T0212066

Lecture Notes in Computer Science 11182

Commenced Publication in 1973
Founding and Former Series Editors:
Gerhard Goos, Juris Hartmanis, and Jan van Leeuwen

More information about this series at http://www.springer.com/series/7412

Jacques Blanc-Talon · David Helbert
Wilfried Philips · Dan Popescu
Paul Scheunders (Eds.)

Advanced Concepts for Intelligent Vision Systems

19th International Conference, ACIVS 2018
Poitiers, France, September 24–27, 2018
Proceedings

 Springer

Editors
Jacques Blanc-Talon
DGA
Bagneux
France

David Helbert
Laboratoire XLIM
Futuroscope Chasseneuil Cedex
France

Wilfried Philips
Ghent University
Ghent
Belgium

Dan Popescu
CSIRO-ICT Centre
Canberra, ACT
Australia

Paul Scheunders
University of Antwerp
Wilrijk
Belgium

ISSN 0302-9743   ISSN 1611-3349   (electronic)
Lecture Notes in Computer Science
ISBN 978-3-030-01448-3   ISBN 978-3-030-01449-0   (eBook)
https://doi.org/10.1007/978-3-030-01449-0

Library of Congress Control Number: 2018955578

LNCS Sublibrary: SL6 – Image Processing, Computer Vision, Pattern Recognition, and Graphics

This Springer imprint is published by the registered company Springer Nature Switzerland AG
The registered company address is: Gewerbestrasse 11, 6330 Cham, Switzerland

Preface

These proceedings gather the selected papers of the Advanced Concepts for Intelligent Vision Systems (ACIVS) Conference, which was held in Poitiers, France, during September 24–27, 2018.

This event was the 19th ACIVS. Since the first event in Germany in 1999, ACIVS has become a larger and independent scientific conference. However, the seminal distinctive governance rules have been maintained:

- To update the conference scope on a yearly basis. While keeping a technical backbone (the classic low-level image processing techniques), we have introduced topics of interest such as video analysis, segmentation, classification, remote sensing, biometrics, deep learning, and image and video compression, restoration and reconstruction. In addition, speakers usually give invited talks on hot issues.
- To remain a single-track conference in order to promote scientific exchanges among the audience.
- To grant oral presentations a duration of 25 minutes and published papers a length of 12 pages, which is significantly different from most other conferences.

The second and third items entail a complex management of the conference; in particular, the number of time slots is rather small. Although the selection between the two presentation formats is primarily determined by the need to compose a well-balanced program, papers presented during plenary and poster sessions enjoy the same importance and publication format.

The first item is strengthened by the notoriety of ACIVS, which has been growing over the years: official Springer records show a cumulated number of downloads on August 1, 2018, of more than 550,000 (for ACIVS 2005–2016 only).

The regular sessions also included a couple of invited talks by Yuliya Tarabalka (Inria Sophia Antipolis, France) and Mihai Datcu (DLR, Germany). We would like to thank all of them for enhancing the technical program with their presentations.

ACIVS attracted submissions from many different countries, mostly from Europe, but also from the rest of the world: Belgium, China, Czech Republic, Finland, France, Germany, India, Italy, Portugal, Korea, Romania, Saudi Arabia, Spain, the UK, the USA, and Vietnam.

From 91 submissions, 36 were selected for oral presentation and 16 as posters. The paper submission and review procedure was carried out electronically and a minimum of two reviewers were assigned to each paper. A large and energetic Program Committee, helped by additional reviewers, as listed on the following pages, completed the long and demanding reviewing process. We would like to thank all of them for their timely and high-quality reviews, achieved in quite a short time and during the summer holidays.

Finally, we would like to thank all the participants who trusted in our ability to organize this conference for the 19th time. We hope they attended a different and stimulating scientific event and that they enjoyed the atmosphere of the various ACIVS social events in the city of Poitiers.

As mentioned, a conference like ACIVS would not be feasible without the concerted effort of many people and the support of various institutions. We are indebted to the local organizers Pascal Bourdon, David Helbert, Mohamed-Chaker Larabi, François Lecellier, Benoit Tremblais, and Thierry Urruty, for having smoothed all the harsh practical details of an event venue.

July 2018

Jacques Blanc-Talon
David Helbert
Wilfried Philips
Dan Popescu
Paul Scheunders

Organization

Acivs 2018 was organized by the University of Antwerp, Belgium.

Steering Committee

Jacques Blanc-Talon	DGA, France
David Helbert	University of Poitiers, France
Wilfried Philips	Ghent University - imec, Belgium
Dan Popescu	CSIRO Data 61, Australia
Paul Scheunders	University of Antwerp, Belgium

Organizing Committee

Pascal Bourdon	XLIM, France
David Helbert	University of Poitiers, France
Mohamed-Chaker Larabi	XLIM, France
François Lecellier	XLIM, France
Benoit Tremblais	XLIM, France
Thierry Urruty	XLIM, France

Program Committee

Hojjat Adeli	Ohio State University, USA
Syed Afaq Shah	The University of Western Australia, Australia
Hamid Aghajan	Ghent University - imec, Belgium
Edoardo Ardizzone	University of Palermo, Italy
Antonis Argyros	University of Crete, Greece
George Bebis	University of Nevada, USA
Fabio Bellavia	University of Florence, Italy
Jenny Benois-Pineau	University of Bordeaux, France
Dominique Béréziat	Université Pierre et Marie Curie, France
Yannick Berthoumieu	Bordeaux INP, France
Janus Bobulski	Czestochowa University of Technology, Poland
Philippe Bolon	University of Savoie, France
Egor Bondarev	Technische Universiteit Eindhoven, The Netherlands
Don Bone	University of Technology Sydney, Australia
Adrian Bors	University of York, UK
Salah Bourennane	Ecole Centrale de Marseille, France
Catarina Brites	Instituto Superior Técnico, Portugal
Vittoria Bruni	University of Rome La Sapienza, Italy
Dumitru Burdescu	University of Craiova, Romania

Vincent Nozick	Université Paris-Est Marne-la-Vallée, France
Danielle Nuzillard	Université de Reims Champagne-Ardenne, France
Rudi Penne	University of Antwerp, Belgium
Fernando Pérez-González	University of Vigo, Spain
Caroline Petitjean	Université de Rouen, France
Hossein Rahmani	Lancaster University, UK
Giovanni Ramponi	University of Trieste, Italy
Florent Retraint	Université de Technologie de Troyes, France
Patrice Rondao Alface	Nokia Bell Labs, Belgium
Florence Rossant	ISEP, France
Luis Salgado	Universidad Politécnica, Spain
Nel Samama	Télécom Sud-Paris, France
Ivan Selesnick	New York University, USA
Wladyslaw Skarbek	University of Technology, Poland
Andrzej Sluzek	Khalifa University, United Arab Emirates
Ferdous Sohel	Murdoch University, Australia
Changming Sun	CSIRO, Australia
Hugues Talbot	Université Paris-Est - ESIEE, France
Attila Tanács	University of Szeged, Hungary
Yuliya Tarabalka	Inria, France
Nadège Thirion-Moreau	SeaTech - Université de Toulon, France
Sylvie Treuillet	Université d'Orléans, France
Florence Tupin	Télécom ParisTech, Université Paris Saclay, France
Cesare Valenti	Università di Palermo, Italy
Marc Van Droogenbroeck	University of Liège, Belgium
Pascal Vasseur	LITIS - Université de Rouen Normandie, France
Peter Veelaert	Ghent University - imec, Belgium
Sergio Velastin	Queen Mary University of London, UK
Nicole Vincent	Université Paris Descartes, France
Domenico Vitulano	National Research Council, Italy
Damien Vivet	ISAE-SUPAERO, France
Shin Yoshizawa	RIKEN, Japan
Gerald Zauner	Fachhochschule Ober Osterreich, Austria
Pavel Zemcik	Brno University of Technology, Czech Republic
Josiane Zérubia	Inria, France
Djemel Ziou	Sherbrooke University, Canada

Additional Reviewers

Syed Afaq Shah	The University of Western Australia, Australia
Hamid Aghajan	Ghent University - imec, Belgium
Roxana Agrigoroaie	U2IS, ENSTA ParisTech, France
Edoardo Ardizzone	University of Palermo, Italy
Antonis Argyros	University of Crete, Greece
George Bebis	University of Nevada, USA
Fabio Bellavia	University of Florence, Italy

Contents

Video Analysis

Biometrics

Deep Learning

Coding and Compression

Image Restoration and Reconstruction

Video Analysis

Improving a Switched Vector Field Model for Pedestrian Motion Analysis

Catarina Barata[✉], Jacinto C. Nascimento, and Jorge S. Marques

Instituto de Sistemas e Robótica, Instituto Superior Técnico,
1049-001 Lisboa, Portugal
ana.c.fidalgo.barata@ist.utl.pt

Abstract. Modeling the trajectories of pedestrians is a key task in video surveillance. However, finding a suitable model to describe the trajectories is challenging, mainly because several of the models tend to have a large number of parameters to be estimated. This paper addresses this issue and provides insights on how to tackle this problem. We model the trajectories using a mixture of vector fields with probabilistic switching mechanism that allows to efficiently change the trajectory motion. Depending on the probabilistic formulation, the motions fields can have a dense or sparse representation, which we believe influences the performance of the model. Moreover, the model has a large set of parameters that need to be estimated using the initialization-dependent EM-algorithm. To overcome the previous issues, an extensive study of the parameters estimation is conducted, namely: (i) *initialization*, and (ii) *priors distribution* that controls the sparsity of the solution. The various models are evaluated in the trajectory prediction task, using a newly proposed method. Experimental results in both synthetic and real examples provide new insights and valuable information how the parameters play an important in the proposed framework.

Keywords: Surveillance · Trajectories · Motion fields
Hidden Markov Models · Expectation maximization

1 Introduction

Activity recognition, movement prediction (see Fig. 1), and detection of abnormal behaviors are critical tasks in surveillance applications, requiring appropriate models to describe them. Since many of the surveillance settings are based on far-field cameras that do not provide detailed information about the pedestrians in a video scene, several of these models are based on the characterization of the trajectories performed by the pedestrians. In fact, trajectories allow us to identify and collect statistics of typical motions, activities, and interactions in a video scene [1]. Consequently, having a reliable description of the possible trajectories is of major importance. However, trajectory modeling is a challenging problem due to its great spatial and temporal variability.

© Springer Nature Switzerland AG 2018
J. Blanc-Talon et al. (Eds.): ACIVS 2018, LNCS 11182, pp. 3–13, 2018.
https://doi.org/10.1007/978-3-030-01449-0_1

Various methods have been proposed to efficiently describe trajectories. Amongst these methods, we are interested in the ones that resort to a generative approach, i.e, the trajectories are assumed to follow a dynamical equation and are governed by known motion patterns. In [2–4] a Gaussian process (GP) regression is used for trajectory modeling under a probabilistic viewpoint. More specifically, in [3] the authors present an unsupervised algorithm in which GP is used to estimate the instantaneous velocity of the pedestrian, while in [2] GP is used to model flow functions. The latter representation also allows for incrementally predicting possible paths and detecting anomalous events from online trajectories. In [4] it is proposed an incremental and unsupervised approach, where the model is updated by receiving new trajectory samples. Dirichlet processes have also been used to model trajectories [6–8], as well as vector fields estimated using the k-means algorithm [5]. The latter work demonstrated the ability of vector fields to describe other types of trajectories besides pedestrian ones (e.g., hurricanes and cell-phone GPS data).

Fig. 1. Pedestrian trajectory: past (yellow) and possible future motions (green). (Color figure online)

Our work is related to [9] where vector fields are used to model pedestrian trajectories. This method assumes that a grid with $\sqrt{n} \times \sqrt{n}$ nodes is defined in the image domain. Each node is characterized by the following parameters: (i) a set of *motion vector fields* (velocity vectors), (ii) a field of isotropic *covariance matrices*, and (iii) a switching *probability matrix* that governs the transition between fields in that node. Only one motion field is active at each trajectory point and the transition between them depends on the spatial location of the pedestrian in the grid. This method has been show to efficiently characterize trajectories in the task of activity recognition [9]. However, it has some limitations: (i) the parameter estimation is performed using the EM algorithm that strongly depends on the initialization, and may encourage wrong estimates; and (ii) the fields were assumed to be dense and estimated in nodes where information was not available. Recently, a new formulation was proposed to deal with the latter issue: imposing a sparsity constraint to the vector fields [10]. However, this work failed to demonstrate the impact of such strategy at describing pedestrian trajectories. Moreover, to the best of our knowledge, none of the vector fields models has ever been used to predict future positions in a trajectory. We believe that this is a critical task in surveillance applications.

This paper provides valuable and comprehensive contributions, namely: (*i*) provides an effective way on how to *initialize* the vector fields in the EM algorithm, (*ii*) evaluates different vector fields formulations, and (*iii*) provides a new methodology for trajectory prediction, using the space-varying dynamical model. For the first item, we perform a systematic study comprising the following three strategies: (a) *random*, (b) *uniform*, and (c) *clustering* initialization, thus not jeopardizing deficient estimates related to poor initializations. For the second item, we compare the vector fields formulations of [9,10]. The proposed prediction strategy is used to evaluate the initialization and type of fields.

2   Switching Dynamical Model

In this work, the trajectory of a pedestrian is assumed to be represented as a sequence of positions $x = (x_1, \ldots, x_L)$, with $x_t \in [0,1]^2$. We also consider that all of the possible motions in a scene may be described using K motion fields $T_k : [0,1]^2 \rightarrow \mathbb{R}^2$, where $k \in \{1, \ldots, K\}$ is the identifier of the field. Only one of these fields is active at a specific time instant, but it is possible to switch between them. Under these assumptions, each position is generated as follows

$$x_t = x_{t-1} + T_{k_t}(x_{t-1}) + w_t, \tag{1}$$

where k_t denotes the active motion field at time t and $w_t \sim N(0, \sigma_{k_t}^2 I)$ is a white random perturbation with zero mean and isotropic covariance.

Switching between active fields depends on the pedestrian position, and it is modeled as a first order Markov process

$$P(k_t = j | k_{t-1} = i, x_{t-1}) = B_{ij}(x_{t-1}), \tag{2}$$

where $B(x) = \{B_{ij}(x), i, j \in \{1, \ldots, K\}\}$ is a stochastic space-varying matrix of switching probabilities (transition matrix) that depends on the position x, *i.e.*, switching probabilities at different positions are usually different.

We specify the motion fields T_k and the transition matrix B using a regular grid on the image $\mathcal{G} = \{g_i \in [0,1]^2, i = 1, \ldots, n\}$. This means that each node g_i of the grid is associated with K velocity vectors T_k^i and a transition matrix B^i. Outside of the nodes, the velocity vector $T_k(x)$ and switching matrix $B(x)$ are obtained by bilinear interpolation, as detailed below

$$T_k(x) = \mathcal{T}_k \Phi(x), \tag{3}$$

$$B(x) = \sum_{i=1}^{n} B^i \phi^i(x), \tag{4}$$

where $\mathcal{T}_k$ is the dictionary of node velocities T_k^i associated with the $k-th$ field, B^i is the switching matrix associated with node g_i, $\Phi(x)$ is a $N \times 1$ normalized sparse vector of known interpolation coefficients, and $\phi^i(x)$ is its $i-th$ element.

If we assume a L-sequence trajectory and a known sequence of active fields $k = (k_1, ..., k_L)$, it is easy to compute the joint probability $p(x, k)$

$$
\begin{aligned}
p(x, k|\theta) &= p(x_1, k_1) \prod_{t=2}^{L} p(x_t|k_t, x_{t-1}) p(k_t|k_{t-1}, x_{t-1}) \\
&= p(x_1) P(k_1) \prod N(x_t|x_{t-1} + T_{k_t}(x_{t-1}), \sigma_{k_t}^2 I) \\
&\quad \times B_{k_{t-1}, k_t}(x_{t-1}),
\end{aligned}
\tag{5}
$$

where $\theta = (\mathcal{T}, B, \sigma^2)$ denotes the complete set of model parameters. From (5), it is clear that the sequence of active fields $k = (k_1, ..., k_L)$ is modeled as a realization of a first order Markov process, with some initial (known) distribution $P(k_1)$ and a space-varying transition matrix, i.e., $B_{k_{t-1}, k_t}(x_{t-1})$. In the following section we address the estimation of the model parameters.

3   MAP Estimation of the Model

Lets assume that we have a set of S observed trajectories, denoted as $\mathcal{X} = \{x^{(1)}, ..., x^{(S)}\}$, where $x^{(j)} = (x_1^{(j)}, ..., x_{L_j}^{(j)})$ is the j-th trajectory with L_j samples. Naturally, the corresponding sequences of active fields $\mathcal{K} = \{k^{(1)}, ..., k^{(S)}\}$ are missing. Given this set, our goal is to estimate the model parameters $\theta = (\mathcal{T}, B, \sigma^2)$ using the MAP formulation

$$
\hat{\theta} = \arg\max_{\theta} \left[\log p(\mathcal{X}|\theta) + \log p(\theta)\right],
\tag{6}
$$

where $\log p(\theta) = \sum_{k=1}^{K} \log p(\theta_k)$ is the log prior distribution. Assuming independence between the various parameters, it is possible to further decompose the prior as follows

$$
\log p(\theta) = \sum_{k=1}^{K} \log p(\mathcal{T}_k) + \sum_{k=1}^{K} \log p(\sigma_k^2) + \log p(B).
\tag{7}
$$

The complete likelihood $\log p(\mathcal{X}, \mathcal{K}|\theta)$ may also be easily defined from (5), by computing a sum over all the trajectories. However, its marginalizing over all the admissible labels $\mathcal{K}$ cannot be computed.

The aforementioned limitation may be addressed using the Expectation-Maximization (EM) method to estimate the model parameters. This amounts to iteratively maximizing the auxiliary function

$$
U(\theta, \theta') = E\left\{\log p(\mathcal{X}, \mathcal{K}|\theta)|\mathcal{K}, \theta^{(t)}\right\} + \log p(\theta)
\tag{8}
$$

with respect to θ and taking into account the available estimates θ' from the previous iteration.

Using (5) and replacing into (8) we obtain the following expression (disregarding the constants)

$$U(\theta, \theta') = - \sum_{s=1}^{S} \sum_{t=2}^{L} \sum_{k=1}^{K} w_k^{(s)}(t) \log N(x_t | x_{t-1} + T_{k_t}(x_{t-1}), \sigma_{k_t}^2 I)$$

$$+ \sum_{s=1}^{S} \sum_{t=2}^{L} \sum_{p,q=1}^{K} w_{p,q}^{(s)}(t) \log B_{k_{t-1}, k_t}(x_{t-1})$$

$$+ \sum_{k=1}^{K} \log p(\mathcal{T}_k) + \sum_{k=1}^{K} \log p(\sigma_k) + \log p(B), \qquad (9)$$

where $w_p^{(s)}(t) = P(k_t^{(s)} = p | x^{(s)}, \theta')$ and $w_{p,q}^{(s)}(t) = P(k_{t-1}^{(s)} = p, k_t^{(s)} = q || x^{(s)}, \theta')$. These weights are computed in the E-step, using the forward-backward algorithm proposed in [12]. The optimization with respect to θ is performed in the M-step, and involves a separate maximization with respect to $\mathcal{T}, B$, and σ^2 [9].

As discussed in [10] the prior $\log p(\theta)$ may be used to incorporate *a priori* knowledge about the parameters, as well as to enforce specific properties. In particular, efforts have been made to define a prior distribution $p(\mathcal{T})$ that could lead to *smooth* and *sparse* fields:

$$\log p(\mathcal{T}_k) = \alpha \|\Delta \mathcal{T}_k\|_2^2 + \beta \|\mathcal{T}_k\|_p^p, \qquad (10)$$

where Δ is an operator that computes all differences between velocities of neighboring nodes and $\|.\|_p$ denotes the pth norm. Here, the first term sets that any neighbor grid nodes $x_{g1}, x_{g2} \in \mathcal{G}$ should have similar estimated velocities, i.e., the difference $\mathcal{T}_{k_t}(x_{g1}) - \mathcal{T}_{k_t}(x_{g2})$ should be small. The second term enforces small velocity values in most of the nodes. These values may be zero if $p = 1$, leading to sparse vector fields where the velocities are only estimated for nodes that are supported by observations [10]. The values chosen for the regularization constants α and β and norm p define the importance of each term, and may influence the estimation of the fields T_k. However, the assessment of their role and relevance has never been performed.

Besides the priors, another criteria that strongly influences the estimation of the model is the initialization of the parameters θ_{init}, since the EM algorithm is very sensitive to initialization [11]. In the following section we propose an experimental framework to evaluate the role of these criteria in the estimation of the motion models.

4   Experimental Setup

The goal of this work is to provide an extensive experimental study on the role of both initialization and the prior $p(\mathcal{T})$ in the estimation of the motion models. To conduct such a study it is necessary to (i) identify the types of initialization and priors to be tested; (ii) select one or more datasets to conduct the experiments;

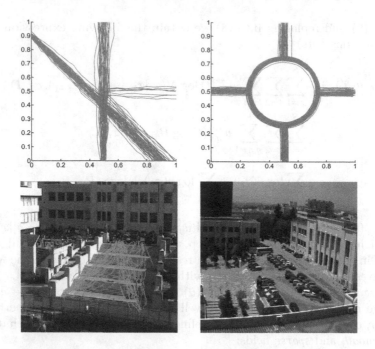

Fig. 2. Experimental datasets: synthetic examples (1st row) and real pedestrian trajectories (2nd row).

and (iii) define an evaluation metric. In this section we address each of these requirements.

(i) Initialization: The EM-algorithm, which is used in the estimation of the parameters, strongly depends on initialization. Thus, we will evaluate three types of initialization: (i) random, where each node of a field is initialized with a random velocity vector; (ii) uniform, where each node of a field is initialized with the same (random) velocity vector; and (iii) clustered, where we perform a preliminary clustering of the velocity values and set the number of centroids to be equal to the number of fields K.

(ii) Prior: Another goal of this paper is to provide a comprehensive study of the role of the fields' prior (10). This means that for each of the aforementioned initializations, we train several models using different combinations of $\alpha \in \{0.5, 1, 2\}$, $\beta \in \{0, 0.05, 0.1, 0.15\}$, and $p \in \{1, 2\}$, leading to a total of 21 models per initialization type.

(iii) Datasets: Our study is conducted using four datasets (two synthetic and two real). The first synthetic case comprises $K = 3$ motion types and 100 randomly generated trajectories, as is exemplified in Fig. 2 (1st row-left). The trajectories were generated according to the following procedure. First, the starting point was randomly chosen between a small region around $[0, 0.9]^T$ and $[0.5, 0]^T$. Conditioned on the selected point, the motion regime can either be a down motion with a slope of $45°$ or an up motion. When the trajectories reach the center of the image there is a probability of switching between these motions and a third one, oriented form left to right.

Table 1. Dataset description. $\bar{L}$ is the average trajectory length and $\bar{v} = \frac{1}{L}\sum_{t=1}^{L}\|x_t - x_{t-1}\|_2^2$ is the average displacement.

Dataset	# Trajectories	$\bar{L}$	$\bar{v}$ ($\times 10^{-3}$)
Synthetic 1	100	27 ± 5	46 ± 4
Synthetic 2	100	77 ± 40	21 ± 0.4
Stairway	86	57 ± 18	12 ± 2
Campus II	1083	50 ± 48	2 ± 0.9

The second synthetic example mimics a roundabout with four entrance/exist points, as exemplified in Fig. 2 (1st row-right). Three motion regimes ($K = 3$) are considered: a circular and counterclockwise one, and two linear movements that represent, respectively, the entrance and exiting of the roundabout. Switching may occur at any of the entrance/exit points. As before, we randomly generate 100 trajectories.

The two real datasets were acquired at the IST university campus in Lisbon (see Fig. 2 (2nd row)), using static far field cameras. The first dataset (2nd row-left), which we call Stairway, comprises 86 trajectories of pedestrians going up and down a stairway, while the second dataset, called Campus II (2nd row-right), comprises more than 900 trajectories of pedestrians crossing the IST campus. Two cameras were used to acquired the Stairway dataset, while the Campus II dataset was acquired using only one. In the first case, two homographies were performed to i) align the images and (ii) correct the distortions cause by the perspective projection. Regarding the Campus II dataset, it was only necessary to correct the distortions. We set the number of motion fields to be estimated to $K = 2$ for the stairway example (up and down), and to $K = 4$ in the Campus II set (roughly north, south, east, and west).

In Table 1 we summarize the properties of each dataset, namely the number of trajectories, the mean trajectory length ($\bar{L}$), and the average displacement between two consecutive trajectory points.

(iv) Evaluation - prediction: Finding an appropriate metric to evaluate the role of the initialization and prior is a challenging task. One could assess the error in the estimation of the model parameters, but such metric is suitable only when we know the ground truth values that generated the trajectories. Such information is only available in the case of the synthetic data. Alternatively, we can assess the ability of the estimated models to predict the future of the trajectory, i.e., given a sequence of past positions $(x(1), ..., x(t_0))$ we want to predict the position $\hat{x}(t_0+\delta)$, δ steps ahead in time using the dynamical model (1).

In our case, this means that we have to predict not only the position, but also the most probable motion field k^* that leads to it (see Fig. 1). To achieve this goal, we apply the forward-backward algorithm [12] used in the E-step of model estimation, namely the forward part, that works as follows. First, lets consider the past, i.e., $t \leq t_0$, and assume the existence of a set of forward variables

$$\alpha_i(t) = Pr\{x(1), ..., x(t), k_t = i|\hat{\theta}\}, \tag{11}$$

such that $\alpha(t) = [\alpha_1(t), ..., \alpha_K(t)]$ is a vector that stores all of the forward probabilities of time instant t. For a given past instant, $\alpha(t)$ is estimated as follows:

$$\alpha(t) = D(x_t)B(x_{t-1})^T \alpha(t-1), \; t = 2, ..., t_0, \; \alpha(1) = D_1\pi, \tag{12}$$

$D(x_t) \in \mathbb{R}^{K \times K}$ is a diagonal matrix with

$$D_{ii} = N\left(x_t; x_{t-1} + T_i(x_{t-1}), \sigma_i^2 I\right), \; i = 1, ..., K. \tag{13}$$

For $t > t_0$ the real position $x(t)$ is unknown. Therefore, matrix D can not be computed and the estimation of the forward probabilities $\hat{\alpha}(t)$ is given by:

$$\hat{\alpha}(t) = B(\hat{x}_{t-1})^T \hat{\alpha}(t-1), \; t = t_0 + 1, ..., t_0 + \delta, \tag{14}$$

where

$$\hat{x}_t = \hat{x}_{t-1} + T_{k_t^*}(\hat{x}_{t-1}) + 0. \tag{15}$$

k_t^* denotes the active field at each instant $t > t_0$, and is defined as: $k^* = \max_k \alpha_k(t)$.

We use the displacement error (DE) as the evaluation metric:

$$DE = \sum_{t_0=1}^{L-\delta} \|x(t_0 + \delta) - \hat{x}(t_0 + \delta)\|_2^2, \tag{16}$$

where $\hat{x}(t_0 + \delta)$ is computed by recursively applying (14) to estimate the best field and (15) to predict the position. We set $\delta = 5$, as proposed in other works (e.g., [13]).

5   Results

In Fig. 3 we show the best performances for each dataset (synthetics 1 and 2, Stairway, and Campus II). These results were obtained using leave-one-trajectory-out cross validation and the performances were computed for each pair (*initialization,prior*). Here dense means that $\beta = 0$, while l_2 and l_1 correspond to $p = 2, 1$, in (10). Table 2 shows the best overall performances, as well as the combination of model parameters and initialization that led to it. These results demonstrate that the proposed prediction strategy is able to successfully the estimate the position of a pedestrian $\delta = 5$ time steps ahead, achieving a low DE in all datasets.

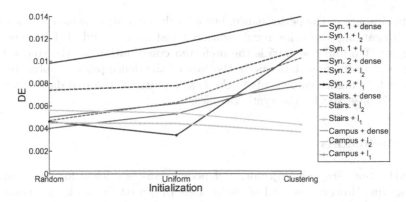

Fig. 3. Experimental results for synthetic examples (red and blue) and real pedestrian trajectories (green and magenta). (Color figure online)

Table 2. Best experimental results out of 63 possible configurations for each dataset.

Dataset	DE	Prior	Best Configuration
Synthetic 1	$5.0 \pm 11.0 \times 10^{-3}$	Dense	Random, $\alpha = 0.5, \beta = 0$
	$4.7 \pm 10.9 \times 10^{-3}$	l_2	Random, $\alpha = 0.5, \beta = 0.1$
	$\mathbf{4.0 \pm 9.0 \times 10^{-3}}$	l_1	**Random, $\alpha = 0.5, \beta = 0.05$**
Synthetic 2	$1.1 \pm 0.80 \times 10^{-2}$	Dense	Random, $\alpha = 1, \beta = 0$
	$7.4 \pm 5.8 \times 10^{-3}$	l_2	Random, $\alpha = 0.5, \beta = 0.15$
	$\mathbf{4.6 \pm 3.0 \times 10^{-3}}$	l_1	**Random, $\alpha = 0.5, \beta = 0.15$**
Stairway	$\mathbf{3.7 \pm 5.8 \times 10^{-3}}$	Dense	**Clustering, $\alpha = 2, \beta = 0$**
	$3.7 \pm 6.4 \times 10^{-3}$	l_2	Clustering, $\alpha = 0.5, \beta = 0.05$
	$4.4 \pm 8.4 \times 10^{-3}$	l_1	Clustering, $\alpha = 1, \beta = 0.05$
Campus II	$\mathbf{2.2 \pm 0.5 \times 10^{-4}}$	Dense	**Clustering, $\alpha = 2, \beta = 0$**
	$2.4 \pm 0.6 \times 10^{-4}$	l_2	Clustering, $\alpha = 0.5, \beta = 0.15$
	$2.4 \pm 0.5 \times 10^{-4}$	l_1	Random, $\alpha = 0.5, \beta = 0.05$

A more thorough analysis of the results suggest that the initialization method has a significant impact in the synthetic and Stairway datasets, while the performance in the Campus II remains almost unchangeable. The role of the initialization is particularly notorious for synthetic example 2 (the roundabout), where velocity clustering leads to a large increase in the DE. This higher error explained by the existence of a circular movement that is inefficiently modeled by velocity clustering.

Enforcing sparsity in the estimation of the fields ($p = 1$) leads to improved prediction results for the synthetic data. However, the opposite is observed for the real datasets, where dense fields achieve the best results. Such disparity may be due to the proportionally higher variability that exists in pedestrian velocities, when compared with the synthetic datasets (see Table 1). Setting $p = 2$ in (10) has some impact in the performance, but it is clearly outperformed by the other priors. Regarding the prior regularization constants, α and β, it seems that setting $\alpha = 0.5$ and $\beta = 0.05$ is the preferred combination for the sparse prior in three out of the four datasets. In the case of the dense prior, we observe that setting α to a higher value leads to the best results for the real datasets, while no trend is observed for the synthetic ones.

6   Conclusions

Efficiently modeling the trajectories of pedestrians is critical for most surveillance setups. However, several of the models proposed to tackle this issue rely on complex methods that require the tunning of multiple parameters. In this paper we addressed this issue, taking as starting point a motion model based on switching motion fields.

Our contributions are two-folded. First we proposed a trajectory prediction strategy for the aforementioned method, which allows a qualitative evaluation of the method. Second, we conducted a thorough study of the model parameters, namely initialization, velocity priors, and constants, in order to obtain insights of their role. Both contributions are original of this paper. Experiments were conducted in four datasets and the obtained results are relevant and promising. Future work should rely on exploring the applicability of the motion fields to other tasks, such as the detection of abnormal trajectories or activities.

Acknowledgments. This work was supported by the FCT project and plurianual funding: [PTDC/EE- IPRO/0426/2014], [UID/EEA/50009/2013].

References

1. Aggarwal, J., Ryoo, M.: Human activity analysis: a review. ACM Comput. Surv. **43**(3), 16 (2011)
2. Kim, K., Lee, D., Essa, I.: Gaussian process regression flow for analysis of motion trajectories. In: Proceedings of IEEE International Conference Computer Vision (ICCV) (2011)
3. Ellis, D., Sommerlade, E., Reid, I.: Modelling pedestrian trajectory patterns with Gaussian processes. In: Proceedings IEEE International Conference Computer Vision (ICCV) Workshops (2009)
4. Bastani, V., Marcenaro, L., Regazzoni, C.S.: Online nonparametric bayesian activity mining and analysis from surveillance video. IEEE Trans. Image Process. **25**(5), 2089–2102 (2016)
5. Ferreira, N., Klosowski, J.T., Scheidegger, C.E., Silva, C.T.: Vector field k-means: clustering trajectories by fitting multiple vector fields. In: Eurographics Conference on Visualization, vol. 32(3), pp. 201–210 (2013)
6. Bastani, V., Marcenaro, L., Regazzoni, C.: Unsupervised trajectory pattern classification using hierarchical Dirichlet process mixture hidden Markov model. In: Proceedings of IEEE International Workshop Machine Learning for Signal Processing (MLSP), p. 16 (2014)

7. Hu, W., Li, X., Tian, G., Maybank, S., Zhang, Z.: An incremental DPMM-based method for trajectory clustering, modeling, and retrieval. IEEE Trans. Pattern Anal. Mach. Intell. **35**(5), 10511065 (2013)
8. Lin, D.: Online learning of nonparametric mixture models via sequential variational approximation. In: Advances in Neural Information Processing Systems, pp. 395–403 (2013)
9. Nascimento, J.C., Figueiredo, M.A.T., Marques, J.S.: Activity recognition using mixture of vector fields. IEEE Trans. Image Process. **22**(5), 1712–1725 (2013)
10. Barata, C., Nascimento, J.C., Marques, J.S.: A sparse approach to pedestrian trajectory modeling using multiple motion fields. In: International Conference on Image Processing (2017)
11. Figueiredo, M.A.T., Jain, A.K.: Unsupervised learning of finite mixture models. IEEE Trans. Pattern Anal. Mach. Intell. **24**(3), 381–396 (2002)
12. Rabiner, L.R.: A tutorial on Hidden Markov Models and selected applications in speech recognition. Proc. IEEE **77**(2), 257–286 (1989)
13. Alahi, A., et al.: Learning to predict human behaviour in crowded scenes. In: Group and Crowd Behavior for Computer Vision, pp. 183–207 (2017)

Matrix Descriptor of Changes (MDC): Activity Recognition Based on Skeleton

Radek Simkanič[✉]

Department of Computer Science, FEECS, VŠB - Technical University of Ostrava,
17. listopadu 15, 708 33 Ostrava - Poruba, Czech Republic
radek.simkanic.st@vsb.cz

Abstract. A new method called Matrix Descriptor of Changes (MDC) is introduced in this work for description and recognition of human activity from sequences of skeletons. The primary focus was on one of the main problems in this area which is different duration of activities; it is assumed that the beginning and the end are known. Some existing methods use bag of features, hidden Markov models, recurrent neural networks or straighten the time interval by different sampling so that each activity has the same number of frames to solve this problem. The essence of our method is creating one or more matrices with a constant size. The sizes of matrices depend on the vector dimension containing the per-frame low-level features from which the matrix is created. The matrices then characterize the activity, even if we assume that certain activities may have different durations. The principle of this method is tested with two types of input features: (i) 3D position of the skeleton joints and (ii) invariant angular features of the skeleton. All kinds of feature types are processed by MDC separately and, in the subsequent step, all the information gathered together as a feature vector are used for recognition by Support Vector Machine classifier. Experiments have shown that the results are similar to results of the state-of-the-art methods. The primary contribution of proposed method was creating a new simple descriptor for activity recognition with preservation of the state-of-the-art results. This method also has a potential for parallel implementation and execution.

Keywords: 3D action feature representation
Human activity recognition · Skeleton joints · Skeleton pose

1   Introduction

Computer vision and image analysis is an interdisciplinary field focusing on processing and obtaining useful information from images. One of the most important and interesting topics is human action recognition. This research has a great potential in many applications such as medical applications, robotics, public surveillance or even entertainment. This field is enriched by depth cameras which provide good quality and real-time capture of depth (or disparity) map

© Springer Nature Switzerland AG 2018
J. Blanc-Talon et al. (Eds.): ACIVS 2018, LNCS 11182, pp. 14–25, 2018.
https://doi.org/10.1007/978-3-030-01449-0_2

and detection of human skeleton pose. Some methods work with a sequence of conventional images (RGB, grayscale, ...). Compared to conventional images, the depth maps are insensitive to changes in lighting conditions and can provide 3D information. The pose of a person in the scene can also be obtained from depth images, and then represented by coordinates of joints of skeleton. The approach based on joints of the skeleton significantly reduces the amount of required information, omitting data from the environment and background.

A new method based on joints of skeleton sequences to solve the problem of human activities recognition is proposed in this paper. The method is designed with a prospect of implementation simplicity and computational efficiency. The method generates a matrix for one time interval of the activity or matrices for some hierarchical time distribution of the activity where each matrix is generated for one time (sub) interval. Motivated by the Self-Similarity Matrix (SSM) [9] method, which generates a symmetric matrix whose size is determined by the activity duration, the focus of this work is to create a symmetric matrix approach where the matrix would have the same size for different durations of activities. The descriptive matrix is created by measuring distances between all pairs of feature vector entries. The matrix or matrices are created for each type of input feature data. The final classification of activities is provided by SVM classifier. The classification is tested on the three datasets that provide information of skeleton joints.

The rest of the paper is organised as follows. In Sect. 2, the related works focused on activity recognition are briefly presented and classified into several types. In Sect. 3, the details of the new method are stated. The experimental results are reported in Sect. 4. The conclusion can be found in Sect. 5.

2   Related Works

With higher availability and low-cost RGBD sensors for public, RGB sensors have also been used in research of human activity recognition. These sensors provide new useful information that was difficult to obtain from conventional cameras, especially about skeleton. Approaches based on the depth maps provide good results and simplify the implementation more than RGB approaches. The Depth Motion Maps (DMM) introduced in [2] can be regarded as a frequently used method. This method has an easy implementation, in which the sum of the differences between the consecutive depth maps is computed. Various other methods are based on the DMM method. They use, for example, Convolution Neural Networks (CNN) [23] or Histogram of Gradients (HoG) [25].

The Space-Time Occupancy Pattern (STOP), which is introduced in [18], is a method based on the space-time probabilistic occupancy grid. The representation based on cloud points was applied to the 4D sub-volumes sampled by weighted sampling scheme in a random occupancy pattern (ROP) in [19].

The method which inspired this work is based on capturing skeleton poses and using the Self-Similarity Matrix (SSM) is presented in [9]. The Recurrent Neural Network (RNN) can model the long-term contextual information of temporal sequences, and it was a reason why it was proposed to use hierarchical RNN for the skeleton based action recognition in [5]. Next approach [3] with capturing the skeleton poses is based on the extraction of key poses to compose a feature vector. The evolutionary algorithm is used to determine the optimal subset of skeleton joints, taking into account the topological structure of the skeleton, which was introduced in article [1]. A skeletal representation that explicitly models the 3D geometric relationships between various body parts using rotations and translations in 3D space was proposed in paper [17]. A local skeleton descriptor that encodes the relative position of Joint quadruples, that leads to a compact 6D view-invariant skeletal feature and using Fisher kernel representation for their representation is described in paper [6]. A descriptor featuring Eigen Joints, which was introduced in [24] also used Accumulated Motion Energy (AME) to remove noisy frames and reduce computational cost.

A combination of input data types is preferred in [20], where the data from the joints of skeleton and depth maps are simultaneously used. The authors observed that their method was resistant to noises and errors in skeleton joint positions that could be caused by severe occlusion.

3 Proposed Method

This section is focused on a detailed description of the new method, which is called the Matrix Descriptor of Changes (MDC). It is partially inspired by the Self-Similarity Matrix (SSM), which is commonly used for action recognition [9], visualisation of musical structure and rhythm [7], video analysis [4], etc. The SSM method generates a graphical representation of similar sequences in data series. The disadvantages of SSM include different sizes of matrices for different actions because every action that should be recognised may have a different time duration, however, we suppose that the beginning and the end of the actions are known. The intervals of activities may be different for each action since the durations of actions are not normalised in the real world either. Moreover, each person may perform the activity in a different time. Somebody can perform the activity quickly, and the other one can do the same activity slowly. The problem with the variable matrix size was an inspiration for developing a new method.

Low level feature vectors with dimension N are obtained from each activity. The proposed method is designed to create symmetric matrices of a constant size $N \times N$ for all activities. For the purpose of detection of interesting time locations, each activity can be divided into several smaller sub-intervals, which is demonstrated in Fig. 1.

The sub-interval $\tau \equiv \{t_b, \ldots, t_e\}$ contains the time sequence from t_b to t_e. The sequence of particular low-level feature values in the time interval τ is represented by the vector $a_{i,\tau} = (a_{i,t_b}, \ldots, a_{i,t_e})$. The symmetric matrix MDC

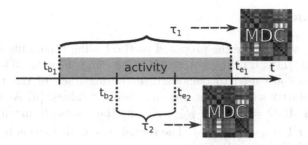

Fig. 1. Generating the MDC matrices from the interval itself (τ_1) and one sub-interval (τ_2). The arrow with dashed line indicate generating the MDC matrices. The green area between t_{b_1} and t_{e_1} demonstrates the duration of activity. (Color figure online)

(Matrix Description of Changes) for the sequence of low-level feature vectors in time interval τ is computed as

$$MDC_\tau = \begin{bmatrix} 0 & d_{\tau,1,2} & \cdots & d_{\tau,1,N} \\ d_{\tau,2,1} & 0 & \cdots & d_{\tau,2,N} \\ \vdots & \vdots & \ddots & \vdots \\ d_{\tau,N,1} & d_{\tau,N,2} & \cdots & 0 \end{bmatrix}, \tag{1}$$

where $d_{\tau,i,j}$ is the distance between two sequences of particular low-level features $a_{i,\tau}$ and $u_{j,\tau}$ in the time interval τ.

The distance between two vectors $a_{i,\tau}$ and $u_{j,\tau}$ in a time interval τ is defined by the formula

$$d_{\tau,i,j} = \sum_{t \in \tau} ||a_{i,t} - a_{j,t}||_2 \cdot g_\tau(t), \tag{2}$$

where $g_\tau(\cdot)$ is a function that represents the weights of the differences from the particular times from sequence τ; and $a_{i,t}$ is a particular low-level feature in the space-time, and it is the value of the i-th low-level feature obtained at a time t. Matrix is computed for each interval τ, where different weight functions $g_\tau(\cdot)$ may be used to compute distances.

It is always necessary to consider (or test) how many matrices are needed for recognition and it is also necessary to determine from which (relative) local time intervals the matrix should be created. This problem is directly dependent on how the actions are different or similar to each other. If the activities are diverse and less complex, then one matrix is sufficient with the original time interval. In the opposite case, more matrices are needed.

One of the suitable solutions to get more local time intervals is creating the hierarchical time pyramid of the time intervals. The first level of the pyramid uses the whole time interval. In each consequent level, the main interval is divided into several smaller sub-intervals that can overlap.

The matrices generated by this algorithm describe the human activity and the recognition can be done by using any algorithm for machine learning; in this case, SVM using intersection kernel was used for the experiments.

4   Experiments

In this section, we evaluate the proposed method using three different datasets: MSR-Action3D [11], UTKinect-Action3D [22] and Florence3D-Action [16]. Three techniques are used to demonstrate the accuracy of the new method: (i) Confusion Matrix with percentages and absolute values, (ii) ACC (accuracy) by the formula $ACC = \frac{TP+TN}{TP+TN+FP+FN}$, (iii) the harmonic mean of precision and sensitivity $F1 = \frac{2 \cdot TP}{2 \cdot TP + FP + FN}$. The robustness of MDC to noisy data is also tested on the MSR-Action3D dataset.

MSR-Action3D dataset [11] was captured using a Microsoft Kinect-like depth sensor. This dataset contains 20 action types, each action is performed by 10 human subjects who repeat the same action 2 or 3 times. Altogether, there are 557 valid action sequences. Each frame in these sequences contains 3D locations of 20 skeleton joints. This is the challenging dataset because the skeleton poses are inaccurate and in some frames do not correspond to real poses and many sequences of actions in different classes are similar to each other.

UTKinect-Action3D dataset [22] was captured by using Microsoft Kinect sensor. It contains 10 action types, each action is performed by 10 human subjects again, every action is repeated twice. Altogether, there are 199 action sequences contained within 20 video sequences. Each frame in each sequence contains 3D locations of 20 skeleton joints representing a skeleton pose. The skeleton poses are inaccurate and, in some frames, do not correspond to real poses and actions contained high intra-class variations.

Florence3D-Action [16] was captured by using Microsoft Kinect sensor. It contains 9 action types, each action is performed by 10 human subjects again, every action is repeated two or three times. Altogether, there are 215 action sequences. Each frame in sequences contains 3D locations of 15 skeleton joints. This is the dataset in which the recognition of some activities is problematic because the actions contain high intra-class variations and many sequences of actions in different classes are highly similar to each other.

Two types of input features are used in the proposed recognition method: (i) 3D positions of the joints of the skeleton, (ii) invariant angular features of the skeleton. In the case of joint coordinates, three entries of the low-level feature vector are used for the coordinates of each joint (i.e. the particular coordinates are placed into the vector). For datasets MSR-Action3D and UTKinect-Action3D, a 60-element vector (20 joints in 3D space) is created and for Florence3D-Action the vector dimension is 45. The invariant angular features of skeletons were published in [14]. The angles are measured between the bones that meet at one joint. This description of skeleton pose is invariant to rotation and its size. When testing the proposed method on the Florence3D-Action, only 3D positions of the joints are used.

Several settings of proposed method are used to perform the experiments in this paper. The first and basic setting is used for all datasets - MDC$_{basic}$, which generates only one matrix, and Gaussian function in this case $g_\tau(\cdot)$ is used to calculate the distance $d_{\tau,ij}$. The Gaussian function is normalised so that it gives the unit value after the integration. This serves to compare the settings with each other and as an indicator of the complexity of activities.

The next settings labelled as MDC_{pyr1-2} and MDC_{pyr1-3} use a (pyramidical) hierarchical binary division of an interval into several new sub-intervals from which matrices are created. These matrices also use the normalised Gaussian function for weight. It is normalised in such a way that it gives the unit value for each level of the pyramid after the integration. The MDC_{pyr1-2} and MDC_{pyr1-3} contain three, respectively seven, matrices per one type of included low level features. The last setting - $MDC_{basic+hg1}$ - contains three matrices per one type of included features. The first matrix is MDC_{basic} and the rest uses half of the main interval. The (left) half of the Gaussian function is used as the weight for the (left) sub-interval. The other half of the Gaussian function is used as the weight for the second (right) sub-interval.

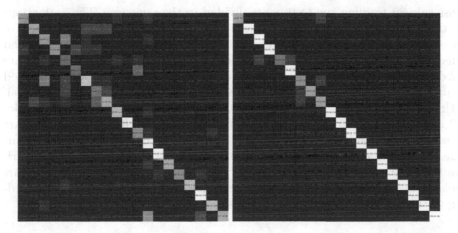

Fig. 2. Confusion matrices of experimental results of protocol Wang et al. [19] on the MSR-Action3D dataset (from left): (i) cross-subject (1:1), (ii) cross-samples (2:1). All is using method MDC_{pyr1-2}.

MSR-Action3D dataset allows for several validation protocols which are summarized in [15]. Several protocols have been used in this article. The first used protocol is introduced in Li et al. [11]. In this standard protocol, the dataset is divided into three action subsets AS1, AS2 and AS3 (8 action types per subset), which may share some action types. The accuracy of proposed method on the dataset is tested by two decompositions: (i) cross-subject and (ii) cross-samples. The challenge is the cross-subject (1:1) testing. One half of the samples of subjects is used for training, the other half is used for validation. The cross-samples contain two types of proportion: (i) 1:2 and (ii) 2:1, where one, respectively two samples are used for training and the rest for testing. The second used protocol is provided in Wang et al. [19]. This protocol is not divided into subsets AS1–AS3, but the test is applied over all action types of the entire dataset. The samples for

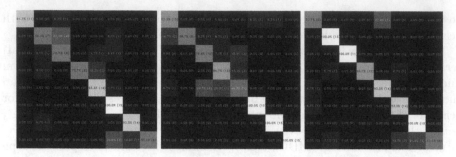

Fig. 3. Confusion matrices of experimental results of cross-subject protocol Li et al. [11] on the MSR-Action3D dataset (from left): (i) AS1 subset, (ii) AS2 subset and (iii) AS3 subset. All is using method $\mathrm{MDC_{pyr1-2}}$.

cross-subject training and testing are divided in the same way. The evaluations by these protocols are shown in Table 1. It can be observed that all subsets of cross-samples of proportions 1:2 and 2:1 achieve high accuracy, precision and sensitivity. In the cases of the cross-subject with proportion 1:1, accuracies are good but worse than in the cases of cross-samples. The best of proposed method configuration is $\mathrm{MDC_{pyr1-2}}$ which has average accuracies of protocol Li et al. [11]: 96.02% (F1 84.09%) for cross-subject, 98.66% (F1 94.63%) for cross-samples 1:2 and 99.32% (F1 97.28%) for cross-samples 2:1. The confusion matrices are shown in Figs. 2 and 3, which demonstrate where erroneous classifications occur. The proposed method is comparable to the state-of-the-art methods. Comparison with the state-of-the-art methods is shown in Table 2. The state-of-the-art skeleton methods achieve approximate accuracy of 65.70–94.49% [15, 21] according to the protocol Li et al. [11] with average accuracies for cross-subject.

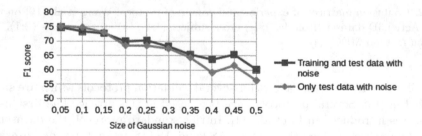

Fig. 4. The progress of F1 score when applying Gaussian noise to the positions of skeleton joints. The x-axis depicts the standard deviation of the position disturbance.

The robustness of the proposed method to noisy data is tested on the MSR-Action3D dataset with the cross-subject protocol. The $\mathrm{MDC_{pyr1-2}}$ setting is used. The Gaussian noise is added to the original coordinates of the skeleton joints. From the skeletons modified in this way, the coordinates and angles are obtained for creating MDCs. The standard deviation of the position disturbance is in range from 0.05 to 0.5, with the step of 0.05. The demonstration of robustness is shown in Fig. 4, where the resulting F1 scores are presented. In this experiment, two cases are used: (i) the noise is applied both to the training and testing data, (ii) the noise is applied to testing data only.

Table 1. The experimental results of proposed method on the MSR-Action3D dataset via cross-subject (1:1) and cross-sample (1:2 and 2:1) decomposition. AS1–AS3 is provided by Li et al. [11] protocol and All is provided by Wang et al. [19] protocol.

Methods	AS1		AS2		AS3		All	
	F1 %	ACC %	F1 %	ACC %	F1 %	ACC %	F1 %	ACC %
Cross-subject 1:1								
MDC$_{basic}$	75.24	93.81	72.32	93.08	89.19	97.30	68.87	96.89
MDC$_{pyr1-2}$	80.95	95.24	83.04	95.76	88.29	97.07	79.93	97.99
MDC$_{basic+hgl}$	79.05	94.76	75	93.75	90.99	97.75	73.63	97.36
Cross-samples 1:2								
MDC$_{basic}$	88.97	97.24	87.58	96.90	93.92	98.48	86.79	98.68
MDC$_{pyr1-2}$	93.79	98.45	92.81	98.20	97.29	99.32	91.64	99.16
MDC$_{basic+hgl}$	90.35	97.59	83.66	95.92	95.95	98.99	85.18	98.52
Cross-samples 2:1								
MDC$_{basic}$	95.78	98.94	94.67	98.67	98.61	99.65	93.96	99 40
MDC$_{pyr1-2}$	97.18	99.30	94.67	98.67	100.00	100.00	96.70	99.67
MDC$_{basic+hgl}$	92.96	98.24	93.33	98.33	95.83	98.96	92.31	99.23

Table 2. Comparison average of AS1–AS3 accuracies provided via cross-subject by protocol Li et al. Note: * - It does not represent method based on the skeleton but depth maps; this is the work of the author of the dataset. [11]

Methods	Accuracy %
Li et al.* [11]	74.7
Yang et al. [24]	83.3
Liu et al. [13]	84.07
Gowayyed et al. [8]	91.26
Du et al. [5, 21]	94.49
MDC$_{pyr1-2}$	96.02

The cross-subject protocol of [27] is used for the UTKinect-Action dataset. In this protocol, half of the subjects is used for training while the other half is used for testing. The evaluation by this protocol is shown in Table 3. The best of the proposed method settings is MDC_{pyr1-3}, which has very high accuracy 98.79% (F1 93.94%) and this is comparable with the state-of-the-art methods. The state-of-the-art skeleton methods achieve approximate accuracy of 90.92–97.00% [10,12,21,22,26]. The comparison of accuracies is shown in Table 4.

Table 3. The experimental results of proposed method on the UTKinect-Action and Florence3D datasets via cross-subject decomposition. The **UTKinect** column presents cross-subject 1:1 validation protocol of [27]. The **Florence3D (cross-subject)** column presents cross-subject protocol. The **Florence3D (leave-one-actor-out)** column presents leave-one-actor-out validation protocol of [16].

Methods	UTKinect		Florence3D (cross-subject)		Florence3D (leave-one-actor-out)	
	F1 %	ACC %	F1 %	ACC %	F1 %	ACC %
MDC_{basic}	84.85	96.97	45.37	87.86	61.40	91.42
MDC_{pyr1-3}	93.94	98.79	58.33	90.74	74.28	94.32
$MDC_{basic+hg1}$	86.87	97.37	55.56	90.12	70.70	93.49

For Florence3D-Action datasets, two evaluation protocols are used: (i) the cross-subject protocol in which half of subjects are used for training, the remaining subjects are used for testing, which is the same as in the protocol used in a part of UTKinect experiments, (ii) the leave-one-actor-out protocol of [16] in which a trained classifier uses all sequences from 9 out of 10 actors and tests on the remaining one and repeats this procedure for all actors and averages the 10 classification accuracy values to the resulting accuracy, in this case, the authors report the accuracy of 90%. The evaluation by this protocols are shown in Table 3.

Table 4. Accuracy comparison for the UTKinect-Action dataset by cross-subject protocol [27].

Methods	Accuracy %
Xia et al. [22]	90.92
Zhang et al. [26]	95.96
Lee et al. [10]	96.97
Jun et al. [12]	97
MDC_{pyr1-3}	98.79

The largest classification errors are observed for actions "answer phone clap" and "read watch". These activities are very similar and, therefore, the proposed method has a bad recognition. Both of these activities are also often confused with the "drink" activity. Other activities are well recognised. All of these are demonstrated on confusion matrices (Fig. 5).

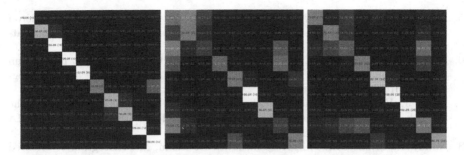

Fig. 5. Confusion matrices of experimental results on the UTKinect-Action3D and Florence3D datasets (from left): (i) UTKinect-Action3D with cross-subject protocol of [27], (ii) Florence3D-Action with cross-subject protocol and (iii) Florence3D-Action leave-one-actor-out protocol of [16]. All are method MDC_{pyr1-3}.

5   Conclusion

This article has presented a new method for the description of space-time data applied to activity recognition. The method generates a matrix or matrices with constant size although the activities have different length of time intervals. The experiments have shown that the new method can be successfully used in this area, and that it is fully comparable with the state-of-the-art methods. If the different activities are too similar, the proposed method may have a problem to classify the correct class. In the future, this proposed method is planned to be used for classification (recognition) of activities from the depth maps and detection of activities from the streams of activities without knowing their beginning and the end.

Acknowledgements. This work was partially supported by Grant of SGS No. SP2018/42, VŠB - Technical University of Ostrava, Czech Republic.

References

1. Chaaraoui, A.A., Flrez-Revuelta, F.: Human action recognition optimization based on evolutionary feature subset selection. In: Proceeding of the Fifteenth Annual Conference on Genetic and Evolutionary Computation Conference - GECCO 2013 (2013)
2. Chen, C., Liu, K., Kehtarnavaz, N.: Real-time human action recognition based on depth motion maps. J. R.-Time Image Process. **12**(1), 155–163 (2016)
3. Cippitelli, E., Gasparrini, S., Gambi, E., Spinsante, S.: A human activity recognition system using skeleton data from RGBD sensors. Comput. Intell. Neurosci. **2016**, 114 (2016)
4. Cooper, M., Foote, J.: Scene boundary detection via video self-similarity analysis. In: Proceedings 2001 International Conference on Image Processing (Cat. No. 01CH37205), pp. 378–381. IEEE (2001)

5. Du, Y., Wang, W., Wang, L.: Hierarchical recurrent neural network for skeleton based action recognition. In: 2015 IEEE Conference on Computer Vision and Pattern Recognition (CVPR) (2015)
6. Evangelidis, G., Singh, G., Horaud, R.: Skeletal quads: human action recognition using joint quadruples. In: 2014 22nd International Conference on Pattern Recognition (2014)
7. Foote, J.: Visualizing music and audio using self-similarity. In: Proceedings of the Seventh ACM International Conference on Multimedia (Part 1) - MULTIMEDIA 1999, pp. 77–80. ACM Press, New York (1999)
8. Gowayyed, M.A., Torki, M., Hussein, M.E., El-Saban, M.: Histogram of oriented displacements (HOD): describing trajectories of human joints for action recognition. In: Proceedings of the Twenty-Third International Joint Conference on Artificial Intelligence, IJCAI 2013, pp. 1351–1357. AAAI Press (2013). http://dl.acm.org/citation.cfm?id=2540128.2540323
9. Junejo, I.N., Dexter, E., Laptev, I., Perez, P.: View-independent action recognition from temporal self-similarities. IEEE Trans. Pattern Anal. Mach. Intell. **33**(1), 172–185 (2011)
10. Lee, I., Kim, D., Kang, S., Lee, S.: Ensemble deep learning for skeleton-based action recognition using temporal sliding LSTM networks. In: 2017 IEEE International Conference on Computer Vision (ICCV) (2017)
11. Li, W., Zhang, Z., Liu, Z.: Action recognition based on a bag of 3D points. In: 2010 IEEE Computer Society Conference on Computer Vision and Pattern Recognition - Workshops, pp. 9–14. IEEE (2010)
12. Liu, J., Shahroudy, A., Xu, D., Wang, G.: Spatio-temporal LSTM with trust gates for 3D human action recognition. CoRR abs/1607.07043 (2016). http://arxiv.org/abs/1607.07043
13. Liu, Z., Zhang, C., Tian, Y.: 3D-based deep convolutional neural network for action recognition with depth sequences. Image Vis. Comput. **55**, 93100 (2016)
14. Ohn-Bar, E., Trivedi, M.M.: Joint angles similarities and HOG2 for action recognition. In: 2013 IEEE Conference on Computer Vision and Pattern Recognition Workshops, pp. 465–470. IEEE (2013)
15. Padilla-López, J.R., Chaaraoui, A.A., Flórez-Revuelta, F.: A discussion on the validation tests employed to compare human action recognition methods using the MSR Action3D dataset. CoRR abs/1407.7390 (2014). http://arxiv.org/abs/1407.7390
16. Seidenari, L., Varano, V., Berretti, S., Bimbo, A.D., Pala, P.: Recognizing actions from depth cameras as weakly aligned multi-part bag-of-poses. In: 2013 IEEE Conference on Computer Vision and Pattern Recognition Workshops (2013)
17. Vemulapalli, R., Arrate, F., Chellappa, R.: Human action recognition by representing 3D skeletons as points in a lie group. In: 2014 IEEE Conference on Computer Vision and Pattern Recognition (2014)
18. Vieira, A.W., Nascimento, E.R., Oliveira, G.L., Liu, Z., Campos, M.F.: On the improvement of human action recognition from depth map sequences using space-time occupancy patterns. Pattern Recogn. Lett. **36**, 221–227 (2014)
19. Wang, J., Liu, Z., Chorowski, J., Chen, Z., Wu, Y.: Robust 3D action recognition with random occupancy patterns. In: Fitzgibbon, A., Lazebnik, S., Perona, P., Sato, Y., Schmid, C. (eds.) ECCV 2012. LNCS, pp. 872–885. Springer, Heidelberg (2012). https://doi.org/10.1007/978-3-642-33709-3_62
20. Wang, J., Liu, Z., Wu, Y., Yuan, J.: Mining actionlet ensemble for action recognition with depth cameras. In: 2012 IEEE Conference on Computer Vision and Pattern Recognition, pp. 1290–1297. IEEE (2012)

21. Wang, P., Li, W., Ogunbona, P., Wan, J., Escalera, S.: RGB-D-based human motion recognition with deep learning: a survey. arXiv e-prints, October 2017
22. Xia, L., Chen, C.C., Aggarwal, J.K.: View invariant human action recognition using histograms of 3D joints. In: 2012 IEEE Computer Society Conference on Computer Vision and Pattern Recognition Workshops, pp. 20–27. IEEE (2012)
23. Yang, R., Yang, R.: DMM-pyramid based deep architectures for action recognition with depth cameras. In: Cremers, D., Reid, I., Saito, H., Yang, M.-H. (eds.) ACCV 2014. LNCS, vol. 9007, pp. 37–49. Springer, Cham (2015). https://doi.org/10.1007/978-3-319-16814-2_3
24. Yang, X., Tian, Y.: Effective 3D action recognition using eigenjoints. J. Vis. Commun. Image Represent. **25**(1), 211 (2014)
25. Yang, X., Zhang, C., Tian, Y.: Recognizing actions using depth motion maps-based histograms of oriented gradients. In: Proceedings of the 20th ACM International Conference on Multimedia - MM 2012 (2012)
26. Zhang, S., Liu, X., Xiao, J.: On geometric features for skeleton-based action recognition using multilayer LSTM networks. In: 2017 IEEE Winter Conference on Applications of Computer Vision (WACV) (2017)
27. Zhu, Y., Chen, W., Guo, G.: Fusing spatiotemporal features and joints for 3D action recognition. In: 2013 IEEE Conference on Computer Vision and Pattern Recognition Workshops (2013)

Person Re-Identification with a Body Orientation-Specific Convolutional Neural Network

Yiqiang Chen[1(✉)], Stefan Duffner[1], Andrei Stoian[2], Jean-Yves Dufour[2], and Atilla Baskurt[1]

[1] Univ Lyon, INSA-Lyon, CNRS, LIRIS, 69621 Villeurbanne, France
yiqiang.chen@insa-lyon.fr
[2] Thales Services, ThereSIS, Palaiseau, France

Abstract. Person re-identification consists in matching images of a particular person captured in a network of cameras with non-overlapping fields of view. The challenges in this task arise from the large variations of human appearance. In particular, the same person could show very different appearances from different points of view. To address this challenge, in this paper we propose an Orientation-Specific Convolutional Neural Network (OSCNN) framework which jointly performs body orientation regression and extracts orientation-specific deep representations for person re-identification. A robust joint embedding is obtained by combining feature representations under different body orientations. We experimentally show on two public benchmarks that taking into account body orientations improves the person re-identification performance. Moreover, our approach outperforms most of the previous state-of-the-art re-identification methods on these benchmarks.

Keywords: Person re-identification · Convolutional neural network
Mixture of experts

1 Introduction

Person re-identification is the problem of identifying people across images that have been captured by different surveillance cameras with non-overlapping views. The task is increasingly receiving attention because of its important applications in video surveillance such as cross-camera tracking, multi-camera behavior analysis and forensic search.

However, this problem is challenging due to the large variations of lightings, poses, viewpoints and backgrounds. The main difficulty is that the pedestrian appearance can be very different with different body orientations under different viewpoints, *i.e.* images of the same person can look quite different and images of different persons can look very similar (see Fig. 1). Moreover, low image resolution and partial occlusion in images make the problem even harder.

© Springer Nature Switzerland AG 2018
J. Blanc-Talon et al. (Eds.): ACIVS 2018, LNCS 11182, pp. 26–37, 2018.
https://doi.org/10.1007/978-3-030-01449-0_3

Fig. 1. Some image examples from a person re-identification dataset. Pedestrian appearance can be very different due to different body orientations.

Most existing approaches consider that pedestrian images come from a single domain. The viewpoint-invariant feature representations are either designed "manually" or learned automatically by a deep neural network. Though, re-identification can be considered as a multi-domain problem, *i.e.* pedestrians with the same body orientation have similar silhouettes and those with different body orientations have dissimilar appearance. Some metric learning approaches, for example, learn to transfer the feature space from one camera to another. But this requires a model for all the combination of cameras. Some other metric learning methods learn to transfer the different view-specific feature spaces to a common subspace where features are discriminative. This addresses the lighting and background variations, but it cannot be generalised to new camera views, and pedestrian images still have variations from different body orientations even if they come from the same camera.

To tackle this issue, we use a multi-task deep Convolutional Neural Network (CNN) to perform body orientation regression in a gating branch, and in another branch separate orientation-specific layers are learned as local experts. The combined orientation-specific CNN feature representations are used for the person re-identification task. Our main contributions are:

- a mixture-of-expert deep CNN to model the multi-domain pedestrian images for person re-identification. We show that learning and combining different feature embeddings of different orientations improves the re-identification performance,
- a novel multi-task CNN framework with combined person orientation estimation and re-identification, where the estimated body orientation is used to steer the orientation specific mixture of experts for re-identification,
- an experimental evaluation showing that our approach outperforms most state-of-the-art methods on the CUHK01 and Market-1501 datasets.

2   Related Work

Existing person re-identification approaches generally build a robust feature representation or learn a distance metric. The features used for re-identification are mainly variants of color histograms, Local Binary Patterns (LBP) or Gabor features. Some approaches use features that are specifically designed to be robust to common appearance variations. For example, Gray *et al.* [6] extract RGB,

YUV and HSV channels and LBP texture histograms in horizontal stripes as feature vector. Liao et al. [13] propose the LOMO features. Color and SILTP histograms are extracted in sliding windows and only the maximal occurance is kept along each horizontal strip. The main metric learning methods include Mahalanobis metrics like KISSME [7], Local Fisher discriminant Analysis (LFDA) [17] and Cross-view Quadratic Discriminant Analysis (XQDA) [13].

With the recent success of deep learning for computer vision applications, many deep convolution neural network approaches have been proposed for person re-identification. For example, Li et al. [11] adopted a filter pairing neural network (FPNN) to model the displacement of body parts for person re-identification. Amed et al. [1] introduced an improved Siamese architecture using the difference of feature maps to measure the similarity. Cheng et al. [4] proposed a variant of the triplet loss function and a CNN network processing parts and the entire body. Varior et al. [22] proposed a Siamese CNN integrating a gate layer to capture effective subtle patterns in the feature map. The deep networks proposed in [8] and [27] learn a body part alignment and localisation and extract body part regions in an unified framework.

Most existing methods for person re-identification focus on developing a robust representation to handle the variations of view. Some methods take into account the view as extra information. For example, Ma et al. [15] divide the data according to the additional camera position information and learn a specific distance metric for each camera pair. Lisanti et al. [14] proposed to apply Kernel Canonical Correlation Analysis which finds a common subspace between the feature space from disjoint cameras. Yi et al. [25] proposed to apply a Siamese CNN to person re-identification. Similar to [14], the weights of two subnetworks are not shared to learn a camera view projection to a common feature space. In these approaches, camera information is used but the body orientation which is only partly due to different camera views is not modelled. That is, in the same camera view, pedestrians can exhibit different orientations and thus largely different appearances in the resulting images.

In order to solve this issue, Bak et al. [2] perform an orientation-driven feature weighting and the body orientation is calculated according to the walking trajectory. some other approaches [18,23] deal with the orientation variations of pedestrian images by using Mixture of Experts. The expert neural networks map the input to the output, while a gating network produces a probability distribution over all experts' final predictions. Verma et al. [23] applied an orientation-based mixture of experts to the pedestrian detection problem. Sarfraz et al. [18] proposed to learn the orientation sensitive units in a deep neural network to perform attribute recognition. Garcia et al. [5] used orientations estimated by a Kalman filter and then trained two SVM classifiers for pedestrian images matching with respectively similar orientations and dissimilar orientations. And the approach of Li et al. [9] learns a mixture of experts, where samples were softly distributed into different experts via a gating function according to the viewpoint similarity.

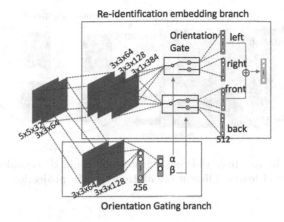

Fig. 2. Overview of the OSCNN architecture.

Sharing the idea of mixture of experts, we propose to build a multi-domain representation in different orientations with deep convolutional neural networks. Intuitively, an orientation-specific model should have a better generalization ability than a camera view-specific model, since we cannot incorporate all possible surveillance camera views. Further, instead of using discrete orientations for the gating activation function, in our method, we use a regressor to estimate an accurate and continuous body orientation. This allows to continuously weight different expert models for re-identification and also avoids combining contradictory orientations.

3 Proposed Method

The overall procedure of our re-identification approach OSCNN is shown in Fig. 2. The network contains an orientation gating branch and a re-identification branch consisting of 4 feature embeddings regarding the 4 main orientations: left, right, frontal and back. The final output feature representation is a linear combination of the four expert outputs and is steered by an orientation gate unit which is a function of the estimated orientation.

3.1 OSCNN Architecture

The proposed neural network architecture consists of two convolution layers shared between an orientation gating branch and a re-identification feature embedding branch. In the re-identification branch, there are 3 further convolution layers followed by 4 separate, parallel fully-connected layers of 512 dimensions, each one corresponding to a local expert. Thus, our network learns different projections to a common feature space, as shown Fig. 3 .

In the orientation regression branch, 2 convolution layers and 2 fully connected layers are connected to the common convolutional layers. The estimated

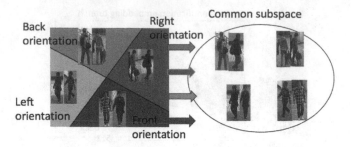

Fig. 3. Pedestrian images from different orientations could be considered as different domains. Our method learns different orientation-specific projections into a common feature space.

orientation output by the orientation gating branch is represented by a two-dimensional Cartesian vector $[\alpha, \beta]$ constructed by projecting the orientation angle on the left-right axis (x) and on the front-back (y) axis and then normalizing it to a unit vector. Based on this vector, the orientation gate selects and weights either the left or the right component and either the front or the back component of the re-identification branch. Thus, we use four different local experts corresponding to left, right, front and back orientations and any orientation can be represented by the combination of these orientations. Let $f_{\{left, right, front, back\}}$ be the output feature vectors of the 4 different orientation branches. The final re-identification output vector is the sum of the left-right component and the front-back component:

$$f_{ouput} = max(\alpha, 0)f_{left} + max(-\alpha, 0)f_{right} + max(\beta, 0)f_{front} + max(-\beta, 0)f_{back}$$
$$(1)$$

Different from the classic mixture of experts approach, our orientation gate is set before the local experts, and we perform a regression in stead of a classification. The advantage of our orientation gate is that it avoids combining contradictory orientations like front and back. Computationally, only two among four orientations are used and combined according to the sign of α and β. This further allows saving computation.

3.2   Training

There are two stages to train the model as shown in Fig. 4. In the first stage, the orientation regressor and a general re-identification feature embedding are both trained in parallel with two separate objective functions. In the second stage, the network is specialized to different orientations. These two steps are detailed in the following.

Multi-task Network Training. We start training the network with pedestrian identity labels and orientation labels respectively. **Identification**: for identification learning, we temporarily add an N-dimensional fully-connected layer to

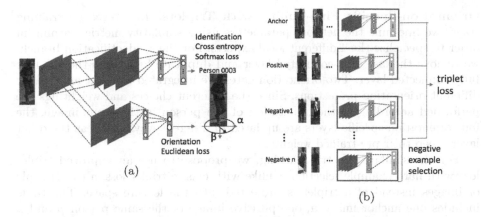

Fig. 4. The two training steps of our method. (a) In the first step, we train the model with identity and orientation labels . (b) Then, we fine-tune the model to train the orientations-specific layers with hard triplets.

the re-identification branch, N being the number of the identities in the training set. The estimated probability of the ith identity is calculated with the softmax function: $p(i) = \frac{exp(z_i)}{\sum_{j=1}^{N} exp(z_j)}$, where $z = [z_1, z_2, ..., z_N]$ is the output of this last fully connected layer. Then, we train the CNN by minimizing the cross-entropy loss:

$$L_{id} = -\sum_{i=1}^{N} log(p(i))l_{id}(i), \qquad (2)$$

where l_{id} is the ground truth one-hot coded identity vector for a given example.

Orientation regression: for the body orientation, we use the Euclidean loss to train the orientation regression of α and β. For a given training example, we have:

$$L_{orien} = \frac{(\alpha - \widehat{\alpha})^2 + (\beta - \widehat{\beta})^2}{2} \qquad (3)$$

where $\widehat{\alpha}, \widehat{\beta}$ are predicted orientation labels of the example. Due to the difficulty in estimating the precise body angle, even for humans, orientation is annotated with 8 discrete labels. For training we convert the orientation class to the vector $[\alpha, \beta]$. To get a more robust orientation learning, we add a uniform random noise of 10 degrees to the orientation labels.

For datasets that have both identity and orientation labels, we train the network with a combined loss $L_{multi-task} = L_{id} + \lambda L_{orien}$. Then, orientation and identification are learned jointly. Otherwise, the two branches are trained separately.

Orientation-Specific Fine-Tuning with Triplets. In the second training stage, we fine-tune the network parameters using similarity metric learning in order to specialize the 4 different local experts. For the re-identification branch, we remove the last fully-connected layer and duplicate four times the the first fully-connected layer. Two orientation gates are integrated to select and weight different orientation projections. Since the different choices and weightings are performed according to the orientation of the person in the input image, the four orientation-specific layers are updated in different ways, whereas the other layers keep their pre-trained weights.

For the similarity metric learning, we propose to use an improved triplet loss with hard example selection. Unlike with classic triplet loss, a (n+2)-tuple of images instead of a triplet is projected into the feature space. The tuple includes one anchor image a, one positive image of the same person p and k negative images of different persons n^j. Training enforces that the projection of the positive example is placed closer to the anchor than the projection of the closest negative example among the k negative examples. This constraint is defined as following:

$$\min_{j=1..k} (\|f(a) - f(n^j)\|_2^2) - \|f(a) - f(p)\|_2^2 > m \tag{4}$$

The negative example that is closest to the anchor is considered the hardest example. The network is thus updated efficiently by pushing the hardest example further away from the anchor. In classic triplet loss, a part of the triplets does not violate the triplet constraint and thus is useless for learning. The selection among k negative examples reduces the number of unused training data and can make the training more efficient. To further enhance the loss function, as [4], we add a term including the distance between the anchor example and the positive example. The loss function for N training examples is defined as follows:

$$E_{triplet} = -\frac{1}{N} \sum_{i=1}^{N} [max(\|f(a_i) - f(p_i)\|_2^2 - \min_{j=1..k} (\|f(a_i^j) - f(n_i^j)\|_2^2) + m, 0)$$

$$+ \gamma \|f(a_i) - f(p_i)\|_2] \tag{5}$$

3.3   Implementation Details

The first convolutional layer has a kernel size of 5×5 and the following have a kernel size of 3×3. All following max-pooling layers have a kernel size of 2×2 except the last one in the re-identification branch which has a kernel size of 3×1 without zero-padding increasing the number of channels and reducing the number of parameters by reducing their size to a single column. Batch normalization and a Leaky ReLU activation function with a slope of 0.2 are applied after the max-pooling layers and fully connected layers. The first fully-connected layers of the re-identification branch and the orientation gating branch output a vector of respectively 512 and 256 dimensions. Dropout is applied to the fully-connected layers to reduce the risk of overfitting. The optimization is performed by Stochastic Gradient Descent with a learning rate of 0.005, a momentum of 0.9 and a batch size of 50. The constant k is set to 5 and γ is set to 0.002 as [4].

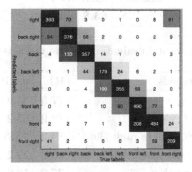

Fig. 5. Orientation confusion matrix on Market-1203.

4   Experiments

4.1   Datasets

The **Market-1501 Dataset** [28] is one of the largest publicly available datasets for human re-identification with 32668 annotated bounding boxes of 1501 subjects. All images are resized to 128 × 48. The dataset is split into 751 identities for training and 750 identities for testing as in [28].

The **Market-1203 Dataset** [16] is a subset of Market-1501 containing 8570 images from 1203 identities under two camera views. 8 body orientations are annotated. We use 601 identities for training and 602 identities for the test. The test on Market-1203 is performed in the way as Market-1501, that means, we pick one image for each identity and each camera view as query (if there's only one image, no image will be picked) and the rest as gallery images. The gallery images from the same identity and the same camera view as the query will be considered as "junk images" which have zero impact on search accuracy. The rank 1 accuracy (R1) and the mean average precision (mAP) are used for performance evaluation.

The **CUHK01 Dataset** [10] contains 971 subjects, each of which has 4 images under 2 camera views. We mannualy annotated each image with 8 body orientations. According to the protocol in [1], the data set is divided into a training set of 871 subjects and a test set of 100 and the extra data from the CUHK03 dataset [11] is also used in training. The CUHK03 dataset is a large person re-identification dataset with 13164 images of 1360 identities. We evaluate in two ways with only CUHK01 data and with CUHK01 plus CUHK03 data in training. The images are all resized to 160 × 60. The Cumulative Match Curve (CMC) is employed as evaluation measure.

Table 1. Experimental evaluation on the market-1203 dataset.

Methods	R1	mAP
Basline	62.0	64.6
OSCNN	**63.8**	**66.4**

Table 2. Experimental evaluation on the market-1501 dataset.

Methods	R1	mAP
Basline	77.3	53.9
OSCNN	78.9	55.2
OSCNN+re-rank [30]	83.9	**73.5**
LOMO+XQDA [13]	43.8	22.2
PersonNet [24]	37.2	18.6
Gated SCNN [4]	65.9	39.6
Divide fues re-rank [26]	82.3	72.4
LSRO [29]	78.1	56.2
DeepContext [8]	80.3	57.5
K-reciprocal re-rank [30]	77.1	63.6
SVDnet [21]	82.3	62.1
JLML [12]	**85.1**	65.5

For all datasets, to reduce over-fitting, we perform data augmentation by randomly flipping the images and by cropping central regions with random perturbation. For the tests on Market-1501 and on CUHK01 with extra data from CUHK03, since only a part of the images has orientation annotations, the re-identification branch and the orientation gating branch are trained separately. For the test on Market-1203 and the one using only the CUHK01 dataset, we perform a joint multi-task training with the combined loss from Sect. 3.2 and $\lambda = 0.01$ determined by a cross-validation.

4.2   Experimental Results

Orientation regression evaluation. We first evaluate the performance of orientation regression. We tested the model after the first training stage on Market-1203 dataset. The confusion matrix is shown in Fig. 5. We calculated also the accuracy rate proposed in [16], *i.e.* result is considered correct if the predicted and true orientation classes are equal or adjacent. Since person appearances obtained in adjacent orientations are very similar, the exact orientation is less important. Thus, this accuracy evaluation criterion is more suitable for the person re-identification problem. On the Market-1203 test set, we can get an accuracy rate of 97.7%.

Orientation gate evaluation. To evaluate the effectiveness of our OSCNN, we set up a baseline method. The baseline performs identity learning with softmax loss, then fine-tuning on hard triplets without the orientation gate. The results on Market-1203, Market-1501 and CUHK01 are respectively shown in Tables 1, 2 and 3. Compared to the baseline, integrating the orientation-based local experts in the CNN framework could achieve a 1.8% point improvement for rank1 on CUHK01, 1.6% and 1.3% points for rank 1 and mAP on Market-1501 and 1.8% and 1.8% points for rank 1 and mAP on Market-1203 . This demonstrates the effectiveness of the orientation gate and the specific projections into a common feature subspace.

Comparison with state-of-the-art. We compared our OSCNN to the state-of-the-art approaches on Market-1501 and CUHK01. Following the test protocol in [1,3,24], we added also the CUHK03 images to the training for the test on the CUHK01 and we compared to the methods only using these two datasets for training. As Table 3 shows, our method is superior to most results in the state-of-the-art. Even without much extra CUHK03 training data, our method shows a competitive performance.

On the Marke-1501 dataset, our OSCNN achieves the same level results as some state-of-the-art methods. Although the result is under the best score of the state-of-the-art, the advantage of our approach is that the model doesn't need a pre-training step with a much larger pre-training dataset composed of ImageNet as [8,21,29,30] . And our model has less complexity (1.15×10^8 FLOPs of our model compared to 1.45×10^9 FLOPs of JLML and to 3.8×10^9 FLOPs of SVDNet). Recently some state-of-the-art approaches show the re-ranking [26,30] which uses information from nearest neighbors in the gallery can significantly improve the performance. As Table 2 shows, our approach can largely benefit from this technique and achieves a state-of-the-art result on Market-1501.

Table 3. Experimental evaluation on the CUHK01 dataset.

Methods	R1	R5	R10	R20
Basline(CUHK01)	76.6	93.8	97.0	98.8
OSCNN(CUHK01)	78.2	94.1	97.3	99.1
OSCNN(CUHK01+03)	**83.5**	96.4	**99.0**	**99.5**
LOMO+XQDA [13]	63.2	83.9	90.1	94.2
ImporvedDL [1]	65.0	88.7	93.1	97.2
PersonNet [24]	71.1	90.1	95	98.1
Deep Embedding [19]	69.4	-	-	-
Norm X-Corr [20]	81.2	-	97.3	98.6
Multi-task [3]	78.5	**96.5**	97.5	-

5   Conclusion

In this paper, we presented a person re-identification approach based on an orientation specific CNN architecture and learning framework. Four orientation-based local experts are trained to project pedestrian images of specific orientations into a common feature subspace. An orientation gating branch learns to predict the body orientation and an orientation gate unit uses the estimated orientation to select and weight the local experts to compute the final feature embedding. We experimentally showed that the orientation gating improves the performance of person re-identification, and our approach outperforms most of the previous state-of-the-art re-identification methods on two public benchmarks.

Acknowledgement. This work was supported by the Group Image Mining (GIM) which joins researchers of LIRIS Lab. and THALES Group in Computer Vision and Data Mining. We thank NVIDIA Corporation for their generous GPU donation to carry out this research.

References

1. Ahmed, E., Jones, M., Marks, T.K.: An improved deep learning architecture for person re-identification. In: CVPR, pp. 3908–3916 (2015)
2. Bak, S., Zaidenberg, S., Boulay, B., Bremond, F.: Improving person re-identification by viewpoint cues. In: IEEE International Conference on Advanced Video and Signal Based Surveillance (AVSS), pp. 175–180. IEEE (2014)
3. Chen, W., Chen, X., Zhang, J., Huang, K.: A multi-task deep network for person re-identification. In: AAAI, pp. 3988–3994 (2017)
4. Cheng, D., Gong, Y., Zhou, S., Wang, J., Zheng, N.: Person re-identification by multi-channel parts-based cnn with improved triplet loss function. In: CVPR, pp. 1335–1344 (2016)
5. García, J., Martinel, N., Foresti, G.L., Gardel, A., Micheloni, C.: Person orientation and feature distances boost re-identification. In: International Conference on Pattern Recognition (ICPR), pp. 4618–4623. IEEE (2014)
6. Gray, D., Tao, H.: Viewpoint invariant pedestrian recognition with an ensemble of localized features. In: ECCV, pp. 262–275 (2008)
7. Koestinger, M., Hirzer, M., Wohlhart, P., Roth, P.M., Bischof, H.: Large scale metric learning from equivalence constraints. In: CVPR, pp. 2288–2295 (2012)
8. Li, D., Chen, X., Zhang, Z., Huang, K.: Learning deep context-aware features over body and latent parts for person re-identification. In: CVPR, pp. 384–393 (2017)
9. Li, W., Wang, X.: Locally aligned feature transforms across views. In: CVPR, pp. 3594–3601 (2013)
10. Li, W., Zhao, R., Wang, X.: Human reidentification with transferred metric learning. In: ACCV (2012)
11. Li, W., Zhao, R., Xiao, T., Wang, X.: Deepreid: deep filter pairing neural network for person re-identification. In: CVPR, pp. 152–159 (2014)
12. Li, W., Zhu, X., Gong, S.: Person re-identification by deep joint learning of multi-loss classification. In: International Joint Conference on Artificial Intelligence (2017)

13. Liao, S., Hu, Y., Zhu, X., Li, S.Z.: Person re-identification by local maximal occurrence representation and metric learning. In: CVPR (2015)
14. Lisanti, G., Masi, I., Del Bimbo, A.: Matching people across camera views using kernel canonical correlation analysis. In: Proceedings of the International Conference on Distributed Smart Cameras, p. 10. ACM (2014)
15. Ma, L., Yang, X., Tao, D.: Person re-identification over camera networks using multi-task distance metric learning. IEEE Trans. Image Process. **23**(8), 3656–3670 (2014)
16. Ma, L., Liu, H., Hu, L., Wang, C., Sun, Q.: Orientation driven bag of appearances for person re-identification. arXiv preprint arXiv:1605.02464 (2016)
17. Pedagadi, S., Orwell, J., Velastin, S., Boghossian, B.: Local fisher discriminant analysis for pedestrian re-identification. In: CVPR, pp. 3318–3325 (2013)
18. Sarfraz, M.S., Schumann, A., Wang, Y., Stiefelhagen, R.: Deep view-sensitive pedestrian attribute inference in an end-to-end model. In: British Machine Vision Conference (BMVC) (2017)
19. Shi, H., et al.: Embedding deep metric for person re-identification: a study against large variations. In: ECCV, pp. 732–748 (2016)
20. Subramaniam, A., Chatterjee, M., Mittal, A.: Deep neural networks with inexact matching for person re-identification. In: NIPS, pp. 2667–2675 (2016)
21. Sun, Y., Zheng, L., Deng, W., Wang, S.: Svdnet for pedestrian retrieval. In: International Conference on Computer Vision (2017)
22. Varior, R.R., Haloi, M., Wang, G.: Gated siamese convolutional neural network architecture for human re-identification. In: ECCV, pp. 791–808 (2016)
23. Verma, A., Hebbalaguppe, R., Vig, L., Kumar, S., Hassan, E.: Pedestrian detection via mixture of CNN experts and thresholded aggregated channel features. In: ICCV Workshops, pp. 163–171 (2015)
24. Wu, L., Shen, C., Hengel, A.v.d.: Personnet: Person re-identification with deep convolutional neural networks. arXiv preprint arXiv:1601.07255 (2016)
25. Yi, D., Lei, Z., Liao, S., Li, S.Z.: Deep metric learning for person re-identification. In: International Conference on Pattern Recognition, pp. 34–39 (2014)
26. Yu, R., Zhou, Z., Bai, S., Bai, X.: Divide and fuse: a re-ranking approach for person re-identification. In: British Machine Vision Conference (BMVC) (2017)
27. Zhao, H., et al.: Spindle net: person re-identification with human body region guided feature decomposition and fusion. In: CVPR, pp. 1077–1085 (2017)
28. Zheng, L., Shen, L., Tian, L., Wang, S., Wang, J., Tian, Q.: Scalable person re-identification: a benchmark. In: International Conference on Computer Vision (2015)
29. Zheng, Z., Zheng, L., Yang, Y.: Unlabeled samples generated by gan improve the person re-identification baseline in vitro. In: International Conference on Computer Vision (2017)
30. Zhong, Z., Zheng, L., Cao, D., Li, S.: Re-ranking person re-identification with k-reciprocal encoding. In: CVPR (2017)

Distributed Estimation of Vector Fields

Ana Portêlo[1]([✉]), Jorge S. Marques[2], Catarina Barata[2], and João M. Lemos[1]

[1] INESC-ID, Instituto Superior Técnico, Universidade de Lisboa, Lisbon, Portugal
ana.i.portelo@gmail.com
[2] Institute for Systems and Robotics, Instituto Superior Técnico,
Universidade de Lisboa, Lisbon, Portugal

Abstract. In many surveillance applications the area of interest is either wide or includes alleys or corners. Thus, the images from multiple cameras need to be combined and this fact motivates the use of distributed optimization approaches. This work proposes three distributed estimation approaches to motion field estimation from target trajectory data: (1) purely decentralized, without communication, (2) distributed estimation based on a cooperative game, and (3) distributed Alternating Direction Method of Multipliers (ADMM). Their performance in estimating different classes of motion fields is important to select the best approach for each application. Experiments using synthetic and real data show that (a) the cooperative game approach is very susceptible to changes in motion direction, and (b) the distributed ADMM approach is the most robust and reliable approach to estimate changing direction motion fields.

Keywords: Distributed optimization · Vector field estimation
Multicamera · Surveillance

1 Introduction

The estimation of motion fields based on target (*e.g.*, pedestrians or vehicles) trajectories provides information about the usual motion flow of targets in a scene [1]. This information can be used in surveillance systems to detect unusual trajectories [2], plan accessibility conditions in cities [3], or for crowd analysis [4].

In many surveillance applications the area of interest is very large or includes alleys or corners. Thus, images from a single camera may not cover the entire area of interest, and the images of two or more cameras need to be combined. In this case, it is reasonable to assume that there is some overlap among the images of the different cameras to make sure the whole area is covered. Multi-camera systems involve several coherence constraints, which motivates the use of distributed optimization algorithms. While previous studies focused on target tracking using image features [5,6], the focus of this work is to ensure coherence among the motion fields estimated using trajectory data from several cameras.

Work supported by FCT and FEDER under contracts PTDC/EEIPRO/0426/2014 (project SPARSIS), UID/CEC/50021/2013 and UID/EEA/50009/2013.

J. Blanc-Talon et al. (Eds.): ACIVS 2018, LNCS 11182, pp. 38–50, 2018.
https://doi.org/10.1007/978-3-030-01449-0_4

In distributed optimization algorithms, several agents estimate a set of variables without the need for a central coordinating agent, nor widespread knowledge of every variable. Distributed algorithms usually involve two steps: the communication step and the computation step. In the communication step, each agent shares its local information (*e.g.*, the new estimates) with its neighbours. In the computation step, each agent minimizes its local cost function using information shared by its neighbours in the prior communication step. Some popular distributed estimation approaches rely on the Alternating Direction Method of Multipliers (ADMM) [7], or on game theory [8].

This work focuses on the estimation of the motion field that describes the observed target trajectories. The proposed distributed estimation approaches are: (1) a purely decentralized approach, without communication among agents, (2) a distributed estimation approach based on a cooperative game, and (3) a distributed version of ADMM. The performance of these approaches in the estimation of different types of motion fields, using different number of trajectories, and with different overlaps among camera images is important to select the best approach for each application. Moreover, it should be relevant to set-up multicamera surveillance systems according to the geometry of the scene, and to optimize existing set-ups. In this framework, the paper contributions are the distributed estimation of motion fields using the three proposed methods and the comparison of their estimation performance considering different classes of motion fields.

This work is organized as follows. Section 2 describes the dynamic model of target trajectories, the parametric motion field representation, and the basic cost function for motion field estimation. Section 3 describes the three proposed distributed estimation approaches. Section 4 presents experiments using synthetic and real target trajectory data, followed by some conclusions in Sect. 5.

2   Dynamic Model and Motion Field Estimation

2.1   Dynamic Model of Target Trajectories

This work assumes that the target trajectories, $\mathbf{x} = (\mathbf{x}_1, \ldots, \mathbf{x}_L)$, on the full image plane (*i.e.*, $[0,1]^2$) are driven by the motion field, $\mathbf{T}(\mathbf{x})$, according to

$$\mathbf{x}(t) = \mathbf{x}(t-1) + \mathbf{T}(\mathbf{x}(t-1)) + \mathbf{w}(t), \tag{1}$$

where $\mathbf{x} \in [0,1]^2$, $\mathbf{T} : [0,1]^2 \to \mathbb{R}^2$, and $\mathbf{w}(t) \sim \mathcal{N}(0, \sigma^2 \mathbf{I})$ is a white random perturbation.

2.2   Motion Field Representation

The motion field is defined only at the grid nodes of an over-imposed uniform grid, $\mathcal{G} = \{\mathbf{g}_i \in [0,1]^2, i = 1, \ldots, N\}$, on the full image plane. This grid contains the open scene and the target trajectories. However, the target trajectories can be defined in any image coordinate, even if it does not correspond to a grid node

($\mathbf{x} \notin \mathcal{G}$). Therefore, it is necessary to use a bilinear interpolation to represent the motion field that drives the trajectories on any coordinate of the full image plane. The bilinear interpolation of the motion field is given by

$$\mathbf{T}(\mathbf{x}) = \sum_{i=1}^{N} \phi_i(\mathbf{x})\, \mathbf{t}_i \,, \tag{2}$$

which is defined in $\mathbf{x} \notin \mathcal{G}$, and where $\phi_i(\mathbf{x})$ are the interpolation coefficients, and $\mathbf{t}_i$ the motion field velocity vectors at the grid nodes.

2.3 Motion Field Estimation

The motion field, $\mathbf{T}$, that rules a set of S collected trajectories $\mathcal{X} = \{\mathbf{x}_1, \ldots, \mathbf{x}_S\}$ on the full image plane is the minimizer of the following cost function

$$f(\mathbf{T}) = \|\mathbf{V} - \mathbf{T}\boldsymbol{\Phi}\|_2^2 + \alpha \, \|\Delta\mathbf{T}\|_2^2 + \beta \, \|\mathbf{T}\|_1 \,, \tag{3}$$

where $\|.\|_p$, $p \in \{1, 2\}$ defines the pth norm of a vector. $\mathbf{T} \in \mathbb{R}^{2 \times N}$, $\mathbf{V} \in \mathbb{R}^{2 \times S\,(L-1)}$, and $\boldsymbol{\Phi} \in \mathbb{R}^{N \times S\,(L-1)}$ are given by

$$\mathbf{T} = \begin{bmatrix} \mathbf{t}_1 \ldots \mathbf{t}_N \end{bmatrix} \,, \tag{4}$$

$$\mathbf{V} = \begin{bmatrix} \mathbf{v}(2) \ldots \mathbf{v}(L_1) \ldots \mathbf{v}(2) \ldots \mathbf{v}(L_S) \end{bmatrix} \,, \tag{5}$$

$$\boldsymbol{\Phi} = \begin{bmatrix} \phi_1(1) \ldots \phi_1(L_1 - 1) \ldots \phi_1(1) \ldots \phi_1(L_S - 1) \\ \vdots \qquad \vdots \qquad \vdots \qquad \vdots \\ \phi_N(1) \ldots \phi_N(L_1 - 1) \ldots \phi_N(1) \ldots \phi_N(L_S - 1) \end{bmatrix} \,, \tag{6}$$

where matrices $\mathbf{V}$ and $\boldsymbol{\Phi}$ respectively consider the velocity, $\mathbf{v}(t) = \mathbf{x}(t) - \mathbf{x}(t-1)$, and the interpolation coefficients per grid node, for each trajectory time point. In (3), the first term is the usual data fidelity criterion, the second term refers to smoothness between pairs of neighbour grid nodes, $(\mathbf{x}_{g1}, \mathbf{x}_{g2}) \in \mathcal{G}$, such that the velocity difference, $\Delta\mathbf{T} = \mathbf{T}(\mathbf{x}_{g1}) - \mathbf{T}(\mathbf{x}_{g2})$, should be small, and the third term refers to sparsity of the motion field.

3   Distributed Estimation Approaches

In this work, the full image plane, *i.e.*, $[0, 1]^2$, is composed of several sub-regions, $\mathcal{R} = \{1, 2, \ldots, R\}$, delimited by the fields of view of the different cameras. Each camera and the respective field of view corresponds to an estimator agent.

There is a set of estimator agents $\mathcal{A} = \{1, 2, \ldots, A\}$. An agent $i \in \mathcal{A}$ has at least one neighbour, $j \in \mathcal{N}_i$. Each agent is composed of sub-regions, $r \in \mathcal{R}_i \subset \mathcal{R}$, that may overlap with neighbour agents. An overlapping sub-region of agent j with its neighbours i is defined as $o \in \mathcal{O}_i^{(j)}$, $j \in \mathcal{N}_i$, where $\mathcal{O}_i^{(j)}$ is the set of overlapping sub-regions of agent j with its neighbours i.

Each estimator agent yields its own motion field estimate. A motion field estimate is the set of estimated velocity vectors sitting on the grid nodes that belong to the set of sub-regions of the respective agent, $i.e.$, $\bar{\mathbf{T}}^{(i)} = \{\mathbf{t}_g\}, g \in \mathcal{R}_i$, in reference to the full image coordinate system. Whenever applicable, $\bar{\mathbf{T}}^{(i)}$ are agent-specific motion field estimates, and $\mathbf{T}_r^{(i)}$, $r \in \mathcal{R}_i$ are sub-region-specific motion field estimates.

3.1   Purely Decentralized Estimation

The first approach is a purely decentralized estimation problem in which communication among agents is not allowed. This approach aims to minimize the local cost functions, f_i defined as in (3). The problem to be solved is

$$\underset{\bar{\mathbf{T}}^{(i)}}{\text{minimize}} \quad f_i(\bar{\mathbf{T}}^{(i)})$$

$$\text{subject to} \quad \mathbf{x}^{(i)}(t) = \mathbf{x}^{(i)}(t-1) + \bar{\mathbf{T}}^{(i)}(\mathbf{x}^{(i)}(t-1)) + \mathbf{w}(t), i \in \mathcal{A}, \quad (7)$$

for each agent i, where $\mathbf{x}^{(i)}$ are the target trajectory data within region $\mathcal{R}_i$. Because there is no communication among agents, (7) can be solved in parallel.

3.2   Distributed Estimation Based on a Cooperative Game

The second approach is based on a distributed cooperative game, which aims to minimize a global cost function. There are three actions, $i.e.$ strategies, that each agent can take to solve the motion field estimation problem. Either keep the previous strategy and share the last best estimate that yielded the global minimizer; cooperate and share an altruist estimate that solves its neighbours motion field estimation problems; or defect and selfishly ask its neighbours to share what it believes to be the best input for its own motion field estimation problem. Thus, in this approach, each agent estimates its motion field, $\bar{\mathbf{T}}^{(i)}$, assuming the neighbour agents share either (a) the last best estimate, $\bar{\mathbf{T}}_b^{(j)}$; (b) the altruist estimate, $\bar{\mathbf{T}}_a^{(j)}$; or (c) the selfish estimate, $\bar{\mathbf{T}}_*^{(j)}$ [9].

The cooperative game algorithm has 2 computation instances, that can be done in parallel, and 2 communication instances. Each agent computes local motion field estimates and cost function values, and shares local motion field estimates with its neighbours and local cost function values with all agents. The problem to be solved is

$$\underset{\bar{\mathbf{T}}^{(1)},...,\bar{\mathbf{T}}^{(A)}}{\text{minimize}} \sum_{i=1}^{A} f_i(\bar{\mathbf{T}}^{(i)}|\mathbf{v}_o^{(i)}, \mathbf{T}_o^{(j)})$$

$$\text{subject to } \mathbf{x}^{(i)}(t) = \mathbf{x}^{(i)}(t-1) + \bar{\mathbf{T}}^{(i)}(\mathbf{x}^{(i)}(t-1)) + \mathbf{w}(t), \, i \in \mathcal{A}, \quad (8)$$

$$\mathbf{v}_r^{(i)}(t) = \mathbf{x}_r^{(i)}(t) - \mathbf{x}_r^{(i)}(t-1), r \in \mathcal{R}_i \setminus \mathcal{O}_j^{(i)},$$

$$\mathbf{T}_o^{(j)} \in \{\mathbf{T}_b^{(j)}, \mathbf{T}_*^{(j)}, \mathbf{T}_a^{(j)}\}, o \in \mathcal{O}_i^{(j)}, j \in \mathcal{N}_i,$$

where $\mathbf{x}_r^{(i)}$ and $\mathbf{v}_r^{(i)}$ are the target trajectory data and respective velocities within agent i sub-region $r \in \mathcal{R}_i \setminus \mathcal{O}_j^{(i)}$, $j \in \mathcal{N}_i$ (where $A \setminus B$ represents the set of elements of set A that are not in set B). In iteration k, each agent i computes its selfish motion field estimate according to

$$\bar{\mathbf{T}}_{*}^{(i),k+1} = \arg\min_{\bar{\mathbf{T}}^{(i)}} f_i(\bar{\mathbf{T}}^{(i)}|\mathbf{v}_r^{(i)}, \mathbf{T}_b^{(j),k}), \qquad (9)$$

which is computed given local data, $\mathbf{v}_r^{(i)}$, $r \in \mathcal{R}_i \setminus \mathcal{O}_j^{(i)}$, and the neighbours best motion field estimates from the previous iteration regarding the overlapping regions, $\mathbf{T}_b^{(j),k}$. Each agent i also computes the altruist motion field estimate it wishes to receive from its neighbours according to

$$\bar{\mathbf{T}}_a^{(j),k+1} = \arg\min_{\bar{\mathbf{T}}^{(j)}} f_i(\bar{\mathbf{T}}^{(j)}|\mathbf{v}_o^{(i)}, \mathbf{T}_*^{(i),k+1}), \qquad (10)$$

which is computed given the data from the overlapping regions, $\mathbf{v}_o^{(i)}$, $o \in \mathcal{O}_i^{(j)}$, $j \in \mathcal{N}_i$, and its new selfish motion field estimate from its local sub-region, $\mathbf{T}_*^{(i),k+1}$.

Then, the agents share their new estimates with their neighbours. This way, each agent can compute its local cost function, defined as in (3), where $\mathbf{V}^{(i)} = \left[\mathbf{v}_r^{(i)} \{\mathbf{T}_o^{(j)}\}_{j\in\mathcal{N}_i}\right]$ considers the neighbours possible new estimates regarding the overlapping regions. The updated local cost function values are also shared. Finally, the decision about which estimate each agent should select is made considering the sums of the local cost functions over all agents, $F = \sum_{i=1}^{A} f_i(\bar{\mathbf{T}}^{(i)}|\mathbf{v}_r^{(i)}, \mathbf{T}_o^{(j)})$, for all the possible combinations of pairs of neighbour agents estimates, $(\bar{\mathbf{T}}^{(i)}, \{\mathbf{T}_o^{(j)}\}_{j\in\mathcal{N}_i}) \in \{\mathbf{T}_b^{(\cdot)}, \mathbf{T}_*^{(\cdot)}, \mathbf{T}_a^{(\cdot)}\}^{|\mathcal{N}_i|+1}$.

3.3 Distributed Estimation Based on ADMM

The third approach is based on the Alternating Direction Method of Multipliers (ADMM) [7]. This work follows the distributed ADMM algorithm [10], adapted to the motion field estimation problem. Each agent i only knows its own cost function, f_i defined as in (3), and the shared motion field estimates from its neighbours, $\mathbf{T}_o^{(j)}$, $o \in \mathcal{O}_i^{(j)}$, $j \in \mathcal{N}_i$. The problem to be solved is

$$\begin{aligned}
\minimize_{\bar{\mathbf{T}}^{(1)},\dots,\bar{\mathbf{T}}^{(A)}} & \sum_{i=1}^{A} f_i(\bar{\mathbf{T}}^{(i)}) \\
\text{subject to } & \mathbf{T}_o^{(i)} = \mathbf{T}_o^{(j)}, o \in \mathcal{O}_i^{(j)}, j \in \mathcal{N}_i, \\
& \mathbf{x}^{(i)}(t) = \mathbf{x}^{(i)}(t-1) + \bar{\mathbf{T}}^{(i)}(\mathbf{x}^{(i)}(t-1)) + \mathbf{w}(t), i \in \mathcal{A},
\end{aligned} \qquad (11)$$

where $\mathbf{x}^{(i)}$ are the target trajectory data within each agent region $\mathcal{R}_i$. The constraints translate in each agent i having a copy of the motion field estimates at its overlapping regions.

The current problem formulation is not yet a distributed optimization problem. A colouring scheme similar to [10] allows the formulation of a distributed version of ADMM by defining

$$\tilde{\mathbf{T}}^c = \begin{cases} \{\bar{\mathbf{T}}^{(i)}\}_{i \in \mathcal{A} \cap \mathcal{C}_c}, & \text{if } j \notin \mathcal{N}_i, (i,j) \in \mathcal{A} \\ \emptyset, & \text{if } j \in \mathcal{N}_i, (i,j) \in \mathcal{A} \end{cases}, \tag{12}$$

where $\tilde{\mathbf{T}}^c$ is the set of $\bar{\mathbf{T}}^{(i)}$ from the agents with colour c. This colouring scheme applied to the constraints of (11), allows their separation into C coupled constraints, such that it is rewritten as

$$\begin{aligned} \underset{\tilde{\mathbf{T}}^1,\dots,\tilde{\mathbf{T}}^C}{\text{minimize}} \quad & \sum_{i \in \mathcal{C}_1} f_i(\bar{\mathbf{T}}^{(i)}) + \dots + \sum_{i \in \mathcal{C}_C} f_i(\bar{\mathbf{T}}^{(i)}) \\ \text{subject to} \quad & \tilde{\mathcal{M}}^1 \tilde{\mathbf{T}}^1 + \dots + \tilde{\mathcal{M}}^C \tilde{\mathbf{T}}^C = 0, \\ & \mathbf{x}^{(i)}(t) = \mathbf{x}^{(i)}(t-1) + \bar{\mathbf{T}}^{(i)}(\mathbf{x}^{(i)}(t-1)) + \mathbf{w}(t), i \in \mathcal{A}, \end{aligned} \tag{13}$$

where $\tilde{\mathcal{M}}^c$ is the diagonal concatenation of the transpose of the neighbour-overlap (*i.e.*, node-arc) incidence matrices $\mathcal{M}_1^c, \mathcal{M}_2^c, \dots, \mathcal{M}_O^c$, over the set of overlapping sub-regions $o \in \mathcal{O}_i^{(j)}$, for all pairs of neighbour agents. This problem can be solved using the multi-block/distributed ADMM [10], where λ_o^{ij} is the dual variable associated to $\mathbf{T}_o^{(i)} = \mathbf{T}_o^{(j)}$, $o \in \mathcal{O}_i^{(j)}$, $j \in \mathcal{N}_i$, and $\gamma = \sum_{i \in \mathcal{C}_c} \sum_{j \in \mathcal{N}_i} \lambda_o^{ij}$. The augmented Lagrangian of (13) is

$$\mathcal{L}_\rho(\tilde{\mathbf{T}}^1, \dots, \tilde{\mathbf{T}}^c; \gamma) = \sum_{c=1}^C \sum_{i \in \mathcal{C}_c} f_i(\tilde{\mathbf{T}}^{(i)}) + \sum_{c=1}^C \gamma^\top \tilde{\mathcal{M}}^c \tilde{\mathbf{T}}^c + \frac{\rho}{2} \left\| \sum_{c=1}^C \tilde{\mathcal{M}}^c \tilde{\mathbf{T}}^c \right\|^2, \tag{14}$$

with penalty parameter, $\rho > 0$. The problem to be solved consists of a sequence of C sub-problems, obtained by minimizing (14) with respect to each block $\tilde{\mathbf{T}}^c$, and of updates of the dual variable γ. The resulting Distributed-ADMM (D-ADMM) algorithm updates are

$$\tilde{\mathbf{T}}^{1,k+1} = \underset{\tilde{\mathbf{T}}^1}{\arg\min} \sum_{i \in \mathcal{C}_1} f_i(\bar{\mathbf{T}}^{(i)}) + \gamma^{k\top} \tilde{\mathcal{M}}^1 \tilde{\mathbf{T}}^1 + \frac{\rho}{2} \left\| \tilde{\mathcal{M}}^1 \tilde{\mathbf{T}}^1 + \sum_{c=2}^C \tilde{\mathcal{M}}^c \tilde{\mathbf{T}}^{c,k} \right\|^2, \tag{15}$$

$$\vdots$$

$$\tilde{\mathbf{T}}^{C,k+1} = \underset{\tilde{\mathbf{T}}^C}{\arg\min} \sum_{i \in \mathcal{C}_C} f_i(\bar{\mathbf{T}}^{(i)}) + \gamma^{k\top} \tilde{\mathcal{M}}^C \tilde{\mathbf{T}}^C + \frac{\rho}{2} \left\| \sum_{c=1}^{C-1} \tilde{\mathcal{M}}^c \tilde{\mathbf{T}}^{c,k+1} + \tilde{\mathcal{M}}^C \tilde{\mathbf{T}}^C \right\|^2, \tag{16}$$

$$\gamma^{k+1} = \gamma^k + \rho \sum_{c=1}^C \tilde{\mathcal{M}}^c \tilde{\mathbf{T}}^{c,k+1}. \tag{17}$$

There are $(C-1)+1$ communication instances in each iteration: (a) the agents of colour c share their new estimates with neighbour agents of colour $c+1$, and (b) all neighbour agents get the new estimates and the updated Lagrange multipliers. Each update $\tilde{\mathbf{T}}^c$ can be decomposed into $|\mathcal{C}_c|$ problems that can be solved in parallel, given that agents of the same colour cannot be neighbours.

4   Experimental Results

This section presents motion field estimation experiments, on the image plane, using synthetic and real trajectory data and the three proposed estimation approaches described above. These experiments consider only 2 estimator agents.

Motion field estimation performance is assessed both regarding the magnitude and the relative angle of the estimated pairs of vectors *via* a vector field evaluation (VFE) diagram [11], which considers the vector similarity coefficient (R_v) and the vector root mean square length (RMSL) defined as,

$$R_v = \frac{1}{N} \sum_{i=1}^{N} \mathbf{A}_i^* \cdot \mathbf{B}_i^*, \tag{18}$$

$$\text{RMSL} = L_V^2 = \frac{1}{N} \sum_{i=1}^{N} \|\mathbf{V}_i\|_2^2, \tag{19}$$

where "." is the inner product between two normalized vector fields $\mathbf{A}_i^*$ and $\mathbf{B}_i^*$, with $\mathbf{V}_i^* = \frac{\mathbf{V}_i}{L_V}$ given $\mathbf{V}_i = (x_{vi}, y_{vi})$, $i = 1, 2, \ldots, N$, and $\|.\|_2$ represents the l_2-norm of a vector. These metrics respectively assess the mean of the inner product of normalized vector pairs (18), and the systematic difference in the mean vector length (19). To facilitate interpretation of the results, RMSL is normalized with respect to L_V, where V is either the known generating motion field (in the synthetic case) or the full image estimated motion field (in the real case) when the focus is accuracy assessment, or the estimated motion field from agent 1 when the focus is consensus assessment. Thus, in the figures below, values closer to the black circumference represent better performance.

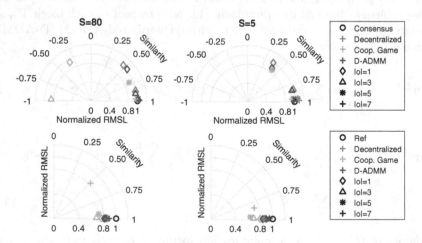

Fig. 1. Vector field diagrams to assess consensus between estimated motion fields of neighbouring agents (top row) and estimation accuracy with reference to the circular motion field (bottom row). The different colours represent each distributed estimation approach. The columns represent different available trajectories ($S = 80, 5$). The symbols represent several widths of the overlapping region ($|o| = 1, 3, 5, 7$).

The values of the cost function parameters α and β on (3) were the minimizers of the motion field estimation problem considering the full image where $\rho = 0$. Then, the D-ADMM penalty parameter, ρ, was selected to yield a good trade-off between accuracy and consensus among neighbour agent estimates. The selected parameter values were $\alpha = 0.2$, $\beta = 0.2$, and $\rho = 1$. The noise variance of the synthetic data was $\sigma^2 = 10^{-4}$. Motion field estimation depended on the number of available trajectories ($S = 5, 80$), and on the width of the overlapping region ($|o| = 1, 3, 5, 7$) within an over-imposed grid of 11×11 nodes.

4.1   Synthetic Target Trajectory Data

Synthetic data consisted on target trajectories of length L generated with different types of motion fields: (a) a circular motion field (changing motion direction); (b) a V-shaped upwards motion field (composed of two upwards diagonal motion fields – to the left or to the right).

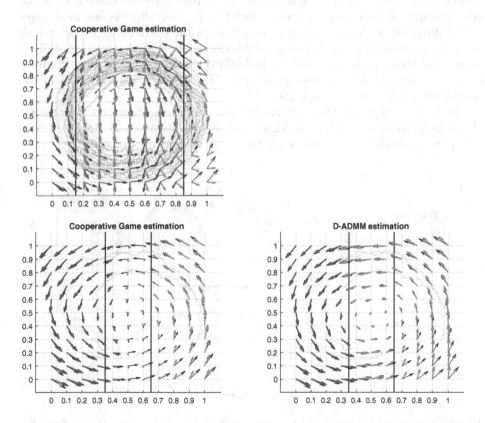

Fig. 2. Circular motion field estimation results (thick arrows). Black thin arrows are the ground truth. *Top*: motion field estimates with $S = 80$, $|o| = 7$ for the Cooperative game approach. *Bottom*: motion field estimates with $S = 5$, $|o| = 3$ for the Cooperative game and D-ADMM approaches. There are two estimator agents: one from the left hand side until the vertical line on the right, and the other from the vertical line on the left until the right hand side.

Regarding the different types of motion field, we expected the circular motion field to highlight the robustness of the D-ADMM approach in estimating motion fields with changing direction within the overlapping region, and the susceptibility of the Cooperative game to local optima. Moreover, because the V-shape upwards motion field was created to mimic the generating motion field of the real target trajectory data, we expected it to provide ground truth results for comparison. Regarding the conditions for motion field estimation, we expected more available trajectories to improve accuracy and consensus error of the estimates. We also expected D-ADMM to be more robust to changes in the width of the overlapping region.

Circular Motion Field. The top row of Fig. 1 shows that higher $|o|$ yields higher angle similarity between pairs of vectors for all approaches. The Cooperative game is the approach that is most affected by changes in $|o|$. The lack of consensus between neighbour agent estimates is mostly due to differences in the relative angle of the estimated vectors (see Fig. 2). This effect is due to changes in the direction of the generating motion field in the overlapping region, which are not easy to extrapolate between neighbour agents. The bottom row of Fig. 1 shows that the Decentralized and the D-ADMM approaches, yield more accurate estimates than the Cooperative game approach. This effect is observed in the magnitude of the estimated vectors.

Moreover, fewer available trajectories also yield less accurate estimates for the Cooperative game and the D-ADMM approaches (see Fig. 2). Regions where there are no available trajectories are the most affected, as expected.

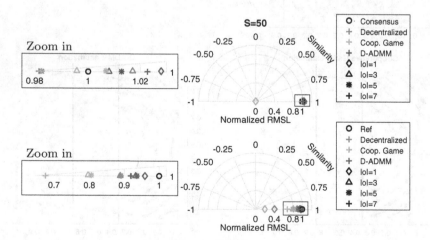

Fig. 3. Assessment of motion field estimation from synthetic data. Vector field diagrams to assess consensus between estimated motion fields of neighbouring agents (top row) and estimation accuracy with reference to the V-shaped upwards motion field (bottom row). The different colours represent each distributed estimation approach. The symbols represent several widths of the overlapping region ($|o| = 1, 3, 5, 7$).

V-Shaped Upwards Motion Field. The zoom in panels of Fig. 3 show that medium $|o|$ yields better matching and more accurate vector field estimates for all approaches. The D-ADMM approach is the least affected by changes in $|o|$, and the one that yields more accurate vector field estimates (see Fig. 4). This experiment, with fixed $S = 50$, serves as ground truth for comparison with the real data experiment.

4.2   Real Target Trajectory Data

The video signal was acquired using a Sony HDR-CX260 video camera with a resolution of 8.9 megapixels per frame and working at a frame rate of 30 frames per second. The targets trajectories were extracted using a tracking algorithm [12]. The trajectories were then sub-sampled at a frame rate of 1 frame per second, and the association errors were corrected. There were $S = 47$ available trajectories (see bottom-right panel of Fig. 4).

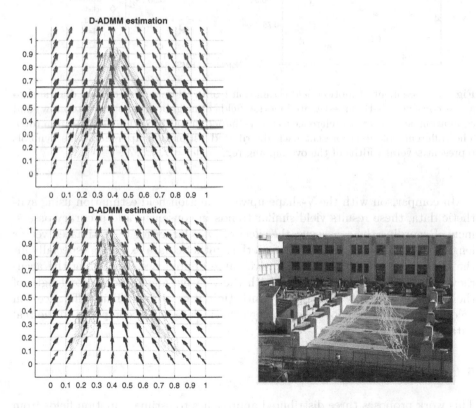

Fig. 4. V-shaped upwards motion field results (thick arrows). Black thin arrows are the ground truth. *Right*: the real data images with over-imposed trajectories. *Left*: V-shaped upwards motion field estimates for the D-ADMM approach with $|o| = 3$, using synthetic (top) and real (bottom) trajectory data. There are two estimator agents: one from the bottom until the upper horizontal line, and the other from the lower horizontal line until the top.

V-Shaped Upwards Motion Field. Figure 5 shows that medium to high $|o|$ yields better matching and more accurate vector field estimates for all approaches. The D-ADMM approach is the least affected by changes in $|o|$, regarding consensus of the estimated vector pairs, and the one that yields more accurate estimates.

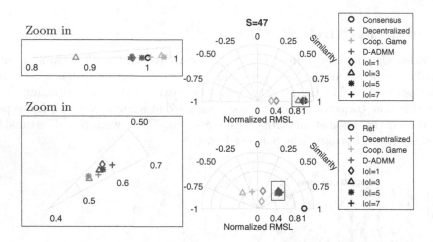

Fig. 5. Assessment of motion field estimation from real data. Vector field diagrams to assess consensus between estimated motion fields of neighbouring agents (top row) and estimation accuracy with reference to the V-shaped upwards motion field (bottom row). The different colours represent each distributed estimation approach. The symbols represent several widths of the overlapping region ($|o| = 1, 3, 5, 7$).

In comparison with the V-shape upwards motion field estimation using synthetic data, these results yield similar trends in consensus and accuracy assessment. Regarding the consensus, the effect of $|o|$ is larger for the estimated vector length of the real data example than in the synthetic data example, especially for the Decentralized and the Cooperative game approaches. Regarding the accuracy, the effect of $|o|$ is larger for both the relative angle and vector length of the real data example than of the synthetic data example. This difference can be due to using low σ^2 for generating synthetic data, when compared to the real data.

5   Conclusions

This work proposes three distributed approaches to estimate motion fields from target (*e.g.*, pedestrians or vehicles) trajectories in an open scene. The motion fields are assumed to rule the target trajectories [1]. This work focuses on surveillance scenarios of very large spaces or including alleys or corners, in which cases the images from a single camera are insufficient to cover the whole area of interest. The first approach considered was a purely decentralized one without communication among estimator agents. The second approach was a distributed estimation approach based on a Cooperative game. In this scenario, the agents estimates do not necessarily converge to the optimum. The third approach was a distributed version of the ADMM algorithm. In this scenario, the goals are to obtain both accurate estimates and consensus among neighbour agents estimates.

The experiments using synthetic target trajectory data show that the D-ADMM approach is the most robust and reliable approach to estimate generating motion fields with changes in direction within the overlapping regions of neighbour agents because only the D-ADMM approach considers a flexible equality constraint among neighbour agents estimates. Contrarily, the Cooperative game approach is very susceptible to changes in motion direction in the overlapping region of neighbour agents.

Regarding the number of available trajectories, the experiments show that fewer available trajectories hamper motion field estimation, as expected. Regarding the overlapping region, the D-ADMM approach is the most robust to different overlapping widths. Finally, the results on the V-shaped upwards motion field estimation showed that, from the proposed approaches, the D-ADMM yields more accurate and consensual estimates.

References

1. Nascimento, J.C., Figueiredo, M.A.T., Marques, J.S.: Trajectory analysis in natural images using mixtures of vector fields. In: ICIP, pp. 4353–4356 (2009)
2. Marques, J.S., Figueiredo, M.A.T.: Fast estimation of multiple vector fields: application to video surveillance. In: ISPA 2011 - 7th International Symposium on Image and Signal Processing and Analysis (2011)
3. Ferreira, N., Klosowski, J.T., Scheidegger, C.E., Silva, C.T.: Vector field k-Means: clustering trajectories by fitting multiple vector fields. In: Eurographics Conference on Visualization (EuroVis), vol. 32, no. 3 (2013)
4. Yao, H., Cavallaro, A., Bouwmans, T., Zhang, Z.: Guest editorial introduction to the special issue on group and crowd behavior analysis for intelligent multicamera video surveillance. IEEE Transa. Circ. Syst. Video Technol. 27(3), 405–408 (2017)
5. Sankaranarayanan, A.C., Veeraraghavan, A., Chellappa, R.: Object detection, tracking and recognition for multiple smart cameras. Proc. IEEE 96(10), 1606–1624 (2008)
6. Taj, M., Cavallaro, A.: Distributed and decentralized multicamera tracking. IEEE Sig. Process. Mag. 28(3), 46–58 (2011)
7. Boyd, S., Parikh, N., Chu, E., Peleato, B., Eckstein, J.: Distributed optimization and statistical learning via the alternating direction method of multipliers. Found. Trends Mach. Learn. 3(1), 1–122 (2010)
8. Li, N., Marden, J.R.: Designing games for distributed optimization. IEEE J. Sel. Topics Sig. Process. 7(2), 230–242 (2013)
9. Maestre, J.M., Muros, F.J., Fele, F., Peña, D.M., Camacho, E.F.: Distributed MPC based on a team game. In: Maestre, J.M., Negenborn, R.R. (eds.) Distributed Model Predictive Control Made Easy. ISCASE, vol. 69, pp. 407–419. Springer, Dordrecht (2014). https://doi.org/10.1007/978-94-007-7006-5_25

10. Mota, J.F., Xavier, J.M., Aguiar, P.M., Puschel, M.: Distributed optimization with local domains: applications in MPC and network flows. IEEE Trans. Autom. Control **60**(7), 2004–2009 (2015)
11. Xu, Z., Hou, Z., Han, Y., Guo, W.: A diagram for evaluating multiple aspects of model performance in simulating vector fields. Geosci. Model Dev. **9**(12), 4365–4380 (2016)
12. Veenman, C.J., Reinders, M.J.T., Backer, E.: Resolving motion correspondence for densely moving points. IEEE Trans. Patt. Anal. Mach. Intell. **23**(1), 54–72 (2001)

Clustering Based Reference Normal Pose for Improved Expression Recognition

Andrei Racoviţeanu, Iulian Felea, Laura Florea[(✉)], Mihai Badea, and Corneliu Florea

Image Processing and Analysis Laboratory,
University Politehnica of Bucharest, Bucharest, Romania
laura.florea@upb.ro

Abstract. In this paper the theme of automatic face expression identification is approached. We propose a robust method to identify the neutral face of a person while showing various expressions. The method consists in separating various images of faces based on expressions with a clustering method and retrieving the neutral face as being in the image closest to the centroid of the dominant cluster. The so found neutral face is used in conjunction with an expression detection method. We tested the method on the Extended Cohn-Kanade database where we identify correctly the neutral face with 100% accuracy and on the UNBC McMaster Pain Shoulder database where the use of the neutral pose leads to an increase of 10% in accuracy thus entering in the range of state of the art in pain detection.

1 Introduction

Multi-modal human-machine communication, like automatic gesture, speech, human expression or emotion recognition has gained popularity in recent years and significant efforts have been made by researchers in these fields of study. However the favorite approach is based on the analysis of the face as most of our information is retrieved as visual data. In this paper, we focus on an automatizing solution for face expression identification. More precisely, we propose a method for face analysis that can help expression recognition algorithms by establishing a reference.

Face expression is one of the most important non-verbal communication methods used by humans in order to understand the mood and the mental state of the interlocutor. Automatic face expression recognition has become an important and challenging area in computer vision. Its applications domain ranges from basic emotion recognition, medical mental evaluation [27] as well as non-psychotic disorders [8], security, automatic counselling systems, face expression synthesis [32], lie detection, music for mood [10], driver fatigue detection [33], etc.

Changes in facial appearance are used for automatic expression recognition. These changes can be minor (e.g. due to wrinkles) or major deformations (like

© Springer Nature Switzerland AG 2018
J. Blanc-Talon et al. (Eds.): ACIVS 2018, LNCS 11182, pp. 51–61, 2018.
https://doi.org/10.1007/978-3-030-01449-0_5

eyes closing, mouth opening, eye-brow rising, etc.). The most used model for studying facial expressions was proposed by Ekman and Friesen and it is called the Facial Action Coding System (FACS) [11,12]. FACS is an anatomically based coding scheme that describes the different combinations of facial muscle movements in terms of 44 unique Action Units (AUs). By means of AUs, the face expression can be encoded in a more simple way. The presence of different combinations of AUs can be interpreted as the subject's emotion, or the presence of pain, etc.

The mentioned deformations are related to facial muscles contraction or relaxation. They may differ according to face shape, appearance, or dynamics. In order to automatically detect these kind of deformations, a reference is desirable. Although there are methods that consider that the neutral pose of the person is a separate expression, there is a plethora of algorithms [6,17,18,29] that use the neutral pose as a reference in order to detect the other expressions occurrence. The use of this neutral pose increases the accuracy in these cases.

In many cases the neutral pose is considered as being known or is extracted from the dataset. Sometimes solutions speculate a particular characteristic of the dataset [6,17,18,29], which allowed them to manually identify a reference image. Alternatively one may actively and automatically try to identify the neutral face. From the more recent examples, Ghosh et al. [15] assumed the mean face averaged over all frames is very similar with the neutral face which is known to be at the beginning of the sequence from the database characteristics. Bishay and Patras [3] trained a dedicated convolutional neural network to substract the subject's mean face. Starting from the assumption that the neutral face is contained in the majority of frames of a person, Baltrusaitis et al. [1] proposed to estimate the neutral expression descriptor by computing the median value of face descriptors in all the images in a sequence of a person.

In some practical, real–life applications, one may extract the neutral pose of the person under expression investigation. For instance, recent high-end smartphones contain a function module for face recognition. Hence, the accumulation of multiple poses for the phone owner is easily achievable; these frames may be analyzed to identify the neutral pose. As other example, in the case of a medical consult, the patient may be filmed before examination, when the pain causing movement may take place; again multiple frames for a person are available (from the first part of the consult) so to identify the neutral pose.

The main contribution of this paper is the introduction of a robust and *automatic* method to identify the neutral pose of a person from a sequence with expressions. The identified image, coupled with an efficient expression detection method is able to increase the performance of simple solution to state of the art level, in the context of pain detection on the UNBC Shoulder Pain database.

The remainder of the paper is organized as follows: in the next section, we recall a few findings from psychology to motivate our solution. In Sect. 3 we describe the two components of the proposed method: the expression detection method and, respectively, the neutral pose identification component. Section 4 presents the used databases, implementation details and results. The paper ends with a short conclusion.

2   Motivation

Our method is motivated from psychological findings. First we need to stress that while temporal analysis over the duration of expression is scarce in the psychological domain, Ekman and Friesen. [13] found that emotional expressions are rather limited in duration, being usually between 0.5 and 4 s in duration. The posed expression varies from 0.25 s to considerably longer than 4s. Krumhuber et al. [20] further found that posed expressions have longer onset and offset durations. Thus, taking into consideration a long enough period of time, one will access plenty of neutral faces or at least plenty of faces with low intensity expressions, near neutral.

Previous attempts on identifying the neutral face were based on the mean face image [15], while very accurate modelling the face dynamic considers Action Units[1]. In this context, two observations need to be made: (1) an individual AU assumes modification in one direction of a specific face part, and (2) there are no opposite AU pairs. If a person performs only, for instance, AU6 "Cheek raiser" with varying intensities, the average image, depending on the meaning of "average" is either a blur over the cheek area, or a moderately Cheek raised, but not a neutral face The second observations refers to the fact that, for instance, while AU1 "Inner brow raiser" and AU4 "Brow lowerer" have opposite directions, they are not the opposite of one another and by averaging them together one will not obtain the neutral pose for the brow.

We hypothesize that due to the fact that expressions represent tension in some facial muscles, which means that the subject invests some effort, from a probabilistic point of view, the most likely to occur is the neutral face which assumes no effort from the subject. However in practical situations, an observer may not be able to record a subject face for sufficiently long time in order to obtain the neutral face as majority, thus we will cluster expressions and we will assume that only the largest cluster (i.e. plurality) is associated with neutral faces. Further more, even if in this cluster expression faces are included, they are placed in marginal positions.

3   Proposed Method

For expression detection we combine a selected ensemble of features extracted from a Pyramid of Histogram of Oriented Gradients with a SVM classifier. For neutral pose identification we rely on a version of Fuzzy C-means clustering. In the next subsections we will review the main features of each method.

---

[1] Visual dynamic examples of the Action Units may be followed at https://imotions. com/blog/facial-action-coding-system/.

First, given an image with a potential face, we use the Caffe based implementation of the multitask cascaded convolutional networks (MTCN) [35] to detect and localize the face. All the following processing is done inside the found gray scale face rectangle.

3.1   Pyramid of Histogram of Oriented Gradients

The Pyramid of Histogram of Oriented Gradients (PIIOG) was proposed for object categorization by Bosch et al. [4]. The idea was to combine the Histogram of Gradient Orientation (HOG) of [7] and the image pyramid representation of [21]. The resulting pyramidal descriptors encode the spatial layout of the computed features at several resolution levels.

A given image is resized into a sequence of decreasing resolutions by halving the number of pixels in each axis direction. Thus the first pyramid level will be the image itself, the second level will contain half the number of pixels on both axes, the third level will contain a fourth of the original number of pixels on each axis and so on. Each level is defined by its corresponding HOG feature vector. All these vectors will be concatenated in the final PHOG feature. The process is illustrated in Fig. 1.

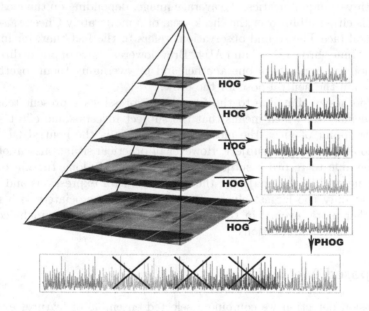

Fig. 1. The Pyramid of Histogram of Oriented Gradients applied on detected faces with expression. At the bottom, we have marked the (2,3,4) levels that have been removed in the Sequential Backward Selection procedure.

3.2   Classifier and Feature Selection

For classification we use a Support Vector Machine (SVM) with a Radial Basis Function (RBF) kernel. The cost C parameter of the SVM learning and the γ of the kernel are found by a grid search optimization on the validation set.

Taking into account the high dimensionality of the initial feature descriptor, a feature selection method has been employed in the form of Sequential Backward Selection (SBS). The objective function is expression detection accuracy on CK+ set achieved with optimal (in the sense of C, γ) SVM. For this test the CK+ database has been split into 70% (of the persons) in training and 30% in testing.

Compared to the standard version of SBS, in our case, the features are not removed independently, but in groups. A group of features is the histogram corresponding to a specific level from the Pyramid of HOG. We have used 5 level description thus, the first iteration of SBS has 5 steps. After 3 eliminations in SBS, the performance no longer increased. The neutral expression identification is based on the histograms from the selected two levels.

3.3   Fuzzy C-Means

To identify the neutral pose, we considered a (unsupervised) clustering method that groups the face descriptors into 8 sets. Our solution is based on Fuzzy C-Means.

We recall that for Fuzzy C-Means (FCM) [2,9], the following objective function has to be minimized:

$$J_{FCM} = \sum_{k=1}^{P}\sum_{i=1}^{N} \nu_{ik}(x_k - v_i)^2, \ s.t. \sum_{i=1}^{N} \nu_{ik} = 1, \forall k \tag{1}$$

where x_k are the image features, P is the number of data instances, N is the total number of clusters and v_i are the centroids/means of the clusters. The solution (ν_{ik}, v_i) is found iteratively once the number of clusters, N is chosen.

To improve the functionality of the standard FCM an adaptation is employed. Iterative FCM convergence has the known drawback of falling into local optima and simulated annealing was proposed to address this aspect [22].

In our case, we have selected all images associated with one person and performed FCM on them. The number of clusters is set to $N = 8$, given as number of fundamental expressions as found by Ekman [11]. A threshold to $\frac{1}{N}$ turns the soft clustering FCM into a hard one and the largest (i.e. with most instances) is selected as representative of the neutral pose. The centroid is taken as the reference image for the neutral pose.

4   Implementation and Results

The implementation has been carried in Python using basic functionalities from the Numpy, SciKit-image and SciKit-learn packages.

Our solution has been developed in the following steps: (1) Initial expression detection on the CK+ database; here, we start by using the full PHOG feature with SVM and optimize via SBS the feature; (2) Neutral pose identification; this is carried using FCM on the reduced PHOG feature set, per person, per database; (3) application: expression detection and respectively pain detection; these use the neutral pose identified in the previous step and are carried on the reduced PHOG set with SVM classifier.

4.1   Evaluation on CK+ Database

We start experimenting with the Extended Cohn-Kanade (CK+) database [25]. The database contains images from 123 subjects. A subject appears in several sequences. A sequence starts with a neutral pose and gradually moves into an expression. However, for a realistic evaluation, prior information about neutral pose and expression pose should not be available. Thus we only take this information into account while evaluating the neutral pose estimation algorithm.

For each subject, FCM was used to select the neutral pose of that person, being given all the images from all the sequences containing that person without any labelling. In *all cases*, a neutral pose (i.e. one of the first two frames from some sequence) was correctly identified. This will be further considered the neutral pose of that person.

As an extended test, we have removed the middle third from all sequences as their membership to either neutral or expression is uncertain. The first third of a sequence is considered (labelled) neutral, while the last is considered (labelled) to contain expression. We sought to detect this correctly assuming, person-wise, separation of 70-30 for training-testing. In this case a binary detector increased its performance from 75% (without using the previously identified neutral face) to 85% (while using the automatically detected neutral image).

4.2   Evaluation on UNBC-Pain Database

The second and more extensive set of experiments is performed on the UNBC McMaster Pain Shoulder database. This database contains face videos of patients suffering from shoulder pain as they perform motion tests of their arms. The movement is either voluntary by the patient itself, or the subjects arm is moved by the physiotherapist. Only one of the arms is affected by pain, but movements of the other arm are recorded as well, to form a control set. 200 sequences from 25 subjects were recorded to a total of 48,398 frames. The pain is computed using Prkchin-Solomon Pain Index metric. Each frame is available with expert annotation. The evaluation is performed while detecting pain as opposed to non-pain frames.

In contrast to CK+ database, where the sequences start by forcible asking for a neutral pose, in the case of UNBC-Pain database, the subjects were not recorded to have a neutral face, and in many cases (sequences), the subject is speaking continually to the doctor, thus the neutral frame is not obvious. The found 25 neutral faces (for the 25 persons) are shown in Fig. 2.

Fig. 2. The found neutral faces for 25 persons from the UNBC McMaster database. While some images are not purely neutral, the expression intensity showed is reduced nonetheless in most of the cases.

Comparison with Alternatives for Neutral Face Identification. As said in the introduction, several solutions have been proposed for the identification of the neutral face, such as Ghosh et al. [15] who assumed the neutral face to be the mean face. Given the face representation (directly, by pixels) or in the feature space, the mean, dimension wise, may not have the meaning of an actual face. We also try the median directly or by aggregated distance as defined in [26]. These results may be followed in Table 1. As one may see, the clustering in FCM allows face in expression to reach other clusters and affect much less the identification of the neutral face.

Using a previously detected neutral face increases the performance of the simple pain detection method. This increase is of nearly 10% if the neutral face is detected with the here proposed method, making the simple method jump closer to state of the art. Other solutions for identifying neutral face also lead to better performance, yet the increase is less consistent.

Table 1. Pain detection performance when various methods for identification of the neutral face are envisaged.

Method	None	Mean image [15]	Median	Aggregated distance	FCM
Accuracy	71.5%	76.5%	77.2%	78.0%	80.2%

Comparison with State of the Art in Pain Detection. From the methods previously used to detect pain, we mention the work of Lucey et al. [23] who use Active Appearance Models (AAM) to align, and track key frames manually selected and then used the same features and a SVM for classifying the frames with pain. The same AAM landmarks with added Local Binary Pattern (LBP) features were used by Chen et al. [5] in order to transfer information from other patients to the current one and enhance the pain detection accuracy. Person specific classifiers were trained by Zen et al. [34] and Sanginieto et al. [30], who also use transfer data for expression detection for pain classification. Irani et al. [19] used temporal features in areas localized by landmarks followed by SVM classification. Pederesen et al. [28] used learned feature selection in the same frame as the previous mentioned solution. Wand et al. [31] used Bag of Words paradigm followed by max pooling for feature selection and SVM for classification. Florea et al. [14] coupled texture description via Histogram of Topographical features with landmarks and ensemble classification for pain detection.

Another category of methods resorted to transfer learning to cope with the limited person variability in UNBC database. Histograms of Local Binary Patterns (LBP) in conjunction with PCA for dimensionality reduction were used as features [5,30,34]. Depending on the method different classifiers and transfer learning procedures were proposed. Chen et al. [5] used AdaBoost (AB) and Transductive Transfer AdaBoost (TTA). Zen et al. [34] report results for Support Vector-based Transductive Parameter Transfer (SVTPT) and also for Selective Transfer Machine (STM). Transductive Parameter Transfer with Density Estimate Kernel were proposed by Sangineto et al. [30].

In Table 2 we present the results reported by the mentioned works comparatively to the performance of the proposed method. As one can see, our method reaches high accuracy managing to outcome multiple methods. In the

Table 2. Comparison with state of the art in pain detection. The explanation for the acronyms is in text.

Method	[24]	Hammal et al. [16]	Irani et al. [19]	Lucey et al. [23]
Accuracy	78.0	75.5	76.5	83.4
Method	[31]	Florea et al. [14]	Chen et al. [5] - AB	Chen et al. [5] - TTA
Accuracy	81.5	80.9	76.9	76.5
Method	Pedersen et al. [28]	Zen et al. [34]	Sangineto et al. [30]	Proposed
Accuracy	86.1	78.4	76.7	**80.2**

more recent years several solutions, [14,23,28,31] managed to reach better performance for pain detection but using much more complex systems. Again, we stress that our system contains descriptors from two levels of PHOG coupled by SVM and uses the automatically found neutral face; we do not use the landmarks on the face or other elaborated transfer learning methods.

5 Discussion

In this paper we have proposed a robust method for automatic identification of the neutral face that can be used as a first step for expression identification. The method is based on clustering the examples of frames from the investigated person. The neutral face is found as the closest to the centroid of the largest resulting cluster. The motivation for choosing the clustering method lies in several psychological results, which indicate a higher (yet not dominant) probability for neutral face while having longer acquisitions with the face of interest.

The correctly identified neutral face in context of expression was first tested on the CK+ database where we coupled a simple method for expression detection with the automatically detected neutral face of the subject and the performance was improved with 10%. With the same simple method for detecting the expression (pain), we manage to reach near state of the art performance on the UNBC database.

Acknowledgment. This work was supported by the Ministry of Innovation and Research, UEFISCDI, project SPIA-VA, agreement 2SOL/2017, grant PN-III-P2-2.1-SOL-2016-02-0002.

References

1. Baltrusaitis, T., Mahmoud, M., Robinson, P.: Cross-dataset learning and person-specific normalisation for automatic action unit detection. In: 2015 11th IEEE International Conference and Workshops on Automatic Face and Gesture Recognition (FG), vol. 6, pp. 1–6 (2015)
2. Bezdek, J.C.: Pattern Recognition with Fuzzy Objective Function Algoritms. Plenum Press, New York (1981)
3. Bishay, M., Patras, I.: Fusing multilabel deep networks for facial action unit detection. In: 2017 12th IEEE International Conference on Automatic Face Gesture Recognition (FG 2017), pp. 681–688 (2017)
4. Bosch, A., Zisserman, A., Munoz, X.: Representing shape with a spatial pyramid kernel. In: Proceedings of the 6th ACM International Conference on Image and Video Retrieval, CIVR 2007, pp. 401–408 (2007)
5. Chen, J., Liu, X., Tu, P., Aragones, A.: Learning person-specific models for facial expression and action unit recognition. Pat. Recog. Lett. **34**(15), 1964–1970 (2013)
6. Chu, W.S., De la Torre, F., Cohn, J.F.: Learning spatial and temporal cues for multi-label facial action unit detection. In: 2017 12th IEEE International Conference on Automatic Face and Gesture Recognition (FG 2017), pp. 25–32 (2017)

7. Dalal, N., Triggs, B.: Histograms of oriented gradients for human detection. In: 2005 IEEE Computer Society Conference on Computer Vision and Pattern Recognition (CVPR 2005), vol. 1, pp. 886–893 (2005)
8. Davies, H., Wolz, I., Leppanen, J., Fernandez-Aranda, F., Schmidt, U., Tchanturia, K.: Facial expression to emotional stimuli in non-psychotic disorders: a systematic review and meta-analysis. Neurosci. Biobehav. Rev. **64**, 252–271 (2016)
9. Dunn, J.C.: A fuzzy relative of the isodata process and its use in detecting compact well-separated clusters. J. Cybern. **3**, 32–57 (1973)
10. Dureha, A.: An accurate algorithm for generating a music playlist based on facial expressions. Int. J. Comput. Appl. **100**(9), 33–39 (2014)
11. Ekman, P., Friesen, W. (eds.): The Facial Action Coding System (FACS): A Technique for the Measurement of Facial Action. Consulting Psychologists Press, Palo Alto (1978)
12. Ekman, P., Rosenberg, E.L. (eds.): What the Face Reveals: Basic and Applied Studies of Spontaneous Expression Using the Facial Action Coding System (FACS). Oxford University Press, New York (2005)
13. Ekman, P., Friesen, W.V.: Felt, false, and miserable smiles. J. Nonverbal behav. **6**(4), 238–252 (1982)
14. Florea, C., Florea, L., Butnaru, R., Bandrabur, A., Vertan, C.: Pain intensity estimation by a self-taught selection of histograms of topographical features. Image Vis. Comput. **56**, 13–27 (2016)
15. Ghosh, S., Laksana, E., Scherer, S., Morency, L.P.: A multi-label convolutional neural network approach to cross-domain action unit detection. In: 2015 International Conference on Affective Computing and Intelligent Interaction (ACII), pp. 609–615 (2015)
16. Hammal, Z., Cohn, J.F.: Automatic detection of pain intensity. In: Proceedings of the 14th ACM International Conference on Multimodal Interaction, pp. 47–52 (2012)
17. Huang, X., Wang, S.J., Zhao, G., Piteikainen, M.: Facial micro-expression recognition using spatiotemporal local binary pattern with integral projection. In: Proceedings of the IEEE International Conference on Computer Vision Workshops, pp. 1–9 (2015)
18. Huang, X., Zhao, G., Hong, X., Zheng, W., Pietikäinen, M.: Spontaneous facial micro-expression analysis using spatiotemporal completed local quantized patterns. Neurocomputing **175**, 564–578 (2016)
19. Irani, R., Nasrollahi, K., Moeslund, T.B.: Pain recognition using spatiotemporal oriented energy of facial muscles. In: Proceedings of the IEEE Conference on Computer Vision and Pattern Recognition Workshops, pp. 80–87 (2015)
20. Krumhuber, E., Manstead, A.S., Cosker, D., Marshall, D., Rosin, P.L.: Effects of dynamic attributes of smiles in human and synthetic faces: a simulated job interview setting. J. Nonverbal Behav. **33**(1), 1–15 (2009)
21. Lazebnik, S., Schmid, C., Ponce, J.: Beyond bags of features: spatial pyramid matching for recognizing natural scene categories. In: 2006 IEEE Computer Society Conference on Computer Vision and Pattern Recognition (CVPR 2006), vol. 2, pp. 2169–2178 (2006)
22. Liu, P., Duan, L., Chi, X., Zhu, Z.: An improved fuzzy c-means clustering algorithm based on simulated annealing. In: Proceedings of FSKD, pp. 39–43 (2013)
23. Lucey, P., Cohn, J., Prkachin, K., Solomon, P., Chew, S., Matthews, I.: Painful monitoring: automatic pain monitoring using the UNBC-McMaster shoulder pain expression archive database. Image Vis. Comp. **30**, 197–205 (2012)

24. Lucey, P., Cohn, J., Lucey, S., Sridharan, S., Prkachin, K.M.: Automatically detecting action units from faces of pain: comparing shape and appearance features. In: 2009 IEEE Computer Society Conference on Computer Vision and Pattern Recognition Workshops, CVPR Workshops 2009, pp. 12–18 (2009)

25. Lucey, P., Cohn, J.F., Kanade, T., Saragih, J., Ambadar, Z., Matthews, I.: The extended cohn-kanade dataset (CK+): a complete dataset for action unit and emotion-specified expression. In: 2010 IEEE Computer Society Conference on Computer Vision and Pattern Recognition Workshops (CVPRW), pp. 94–101 (2010)

26. Lukac, R., Plataniotis, K.N.: A taxonomy of color image filtering and enhancement solutions. Adv. Imaging Electr. Phys. **140**, 188 (2006)

27. Mandal, M., Pandey, R., Prasad, A.: Facial expressions of emotions and schizophrenia: a review. Schizophrenia Bull. **24**, 399–412 (1998)

28. Pedersen, H.: Learning appearance features for pain detection using the UNBC-McMaster shoulder pain expression archive database. In: Nalpantidis, L., Krüger, V., Eklundh, J.-O., Gasteratos, A. (eds.) ICVS 2015. LNCS, vol. 9163, pp. 128–136. Springer, Cham (2015). https://doi.org/10.1007/978-3-319-20904-3_12

29. Qu, F., Wang, S.J., Yan, W.J., Li, H., Wu, S., Fu, X.: CAS (ME)‸ 2: a database for spontaneous macro-expression and micro-expression spotting and recognition. IEEE Trans. Affect. Comput. (2017)

30. Sangineto, E., Zen, G., Ricci, E., Sebe, N.: We are not all equal: personalizing models for facial expression analysis with transductive parameter transfer. In: ACM Multimedia, pp. 357–366 (2014)

31. Wang, X., Wang, L.M., Qiao, Y.: A comparative study of encoding, pooling and normalization methods for action recognition. In: Lee, K.M., Matsushita, Y., Rehg, J.M., Hu, Z. (eds.) ACCV 2012. LNCS, vol. 7726, pp. 572–585. Springer, Heidelberg (2013). https://doi.org/10.1007/978-3-642-37431-9_44

32. Xie, W., Shen, L., Yang, M., Jiang, J.: Facial expression synthesis with direction field preservation based mesh deformation and lighting fitting based wrinkle mapping. Multimed. Tools Appl. **77**, 7565–7593 (2018)

33. Yang, Q., Li, C., Li, Z.: Application of FTGSVM algorithm in expression recognition of fatigue driving. J. Multimedia **9**(4), 527–533 (2014)

34. Zen, G., Sangineto, E., Ricci, E., Sebe, N.: Unsupervised domain adaptation for personalized facial emotion recognition. In: ACM International Conference on Multimodal Interaction, pp. 128–135 (2014)

35. Zhang, K., Zhang, Z., Li, Z., Qiao, Y.: Joint face detection and alignment using multitask cascaded convolutional networks. IEEE Sig. Process. Lett. **23**(10), 1499–1503 (2016)

Detecting and Recognizing Salient Object in Videos

Rahma Kalboussi[1,2]($\boxtimes$), Mehrez Abdellaoui[1], and Ali Douik[1]

[1] Networked Objects Control and Communication Systems Laboratory (NOCCS), University of Sousse, Pôle technologique de Sousse, Route de Ceinture Sahloul, 4054 Sousse, Tunisia
rahma.kalboussi@gmail.com
[2] Higher Institute of Computer Sciences and Communications Technology (ISITCom), Hammam Sousse, Sousse, Tunisia

Abstract. Saliency detection has been an interesting research field. Some researchers consider it as a segmentation problem some others treat it differently. In this paper, we propose a novel video saliency framework that detects and recognizes the object of interest.

Starting from the assumption that spatial and temporal information of an input video frame can provide better saliency results than using each information alone, we propose a spatio-temporal saliency model for detecting salient objects in videos. First, spatial saliency is measured at patch-level by fusing local contrasts with spatial priors to label each patch as a foreground or a background one. Then, the newly proposed motion distinctiveness feature and temporal gradient magnitude measure are used to obtain the temporal saliency maps. Spatial and temporal saliency maps are fused together into one master saliency map.

Object classification framework contains training and testing stage. On the training phase, we use a convolutional neural network to extract features of the proposed training set. Then, deep features are fed into a Support Vector Machine classifier to produce a classification Model. This model will be used to predict the class of the salient object.

Despite the framework is simple to implement and efficient to run, it has shown good performances and achieved good results.

Experiments on two standard benchmark datasets for video saliency have shown that the proposed temporal cues improve saliency estimation results. Results are compared to six state-of-the-art methods on two benchmark datasets.

1 Introduction

As Humans, we are able to quickly identify the visually most attractive object in the scene known as salient object, and adaptively focus our attention on such important region. However, developing a computational model that detects automatically and efficiently the salient object which matches the human annotators behaviour is very desirable.

© Springer Nature Switzerland AG 2018
J. Blanc-Talon et al. (Eds.): ACIVS 2018, LNCS 11182, pp. 62–73, 2018.
https://doi.org/10.1007/978-3-030-01449-0_6

Considering the instantaneous ability of defining the technical support to extract features, decide which one belongs to the object of interest and segment it from the confusing background, it remains a challenging problem.

Unlike segmentation problems which aim to divide the input frame into coherent properties, salient object detection approaches segment the salient foreground object from its background to produce a map called saliency map which highlight only the salient object. Generally, saliency map computation starts from selecting the salient object then segmenting it from the whole scene. In the saliency map, every pixel's value represents its saliency degree (its probability to belong to the salient object).

Since the visual world is well structured, the human visual system selects the salient region according to the different stimuli presented in the observed image. So, a very interesting question can be asked: How does our brain order our gaze? The stimuli properties play a fundamental role in salient regions selection. When observing a natural image, gaze is moved to regions with high local contrast like edges and borders, and warm colors.

While saliency detection methods for still images are numerous, video saliency detection is a promising research field (methods are still in their early stages).

Visual saliency methods can be divided into two groups: local and global methods. Local approaches measure rarity of a region over its neighborhoods. [11,16]. In contrast, global approaches are based on the rarity and uniqueness of image regions with respect to the whole scene [4,16,24] Most of the saliency methods are theoretically based on the feature integration theory of visual attention of Treisman and Gelade [27].

Lately, Zhang et al. [30] and Qi et al. [21] used the Boolean Map theory for visual attention [9] to detect salient object in still images and Kalboussi et al [13] proposed a similar framework based on boolean map theory that detects salient object in videos where temporal information is used.

In this paper, we propose a salient object detection and recognition framework. For saliency detection we propose a spatio-temporal saliency detection method. Temporal saliency estimation is performed using our newly proposed motion features: motion distinctiveness feature which is an inter-frame motion estimation to measure motion coherence between each pair of frames and highlights the special movement that occurs in the current frame taking account the previous frame (under the assumption that surprising motion catch human's attention) and the temporal gradient magnitude as an intra-frame motion measure to ensure more motion consistency. Spatial saliency estimation is performed at patch level using classic saliency cues including surrounding contrast measure, and the surrounding contrast with consistency proposed by [29]. Spatial and temporal saliency information are fused together to provide final master saliency map that highlight only the object that attracts attention.

Once the salient object is detected, it is well desired to recognize it for different tasks including video-camera surveillance, military, object tracking, etc.

For these reasons, we propose an additional contribution which consists on an object classification framework using Support vector machine classifier and deep features extracted from a convolutional neural network. At training phase, we extract features from the training dataset using a convolutional neural network to provide deep features rich of semantic information. Those deep features are fed into a SVM classifier to train a classification model.

At the testing phase, the detected salient object using our spatio-temporal saliency estimation is cropped from the original RGB frame and use the newly cropped image as an input to the SVM model to predict the class of the object (see Fig. 1).

As a final contribution, we propose a salient object recognition training dataset that contains eight classes of different objects.

We evaluate the performance of the proposed approach on two standard benchmark datasets for video saliency namely **SegTrack v2** [17] **and Fukuchi** [6] against six state-of-the-art methods. Despite its simplicity, our method consistently achieves state-of-the-art performance on all datasets and is the only visual saliency method that includes salient object recognition framework.

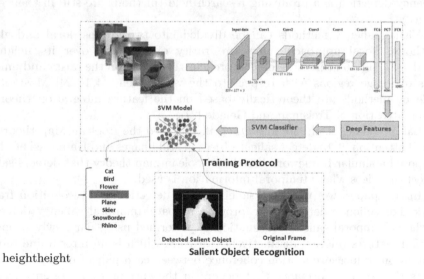

heightheight

Fig. 1. Method framework

2 Related Work

When an observer watches a video, he does not have enough time to examine the whole scene, so his gaze is always focused on the region that contains the dynamic object. From that assumption different video saliency models that include temporal and spatial cues are developed. Here we only review the main methods on video saliency, for an excellent review of saliency methods in still images we refer to [1,2].

Video saliency detection is a promising research field. Itti et al. [10] defined a salient event as an important event that can stimulate attention. Rahtu et al. [23] proposed a saliency model which combines saliency measure formulated by using local features and statistical framework, with a conditional random field model. Mancas et al. [19] introduced a motion selection in crowd model where optical flow is used to determine the motion region. A spatio-temporal visual saliency model was proposed where the resultant saliency map is the fusion of static and dynamic saliency maps [22]. Goferman et al. [7] proposed a model that considers local and global surroundings of an object to compute its saliency level, so that the whole context of the input image is considered. Wang et al. [28] propose a video saliency object segmentation model based on geodesic distance where both spatial edges and temporal motion boundaries are used as foreground indicators. Also, Wang et al. [28] proposed a video saliency object segmentation model based on geodesic distance where both spatial edges and temporal motion boundaries are used as foreground indicators. Singh et al. [26] presented a method that extracts salient objects in video by integrating color dissimilarity, motion difference, objectness measure, and boundary score feature, they used temporal superpixels to simulate attention to a set of moving pixels. Mauthner et al. [20] proposed an approach based on the Gestalt principle of figure-ground segregation for appearance and motion cues. Kalboussi et al. [14] proposed a video saliency detection method using supervoxels uniqueness measure. A local spatio-temporal model is proposed by Kalboussi et al. [15] where spatial information is measured through surrounding contrast metric and temporal information is issued from the magnitude and orientation of the optical flow at patch level.

3   Salient Object Detection

3.1   Spatial Saliency Estimation

Surrouding Contrast Measure. Salient region is usually distinctive from the rest of the scene. Previous works on saliency detection have proved that a change in contrast is the main cue to highlight the object of interest [5]. Given a frame divided into patches, local contrast is defined as the brutal change of color independently of the spatial distance between considered patches. While salient patches are generally spatially grouped, spatial distance is considered for a good local contrast representation. In this context, surrounding contrast cue is presented by [7] which assumes that not only color distinctiveness is necessary for saliency detection but also the surrounded patches characteristics. To do so, local contrast distinctiveness for each patch P_i is defined as follow

$$LC(p_i) = \sum_{p_j \forall j} \frac{d_c(p_i, p_j)}{1 + \alpha.d_p(p_i, p_j)} \qquad (1)$$

where α is a control parameter of the color by spatial distance rate, $d_c(p_i, p_j)$ is the euclidean distance between p_i and p_j in the CIE L*a*b color space and $d_p(p_i, p_j)$ is the Euclidean distance between p_i and p_j positions.

3.2   Consistent Surrounding Contrast

Local Contrast LC can be considered as one of the strongest object boundaries detection in color images. This detector is not enough to produce a good saliency estimation in still images and highlight the whole salient object. To solve this problem, Yeh et al. [29] have proposed a novel metric that considers not only the contrast of the current patch but compares its color-degrees to the surrounding patches. They define the contrast consistency as:

$$CC(p_i, p_j) = \frac{exp(-d_c(p_i, p_j)}{\sum\limits_{p_h \in N_q} exp(-d_c(p_i, p_h))} \tag{2}$$

N_q is the set of neighboring patches of the current patch p_i. The final consistent surrounding contrast is defined as

$$SCC(p_i) = \sum_{p_j \in N_q} LC(p_j) \times CC(p_i, p_j) \tag{3}$$

3.3   Static Saliency Map Generation

In general, patches with higher contrast values arouse the attention. Thus, the consistent surrounding contrast measure will be used to select foreground and background patches. The first thing to do is to sort the SCC values in the ascending order, then patches are ranked according to their SCC degree, where patches with high SCC degree are marked as foreground patches and patches with lowe SCC degree are marked as background patches. More precisely, the P_f are the first 10% patches and the P_b are the last 70% patches.

Given foreground and background patches sets (F, B), we can define foreground and background probabilities of a given patch as a superimposed mixture distribution

$$Pr(P|F) = \frac{SCC(P)}{|F|} \sum_{Y \in F} exp(-\frac{d_c(P,Y)}{\sigma_c}) exp(-\frac{d_p(P,Y)}{\sigma_p}) \tag{4}$$

and

$$Pr(P|B) = \frac{1 - SCC(P)}{|B|} \sum_{X \in B} exp(-\frac{d_c(P,X)}{\sigma_c}) exp(-\frac{d_p(P,X)}{\sigma_p}) \tag{5}$$

Foreground and background probabilities of a given patch P_i depend on the distance in the space and color domains regarding the other patches of the whole frame and on the uniform contrast measure. The final static saliency map will be refined according to the following equation

$$S(P) = \frac{Pr(P|F)}{Pr(P|F) + Pr(P|B)} \tag{6}$$

4   Temporal Saliency Estimation

4.1   Motion Distinctiveness

In this paper we define a salient region as a region that have a distinctive motion compared to the previous frame. Therefore, we define a new metric to quantify the motion distinctiveness at pixel level.

Here the concept of motion distinctiveness is defined as a region that have a low motion commonality compared to the previous.

The pixel-level motion vectors mv are calculated for each frame using the optical flow estimation method proposed by [3]. Motion vector provides the information about the motion activity of objects in the current frame. A uniform motion vector at frame-level when moving from frame f_t to frame f_{t+1} shows that there is no new motion activity in the frame, while a fluctuating motion denotes that it contains a new or distinctive moving object. In general, newly appearing moving object catch attention.

Therefore, for every video frame, we exploit correlation measure between current and previous frames. The sum of squared difference (SSD) was computed to measure the similarity between a pair of frames in different previous works [25]. However, after testing different similarity measures, we have found that Pearson correlation measure is more adequate to compute the motion distinctiveness.

Unlike the Euclidean Distance score which is scaled to vary between 0 and 1, Pearson correlation measure $\rho_i(f_t, f_{t-1})$ is scaled between 1 and -1 is given by Eq. 7

$$\rho_i(f_t, f_{t-1}) = \frac{cov_{mv_t, mv_{t-1}}}{\sigma_{mv_t} \times \sigma_{mv_{t-1}}} \tag{7}$$

where mv_t is the motion vector at frame f at instant t, and $cov_{mv_t, mv_{t-1}}$ is given by

$$cov_{mv_t, mv_{t-1}} = \frac{\sum_{i=1}^{N}(mv_t - \overline{mv_t})(mv_{t-1} - \overline{mv_{t-1}})}{N - 1} \tag{8}$$

and σ_{p_i} is defined as

$$\sigma_{mv_t} = \sqrt{\frac{\sum_{i=1}^{N}(mv_t - \overline{mv_t})^2}{N - 1}} \tag{9}$$

The Pearson measure indicates how two variables are correlated and is varied from -1 to 1, where a value of 1 indicates that both patches are similar and a score of -1 indicates that the two patches are not correlated and are totally distinctive. Since we are interested in measuring the motion distinctiveness score, we propose a new metric defined as follows

$$M_d^t = \frac{1}{\alpha} exp(-\frac{\rho_i(f_t, f_{t-1}) - 1}{2}) \tag{10}$$

where α is a parameter equal to 0.5. M_d is the motion distinctiveness measured at fame-level and is used as an inter-frame saliency indicator.

4.2 Temporal Gradient Magnitude

While contrast measure is a good saliency indicator in still images, it can not be discriminative in a complex scene with a high textured background. In the last section, we introduced a new measure to quantify the inter-frame distinctive motion. To ensure the motion consistency, we have found that integrating intra-frame motion information into the same framework can be more effective.

To compute the optical flow, we use the large displacement motion estimation algorithm [3] Given a video frame f_i, let mv_i be its optical flow field. We propose a temporal gradient field which uses an exponential function to highlight the optical flow gradient magnitude $||\nabla mv_i||$ and eliminate noise

$$M_i = 1 - exp(-\lambda||\nabla mmv_i||) \tag{11}$$

λ is used to scale the exponential function and is set to the value 1.

The temporal gradient magnitude reveals the boundaries of the moving object.

Motion distinctiveness and temporal gradient magnitude are fused together into one temporal measure T defined as:

$$T(x,y) = M_i(x,y) + M_d(x,y) \tag{12}$$

The measure T highlights the salient object that have distinctive motion compared to the rest of the scene.

5 Final Saliency Map Generation

The final saliency map is the fusion of the static and dynamic saliency maps. The combination is performed to modify static saliency maps with the corresponding dynamic saliency value.

According to previous works on video saliency, [7], locations that are distant from the region of attention are less attractive than those which are around. Which means that, pixels that are closer to the object of interest get higher saliency scores than further ones.

Hence, the saliency at location $X = (x,y)$ can be defined as

$$SM(x,y) = (S_m(x,y))(1 - d(X,C)) \tag{13}$$

where $d(x,y)$ is the euclidean distance between $X = (x,y)$ and the center $C = (x_c, y_c)$, $S_m(x,y)$ is the saliency values at location (x,y) and is given by

$$S_m(x,y) = N(S(x,y)) \times exp(T(x,y)) * I_{k*k} \tag{14}$$

the exponential function is used to widen the contrast of the dynamic saliency weights and $N(S(x,y)$ is a normalization operation used to normalize the values of $S(x,y)$ to the range of $[0,1]$. $S(x,y)$ is the saliency degree at location (x,y) from the patch P. To minimize noise caused by camera motion we use a 2D Gaussian low-pass filter I_{k*k} k is the kernel value equal to 5.

6   Salient Object Recognition

Recognizing the object that attracts human gaze is a very important task that can be used in different applications including video surveillance applications.

Machine learning techniques have been used for decades to detect and recognize objects. They use the extracted features of the input image to train a model which then will be used to predict the object class. The traditional machine learning application usually use the hand-crafted features. In standard situations, hand-crafted features can achieve desired goals but, are not enough effective in some challenging situations like complex background or overlapping between background and foreground object. To solve this problem, recently, convolutional neural networks (CNNs) have been employed. Since these CNNs are essentially pre-trained for visual recognition, they contain high-level semantic information. Therefore, features issued from a CNNs are rich of semantic and produce good recognition rates.

In this paper, we use the CNN classifier **Alexnet** that is pre-trained on the challenging ImageNet dataset, to extract features. The recognition algorithm is simple and efficient.

To train a SVM (support vector machine) machine we need a training data. First, we collect a set of images from internet which will be classified into eight classes of different objects. Then, we use the alexnet CNN to extract deep features of each image from the training set of images. CNNs are constructed by using the interconnected neurons or layers of non linear elements. The first layers situated at the beginning of the network detect basic image features, like edges and blobs. Then, "primitive" features are processed by the deeper network layers, which combine the early features to produce higher-level image features. These higher-level features are better suited for recognition tasks because they combine all the primitive features into a richer image representation.

Identifying the exact layer that is the most suitable for image feature extraction may be impossible. For this project we tried different tests to realize that the layer before the classification layer is the best to extract input's features. While convolutional neural networks are very used for object classification, we opt for the support vector machine (SVM) classifier to generated classification model for different reasons including the strong founding theory, the less prone to over-fitting and specially the less memory requirements to store the predictive model.

After detecting the salient object and producing the saliency map, we use a bounding box to encircle the object of interest. Then, we use the bounding box's coordinates to crop a region that contains the salient object from the original RGB frame. The cropped image will be used as a testing sample for our SVM trained model. Our classification framework showed good results on the video saliency dataset **Fukuchi** [6] (see confusion matrix in Fig. 4).

7   Experiments

Our method detects and recognizes automatically the object that attract human gaze in video sequences. In this section, we compare our video saliency method against state of the art methods on the Segtrack v2 [17] and Fukuchi [6] datasets. In our proposed method, we utilize the spatial and temporal information of the input frame at pixel and frame levels to decide saliency probability of each pixel. Spatial information includes color and contrast measure, while temporal information makes use of motion distinctiveness and magnitude gradient flow field. The fusion of spatial and temporal saliency maps leads to a saliency map which highlights region of interest and segments salient object from the background.

We compare our video saliency approach to six state of the art methods namely CBS [12], GB [8], GVS [28], ITTI [10], RR [19] and RT [23].

On both datasets Segtrack v2 and Fukuchi datasets we clearly obtain good results in terms of F-score measure. The **F-measure** is defined as:

$$F_\beta = \frac{(1 + \beta^2) \cdot \text{precision} \cdot \text{recall}}{\beta^2 \cdot \text{precision} + \text{recall}} \tag{15}$$

where we use $\beta^2 = 0.3$ following [18].

Comparing our results in terms of F-score values we outperform other methods on both sequences from the fukuchi and segtrack v2.

Figure 2 plots the precision recall curves of two challenging video sequences of the benchmark datasets. From left to right, the first is dedicated to the bird sequence from the fukuchi dataset. This sequence contains a moving bird. The problem is that the bird is moving slowly, so motion information alone could not provide good saliency estimation. Our proposed method which combines spatial and temporal information achieved better results comparing to the other methods. The figure on the right plots the precision recall curves of the fallen bird video sequence from the segtrack v2 dataset which is a challenging video sequence since it contains a fallen bird that has an almost similar color with the background. It such a case, spatial information will surely fail to detect the salient object and our motion features will succeed to highlight the salient object and provide good results (Table 1).

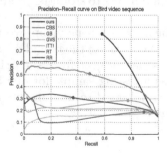 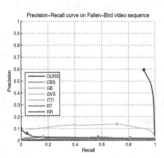

Fig. 2. PR curves on Fukuchi and Segtrack v2 datasets

Table 1. Different values of F-measure on two sequences from the bechmark datasets

Sequence	GVS	GB	RR	RT	ITTI	CBS	Ours
Fallen bird	0.168	0.025	0.055	0.012	0.038	0.354	**0.647**
Bird	0.215	0.333	0.32	0.23	0.259	0.504	**0.763**

We added qualitative comparison on different challenging cases in Fig. 3 and in each situation, our method outperforms other methods. Saliency maps provided by **GB** [8] and **ITTI** [10] and do not show the exact location of the salient object because of lack of motion information specially with complex backgrounds. **RT** [23] is quite good, the salient object is correctly detected but the background gets high saliency probability. While optical flow is one of the most used techniques to detect moving objects, it can not be a good saliency estimator. The performance of video saliency detector **RR** [19] based on optical flow, assign low saliency probability to static pixels which belong to salient object (see fifth row). In most cases, **CBS** [12] and **GVS** [28] are able to locate the salient object even in complex situations where foreground-background colors are similar (first and second rows) since their motion information is very informative.

Results of a moving object with higher speed and a static camera are shown in the third row, and produces a good saliency map. In case of an object with high speed and a moving camera, (last row) our proposed motion feature highlights only the moving object. Based on the aforementioned analysis, two main conclusions can be drawn. First, to detect salient object in videos, it is essential to examine motion information. Second, developing a method that depends only on motion information is not an excellent idea. Combining spatial and temporal information into a video saliency framework leads to the best results.

The confusion matrix in Fig. 4 shows that our object recognition framework can also be effective. All the object classes has a 100 per cent recognition rate except cat, skier and snow-border which can be explained by the big similarity between snowboarding and skying.

8   Conclusions and Perspectives

In this paper, we proposed a saliency detection and recognition framework which detects the object of interest and recognize it in videos. To detect salient objects, we proposed two motion features to quantify motion distinctiveness inter video frames and highlight the shape of the dynamic object intra each video frame. Motion features are combined with spatial features into one final and master saliency map. The highlighted object in the saliency map is cropped from the original RGB video frame and fed into the proposed SVM classification model to be recognized. Our proposed video saliency method is compared to six state-of-the-art methods and shows great performance on two benchmark datasets.

In the future work, we plan to work with sound videos to see the impact of sound on human gaze, and to figure out how to incorporate sound information into our framework.

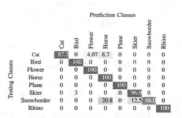

Prediction Classes

Testing Classes	Cat	Bird	Flower	Horse	Plane	Skier	Snowborder	Rhino
Cat	87.2	0	4.07	8.7	0	0	0	0
Bird	0	100	0	0	0	0	0	0
Flower	0	0	100	0	0	0	0	0
Horse	0	0	0	100	0	0	0	0
Plane	0	0	0	0	100	0	0	0
Skier	0	3	0	0	0	96.9	0	0
Snowborder	0	0	0	20.8	0	12.5	66.0	0
Rhino	0	0	0	0	0	0	0	100

Fig. 3. Visual comparison of saliency maps

Fig. 4. Confusion Matrix

References

1. Borji, A.: What is a salient object? a dataset and a baseline model for salient object detection. IEEE Trans. Image Process. **24**(2), 742–756 (2015)
2. Borji, A., Cheng, M.M., Jiang, H., Li, J.: Salient object detection: a survey. arXiv preprint arXiv:1411.5878 (2014)
3. Brox, T., Bregler, C., Malik, J.: Large displacement optical flow. In: 2009 IEEE Conference on Computer Vision and Pattern Recognition, CVPR 2009, pp. 41–48. IEEE (2009)
4. Bruce, N., Tsotsos, J.: Saliency based on information maximization. In: Advances in Neural Information Processing Systems, pp. 155–162 (2005)
5. Cheng, M.M., Mitra, N.J., Huang, X., Torr, P.H., Hu, S.M.: Global contrast based salient region detection. IEEE Trans. Patt. Anal. Mach. Intell. **37**(3), 569–582 (2015)
6. Fukuchi, K., Miyazato, K., Kimura, A., Takagi, S., Yamato, J.: Saliency-based video segmentation with graph cuts and sequentially updated priors. In: 2009 IEEE International Conference on Multimedia and Expo, pp. 638–641. IEEE (2009)
7. Goferman, S., Zelnik-Manor, L., Tal, A.: Context-aware saliency detection. IEEE Trans. Patt. Anal. Mach. Intell. **34**(10), 1915–1926 (2012)
8. Harel, J., Koch, C., Perona, P., et al.: Graph-based visual saliency. In: NIPS, vol. 1, p. 5 (2006)
9. Huang, L., Pashler, H.: A boolean map theory of visual attention. Psychol. Rev. **114**(3), 599 (2007)
10. Itti, L., Baldi, P.: A principled approach to detecting surprising events in video. In: 2005 IEEE Computer Society Conference on Computer Vision and Pattern Recognition (CVPR 2005), vol. 1, pp. 631–637. IEEE (2005)
11. Itti, L., Koch, C., Niebur, E.: A model of saliency-based visual attention for rapid scene analysis. IEEE Trans. Patt. Anal. Mach. Intell. **20**(11), 1254–1259 (1998)
12. Jiang, H., Wang, J., Yuan, Z., Liu, T., Zheng, N., Li, S.: Automatic salient object segmentation based on context and shape prior. In: BMVC, vol. 6, p. 9 (2011)
13. Kalboussi, R., Abdellaoui, M., Douik, A.: Video saliency detection based on boolean map theory. In: Battiato, S., Gallo, G., Schettini, R., Stanco, F. (eds.) ICIAP 2017. LNCS, vol. 10484, pp. 119–128. Springer, Cham (2017). https://doi.org/10.1007/978-3-319-68560-1_11

14. Kalboussi, R., Abdellaoui, M., Douik, A.: Video saliency using supervoxels. In: De Pietro, G., Gallo, L., Howlett, R.J., Jain, L.C. (eds.) KES-IIMSS 2017. SIST, vol. 76, pp. 544–553. Springer, Cham (2018). https://doi.org/10.1007/978-3-319-59480-4_54
15. Kalboussi, R., Azaza, A., Abdellaoui, M., Douik, A.: Detecting video saliency via local motion estimation. In: 2017 IEEE/ACS 14th International Conference on Computer Systems and Applications (AICCSA), pp. 738–744. IEEE (2017)
16. Kim, J., Han, D., Tai, Y.W., Kim, J.: Salient region detection via high-dimensional color transform. In: Proceedings of the IEEE Conference on Computer Vision and Pattern Recognition, pp. 883–890 (2014)
17. Li, F., Kim, T., Humayun, A., Tsai, D., Rehg, J.M.: Video segmentation by tracking many figure-ground segments. In: Proceedings of the IEEE International Conference on Computer Vision, pp. 2192–2199 (2013)
18. Li, Y., Hou, X., Koch, C., Rehg, J.M., Yuille, A.L.: The secrets of salient object segmentation. In: Proceedings of the IEEE Conference on Computer Vision and Pattern Recognition, pp. 280–287 (2014)
19. Mancas, M., Riche, N., Leroy, J., Gosselin, B.: Abnormal motion selection in crowds using bottom-up saliency. In: 2011 18th IEEE International Conference on Image Processing, pp. 229–232. IEEE (2011)
20. Mauthner, T., Possegger, H., Waltner, G., Bischof, H.: Encoding based saliency detection for videos and images. In: Proceedings of the IEEE Conference on Computer Vision and Pattern Recognition, pp. 2494–2502 (2015)
21. Qi, S., Ming, D., Ma, J., Sun, X., Tian, J.: Robust method for infrared small-target detection based on boolean map visual theory. Appl. Opt. 53(18), 3929–3940 (2014)
22. Rahman, A., Houzet, D., Pellerin, D., Marat, S., Guyader, N.: Parallel implementation of a spatio-temporal visual saliency model. J Real-Time Image Process. 6(1), 3–14 (2011)
23. Rahtu, E., Kannala, J., Salo, M., Heikkilä, J.: Segmenting salient objects from images and videos. In: Daniilidis, K., Maragos, P., Paragios, N. (eds.) ECCV 2010. LNCS, vol. 6315, pp. 366–379. Springer, Heidelberg (2010). https://doi.org/10.1007/978-3-642-15555-0_27
24. Scharfenberger, C., Wong, A., Fergani, K., Zelek, J.S., Clausi, D.A.: Statistical textural distinctiveness for salient region detection in natural images. In: Proceedings of the IEEE Conference on Computer Vision and Pattern Recognition, pp. 979–986 (2013)
25. Shi, J., Malik, J.: Normalized cuts and image segmentation. IEEE Trans. Patt. Anal. Mach. Intell. 22(8), 888–905 (2000)
26. Singh, A., Chu, C.H.H., Pratt, M.: Learning to predict video saliency using temporal superpixels. In: 4th International Conference on Pattern Recognition Applications and Methods, pp. 201–209 (2015)
27. Treisman, A.M., Gelade, G.: A feature-integration theory of attention. Cogn. Psychol. 12(1), 97–136 (1980)
28. Wang, W., Shen, J., Porikli, F.: Saliency-aware geodesic video object segmentation. In: Proceedings of the IEEE Conference on Computer Vision and Pattern Recognition, pp. 3395–3402 (2015)
29. Yeh, H.H., Liu, K.H., Chen, C.S.: Salient object detection via local saliency estimation and global homogeneity refinement. Patt. Recogn. 47(4), 1740–1750 (2014)
30. Zhang, D., Javed, O., Shah, M.: Video object segmentation through spatially accurate and temporally dense extraction of primary object regions. In: Proceedings of the IEEE Conference on Computer Vision and Pattern Recognition, pp. 628–635 (2013)

Directional Beams of Dense Trajectories
for Dynamic Texture Recognition

Thanh Tuan Nguyen[1,2,3]([⊠]), Thanh Phuong Nguyen[1,2], Frédéric Bouchara[1,2],
and Xuan Son Nguyen[4]

[1] Université de Toulon, CNRS, LIS, UMR 7020, 83957 La Garde, France
tnguyen108@etud.univ-tln.fr
[2] Aix-Marseille Université, CNRS, ENSAM, LIS, UMR 7020, 13397 Marseille, France
[3] HCMC University of Technology and Education, Faculty of IT,
HCM City, Vietnam
[4] Université de Caen, CNRS, GREYC, UMR 6072, 14000 Caen, France

Abstract. An effective framework for dynamic texture recognition is
introduced by exploiting local features and chaotic motions along beams
of dense trajectories in which their motion points are encoded by using
a new operator, named LVP$_{full}$-TOP, based on local vector patterns
(LVP) in full-direction on three orthogonal planes. Furthermore, we also
exploit motion information from dense trajectories to boost the discriminative power of the proposed descriptor. Experiments on various benchmarks validate the interest of our approach.

Keywords: Dynamic texture · Local feature · Dense trajectory
LBP · LVP

1 Introduction

Dynamic texture (DT) is a string of textures moving in the temporal domain
such as fire, clouds, trees, waves, foliage, blowing flag, fountain, etc. Analysis for *"understanding"* DTs is one of fundamental issues in computer vision
tasks. Various approaches have been proposed for DT description. In general,
existing methods can be classified into six groups as follows. First, *optical-flow-
based methods* [1] are natural approaches for DT recognition thanks to their
efficient computation and describing videos in effective ways. Second, *model-
based methods* [2,3] have been widely used for DT since the typical model Linear Dynamical System (LDS) [2] was introduced. Third, *filter-based methods*
have been also utilized for handling DT recognition. Different filtering operations have been addressed for encoding dynamic features: Binarized Statistical Image Features on Three Orthogonal Planes (BSIF-TOP) [4], Directional
Number Transitional Graph (DNG) [5]. Fourth, various *geometry-based methods*
have been presented using fractal analysis techniques in which fractal dimension
and other fractal characteristics are taken into account for DT representation:

© Springer Nature Switzerland AG 2018
J. Blanc-Talon et al. (Eds.): ACIVS 2018, LNCS 11182, pp. 74–86, 2018.
https://doi.org/10.1007/978-3-030-01449-0_7

dynamic fractal spectrum (DFS) [6], Multi-fractal spectrum (MFS) [7], wavelet-based MFS descriptor [8]. Fifth, owing to outperforming results, *learning-based methods* have recently attracted researchers with promising techniques coming from recent advances in deep learning: Transferred ConvNet Features (TCoF) [9], PCA convolutional network (PCANet-TOP) [10], Dynamic Texture Convolutional Neural Network (DT-CNN) [9]. Lately, dictionary-learning-based methods [11,12] have also become more popular in which local DT features are figured out by kernel sparse coding. Sixth, *local-feature-based methods* have been also considered with different LBP-based variants owing to their simplicity and efficiency since Zhao et al. [13] proposed two LBP-based variants for DT depiction: Volume LBP (VLBP) and LBP on three orthogonal planes (LBP-TOP). Lately, several efforts based on extensions of these typical operators are addressed to enhance the discriminative power of DT description [14–18].

This paper addresses a new efficient framework using directional beams of dense trajectories for DT representation, in which local feature patterns of motion points are encoded along their trajectories in conjunction with directional features of their neighbors structured by the proposed LVP_{full}-TOP operator with full-direction on three orthogonal planes of sequences. Furthermore, the motion information extracted from dense trajectories is conducted as a complement component to enhance the recognition power of DT descriptor. It could be seen that the advantages of both *optical-flow-based* and *local-feature-based* methods are consolidated into our approach to construct an effective DT descriptor.

2   Related Works

LBP-based Variants for Dynamic Texture: An efficient operator, called Local Binary Pattern (LBP), has been introduced in [19] to encode local features of a texture image as a binary chain by regarding relations between the center pixel and its surrounding neighbors interpolated on the neighboring circle centered at this pixel. In order to reduce effectively the dimensionality, different mappings have been proposed to select representative or important patterns: uniform patterns $u2$, $riu2$ [19], topological patterns [20], etc. Inherited by the benefits of LBP for still images, various LBP-based variants have been proposed to inspect DT recognition. At first, Zhao et al. [13] introduced VLBP considering three consecutive frames to form a $(3P+2)$-bit pattern for each voxel. An another variant, called LBP-TOP [13], has been also presented to overcome the curse of dimensionality of VLBP by addressing LBP on three orthogonal planes. Various extensions based on two above works have been then proposed to advance the discriminative power: CVLBC [21], CVLBP [15], CLSP-TOP [14], HLBP [16].

Directional LBP-based Patterns: The classical LBP captures only the first-order derivative variations. Thus, exploiting higher-order derivative variations is one important approach to develop LBP-based variants for different applications [22–24]. Zhang et al. [23] introduced Local Derivative Pattern (LDP), a directional extension of LBP, by taking into account local high-order derivative variations to encode directional patterns of voxels for capturing more robust features. To obtain potential information between derivative directions eliminated in the LDP, Fan et al. [25] proposed Local Vector Pattern (LVP) by regarding the pairwise of directional vectors to remedy the shortcomings remaining in local pattern representation. As adopting LVP as a component in our framework, we recall LVP in detail hereafter. Let I denote a sub-region of a 2D image. The first-order LVP of the center pixel $\mathbf{q}_c$ conducted by a direction α (in practice, 4 directions are considered, i.e. $\alpha = \{0°, 45°, 90°, 135°\}$) is calculated as follows.

$$\text{LVP}_{P,R,\alpha}(\mathbf{q}_c) = \left\{ h\big(V_{\alpha,D}(\mathbf{q}_c), V_{\alpha+45°,D}(\mathbf{q}_c), V_{\alpha,D}(\mathbf{q}_i), V_{\alpha+45°,D}(\mathbf{q}_i)\big) \right\}_{1 \leq i \leq P} \tag{1}$$

where $V_{\alpha,D}(\mathbf{q}) = I(\mathbf{q}_{\alpha,D}) - I(\mathbf{q})$, known as the first-order LVP, means the directional value of a vector obtained by concerning the current pixel $\mathbf{q}$ with its adjacent neighbor $(\mathbf{q}_{\alpha,D})$ in direction of α; $D = \{1, 2, 3\}$ presents the distance of the considered pixel with its contiguous points, and the function $h(.)$, called Comparative Space Transform (CST), with four parameters corresponding to four directional vectors in (1) is defined as

$$h(x, y, z, t) = \begin{cases} 1, & \text{if } t - \dfrac{(y * z)}{x} \geq 0 \\ 0, & \text{otherwise.} \end{cases} \tag{2}$$

Other formulations of LVP and samples of encoding LVP-based patterns for texture images are clearly discussed in [25].

Dense Trajectories: Wang et al. [26] extracted dense trajectories in videos by utilizing a dense optical flow field to sample and track the motion paths of points. For a point $\mathbf{q}_f = (x_f, y_f)$ at the fth frame, its position is tracked into the $(f + 1)^{th}$ frame by interpolating with a median filter on an optical flow $\omega_f = (u_f, v_f)$, in which u_f and v_f refer to horizontal and vertical of the optical flow component. The new position of $\mathbf{q}_f$, i.e. $\mathbf{q}_{f+1}$, at the adjacent frame is inferred as

$$\mathbf{q}_{f+1} = (x_{f+1}, y_{f+1}) = (x_f, y_f) + (M * \omega_f)|_{(\overline{x}_f, \overline{y}_f)} \tag{3}$$

where $(\overline{x}_f, \overline{y}_f)$ means the rounded position value of $\mathbf{q}_f$, M denotes a median filter kernel of 3×3 pixels. Finally, a trajectory $t = \{\mathbf{q}_f, \mathbf{q}_{f+1}, ..., \mathbf{q}_{f+L-1}\}$ with length of L is formed by concatenating points of consecutive frames. In our framework, we use the latest version (1.2) of dense trajectories[1] as a tool to extract motion paths of dynamic textures for video representation.

3   Proposed Method

3.1   Overview

We introduce an efficient framework taking into account the advantages of two well-known approaches: *optical-flow-based* and *local-feature-based* for an effec-

---

[1] http://lear.inrialpes.fr/people/wang/dense_trajectories.

tive DT representation. The main idea is to exploit local features by using the proposed LVP$_{full}$-TOP operator along dense trajectories together with motion information extracted from them. Figure 1 graphically illustrates the proposed framework. Our main contribution is four-fold. First, dense trajectories are used to exploit motions from dynamic textures instead of typical optical-flow-based approaches. Second, an effective operator, called LVP$_{full}$-TOP, is introduced to capture more second-order derivative variations on three orthogonal planes. Third, a local-feature-based descriptor is presented by addressing LVP$_{full}$-TOP along a beam of directional trajectories. Fourth, motion angle patterns are utilized to capture more chaotic motions of DT from dense trajectories. In the following, we then detail the proposed method for DT representation.

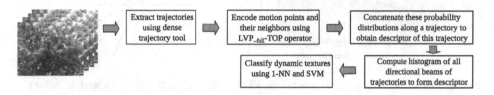

Fig. 1. Illustration of different steps of our proposed framework.

3.2 Components of the Proposed Descriptor

LVP$_{full}$-TOP Operator: Because only 4 derivative directions of the center pixel are considered in LDP and LVP operators, the relations between the center and other neighbors are less exploited. To remedy this problem, we extend encoding of LVP on full-direction, called LVP$_{full}$ with 8 directions of $\alpha = \{0°, 45°, 90°, 135°, 180°, 225°, 270°, 315°\}$, to capture full spatial relations between a pixel $\mathbf{q}_c$ and its neighbors as follows.

$$\text{LVP}_{P,R,full}(\mathbf{q}_c) = \sum_{i=0}^{P-1} h(.)2^i \Big|_{\alpha} \qquad (4)$$

where $h(.)$ is referred by Eqs. 1 and 2, P is neighbors of the considered pixel $\mathbf{q}_c$ sampled on a circle with radius R. Due to full-directional encoding, each first-order LVP$_{full}$ code of a pixel has a 64-bit binary pattern in total. It leads to be impossible in real implementation. Therefore, we apply the concept of $u2$ mapping of the typical LBP [19] for LVP^{u2} on each direction to reduce the dimension of descriptor, i.e. $8 \times (P(P-1) + 3)$ bins for full-direction where P is the sampled neighbors.

Furthermore, based on the idea of LBP-TOP [13], we investigate the proposed LVP$_{full}$ on three orthogonal planes to form a new operator, LVP$_{full}$-TOP which is able to effectively obtain spatio-temporal structures of dynamic features.

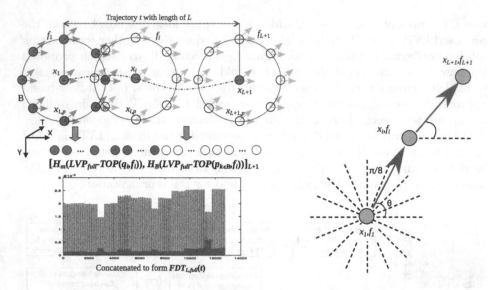

Fig. 2. (Best viewed in color) Encoding FDT with the proposed LVP$_{full}$-TOP in direction $\alpha = 0°$.

Fig. 3. (Best viewed in color) Computing motion angle patterns for a trajectory. (Color figure online)

Features of Directional Trajectory: Let $t = \{\mathbf{q}_1, \mathbf{q}_2, ..., \mathbf{q}_L, \mathbf{q}_{L+1}\}$ be a trajectory with length of L which is formed by $L + 1$ motion points corresponding to $L + 1$ consecutive frames $F = \{f_1, f_2, ..., f_L, f_{L+1}\}$. Figure 2 graphically demonstrates the proposed method to encode features of directional trajectory (FDT) t in direction $\alpha = 0°$. Accordingly, we consider movements of the motion point $\mathbf{q}_i \in t$ and its neighbors positioned by a vicinity of B in order to compute probability distributions for chaotic motions as well spatial features of $\mathbf{q}_i$ along trajectory t using the proposed LVP$_{full}$-TOP operator with full-direction. Finally, the gained histograms are concatenated to shape the FDT of t as follows.

$$\text{FDT}_{L,full}(t) = \sum_{i=1}^{L+1} \Big[H_m\big(\text{LVP}_{full}\text{-TOP}(\mathbf{q}_i, f_i)\big), H_B\big(\text{LVP}_{full}\text{-TOP}(\mathbf{p}_{k \in B}, f_i)\big) \Big]$$

(5)

where LVP$_{full}$-TOP$(\mathbf{q}_i, f_i)$ is local vector pattern of a pixel $\mathbf{q}_i$ at the frame f_i based on full-direction in three orthogonal planes; $\mathbf{p}_k$ refers to the kth neighbor of $\mathbf{q}_i$ in the vicinity B; $H_m(.)$ and $H_B(.)$ are respectively the distributions of the motion point $\mathbf{q}_i$ and its neighbors placed in B.

Motion Angle Patterns (MAP): To capture motion information from dense trajectories, we extend the idea of [27]. Accordingly, two angles of $\beta = \pi/4$ and $\beta = \pi/8$ can be considered to form 8 and 16 bins of directional motion angle pattern for each pair of motion points respectively. Figure 3 graphically illustrates a sample of this calculation for a trajectory t, i.e. $\text{MAP}_\beta(t)$, using angle of $\beta = \pi/8$ with its length of L located by $L+1$ motion points in corresponding consecutive frames. These obtained motion angle patterns at point level are concatenated to shape MAP histogram of the whole trajectory.

Proposed Video Descriptor: Let $T = \{t_1, t_2, ..., t_n\}$ be a set of trajectories with the same length of L extracted from a video $\mathcal{V}$ using a tool of dense trajectories introduced in [26]. To construct a robust descriptor for DT recognition, we combine and normalize the features of directional beams of T trajectories with the structures of their motion angle patterns, named FD-MAP and defined as follows.

$$\text{FD-MAP}_{L,\beta}(\mathcal{V}) = \sum_{i=1}^{n} \Big[\text{FDT}_{L,full}(t_i), \text{MAP}_\beta(t_i) \Big] \qquad (6)$$

To investigate the advantage of the angle vector complement component, we also evaluate recognition of $\text{FDT}_{L,full}(T)$ descriptor on the benchmark DT datasets.

3.3 Classification

For DT recognition, we adopt two classifying algorithms as follows.

Support Vector Machines (SVMs): We utilize a linear SVM algorithm provided in the LIBLINEAR library [28] to implement a multi-class classifier. All our experiments in this work are inspected by the latest version (2.20) with the default parameters of this tool.

k-nearest neighbors (k-NN): To be comparable with results of existing methods [14–16], we employ the simple of k-nearest neighbors, i.e. $k = 1$ (1-NN), in which chi-square (χ^2) measure as dissimilarity measure.

4 Experiments

4.1 Experimental Settings

Settings for Extracting Trajectories: Because most of chaotic motion points in DT sequences are usually *"living"* in a short period, we investigate dense trajectories with length of $L = \{2, 3\}$ using the tool presented in [26]. Since the default parameters of this tool are set for recognition of human actions in videos, to be suitable for the particular DT characteristics, we changed the original parameter of rejecting trajectory $min\_var = 5 \times 10^{-5}$ in order to obtain *"weak"* directional trajectories of turbulent motion points. Figure 4 graphically illustrates trajectories extracted from the corresponding sequences with the customized settings. Empirically, for datasets (like DynTex++) which are built by splitting from other original videos, some of cropped sequences have number of obtained trajectories that are not sufficient for DT representation (see Fig. 4(c)). In this case, a few

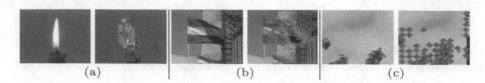

(a)   (b)   (c)

Fig. 4. (Best viewed in color) Samples (a), (b), (c) of directional trajectories extracted from the corresponding videos in UCLA, DynTex, and DynTex++ respectively in which green lines show paths of motion points through the consecutive frames. (Color figure online)

tracking parameters should be decreased to boost the quantity of trajectories in our framework as $min\_distance = 1$ and $quality = 10^{-8}$ [2].

Parameter Settings for Descriptors: We use the first-order LVP$_{full}$-TOP with $D = 1, P = 8, R = 1$ to structure local vector patterns of dynamic features in full-direction on three orthogonal planes. To be compliant with LBP-based methods, encoding spatial patterns of each motion point is adopted with its neighbors $P_B = 8$ circled by radius $R_B = 1$, i.e. $B = \{P_B, R_B\} = \{8, 1\}$ (see Fig. 2). In this case, the FDT$_{L,full}$ descriptor has $3 \times 9 \times 8 \times (P(P-1) + 3)$ dimensions with $u2$ mapping utilized for the proposed operator. Furthermore, the angle vector complement with $L \times |\text{MAP}_\beta|$ bins is also employed to form the FD-MAP$_{L,\beta}$ descriptor with $|\text{FDT}_{L,full}| + L \times |\text{MAP}_\beta|$ bins, where L is length of the considered trajectory, $|.|$ is the size of the descriptor. Regarding the configuration for the DT representation, the best parameter setting is selected as follows to compare with existing approaches: FD-MAP$_{L,\beta}$ with $L = 2$ and $\beta = \pi/4$, which will be further provided to the classifiers.

4.2 Datasets and Experimental Protocols

UCLA Dataset: UCLA dataset [2] consists of 50 categories with 200 different DT videos, corresponding to four sequences per class, which demonstrate the moving of dynamic textures such as fire, boiling water, fountain, waterfall, flower, and plant. Each original sequence is captured in 75 frames with dimension of 110×160 for each frame. The categories are arranged in varied ways to compose more challenging sub-datasets as follows.

- *50-class:* Original 50 classes are utilized using two experimental protocols: *leave-one-out* (50-LOO) [4,17,29] and *4-fold cross validation* (50-4fold) [14,16].
- *9-class and 8-class:* Original 50 classes of sequences are divided into 9 semantic categories [6,29] consisting of "boiling water" (8), "fire" (8), "flowers" (12), "fountains" (20), "plants" (108), "sea" (12), "smoke" (4), "water" (12), and "waterfall" (16), where the numbers in parentheses take account of sequences in each class. The "plants" category is eliminated from 9-class

---

[2] Please see [26] for more details about these above parameters.

to form more challenging 8-class scheme [6, 29]. For these two schemes, following [14, 29, 30], a half of DTs is randomly selected for training and the remaining for testing work. The final evaluation of recognition is estimated by the average of rates in 20 runtimes.

DynTex Dataset: DynTex dataset [31] consists of more than 650 high-quality DT sequences recorded in various conditions of environment. Following [4, 13, 16], we use a common version of this dataset containing 35 categories (DynTex35) in which each sequence is randomly cropped into 8 non-overlapping sub-videos that splitting points are not in half of the X, Y, and T axes. In addition, two more sub-sequences are also obtained for the experiment by cutting along the temporal axis of the original sequence. Consequently, 10 sub-DTs with different spatial-temporal dimension split from each sequence make classification tasks more challenging.

Other popular schemes stated as benchmark sub-datasets for DT recognition are compiled from [31] using leave-one-out as experimental protocol [4, 10].

- *Alpha* consists of 60 DT videos equally divided into three categories, i.e. "sea", "grass", and "trees", with 20 sequences in each of them.
- *Beta* contains 162 DT videos grouped into 10 classes with various numbers of sequences for each: "sea", "vegetation", "trees", "flags", "calm water", "fountains", "smoke", "escalator", "traffic", and "rotation".
- *Gamma* comprises 10 classes with 264 DT videos in total: "flowers", "sea", "naked trees", "foliage", "escalator", "calm water", "flags", "grass", "traffic", and "fountains". Each of which includes a sample of diverse sequences.

DynTex++ Dataset: From more than 650 sequences of the original DynTex, Ghanem et al. [30] filtered 345 raw videos to build DynTex++ in which the filtered videos only contain the main dynamic texture, not consist of other DT features such as panning, zooming, and dynamic background. They are finally divided into 36 classes in which each class has 100 sequences with fixed dimension of $50 \times 50 \times 50$, i.e. 3600 dynamic textures in total. Similarity to [4, 30], training set is formed by randomly selecting a half of DTs from each class and the rest for testing. This is repeated 10 times to obtain the average rate as the final result.

4.3   Experimental Results

Estimations of our method on the datasets using the proposed descriptors are presented in Tables 1 and 2 respectively, in which the highest classification rates corresponding to the settings are in bold. It is clear from those tables that the combination between the descriptor of directional trajectories FDT and the complement component of motion angle patterns MAP outperforms significantly in comparison with only using FDT descriptor on these datasets. The obtained rates from the observations are then compared with the state-of-the-art approaches through Table 3, in which the highest rates are in bold and results of VLBP [13], LBP-TOP [13] operators are referred to the evaluations of [16, 32] in the meanwhile the remains are taken from the original approaches. In general, our proposed FD-MAP descriptor achieves outstanding results in comparison with most of competitive LBP-based approaches on various DT datasets for recognition problems. For comparison with deep learning approaches, our method outperforms significantly on UCLA dataset (see Table 3), but not better on other datasets since deep-learning-based methods use sophisticated learning techniques with a gigantic cost of computation in the meanwhile our framework only concentrates on encoding directional trajectories of DT sequences.

UCLA Dataset: It can be observed from Tables 1 and 3 that the proposed descriptor achieves promising results compared to competing methods (both LBP-based and deep-learning-based methods) on all subsets.

Table 1. Classification rates (%) on UCLA using $FDT_{L,full}$ and FD-MAP$_{L,\beta}$

	50-LOO		50-4fold		9-class		8-class	
Descriptor	1-NN	SVM	1-NN	SVM	1-NN	SVM	1-NN	SVM
FDT_2	97.00	99.00	98.50	99.00	97.50	98.75	98.48	98.59
FDT_3	97.00	98.50	98.50	99.00	97.60	97.70	98.91	99.35
FD-MAP$_{2,\pi/4}$	97.00	**99.50**	98.50	99.00	97.30	**99.35**	99.02	**99.57**
FD-MAP$_{2,\pi/8}$	97.00	**99.50**	98.50	**99.50**	97.25	99.15	98.70	99.13
FD-MAP$_{3,\pi/4}$	**97.50**	99.00	**99.00**	99.00	**98.00**	99.00	98.59	99.35
FD-MAP$_{3,\pi/8}$	**97.50**	98.50	**99.00**	99.00	97.35	99.00	**99.13**	99.13

Table 2. Rates (%) on DynTex and Dyntex++ with $FDT_{L,full}$ and FD-MAP$_{L,\beta}$

	DynTex35		Alpha		Beta		Gamma		DynTex++	
Descriptor	1-NN	SVM	1-NN	SVM	1-NN	SVM	1-NN	SVM	1-NN	SVM
FDT_2	95.71	**98.86**	**93.33**	**98.33**	**84.57**	92.59	80.30	**91.67**	92.63	95.66
FDT_3	**96.00**	**98.86**	91.67	**98.33**	83.95	**93.21**	80.68	**91.67**	92.62	95.31
FD-MAP$_{2,\pi/4}$	95.71	**98.86**	91.67	**98.33**	**84.57**	92.59	80.30	**91.67**	92.87	**95.69**
FD-MAP$_{2,\pi/8}$	95.71	98.57	91.67	**98.33**	**84.57**	92.59	80.30	**91.67**	**93.07**	95.56
FD-MAP$_{3,\pi/4}$	**96.00**	98.57	91.67	**98.33**	83.95	**93.21**	**81.06**	91.29	92.81	95.26
FD-MAP$_{3,\pi/8}$	**96.00**	98.57	91.67	**98.33**	83.95	92.59	**81.06**	**91.67**	92.55	95.47

50-class: It can be verified that our obtained results (99.5% and 99%) are similar to methods' using deep learning techniques, i.e. PCANet-TOP [10] and DT-CNN [9]. Although our descriptor takes dimension of 12,760 bins, larger than MBSIF-TOP's [4] with 7-scale (5,376 bins), to obtain the same accuracy (99.5%), it is clear that our method is more efficient in other DT datasets (DynTex, DynTex++) in which MBSIF-TOP [4] achieves those with different multi-scale settings. DFS [6], a geometry-based method, gains rate of 100% in 4-fold cross validation scenario, but not perform better ours in other DT datasets.

9-class: Our proposed framework gains the best result of 99.35% compared to all existing approaches including deep-learning-based methods, DT-CNN [9] with rates of 98.05% and 98.35% using AlexNet and GoogleNet architectures respectively, (except DNGP [5], just 0.25% higher than ours).

8-class: On the more challenging 8-class scheme, our method demonstrates outstanding performance with 99.57%, the highest evaluation in comparison with the state-of-the-art approaches while the best rate of DT-CNN [9] just achieves 99.02% by the GoogleNet framework. A geometry-based method, 3D-OTF [7], also obtains nearly same ours but it failed in testing on other DT datasets.

DynTex Dataset: It could be seen from Table 3 that our descriptor performs better than all competing LBP-based methods. Furthermore, it is also comparative to deep-learning-based methods such as PCANet-TOP [10], st-TCoF [32].

DynTex DynTex35: Our method obtains the best performance on this dataset with 98.86% rate of classification with the selected settings for comparison (see Table 2) compared to most of LBP-based approaches except MEWLSP [18] with 99.71% but it is not efficient on UCLA as well not verified on other challenging DynTex variants (i.e. *Alpha, Beta, Gamma*).

DynTex Alpha: Our proposal achieves the highest rate of 98.33% compared to all LBP-based methods; and even outperforms deep learning methods (i. e. PCANet-TOP [10], st-TCoF [32] with rate of 96.67%, 98.33% respectively).

DynTex Beta and Gamma: In the best configurations formed for comparison, our result of 92.59% and 91.67% on DynTex beta and gamma respectively shows that our method performs the best compared to all existing approaches except deep learning methods, i.e. st-TCoF [32] and DT-CNN [9] utilizing a complicated computation in training step.

Dyntex++ Dataset: It is evident from Table 3 that our proposal gained comparative results in comparison with LBP-based approaches. More specifically, our best result of recognition rate on this scheme is 95.69% (see Table 2), nearly same DDLBP with MJMI [33] (95.8%), the highest rate of recognition on this scheme using SVM classifier. With 92.87% using 1-NN, our descriptor outperforms DNGP [5] and CVLBC [21] over 2% and 1% respectively in the meanwhile it is not better than other LBP-based methods. This may be because DT sequences in DynTex++ dataset, which includes sub-videos split from the original DynTex dataset, comprise lack of directional trajectories (see Fig. 4(c)) for encoding although we reduced the *min_distance* parameter of the tool to minimum value for extracting more trajectories (see more detail in 4.1). In LBP-based

Table 3. Comparison on UCLA, Dyntex, and Dyntex++ datasets.

Dataset	UCLA					DynTex			
Method	50-LOO	50-4fold	9-class	8-class	DynTex35	Alpha	Beta	Gamma	DynTex++
VLBP [13]	-	89.50	96.30	91.96	81.14	-	-	-	94.98
LBP-TOP [13]	-	94.50	96.00	93.67	92.45	96.67	85.80	84.85	94.05
DFS [6]	-	100^s	97.50^s	99.00^s	97.16^s	85.24^s	76.93^s	74.82^s	91.70^s
3D-OTF [7]	-	87.10^s	97.23^s	99.50^s	-	82.80^s	75.40^s	73.50^s	89.17^s
CLSP-TOP [14]	99.00	99.00	98.60	97.72	98.29	95.00	91.98	91.29	95.5
MBSIF-TOP [4]	**99.50**	99.50	98.75	97.80	98.61	90.00	90.70	91.30	97.12
MEWLSP [18]	96.50	96.50	98.55	98.04	**99.71**	-	-	-	98.48
CVLBP [15]	-	93.00	96.90	95.65	85.14	-	-	-	-
HLBP [16]	95.00	95.00	98.35	97.50	98.57	-	-	-	96.28
DNGP [5]	-	-	$\mathbf{99.60^s}$	99.40^s	-	-	-	-	93.80^s
DDLBP with MJMI [33]	-	-	-	-	-	-	-	-	95.80^s
WLBPC [17]	-	96.50	97.17	97.61	-	-	-	-	95.01
CVLBC [21]	98.50	99.00	99.20	99.02	98.86	-	-	-	91.31
Chaotic vector [3]	-	-	85.10	85.00	-	-	-	-	69.00
Orthogonal Tensor DL [12]	-	-	-	-	99.00	87.80^s	76.70^s	74.80^s	94.70^s
Equiangular Kernel DL [11]	-	-	-	-	-	88.80^s	77.40^s	75.60^s	93.40^s
PCANet-TOP [10]	-	99.50^d	-	-	-	96.67^d	90.74^d	89.39^d	-
st-TCoF [32]	-	-	-	-	-	98.33^d	98.15^d	98.11^d	-
DT-CNN-AlexNet [9]	-	99.50^d	98.05^d	98.48^d	-	$\mathbf{100^d}$	99.38^d	$\mathbf{99.62^d}$	98.18^d
DT-CNN-GoogleNet [9]	-	99.50^d	98.35^d	99.02^d	-	$\mathbf{100^d}$	$\mathbf{100^d}$	$\mathbf{99.62^d}$	$\mathbf{98.58^d}$
Ours	$\mathbf{99.50^s}$	99.00^s	99.35^s	$\mathbf{99.57^s}$	98.86^s	98.33^s	92.59^s	91.67^s	95.69^s

Note: Superscript "d" indicates deep-learning methods, "s" is for results using SVM, otherwise using 1-NN; "-" means "not available".

approaches, MEWLSP [18] points out the highest rate of 98.48% in this scheme, even higher than the rate of DT-CNN [9] (98.18%) using AlexNet for learning patterns. However, it is not better than ours on UCLA dataset as well has not been tested on other challenging DynTex variants (i.e. *Alpha, Beta, Gamma*). Deep learning methods [9] have outstanding results but they take a long time to learn features with a huge complex computation.

5   Conclusions

An effective framework for DT representation has been proposed in this paper in which directional trajectories extracted from DT sequences are encoded by the proposed operator LVP_{full}-TOP, an extension of local vector pattern operator to full-direction on three orthogonal planes in order to exploit the entire reactions between motion points and their neighbors. In addition, we also introduced motion angle patterns of directional trajectories to capture the angle vector of pairs of their motion points as a complementary feature for texture representation in order to make the proposed descriptor more robust. Evaluations on different DT datasets have demonstrated that the proposed framework significantly outperforms recent state-of-the-art results. A combination between LVP and a filtering technique [34] will be the subject of a future work.

References

1. Peh, C., Cheong, L.F.: Synergizing spatial and temporal texture. IEEE Trans. IP **11**(10), 1179–1191 (2002)
2. Saisan, P., Doretto, G., Wu, Y.N., Soatto, S.: Dynamic texture recognition. In: CVPR, pp. 58–63 (2001)
3. Wang, Y., Hu, S.: Chaotic features for dynamic textures recognition. Soft Comput. **20**(5), 1977–1989 (2016)
4. Arashloo, S.R., Kittler, J.: Dynamic texture recognition using multiscale binarized statistical image features. IEEE Trans. Multimedia **16**(8), 2099–2109 (2014)
5. Rivera, A.R., Chae, O.: Spatiotemporal directional number transitional graph for dynamic texture recognition. IEEE Trans. PAMI **37**(10), 2146–2152 (2015)
6. Xu, Y., Quan, Y., Zhang, Z., Ling, H., Ji, H.: Classifying dynamic textures via spatiotemporal fractal analysis. Pattern Recogn. **48**(10), 3239–3248 (2015)
7. Xu, Y., Huang, S.B., Ji, H., Fermüller, C.: Scale-space texture description on sift-like textons. CVIU **116**(9), 999–1013 (2012)
8. Ji, H., Yang, X., Ling, H., Xu, Y.: Wavelet domain multifractal analysis for static and dynamic texture classification. IEEE Trans. IP **22**(1), 286–299 (2013)
9. Andrearczyk, V., Whelan, P.F.: Convolutional neural network on three orthogonal planes for dynamic texture classification. Pattern Recogn. **76**, 36–49 (2018)
10. Arashloo, S.R., Amirani, M.C., Noroozi, A.: Dynamic texture representation using a deep multi-scale convolutional network. JVCIR **43**, 89–97 (2017)
11. Quan, Y., Bao, C., Ji, H.: Equiangular kernel dictionary learning with applications to dynamic texture analysis. In: CVPR, pp. 308–316 (2016)
12. Quan, Y., Huang, Y., Ji, H.: Dynamic texture recognition via orthogonal tensor dictionary learning. In: ICCV, pp. 73–81 (2015)
13. Zhao, G., Pietikäinen, M.: Dynamic texture recognition using local binary patterns with an application to facial expressions. IEEE Trans. PAMI **29**(6), 915–928 (2007)
14. Nguyen, T.T., Nguyen, T.P., Bouchara, F.: Completed local structure patterns on three orthogonal planes for dynamic texture recognition. In: IPTA, pp. 1–6 (2017)
15. Tiwari, D., Tyagi, V.: Dynamic texture recognition based on completed volume local binary pattern. MSSP **27**(2), 563–575 (2016)
16. Tiwari, D., Tyagi, V.: A novel scheme based on local binary pattern for dynamic texture recognition. CVIU **150**, 58–65 (2016)
17. Tiwari, D., Tyagi, V.: Improved weber's law based local binary pattern for dynamic texture recognition. MTA **76**(5), 6623–6640 (2017)
18. Tiwari, D., Tyagi, V.: Dynamic texture recognition using multiresolution edge-weighted local structure pattern. Comput. Electr. Eng. **62**, 485–498 (2017)
19. Ojala, T., Pietikäinen, M., Mäenpää, T.: Multiresolution gray-scale and rotation invariant texture classification with local binary patterns. IEEE Trans. PAMI **24**(7), 971–987 (2002)
20. Nguyen, T.P., Manzanera, A., Kropatsch, W.G., N'Guyen, X.S.: Topological attribute patterns for texture recognition. Pattern Recogn. Lett. **80**, 91–97 (2016)
21. Zhao, X., Lin, Y., Heikkilä, J.: Dynamic texture recognition using volume local binary count patterns with an application to 2d face spoofing detection. IEEE Trans. Multimedia **20**(3), 552–566 (2018)
22. Nguyen, X.S., Nguyen, T.P., Charpillet, F., Vu, N.S.: Local derivative pattern for action recognition in depth images. MTA **77**(7), 8531–8549 (2018)
23. Zhang, B., Gao, Y., Zhao, S., Liu, J.: Local derivative pattern versus local binary pattern: face recognition with high-order local pattern descriptor. IEEE Trans. IP **19**(2), 533–544 (2010)

24. Nguyen, X.S., Mouaddib, A.I., Nguyen, T.P., Jeanpierre, L.: Action recognition in depth videos using hierarchical gaussian descriptor. MTA (2018)
25. Fan, K., Hung, T.: A novel local pattern descriptor - local vector pattern in high-order derivative space for face recognition. IEEE Trans. IP **23**(7), 2877–2891 (2014)
26. Wang, H., Kläser, A., Schmid, C., Liu, C.: Dense trajectories and motion boundary descriptors for action recognition. IJCV **103**(1), 60–79 (2013)
27. Nguyen, T.P., Manzanera, A., Garrigues, M., Vu, N.: Spatial motion patterns: action models from semi-dense trajectories. IJPRAI **28**(7), 1460011 (2014)
28. Fan, R., Chang, K., Hsieh, C., Wang, X., Lin, C.: LIBLINEAR: a library for large linear classification. J. Mach. Learn. Res. **9**, 1871–1874 (2008)
29. Ravichandran, A., Chaudhry, R., Vidal, R.: View-invariant dynamic texture recognition using a bag of dynamical systems. In: CVPR, pp. 1651–1657 (2009)
30. Ghanem, B., Ahuja, N.: Maximum margin distance learning for dynamic texture recognition. In: Daniilidis, K., Maragos, P., Paragios, N. (eds.) ECCV 2010. LNCS, vol. 6312, pp. 223–236. Springer, Heidelberg (2010). https://doi.org/10.1007/978-3-642-15552-9_17
31. Péteri, R., Fazekas, S., Huiskes, M.J.: Dyntex: a comprehensive database of dynamic textures. Pattern Recogn. Lett. **31**(12), 1627–1632 (2010)
32. Qi, X., Li, C.G., Zhao, G., Hong, X., Pietikainen, M.: Dynamic texture and scene classification by transferring deep image features. Neurocomputing **171**, 1230–1241 (2016)
33. Ren, J., Jiang, X., Yuan, J., Wang, G.: Optimizing LBP structure for visual recognition using binary quadratic programming. IEEE Signal Process. Lett. **21**(11), 1346–1350 (2014)
34. Nguyen, T.P., Vu, N., Manzanera, A.: Statistical binary patterns for rotational invariant texture classification. Neurocomputing **173**, 1565–1577 (2016)

Intrinsic Calibration of a Camera to a Line-Structured Light Using a Single View of Two Spheres

Yu Liu[1,2], Jiayu Yang[2], Xiaoyong Zhou[3], Qingqing Ma[3], and Hui Zhang[2(✉)]

[1] Department of Computer Science, Hong Kong Baptist University, Kowloon Tong, Hong Kong SAR, China
[2] Department of Computer Science and Technology, United International College, BNU-HKBU, Zhuhai, China
[3] Bomming Vision Technology Co., Ltd., Zhuhai, China
amyzhang@uic.edu.hk

Abstract. This paper proposes a novel approach to calibrate the intrinsic camera parameters from a single image, which includes the silhouette of two spheres and two ellipses generated by the intersection between the line-structured laser light and the two spheres. This approach uses the vanishing line of a plane and its normal direction to calculate the orthogonal constraints on the image of absolute conic (IAC). And this plane is formed by the camera center and two sphere centers. In addition, the pair of the circular points of the light plane is calculated from the generalized eigenpairs from the intersection between the light plane and the spheres. The intrinsic parameters of the camera can then be recovered from the derived orthogonal constraint and the pair of circular points on the IAC. Furthermore, the 3D positions of these two sphere centers under the camera coordinate can be recovered from the camera intrinsic matrix and then used to evaluate the accuracy of the camera intrinsic matrix. Experiment results on both synthetic and real data show the accuracy and the feasibility of the proposed approach.

Keywords: Intrinsic camera calibration · The line-structure light Spheres · Generalized eigenvectors

1 Introduction

Camera calibration is a vital step for the task of 3D reconstruction in computer vision or photogrammetry in machine vision. The camera intrinsic parameters indicate the projective geometry from 3D objects under the camera coordinate to a 2D plane under the image coordinate system. The camera extrinsic parameters describe the rotation and translation information of a camera relative to the world coordinate system. Many approaches have been developed according to the dimension of the calibration objects [1], such as 3D objects [9,10,14–16], 2D plane [5,13,20], one-dimensional objects [1], or even without calibration objects

© Springer Nature Switzerland AG 2018
J. Blanc-Talon et al. (Eds.): ACIVS 2018, LNCS 11182, pp. 87–98, 2018.
https://doi.org/10.1007/978-3-030-01449-0_8

[10], i.e. self-calibration. Due to its high accuracy and convenience, the most widely used calibration object is the 2D chessboard plane [13]. However, since a single plane can only provide two linearly independent constrains on calibration, pictures of multiple poses are required to be taken to fully calibrate the camera intrinsic parameters. Such procedures make the process of calibration manually cumbersome and time consuming. In order to overcome these problems, some common objects such as spheres or surfaces of revolutions (SOR) were proposed to replace the calibration board. In [16], the camera intrinsic parameters were calibrated with a single image which includes at least three spheres. In [10], the surface of revolution was used to recovering the intrinsic of a camera. Compared with the 2D calibration patterns, the images of 3D objects such as spheres or SOR can provide more constraints for intrinsic calibration.

Recently, these 3D calibration objects have been employed in calibrating the system with line-structured light vision sensors [2,4,18]. In [18], four balls placed on a flat board are used to calibrate the structured light plane. In [2], a method with a single object was proposed to calibrate the line-structured system. However, these methods generally involve calibrating the camera extrinsic parameters. A 2D calibration board was required to be firstly used to recover the intrinsic matrix in a separated step. 3D calibration objects were then introduced to estimate the light plane parameters [2].

This paper proposes a method to calibrate the intrinsic parameters from the intersection of a laser light plane with two spheres. Inspired by [16], the vanishing line of a plane and its normal direction were derived from the eigenvectors of the dual of the conic homograph and conic homograph generated from two sphere silhouettes, which formed the orthogonal constraint on the IAC. Besides, the intersection of the two spheres and the light plane form two co-planar circles. From generalized eigenvectors of their images can be broken down into constraints on the dual conic of the circular points from generalized eigenvectors. Together with the orthogonal constraints from the sphere silhouettes, the camera intrinsic parameters can then be recovered. From the camera intrinsic, the position of the sphere center can also be obtained.

The paper is organized as follows. Section 2 presents the fundamental theory of the image of the absolute conic. Section 3 introduces the orthogonal constraints from the silhouettes of two spheres. Section 4 relates the pair of circular points to the images of the intersection circles between the line-structured light plane and the spheres. The camera intrinsic parameters can then be estimated. Section 5 explains the process of recovering the 3D position of the sphere center. Experimental results and conclusions are then given in Sects. 6 and 7, respectively.

Throughout this paper, we use italic font (e.g., f) to represent a scale value, and use bold font (e.g., $\mathbf{v}$, $\mathbf{M}$) to represent a vector or a matrix.

2   The Image of the Absolute Conic

The absolute conic (AC) was firstly introduced in camera calibration by Faugeras et al. [6], and its projection in image is IAC which is an imaginary points conic, without real points lying on it. IAC is a 2D projection on the plane at infinity which is invariant under Euclidean transformations and can be written as

$$\omega = \mathbf{K}^{-T}\mathbf{K}^{-1} = \begin{pmatrix} \omega_1 & \omega_2 & \omega_3 \\ \omega_4 & \omega_5 & \omega_6 \\ \omega_7 & \omega_8 & \omega_9 \end{pmatrix}, \tag{1}$$

where $\mathbf{K}$ is the camera intrinsic matrix

$$\mathbf{K} = \begin{pmatrix} \alpha f & s & u_0 \\ 0 & f & v_0 \\ 0 & 0 & 1 \end{pmatrix}. \tag{2}$$

Here f is the focal length, α is the aspect ratio, (u_0, v_0) is the principle point and s is the skew. Generally, in order to calibrate the camera intrinsic, the image of the absolute conic (IAC) was firstly recovered, and then the camera intrinsic matrix $\mathbf{K}$ can be easily obtained by the Cholesky decomposition [7] of IAC.

3 Orthogonal Constraint from the Silhouettes of Two Spheres

The setup of a camera and a line-structured light system is shown in Fig. 1. The camera locates outside of the light plane. Two spheres, as the calibrate objects, are placed in front of the camera so that they are both visible to the camera. The line-structured light hits on both spheres and the intersection curves are partially visible to the camera. In Fig. 1, the solid part of the intersection curves are the visible outline to the camera, and the dashed curves are invisible part to the light source and cannot be seen by the camera.

Let the spheres be $\mathbf{x}^T \mathbf{B}_i \mathbf{x} = 0$ ($i = 1, 2$), $\mathbf{x} = [x\ y\ z\ 1]^T$ is any point on the spheres. $\mathbf{B}_i$ is a 4×4 matrix

$$\mathbf{B}_i = \begin{pmatrix} \mathbf{I}_{3\times3} & -\mathbf{o}_i \\ -\mathbf{o}_i^T & \mathbf{o}_i^T\mathbf{o}_i - r_i^2 \end{pmatrix}, \tag{3}$$

where the radius of the sphere is r_i and the sphere center is $\mathbf{o}_i$. After projection, we can get two conics $\mathbf{C}_1$ and $\mathbf{C}_2$ in the image (see Fig. 2 for details), which are the outlines of these two spheres. These two conics are both 3×3 matrices and satisfied $\mathbf{m}^T \mathbf{C}_i \mathbf{m} = 0$ ($i = 1,2$), where $\mathbf{m} = [u\ v\ 1]^T$.

The vanishing line $\mathbf{l}$ of the plane formed by the camera and the two sphere centers can be recovered as one of the eigenvectors of the "dual of the conic homograph" $\mathbf{H}_c$ [3]

$$\mathbf{H}_c = \mathbf{C}_2\mathbf{C}_1^{-1}, \tag{4}$$

which passes through both sphere silhouettes C_1 and C_2 (see Fig. 2 for details). The other two eigenvectors are discarded since they only pass through one conic. Note that the eigenpairs of H_c are as same as the generalized eigenpairs of (C_1^{-1}, C_2^{-1}). Similarly, the vanishing point v of the plane normal can be obtained directly as one of the eigenvectors of the "conic homograph" H_d [3]

$$H_d = C_2^{-1}C_1, \tag{5}$$

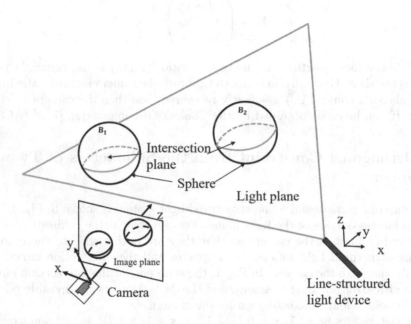

Fig. 1. The system setup. The two spheres B_1 and B_2, as the calibration objects, are placed in front of the camera so that they are both visible to the camera. The line-structured light hits on both spheres. The solid part of the intersection circles are the visible outline to the camera, and the dashed curves are invisible to the light source and cannot be seen by the camera.

which is lying outside of both conics. The other two eigenvectors are discarded since they lie within both of the silhouettes. The orthogonal constraints between the vanishing point v and the vanishing line l provides the pole-polar relationship with respect to the IAC ω [3]

$$l = \omega v. \tag{6}$$

Here the point v is the pole and the line l is the polar. Thus, two linear independent constraints on the IAC are provided by an image of two spheres. To fully calibrate the camera we still need to find other constraints from the system.

4   Circular Points from the Intersection Curves

Let the plane of the line-structured light in the world coordinate frame be

$$l_{light}^T \mathbf{P} = 0, \tag{7}$$

where $l_{light} = [a\ b\ c\ d]^T$ is the coefficient vectors of the light plane, $\mathbf{p} = [x\ y\ z1]$ is any point on this plane. The projection of the laser line onto the two spheres is the front half of the intersection between the light plane and the two spheres, which can be fitted to two complete conics $\mathbf{x}^T \mathbf{D}_1 \mathbf{x} = 0$ and $\mathbf{x}^T \mathbf{D}_2 \mathbf{x} = 0$ on the image, where $\mathbf{D}_1, \mathbf{D}_2$ are two 3×3 matrices. Since $\mathbf{D}_1$ and $\mathbf{D}_2$ are the projection of two co-planar circles, and intersect the vanishing line l_∞ of the light plane at two invariant conjugate complex points, which are the circular points $\mathbf{i}$ and $\mathbf{j}$ [8] (see Fig. 3 for details). In generally, when we know the conic $\mathbf{D}_1$ and $\mathbf{D}_2$, we can get these two intersection points by solving quadratic equations. However, the process of solving two quadratic equations is time consuming.

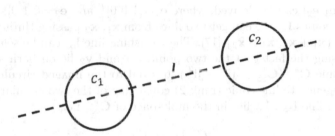

Fig. 2. The vanishing line l of the plane, which is pass through the camera center and the center of two spheres. The vanishing point $\mathbf{v}$ of the plane normal and vanishing line l, satisfied the pole-polar relationship and provide two linear independent constraints on the IAC.

In order to get the circular points $\mathbf{i}$, $\mathbf{j}$ in a linear way, the generalized eigen-pairs $(\lambda_i, \mathbf{v}_i)$ $(i = 1, 2, 3)$ of conic $\mathbf{D}_1$ and $\mathbf{D}_2$ are firstly obtained, where λ_i are the eigenvalues, $\mathbf{v}_i$ are the eigenvectors. Similar to Sect. 3, being regarded as points, one of the eigenvectors (let it be $\mathbf{v}_3$) lies outside of both conics, on the vanishing line l_∞. Its corresponding eigenvalue λ_3 has the smallest absolute value. The other two eigenvectors ($\mathbf{v}_1$ and $\mathbf{v}_2$) lie inside of either one of the conic $\mathbf{D}_1$ and $\mathbf{D}_2$. Secondly, each of the degenerated conic $\mathbf{L}_i = \mathbf{D}_1 - \lambda_i \mathbf{D}_2$ $(i = 1, 2, 3)$ composes two lines which can be obtained by singular value decomposition (**SVD**) [17]

$$\mathbf{L}_i = \mathbf{U}_i \mathbf{S}_i \mathbf{U}_i^T = \begin{pmatrix} \mathbf{u}_{i1} & \mathbf{u}_{i2} & \mathbf{u}_{i3} \end{pmatrix} \begin{pmatrix} s_{i1} & 0 & 0 \\ 0 & s_{i2} & 0 \\ 0 & 0 & 0 \end{pmatrix} \begin{pmatrix} \mathbf{u}_{i1}^T \\ \mathbf{u}_{i2}^T \\ \mathbf{u}_{i3}^T \end{pmatrix}, \tag{8}$$

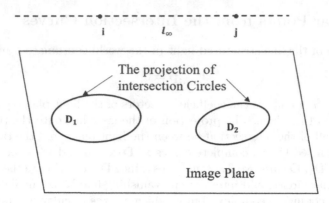

Fig. 3. Co-planer conics $\mathbf{D}_1$ and $\mathbf{D}_2$ intersect the vanishing line $\mathbf{l}_\infty$ at two imaged circular point $\mathbf{i}$ and $\mathbf{j}$. The pair of circular points provide the other two linear independent constraints on the IAC.

where the diagonal matrix $\mathbf{S}_i$ is with the singular values $s_{i1} > s_{i2} \geqslant 0$, and the $\mathbf{U}_i \in \mathbf{R}^{3\times3}$ is a orthogonal matrix. From each $\mathbf{S}_i$ and $\mathbf{U}_i$, two vectors $[\mathbf{x}_{i1}, \mathbf{x}_{i2}]$ $\equiv \mathbf{U}_i\, \mathbf{S}_i^{1/2}\, [\mathbf{e}_1\ \mathbf{e}_2]$ can be derived, where $\mathbf{e}_1=[1\ 0\ 0]^T$ and $\mathbf{e}_2=[0\ 1\ 0]s^T$. For the degenerated conics $\mathbf{L}_3$, we can get two lines from $\mathbf{x}_1, \mathbf{x}_2$ passing through $\mathbf{v}_3$, i.e., $\{\mathbf{m}_3, \mathbf{n}_3\} = \{\mathbf{x}_1 + \mathbf{x}_2, \mathbf{x}_1 - \mathbf{x}_2\}$ [17]. The vanishing line $\mathbf{l}_\infty$ can be selected from $\{\mathbf{m}_3, \mathbf{n}_3\}$ using the fact that the two points $\mathbf{v}_1$ and $\mathbf{v}_2$ lie on both side of $\mathbf{l}_\infty$. Since the conic $\mathbf{C}_\infty^*$ ($\mathbf{C}_\infty^* = \mathbf{ij}^T + \mathbf{ji}^T$) [6] dual to the imaged circular points $\mathbf{i}$ and $\mathbf{j}$ is a degenerate line conic (rank 2) consisting of the two circular points on the vanishing line $\mathbf{l}_\infty$, $\mathbf{l}_\infty$ lies in the null space of $\mathbf{C}_\infty^*$, i.e.,

$$\mathbf{C}_\infty^* \mathbf{l}_\infty = 0. \tag{9}$$

This provides two constraints on $\mathbf{C}_\infty^*$. Besides, each of the other two degenerated conics $\mathbf{L}_1$ and $\mathbf{L}_2$ composes two complex conjugate lines passing through the circular points, which provides further constraints on $\mathbf{C}_\infty^*$ [17],

$$\mathbf{x}_{i1}^T \mathbf{C}_\infty^* \mathbf{x}_{i2} = \mathbf{0} \ \ and \ \ \mathbf{x}_{i1}^T \mathbf{C}_\infty^* \mathbf{x}_{i1} - \mathbf{x}_{i2}^T \mathbf{C}_\infty^* \mathbf{x}_{i2} = 0\,(where\ i=1,2,3). \tag{10}$$

Finally, from (9) and (10), the conic $\mathbf{C}_\infty^*$ dual to the imaged circular points can be recovered and it can then be broken down to the circular points $\mathbf{i}, \mathbf{j}$ by **SVD**.

The circular points $\mathbf{i}, \mathbf{j}$ are lying on $\mathbf{l}_\infty$ in pairs (see Fig. 3 for details), and are also lying on the IAC [3], i.e.,

$$\mathbf{i}^T \boldsymbol{\omega} \mathbf{i} = \mathbf{j}^T \boldsymbol{\omega} \mathbf{j} = 0. \tag{11}$$

This provides another two linear independent constraints on IAC additional to (6). The system resulting from (6) and (11) provides four linear constraints on $\boldsymbol{\omega}$, hence a natural camera with only four parameters, i.e., a zero skew camera, can be recovered.

5   The Recovery of Sphere Center 3D Position

The position of the sphere center can be estimated from the camera intrinsic matrix. In conventional methods, triangularization is generally used to reconstruct the position from two images captured at two different viewpoints. This method is accurate, but exists ambiguity when decomposing the essential matrix $\mathbf{E}$, and needs at least two images and known some previous information.

To obtain the 3D position of the sphere center just from one image, we can use the eigenpairs from the right-circular cone of spheres' conic under camera coordinate system. From Sects. 3 and 4, we have already know the intrinsic matrix $\mathbf{K}$ and detected two spheres' outlines $\mathbf{C}_i$ $(i = 1, 2)$ from image. Then, we can recovery the right-circle cone $\mathbf{Q}_i$ for each sphere circle from back perspective projection model.

$$\mathbf{Q}_i = \mathbf{K}^T \mathbf{C}_i \mathbf{K}. \tag{12}$$

We apply matrix decomposition on the $\mathbf{Q}_i$, and get eigenvectors $(\mathbf{e}_{i1}, \mathbf{e}_{i2}, \mathbf{e}_{i3})$ together with the corresponding eigenvalues $(\lambda_{i1}, \lambda_{i2}, \lambda_{i3})$, $\lambda_{i1}, \lambda_{i2} > 0$, $\lambda_{i3} < 0$, where $i = 1, 2$. Once we get the measured sphere radius R, we can get the sphere center position under the camera coordinate system from [19]

$$[X_i\, Y_i\, Z_i]^T = \lambda_i' [\mathbf{e}_{i3}], \tag{13}$$

where

$$\lambda_i' = R\sqrt{\frac{|\lambda_{i1}|\,(|\lambda_{i2}| + |\lambda_{i3}|)}{|\lambda_{i3}|\,(|\lambda_{i1}| + |\lambda_{i3}|)}}, (i = 1, 2). \tag{14}$$

6   Experiments and Results

6.1   Synthetic Data

The synthetic camera has fixed intrinsic parameters, with focal length $f = 1000$, $\alpha = 1.2$, $s = 0$, and principle point $(u_0, v_0) = (320,240)$. The points on the silhouette of each sphere and each laser ellipse are corrupted with Gaussian noises at different levels from 0 to 5 pixels with 0.1 pixel step, and then two conics are fitted on the noisy points [11].

The synthetic experiment is to calibrate the camera intrinsic parameters using our proposed method under different noise levels. For each level, 100 independent tests are carried out. And, then computing the RMS error in each noise step. Figure 4 shows the relative error between the ground truth and RMS (root mean square) errors of the focal length f, the aspect ratio α, and the principle point (u_0, v_0). From Fig. 4, it can be seen the noise is slightly effect the four parameters in intrinsic matrix. Even for the worst error of the principle point v_0 is less than 15%.

Furthermore, we compare the proposed method with the orthogonal method mentioned in [16]. Under the same condition, the average intrinsic matrix parameters is showed in Table 1. It can be seen that the result of proposed method is better than the orthogonal method for a natural camera. Besides, the orthogonal method requires two images, which contain only the silhouettes of the two spheres to calibrate the camera intrinsic parameters.

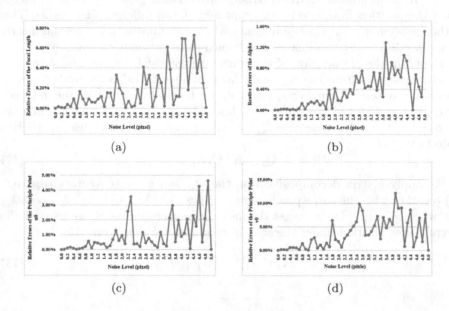

Fig. 4. The relative error between the RMS errors and ground truth at different noise levels below 5 pixels with 0.1 pixel step. And for each step, 100 independent tests are carried out. (a) The relative errors for the focal length f. (b) The relative errors for the aspect ratio α. (c) The relative errors for u_0. (d) The relative errors for v_0.

To verify the correctness of the estimated camera intrinsic parameters, we can obtain the positions of the two sphere centers and compare the estimated distance with the ground truth value. In the synthetic experiment, the sphere radius and the distance of sphere centers are given. Then, we also can use the same way to assess the stability of intrinsic matrix under different noise levels. Figure 5 shows the RMS error of the distance of two sphere centers under the noise level below 5 pixels. We can find the distance is weakly effected by the noise. The biggest RMS error is still below 1.00%, which means the evaluating the accuracy of camera intrinsic parameters by sphere center distance is reliable.

6.2   Real Scene

In the real scene experiment, an image of two connected ceramic balls is taken by an industrial camera. The setup of the real system is shown in Fig. 6(a). The two spheres are hit by the laser light and the detected light are fit to two conics (see Fig. 6(b) for details). The distance between these two sphere centers is 74 mm. The radius of both spheres is 12.7 mm. The image resolution is 1280×960. The

Table 1. Estimated camera parameters by the proposed method and the orthogonal method under the noise lever below 5 pixels.

Approach	α	f	u_0	v_0
Ground truth	1.2	1000	320	240
Proposed method	1.2056	997.408	318.6277	242.0497
Error (%)	0.46	0.21	0.43	0.85
Orthogonal method (2 images)	1.1748	986.1887	322.1086	264.2599
Error (%)	2.1	1.38	0.66	10.11

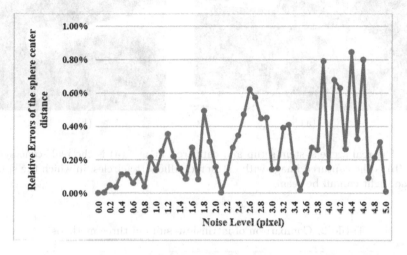

Fig. 5. The relative errors between RMS error and ground truth for the sphere center distance at each noise level. The noise level is below 5 pixels with 0.1 pixel step. And for each step, 100 independent tests are carried out.

Canny edge detector [12] is firstly applied to find the points on the silhouette of the spheres to which conics are fitted with a least square approach [11]. The intersection conics between the light plane and the spheres are obtained by the following steps. First, the Hessian matrix was used to compute the center curve of the laser light rays in image, and then the center curve was fitted to an ellipse with a least square approach [11]. By using proposed methods, the estimated intrinsic parameters are then obtained and listed in Table 2. For the purpose of comparison, the result from the classical method of Zhang [13] is taken as the ground truth. From Table 2, we can see that the result of proposed method is much close to the ground truth, except the vertical coordinate of the principle points. This may due to the errors induced from the vanishing point in (6). Compared with Zhang's method [13], this proposed method can just use one single image to recovery the camera intrinsic parameters linearly and get the result almost accurate, which means the proposed method is correct and reliable. Besides, we cannot use orthogonal method [16] to obtain the camera intrinsic matrix from one image which include only two spheres. Two images can provide four linear independent constraints on the camera intrinsic parameters. Table 2 also shows the results of the proposed method and the orthogonal method based on two same images. From the result, we can see the focal length is better than that of the proposed method. The principle point is worse own to the silhouettes extraction. The result of the orthogonal method is worse and own to the vanishing point in (6) derived far away.

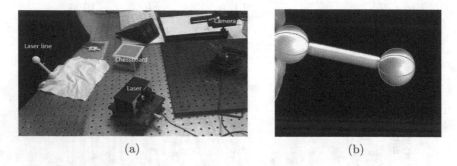

(a)   (b)

Fig. 6. The real scene system setup and captured image. (a) is the real scene system setup. (b) is the captured image with extracted silhouette circles, in which the spheres and laser light can all be seen.

Table 2. Comparison of intrinsic results in three methods

Method	α	f	(u_0, v_0)
Zhang	1.00689	6659.2	(592.9, 460.6)
Proposed method	1.00035	6555.2	(594.7, 418.7)
Error(%)	0.65	1.56	(-0.3,9.1)
Proposed method (2 images)	1.0042	6565	(431.1, 398.5)
Orthogonal method (2 images)	0.99	7192.9	(241.3, 158.5)

After we get the intrinsic matrix and the sphere conics from the image, we can use the method mentioned in Sect. 5 to estimate the distance between the sphere centers. Please note, at this moment, the sphere center is under the camera coordinate system. And based on the projection property, the distance between two sphere centers will not change. Hence, we can compare the estimated distance with the ground truth to assessment accuracy of the proposed method. The results of the distance of the sphere centers are list in Table 3. The camera intrinsic matrix calibrated by the proposed method is accepted. The difference from the ground truth is caused by the silhouette and extracted laser center line.

Table 3. Comparison of the distance of sphere centers between the estimated value and ground truth

Estimated distance (mm)	Ground truth (mm)	Error (%)
74.899	74	1.21

7   Conclusion

This paper has proposed an approach to calibrate the camera intrinsic parameters using a single image which includes the outline of two spheres and two ellipses generated by the intersection between the line-structured laser light plane and the two spheres. Then, the distance of two sphere centers can be recovered using the estimated intrinsic matrix $\mathbf{K}$. For the intrinsic matrix calibration, a plane is formed by the camera center and two sphere centers. The orthogonal constraint on the silhouettes of two spheres can be generated from the vanishing line of the plane and its normal direction. Besides, the pair of the circular points of the light plane is calculated from the generalized eigenpairs from the intersection between the light plane and the spheres. Four linear constraints derived from the orthogonal constraint and the pair of circular points. Thus, the camera intrinsic parameters $\mathbf{K}$ can be recovered. Furthermore, the 3D positions of these two sphere centers under the camera coordinate can be recovered from the camera intrinsic matrix. After that, on the strength of the distance invariability under the camera coordinate system, we can further obtain the distance of two sphere centers. And then using this distance to assess the accuracy of the camera intrinsic matrix. Both synthetic and real experiments show the feasibility of the proposed approach.

Acknowledgment. The work described in this paper was supported by the Major Innovation Projects of University - Industry Collaboration (Project no. 201604020095) and Province Natural Science Fund of Guangdong, China (Project no. 2017A030313362).

References

1. Zhang, Z.Y.: Camera calibration with one-dimensional objects. IEEE Trans. Pattern Anal. Mach. Intell. **26**(7), 892–899 (2004)
2. Liu, Z., Li, X.J., Li, F.J., Zhang, G.J.: Calibration method for line-structured light vision sensor based on a single ball target. Optics Lasers Eng. **69**, 20–28 (2015)
3. Zhang, H., Zhang, G., Wong, K.-Y.K.: Camera calibration with spheres: linear approaches. In: The Process of International Conference on Image Processing, Genova, vol. (II), pp. 1150–1153 (2005)
4. Colombo, C., Comanducci, D., Del Bimbo, A.: A desktop 3D scanner exploiting rotation and visual rectification of laser profiles. In: The Process of IEEE International Conference on Computer Vision Systems (2006)

5. Chen, D., Sakamoto, R., Chen, Q., Wu, H.Y.: Extrinsic camera parameters estimation from arbitrary co-planar circles. In: The Process of SIGGRAPH Asia Posters (2014)

6. Faugeras, O.D., Luong, Q.-T., Maybank, S.J.: Camera self-calibration: theory and experiments. In: The Process of European Conference on Computer Vision, pp. 321–334 (1992)

7. James, E.: Gentle: Numerical Linear Algebra for Applications in Statistics. Springer, New York (1998)

8. Hartley, R., Zisserman, A.: Multiple View Geometry in Computer Vision. Cambridge University Press, Cambridge (2003)

9. Colombo, C., Del Bimbo, A., Pernici, F.: Metric 3D reconstruction and texture acquisition of surfaces of revolution from a single uncalibrated view. IEEE Trans. Pattern Anal. Mach. Intell. 27(1), 99–114 (2005)

10. Zhang, H., Wong, K.-Y.K., Mendonca, P.R.S.: Reconstruction of surface of revolution from multiple uncalibrated views: a bundle-adjustment approach. In: Proceeding of the Asian Conference on Computer Vision, pp. 378–383 (2004)

11. Fitzgibbon, A., Pilu, M., Fisher, R.B.: Direct least square fitting of ellipses. IEEE Trans. Pattern Anal. Mach. Intell. 21(5), 476–480 (1999)

12. Canny, J.: A computational approach to edge detection. IEEE Trans. Pattern Anal. Mach. Intell. 8(6), 679–698 (1986)

13. Zhang, Z.Y.: A flexible new technique for camera calibration. IEEE Trans. Pattern Anal. Mach. Intell. 22(11), 1330–1334 (2000)

14. Zhang, H., Wong, K.-Y.K.: Self-calibration of turntable sequences from Silhouettes. IEEE Trans. Pattern Anal. Mach. Intell. 31(1), 5–14 (2008)

15. Wong, K.-Y.K., Mendonca, P.R.S., Cipolla, R.: Camera calibration from surfaces of revolution. IEEE Trans. Pattern Anal. Mach. Intell. 25(2), 147–161 (2003)

16. Zhang, H., Wong, K.-Y.K., Zhang, G.: Camera calibration from images of spheres. IEEE Trans. Pattern Anal. Mach. Intell. 29(3), 499–503 (2007)

17. Gurdjos, P., Sturm, P., Wu, Y.H.: Euclidean structure from N⩾2 parallel circles: theory and algorithms. In: The Process of European Conference on Computer Vision, pp. 238–252 (2006)

18. Xu, J., Doubet, J., Zhao, J.G., Song, L.B., Chen, K.: A simple calibration method for structured light-based 3D profile measurement. Optics Lasers Eng. 48, 187–193 (2013)

19. Shiu, Y., Ahmad, S.: 3D location of circular spherical features by monocular model-based vision. In: The Process of the IEEE Conference on System, Man and Cybernetics, Cambridge, Mass, USA, pp. 576–581 (1989)

20. Chen, Q., Wu, H.Y., Higashino, S., Sakamoto, R.: Camera calibration by recovering projected centers of circle pairs. In: The Process of SIGGRAPH: Posters, Article No. 39 (2016)

3D Object-Camera and 3D Face-Camera Pose Estimation for Quadcopter Control: Application to Remote Labs

Fawzi Khattar[1,3]([✉]), Fadi Dornaika[3,4], Benoit Larroque[2], and Franck Luthon[1]

[1] UNIV PAU & PAYS ADOUR/E2S UPPA, LIUPPA Lab, Anglet, France
{fawzi.khattar,franck.luthon}@univ-pau.fr
[2] UNIV PAU & PAYS ADOUR/E2S UPPA, SIAME Lab, Anglet, France
benoit.larroque@univ-pau.fr
[3] University of the Basque Country UPV/EHU, San Sebastian, Spain
fadi.dornaika@ehu.eus
[4] IKERBASQUE, Basque Foundation for Science, Bilbao, Spain

Abstract. We present the implementation of two visual pose estimation algorithms (object-camera and face-camera) with a control system for a low cost quadcopter for an application in a remote electronic laboratory. The objective is threefold: (i) to allow the drone to inspect instruments in the remote lab, (ii) to localize a teacher and center his face in the image for student-teacher remote communication, (iii) and to return back home and land on a platform for automatic recharge of the batteries. The object-camera localization system is composed of two complementary visual approaches: (i) a visual SLAM (Simultaneous Localization And Mapping) system, and (ii) a homography-based localization system. We extend the application scenarios of the SLAM system by allowing close range inspection of a planar instrument and autonomous landing. The face-camera localization system is based on 3D modeling of the face, and a state of the art 2D facial point detector. Experiments conducted in a remote laboratory workspace are presented. They prove the robustness of the proposed object-camera visual pose system compared to the SLAM system, the performance of the face-camera visual servoing and pose estimation system in terms of real-time, robustness and accuracy.

1 Introduction

Remote labs constitute an interesting and novel way of doing labs. Anywhere and at anytime the student can access the lab equipment and do his labwork. This new way of distance learning can be used to increase the motivation of nowadays students [1]. Quadcopters equipped with a camera can be used in these laboratories in order to mimic the student behavior in traditional lab and increase motivation for learning. It can be an interesting way to make the lab experience immersive and ludic. Specifically, in remote electronics laboratories it can fly and move in 3D space to inspect electrical instruments, consequently sending

© Springer Nature Switzerland AG 2018
J. Blanc-Talon et al. (Eds.): ACIVS 2018, LNCS 11182, pp. 99–111, 2018.
https://doi.org/10.1007/978-3-030-01449-0_9

direct visual feedback of the results of an experiment to the student. Furthermore it can also search for a teacher in the lab and move towards him, centering his face in the image to allow remote student-teacher interaction. In this way the student can achieve his lab-work and can also ask questions to the teacher in case he needs to. To achieve this double objective two systems are needed. First, a localization system that can estimate the position and orientation of the quadcopter in 3D space with respect to an object of interest (front panel of an electrical instrument or face of the teacher in our case). Second, a control system that sends appropriate commands to the quadcopter in order to reach a reference relative or absolute 3D position. Many localization systems and sensors can be used in order to estimate quadcopter position and orientation in 3D space. For outdoor environments, GPS sensors are the best solution to localize a quadcopter. For indoor environments, artificial markers can be placed in the scene to facilitate the task of localization [2]. These markers can also be reflective and detected by an external localization system that gives accurate position estimate. However, the challenge in these applications is to use only available on-board sensors. Stereo rig cameras [3] and RGBD cameras [4] have been investigated. They allow for absolute pose estimate however this comes with an additional weight and power consumption. In this work we use the Parrot AR Drone 2.0 [5], a low cost quadcopter equipped with two monocular cameras facing forwards and downwards, in addition to pressure, ultrasound and inertial sensors. Using the monocular cameras available on-board constitutes a good trade-off between weight and information recovery from the environment (3D localization, environment recognition, etc.). However, a monocular camera alone cannot give absolute scale pose estimate due to the well known scale ambiguity rising from the perspective projection of 3D world into 2D images. Despite this fact, combining the visual information with some prior knowledge of 3D world or other sensors, that can give partial but absolute pose estimate, can overcome this issue and allow for absolute 3D pose estimate. The well known SLAM algorithm PTAM (Parallel tracking and mapping) [6] combined with inertial and ultrasound sensors readings in order to get the absolute 3D pose estimate is used in [7]. All the available data is processed in a Kalman filter allowing for information fusion and delay compensation. Here, starting from [7] we propose the object-camera pose system and use it for 3D pose estimation when the quadcopter is exploring the 3D world to search for an instrument or for a teacher. However, since the visual SLAM relies on corresponding points detected in the flow of images, it will fail to give 3D pose estimate if the quadcopter is asked to inspect an object of interest, since these points will disappear when the object of interest occupies the majority of the image. To overcome this limitation, this system is extended by using the object of interest as a landmark. In this way two localization modules are available: a visual SLAM module (for localization w.r.t. an arbitrary world coordinate system) and a localization module that relies on detecting and localizing a planar object with respect to the quadcopter and to the arbitrary world coordinate system (needed for controlling the remote lab activities and sending visual feedback to the student). The first module is suitable while exploring the environment whereas the latter is suitable when the drone is in a short distance range from the object of interest or when it needs to land on the electrical recharge platform. To be able to control the drone to center the face of a teacher in the image, the position of the drone with respect to the face of the teacher must be estimated. For this purpose, we use a deformable 3D model of the face and fit it to the image. The paper is organised as follows: In the second section we present an overview of the remote electronic lab LaboREM. In Sect. 3 we explain how detection and 3D localization can be carried out for the purpose of remote instrument inspection. In Sect. 4 we present the face-camera pose estimation approach. In Sect. 5, the 3D pose visual servoing of the quadcopter is explained. In the final section, we present our experiments and results.

2 Review of the Remote Electronics Lab

LaboREM is a remote laboratory in electronics developed for first year undergraduate students in engineering. The learning objective of LaboREM is to enable students to wire and test remotely electronic circuits, make measurements and characterize each circuit by its time or frequency response. The electronic circuits consist of operational amplifiers, active filters and oscillators. Its design is based on a classic client-server architecture [8]. The student calls for a lab session by simple URL addressing. A first-in first-out strategy is adopted to give access to the remote lab to one client (student) at a time. The remote lab application is developed using NI-LabVIEW software and the easy-to-use RFP protocol to pilot the remote devices. The hardware setup includes: (i) a robotic arm that mimics the student's hand for placing electronic components equipped with magnets on an electronic breadboard, (ii) measurement instruments and data acquisition system (DAQ), (iii) a webcam with zoom control that mimics the student's eye in order that the student doesn't feel so far away from what is actually happening in the lab, (iv) a quadcopter (AR quadcopter 2.0) with the role of flying in the lab for exploring the environment, inspecting electrical instruments and interacting with a teacher in order to increase student immersion and motivation.

3 Object-Camera Pose Estimation

3.1 Quadcopter-Object Relative 3D Pose: A Real-Time and Marker Free Solution

Planar objects are a well defined type of objects that are widely available in human made environments. Incorporating the information that the object of interest is planar is of great benefit for object-camera pose estimation. The homography matrix is a matrix that relates 3D points lying on a plane to their 2D projections. Given this transform one can directly calculate the rotation and translation matrix as done in [9]. In order to estimate the homography that maps any plane into another plane by means of perspective projections several methods can be used. These methods are usually classified into local (feature-based) and global (featureless) methods. Given a template image of the planar object, local methods extract local keypoints and attribute a descriptor to each of them both in the template image and the current image. After this step, keypoints (at least four keypoints) in both images are matched according to a similarity metric performed on the descriptors. Given the point correspondences, the homography matrix is estimated using robust methods like RANSAC (Random Sample Consensus) in order to deal with the presence of outlier correspondences. Local methods can work well with no prior information on the homography parameters. However, in some cases the robust computation may be computationally expensive and do not work in real time. A survey about keypoint detectors and descriptors can be found in [10]. On the other hand, the global methods use all the information in the image and attempt to find the homography matrix that best aligns the template patch to the test image. This process however gives rise to NLM (non linear minimization) problems that can be solved using iterative algorithms like gradient descent or LM (Levenberg-Marquardt). Thus, a good initialization is necessary to guarantee the convergence of those algorithms. Different similarity functions exist to measure the degree to which two patches are aligned, the most used ones being the sum of squared distance (SSD) and the enhanced correlation coefficient (ECC) [11]. In practice the first one uses a brightness model in order to cope with variation of additive and multiplicative change of illumination [12] whereas the latter is by definition insensitive to those illumination changes. These methods have the advantage that they can run in real time and give good results if a coarse estimate of the homography parameters is known. Thus the two families of methods are complementary. The first one is robust with no priors needed but computationally expensive, while the second is fast and works well if a prior is available. Here, both approaches are used in order to estimate the 3D pose of the quadcopter with respect to the object of interest. The first approach is used for detecting the object of interest as well as for recovering from a tracking loss. The second is used in the tracking process. The approach is divided into two steps: detection and tracking [13]. In the detection step, template matching on a pyramid of the image is used to search for the desired object. If the normalized correlation coefficient is greater than α the object is declared detected. Once the detection is done, a homography transformation is computed by using the bounding box of the detected object to determine the object-camera relative pose. A command is then sent to the drone in order to move it closer to the object. Template matching is used to allow successful detection of the object despite its distance from the camera and its size, as keypoints detector fails to detect and put in correspondences keypoints if the object of interest doesn't occupy a certain amount of image pixels. However once the distance between the camera of the drone and the instrument is less than a threshold λ, the SIFT descriptor [14] is used to allow more robustness to orientation changes. The object is declared detected if the number of matched keypoints is greater than N. Once the object is detected, the tracking stage begins. As a rough estimation of the homography matrix is available from the detection stage, it is used as an initial solution for the next frame and the ECC algorithm is applied to estimate the homography in this frame. The homography estimation is propagated in this way from a frame to the next one, and used as a prior for the ECC algorithm. However, sometimes the ECC algorithm will fail to converge due to several reasons. For example, communication problems between the quadcopter and the computer makes the last estimated homography not close enough to the real solution of the current frame, which prevents algorithm convergence. Besides, the image quality can be degraded by motion blur or decoding/encoding problems. In this work, tracking loss is declared if the ECC algorithm is unable to converge or if it converges to a clearly unrealistic estimation. At each frame, we compute the 3D pose of the quadcopter with respect to the planar object. By monitoring the estimated traveled distance between two consecutive frames and comparing it to a threshold D, we can detect a loss of tracking. Another threshold β is also imposed on the difference of each angle of orientation (yaw, roll, pitch). If the tracking fails, we resort back to the local method (SIFT) if the quadcopter-object distance is relatively small, or to the template matching method in the other case, to reinitialize the ECC tracker as shown in Fig. 1. This pose estimate is fused with inertial measurements sent by the drone in a Kalman filter framework in order to smooth this estimate, and to provide robustness when the visual tracker fails. The Kalman filter is used also to compensate for time delays as done in [7]. The homography estimation process is shown in Fig. 1 while the feedback control loop is shown in Fig. 2.

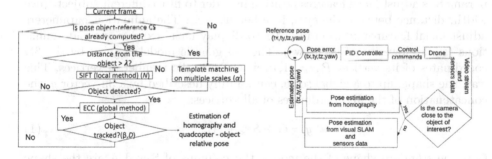

Fig. 1. Homography estimation diagram.   **Fig. 2.** Feedback control loop.

4   Face-Camera Pose Estimation

Human-drone interaction is an interesting way of controlling drones. In [15] authors use face pose and hand gestures in order to allow human-drone interaction. Their face pose estimation process is based on the Viola and Jones face detector [16]. They compute a face score vector by applying frontal and side face detector on the flipped and the original image. Using this face score and a machine learning technique they estimate the *yaw* angle of the face pose. The distance from the face is estimated by the size of the face bounding box. Hand gestures are used to give order to the drone to move to an orientation while maintaining the distance from the face. In [17], authors also used hand gestures and face localization for drone-human interaction. Their approach is unique for the fact that it allows the drone to approach a human that is 20 meters away, by detecting periodic hand gestures. The drone then approach the target by tracking its appearance. Once at a short distance, the drone centers the face of the subject and detects hand gestures in order to take a picture. However, the orientation of the face is not estimated and the user has to be facing the camera in order to take a frontal photo. In this work, we adopt a 3D approach that models the human face in 3D and subsequently uses full perspective projection in order to recover the 3D face pose parameters. By using this modeling and matching it with image specific data related to the face, all the 6 pose parameters are inferred. The 3D modeling is based on the CANDIDE deformable 3D face model.

4.1 CANDIDE 3D Model

CANDIDE is a parameterized 3D face model specifically developed for model-based coding of human faces. CANDIDE is controlled by 3 sets of parameters: global, shape and animation parameters. The global parameters correspond to the pose of the face with respect to the camera. There exist 6 global parameters: 3 Euler angles for the rotation and 3 for the translation (t_x, t_y, t_z). The shape parameters adjust facial features position in order to fit to different subjects (eye width, distance between the eyes, face height etc.). The animation parameters adjust facial features position in order to display facial expressions and animations (smile, lowering of eyebrows). The 3D generic model is given by the 3D coordinates of its vertices $P_i, i = 1, n$. where n is the number of vertices. This way, the shape, up to a global scale, can be fully described by a 3n-vector g, the concatenation of the 3D coordinates of all vertices:

$$g = G + S\tau_s + A\tau_a \tag{1}$$

G is the standard shape of the model, the columns of S and A are the shape and animation units, $\tau_s \in \mathbb{R}^m$ and $\tau_a \in \mathbb{R}^k$, are the shape and animation control vectors, respectively.

4.2 Inferring Pose Parameters

In order to determine the pose from the 3D model, we have to fit this model to the face data available in the image. Fitting the model means determining its different parameters: pose, shape and animation parameters. In this work only pose and shape parameters are of interest for us, however recovering the animation parameters can be an interesting way to allow human-drone interaction based on facial expressions. Different approaches attempt to adapt the model in different ways. However, the majority of them follow a step by step approach, starting by estimating the shape parameters τ_s in order to adapt the 3D model to different face anatomy and then estimating the pose and animation parameters. From the face image, many face related data can be used to fit the 3D model. In [18], the authors use the gray scale appearance of the image to adapt the 3D model after estimating its shape parameter off-line. In our work, we make use of the advancement in facial landmark detection and use these landmarks to adapt the model and recover the 3D face pose from a set of 3D-to-2D correspondences. The shape parameters are estimated using a frontal picture of the subject following the method described in [19]. We use the facial point detector in [20], that can detect 68 2D landmarks on a face in one millisecond by a pre-trained ERT (Ensemble of Regression Trees), given that a face image patch is available. However, since the algorithm needs a region of interest that contains a face, the total time for its execution from the detection of the face to the detection of the landmarks is more than one millisecond due to the computationally expensive face detection step. One way to reduce this time and make the process working at more than 30 fps (frame per second) is to perform a search for the face around the last detected bounding box of the face instead of looking for the face in the whole image. We make use of only 46 points from the 68 points given by the landmark detector. The points were chosen to be semantic and mostly rigid thus eliminating points along face contour. Once the 2D landmarks are detected in the image we use state of the art pose estimation algorithms that are based on 3D-2D point correspondences to recover the pose. This problem is known in the literature as PnP (Perspective n Point). Many algorithms attempt to solve this problem. The P3P algorithm [21] (perspective 3 point) can estimate the pose using only 3 point correspondences. Other algorithms like EPNP [22] (Efficient Perspective N Point) can handle any number of points. Another approach is to use non-linear minimization techniques to recover the pose that best minimizes the distance between the projected 3D points and the 2D points. However, this method requires an initial guess of the pose parameters in order to converge to the global minimum. This initial guess can be made available using the estimated pose from the previous frame or using any closed-form solution like EPNP, P3P, etc. in case it is not available. We propose to use the Levenberg-Marquardt technique as it gives good results and fast execution time. The face-camera pose estimation process is shown in Fig. 3. The pose used to control the drone is computed by fusing the visual pose from the 3D model with inertial and ultrasound measurements in a Kalman filter as done in [7].

5   Visual Control of the Quadcopter

In order to control the 3D position and orientation of the quadcopter for the purpose of instrument inspection, a closed-loop control is used taking as a feedback the pose derived from the SLAM or the homography. The control loop is shown in Fig. 2. If the objective is to control the drone to maintain a relative position from the face, the algorithm explained in Sect. 4 is used as a feedback sensor for the control loop. The controlled degrees of freedom associated with the quadcopter are the 3D translation vector and the yaw angle. Each degree is controlled by a closed loop control system with a traditional PID controller.

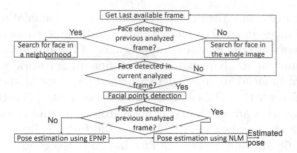

Fig. 3. Face-camera pose estimation.

6   Experiments

Before presenting the different scenarios for evaluation, we begin with an evaluation of the performance of the Face-Camera pose estimation. We compare the accuracy of different techniques and their execution time on a database for pose estimation (UPNA head pose database) [23]. The UPNA database contains 120 videos corresponding to 10 different subjects, 12 videos each, in which the subject changes its head pose by following guided and free movement. The ground-truth relative 3D face motion is known for all frames in all videos. We conclude that all the techniques converge to the same optimal solution if followed by a non-linear minimization method. As shown in Table 1 the method that yields the best pose estimate and exccecution time is the Levenberg- Marquardt method that tracks the 3D face from a frame to the next one based on an initial estimate. In order to evaluate the proposed implementation of 3D pose estimation and 3D pose-based servoing, we design three different scenarios. These scenarios are the following: behavior of the system in response to perturbations when asked to inspect an object, autonomous visual inspection of planar object, and drone-face visual servoing.

First Scenario: The first experiment aims to test the quality of the homography based visual feedback control system. To this end, we control the quadcopter in such a way that the reference 3D pose of its on-board camera is fronto-parallel to the planar object with a translation vector allowing a centered view. The pose used for the visual feedback control is the homography based pose. Since the servoing objective is to maintain a rigid link between the quadcopter and the object of interest, any motion induced to the object will force the quadcopter to compensate for it. We can induce such motion by a walking person that carries the object or by giving manual kicks to the quadcopter. The quadcopter then follows the object, centering it in the image. Figure 4 shows the response of the system facing manual perturbations applied to the drone (5 perturbations to the x position, 2 for y, 2 for z and 3 for the angle yaw). The objective is to see the behavior of the system when the drone is pushed away from the reference pose causing the visual tracking to fail. Despite the loss of tracking (red curves in Fig. 4) due to the fast kicks applied to the drone, it is always able to return to a position that allows the tracking to restart. This is done by controlling the drone based on the pose estimation procured by the Kalman filter [7] that fuses the inertial and ultrasound measurements to have an estimate of the current position of the drone. Two videos of this scenario are available online [24, 25].

Second Scenario: In remote lab context, an interesting scenario is the following: the remote student will send a command to the quadcopter to go and inspect an electrical device. After receiving this command, the server tells the quadcopter to carry out the following tasks: it should first take off, initialize the SLAM algorithm (by following a vertical path in order to change the height and correctly estimate the scale of the SLAM map [7]), and initiate a search procedure for the required instrument. After object detection, the drone moves towards the instrument and the feedback loop uses the 3D pose based on the homography for control. In this way the quadcopter is able to fly to inspect an electrical instrument maintaining its position with respect to the instrument. After the mission is over the drone is sent back home by using the pose derived from the SLAM. The position of the landing platform is estimated by using the homography-based pose estimation applied on the video of the bottom camera of the AR Drone 2.0. The drone hover then at a certain altitude above the landing platform preparing for landing. A quantitative comparison between the approach proposed here for pose estimation when inspecting an object at close range with the SLAM algorithm used in [7] is shown in Fig. 5. It shows the superiority of our approach and the drift of the SLAM approach due to the reasons explained in the introduction. Figure 6 shows the first detection and the tracking of the object of interest and its robustness in spite of problems of tracking failure. The green cross represents the center of the image while the red cross represents the projection of the center of the object on the image. The objective is to control the position of the drone so that the two crosses are as close as possible. Videos of this scenario are available in [26, 27].

Third Scenario: In this scenario we test the performance of the face-camera visual servoing system explained in Sect. 4. The drone has to fly, detect a face, align its line of sight with that of the subject face, and centering the face in the image while maintaining a fixed distance with it. In this experiment the user is in motion in order to induce perturbation to the control system. The drone has

Table 1. Average pose errors and computation time for different face pose estimation methods. t_x, t_y, t_z are in millimeters, *roll*, *yaw*, *pitch* in degrees, time in milliseconds.

Method	t_x	t_y	t_z	roll	yaw	pitch	time
EPNP	11,84	7,11	12,67	0,55	3,74	2,39	0.117
Ransac P3P	12,50	7,79	18,34	1,56	6,52	6,11	0.898
EPNP + NLM	11,51	7,23	13,38	0,56	2,28	1,45	0.363
P3P Ransac + NLM	11,51	7,23	13,38	0,56	2,28	1,45	1.138
NLM	**11,51**	**7,23**	**13,38**	**0,56**	**2,28**	**1,45**	**0.234**

to correct for the user displacement and the out-of-plane orientation of his face. Figure 7 shows the experiment seen from the camera of the drone and from an external camera. A video of the experiment is available on [28].

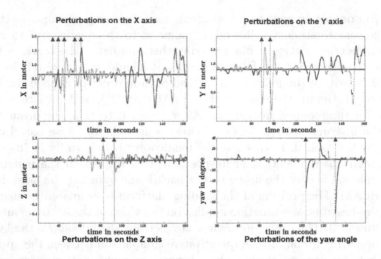

Fig. 4. Quadcopter control and estimated pose facing perturbations. For each control loop: In green the estimation from the homography when the tracking is good, in red estimation based solely on navigation data when the tracking fails and in light blue the reference 3D pose. Vertical arrows indicate the time each perturbation was applied. (Color figure online)

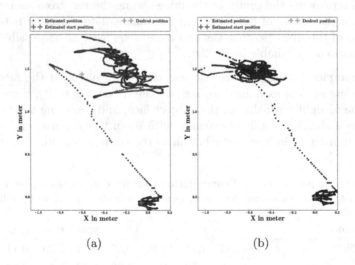

(a)   (b)

Fig. 5. Estimated pose in the XY plane. (a) with [7], (b) with the proposed approach.

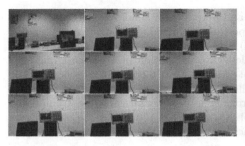

Fig. 6. Detection and tracking of the instrument. (Color figure online)

Fig. 7. Third scenario experiment: Face tracking.

7 Conclusion

This paper presents the implementation of a visual servoing system of a quadcopter in a remote lab environment to increase student immersion in the lab and hence his motivation. The objective is to allow remote instrument inspection and remote human-teacher communication. The proposed localization system for the first objective is proven to outperforms the SLAM system in [7] through qualitative and quantitative experiments, allowing the quadcopter to inspect an object and return to its base autonomously. The approach uses only the on-board sensors available on the low cost drone. The localization system for face-camera servoing is based on 3D modelling of the face and a state-of-the art 2D facial point detector. The approach controls all 4 degrees of freedom (3° for translation as well as the orientation of the face). It is shown robust, accurate and working at frame rate through qualitative and quantitative experiments.

References

1. Luthon, F., Larroque, B., Khattar, F., Dornaika, F.: Use of gaming and computer vision to drive student motivation in remote learning lab activities. In: 10th Annual International Conference of Education, Research and Innovation, ICERI 2017, pp. 2320–2329 (2017)
2. Eberli, D., Scaramuzza, D., Weiss, S., Siegwart, R.: Vision based position control for MAVs using one single circular landmark. J. Intell. Robot. Syst. **61**(1–4), 495–512 (2011)
3. Schauwecker, K., Zell, A.: On-board dual-stereo-vision for the navigation of an autonomous MAV. J. Intell. Robot. Syst. **74**(1–2), 1–16 (2014)
4. Flores, G., Zhou, S., Lozano, R., Castillo, P.: A vision and GPS-based real-time trajectory planning for a MAV in unknown and low-sunlight environments. J. Intell. Robot. Syst. **74**(1–2), 59–67 (2014)
5. https://jpchanson.github.io/ARdrone/ParrotDevGuide.pdf

6. Klein, G., Murray, D.: Parallel tracking and mapping for small AR workspaces. In: 6th IEEE and ACM International Symposium on Mixed and Augmented Reality, ISMAR 2007, pp. 225–234. IEEE (2007)
7. Engel, J., Sturm, J., Cremers, D.: Scale-aware navigation of a low-cost quadro-copter with a monocular camera. Robot. Auton. Syst. **62**(11), 1646–1656 (2014). Special Issue on Visual Control of Mobile Robots
8. Luthon, F., Larroque, B.: LaboREM a remote laboratory for game-like training in electronics. IEEE Trans. Learn. Technol. **8**(3), 311–321 (2015)
9. Medioni, G., Kang, S.B.: Emerging Topics in Computer Vision. Prentice Hall PTR, Upper Saddle River (2004)
10. Krig, S.: Interest Point Detector and Feature Descriptor Survey. In: Computer Vision Metrics. Apress, Berkeley (2014). https://doi.org/10.1007/978-1-4302-5930-5_6
11. Evangelidis, G.D., Psarakis., E.Z.: Parametric image alignment using enhanced correlation coefficient maximization. IEEE Trans. Pattern Anal. Mach. Intell. **30**(10), 1858–1865 (2008)
12. Dornaika, F.: Registering conventional images with low resolution panoramic images. In: The 5th International Conference on Computer Vision Systems (2007)
13. Fawzi, K., Fadi, D., Franck, L., Benoit, L.: Quadcopter control using onboard monocular camera for enriching remote laboratory facilities. In: 2018 IEEE International Conference on Automation, Quality and Testing, Robotics (AQTR). IEEE (2018)
14. Lowe, D.G.: Object recognition from local scale-invariant features. In: The proceedings of the Seventh IEEE International Conference on Computer Vision, vol. 2, pp. 1150–1157. IEEE (1999)
15. Nagi, J., Giusti, A., Di Caro, G.A., Gambardella, L.M.: Human control of UAVS using face pose estimates and hand gestures. In: Proceedings of the 2014 ACM/IEEE International Conference on Human-Robot Interaction, pp. 252–253. ACM (2014)
16. Paul, V., Jones, M.J.: Robust real-time face detection. Int. J. Comput. Vis. **57**(2), 137–154 (2004)
17. Monajjemi, M., Mohaimenianpour, S., Vaughan, R.: UAV, come to me: end-to-end, multi-scale situated HRI with an uninstrumented human and a distant UAV. In: 2016 IEEE/RSJ International Conference on Intelligent Robots and Systems (IROS), pp. 4410–4417. IEEE (2016)
18. Dornaika, F., Davoine, F.: On appearance based face and facial action tracking. IEEE Trans. Circuits Syst. Video Technol. **16**(9), 1107–1124 (2006)
19. Unzueta, L., Pimenta, W., Goenetxea, J., Santos, L.P., Dornaika, F.: Efficient deformable 3d face model fitting to monocular images (2016)
20. Kazemi, V., Josephine, S.: One millisecond face alignment with an ensemble of regression trees. In: 27th IEEE Conference on Computer Vision and Pattern Recognition, CVPR 2014, Columbus, United States, 23 June 2014 through 28 June 2014, pp. 1867–1874. IEEE Computer Society (2014)
21. Gao, X.-S., Hou, X.-R., Tang, J., Cheng, H.-F.: Complete solution classification for the perspective-three-point problem. IEEE Trans. Pattern Anal. Mach. Intell. **25**(8), 930–943 (2003)
22. Lepetit, V., Moreno-Noguer, F., Fua, P.: EPNP: an accurate O(n) solution to the PnP problem. Int. J. Comput. Vis. **81**(2), 155 (2009)
23. Ariz, M., Bengoechea, J.J., Villanueva, A., Cabeza, R.: A novel 2d/3d database with automatic face annotation for head tracking and pose estimation. Comput. Vis. Image Underst. **148**, 201–210 (2016)

24. https://youtu.be/42nZTCsfQjE
25. https://youtu.be/Kr6TnjoByZ0
26. https://youtu.be/kXZH9uz9Hkc
27. https://youtu.be/PTMVeJizjF8
28. https://youtu.be/Xytlz0UdaDk

Orthogonally-Divergent Fisheye Stereo

Janice Pan[1]([⊠]), Martin Mueller[2], Tarek Lahlou[2], and Alan C. Bovik[1]

[1] The University of Texas at Austin, Austin, USA
janicepan@utexas.edu
[2] Texas Instruments Incorporated, Dallas, USA

Abstract. An integral part of driver assistance technology is surround-view (SV), a system which uses four fisheye (wide-angle) cameras on the front, right, rear, and left sides of a vehicle to completely capture the surroundings. Inherent in SV are four wide-baseline orthogonally-divergent fisheye stereo systems, from which, depth information may be extracted and used in 3D scene understanding. Traditional stereo approaches typically require fisheye distortion removal and stereo rectification for efficient correspondence matching. However, such approaches suffer from loss of data and cannot account for widely disparate appearances of objects in corresponding views. We introduce a novel method for computing depth from fisheye stereo that uses an understanding of the underlying lens models and a convolutional network to predict correspondences. We also built a synthetic database for developing and testing fisheye stereo and SV algorithms. We demonstrate the performance of our depth estimation method on this database.

Keywords: Fisheye · Stereo · Orthogonal · Diverging · Surround-view

1 Introduction

A common feature of Advanced Driver Assistance Systems (ADAS) is surround-view (SV), which uses four outward-facing fisheye cameras placed on the front, right, rear, and left sides of a vehicle [2,8,21,23,26]. By exploiting the wide Field-of-View (FoV) of fisheye lenses, such a configuration allows for the generation of a complete 360° birdseye-view image, also referred to as the *surround-view* or *top-view* output. An example set of synthetically generated fisheye images along with a "true" simulated SV image are shown in Fig. 1. Our motivation is to estimate the unknown ideal SV image from the four captured fisheye images. This typically entails fisheye correction, followed by perspective transformation for geometric alignment, stitching of the images, and finally, photometric correction to reduce visual artifacts [2,21,23,26].

The type of SV system we are interested in deploys four fisheye stereo systems comprised of four camera pairs (front/right, right/rear, rear/left, left/front). The use of wide-angle lenses ensures significant overlap between the camera FoVs. Within these regions, information regarding the scene structure or objects

© Springer Nature Switzerland AG 2018
J. Blanc-Talon et al. (Eds.): ACIVS 2018, LNCS 11182, pp. 112–124, 2018.
https://doi.org/10.1007/978-3-030-01449-0_10

present may be analyzed to notify drivers of obstacles or to improve top-view rendering by reducing artifacts. In the type of SV we are considering, each stereo system is assumed to have a wide baseline (>1 m) with camera axes oriented approximately 90° apart.

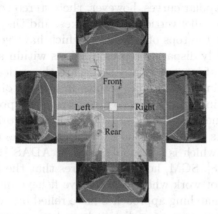

Fig. 1. Example set of fisheye captures and the associated true SV

Stereo systems are typically comprised of a pair of rectilinear cameras with parallel camera axes and narrow ($\sim$10 cm) baselines. The use of fisheye lenses enables stereo configurations to use wider baselines and non-parallel camera axes, since they capture larger FoVs that may still overlap under disparate camera positions and orientations. However, using fisheye lenses for stereo applications is inherently more challenging because wide-angle lenses produce spatial image distortions that get more severe closer to image edges, which are coincidentally where FoVs overlap and where one would want to exploit stereo vision.

We train a neural network to conduct correspondence prediction and use it along with the stereo camera parameters to predict a depth map, given a fisheye stereo pair. The proposed method does not require the removal of fisheye distortion, nor does it require stereo rectification. The trained network searches along epipolar curves to find correspondences, then triangulates and estimates depths under smoothness constraints.

2   Background

Past work exploring fisheye stereo has typically involved fisheye correction, or some form of spatial distortion removal, followed by stereo rectification [3,5,6,9,13,18,20,22,27]. After the removal of fisheye, the transformed view is usually cropped so that the new image looks rectilinear, which results in a loss of information at the FoV boundaries. In cases where the 'undistorted' view is not cropped, the edge content may still suffer from horizontal distortions and lower resolution due to spatial spreading of the original fisheye pixels.

Few methods do not remove fisheye distortion prior to stereo computation [12,16,17], but the work in [12] used stereo images designed to be densely textured with a correspondence method based on template-matching. The authors in [16,17] take an approach more similar to ours, based on an analytical model of the fisheye lens and the stereo camera geometry. They perform stereo matching along computed epipolar curves, however, their stereo configuration involves fisheye cameras with parallel vertical camera axes, and the points they attempt to match are primarily the tops of buildings, which have very large depths and do not suffer from widely disparate representations within a stereo pair.

In SV, the regions of interest in any scene are those closest in depth. Additionally, having diverging camera axes results in imaged objects having widely disparate appearances in a stereo pair, so the stereo correspondence task is much more challenging. Some traditional stereo-matching algorithms rely on local [24] and/or global costs [10,11], a good example being the popular Semi-global Matching (SGM) [10] which is commonly used in ADAS [19] and autonomous driving applications [28]. SGM, however, requires that the input image pair be rectified, which does not work when using native fisheye image pairs. A variety of more recent stereo matching approaches have relied on neural networks. Networks trained for correspondence prediction have been proven to perform better [14,15,25] than traditional methods, and even networks trained on synthetic data have performed well on real-world data [15]. We have taken a previously-established model architecture [25] that was originally developed for evaluating correspondences between rectilinear stereo images, and retrained it on image patches extracted from the new database.

3   Database

No openly available databases of real-world surround-view fisheye inputs with birdseye-view outputs currently exist, so the development of SV algorithms has necessarily relied on the collection of private datasets [26]. Gathering a large database of well-calibrated SV image sets (four fisheye images and one birdseye-view output) is a very difficult, time-consuming, and costly problem. As a practical alternative, we created a synthetic database using the graphics rendering software Blender [1]. Past work has shown that methods developed on synthetic data have transferred well to real data [15], so we aimed to generate photo-realistic synthetic data representing many real-world geometries and structures to help us extend our methods to real data later. To gather the synthetic data, we built a realistic city environment and populated it with vehicles, motorcycles, bikes, pedestrians, and sidewalk furniture (e.g., trash cans and benches).

The proposed database includes 222 unique manually selected scenes captured within the city, each of which was captured 21 times by the four fisheye cameras and true birdseye-view camera, and each capture has an associated ground-truth depth image, also acquired in Blender. Each of the 21 captures is a different augmentation, chosen by randomly rotating about the vertical axis the 5-camera configuration, which includes the four fisheye cameras and an overhead rectilinear camera to capture a true virtual SV image. The data was also diversified by randomly applying different lighting conditions to the scene. Noise and other image degradations were not added to the data, but they can easily be applied for developing more robust algorithms in the future. Overall, the database includes 4,662 surround-view image sets, each of which includes the front, right, rear, left, and overhead captures along with corresponding depth maps for each captured image. So that users can attempt fisheye stereo, which can potentially be used to augment SV images, the database contains $18,648$ unique stereo pairs. The true SV image is not used in this work but was captured for future SV algorithm development. The database also includes sequences of captures that enable temporal testing of SV algorithms. Figure 1 shows an example set of fisheye captures along with the corresponding top-view capture.

Due to the large size of the database, we make available the files used to generate the database here: https://github.com/janicepan/sv-db.git. We include instructions for obtaining the Blender file containing the city scene, the files used to capture and process the data, and the calibration data used for synthesizing the output of the SV system.

4 Depth Estimation from Divergent Fisheye Stereo

We now present our method for predicting depth from fisheye stereo. The pipeline is shown in Fig. 2. Given the stereo pair as input, the search space in each image for finding correspondences, in Step 2, is limited by the patch size and by the spatial extent of the fisheye FoV projected on the image plane. Patches are extracted in Step 3 according to these constraints. Point correspondences in Step 4 are computed by taking the best match along epipolar search curves, as predicted by a trained network. The epipolar curves are determined by the camera and stereo parameters, so we can pre-compute them in an offline calibration step and then input them in a LUT to more efficiently compute matching points. When triangulating depth (Step 5) for any reference pixel, the Euclidean distance between the world point and the left camera center is the estimated depth. Hence, depth is a measurement on all three 3D space coordinates.

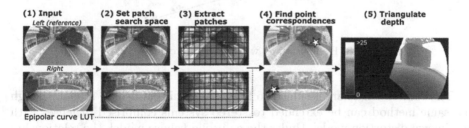

Fig. 2. Algorithm pipeline for computing depth image

We begin in Sect. 4.1 by discussing the underlying geometry of the fisheye stereo system, which we use to triangulate depth from an estimated pair of corresponding pixels. Without first removing the radial distortions caused by the fisheye lenses, the captured stereo images cannot be rectified, and therefore, we necessarily rely on an understanding of the system geometry and lens models to triangulate depth from estimated correspondences. Then, in Sect. 4.2, we discuss the network we trained using patches extracted from image pairs in the new database to predict these correspondences. Simple patch similarity metrics, such as sum of absolute or squared differences, fall short in reliably evaluating how well patches match, which we discuss in Sect. 5, so we turn to a data-driven learning-based approach in training a correspondence predictor.

4.1 Divergent Fisheye Stereo Geometry

Because our model stereo system uses fisheye lenses instead of rectilinear lenses, each camera can be rotated about its vertical axis away from the other and still maintain FoVs that overlap. Figure 3 depicts a diverging fisheye stereo system, where the left and right cameras are symmetrically rotated by angles α and $-\alpha$, respectively, about their Y-axes. In the context of SV, usually $\alpha \approx 45°$. When computing SV, each camera is also usually angled slightly downward, i.e., rotated about their X-axes, to capture more of the ground plane. The 'left' and 'right' cameras in a stereo pair may be any of the front/right, right/rear, rear/left, or left/front camera pairs (refer to Fig. 1).

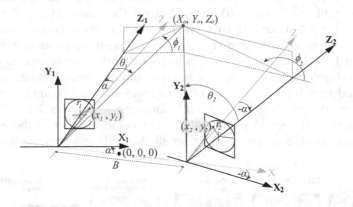

Fig. 3. Fisheye stereo with diverging camera axes

The camera lenses are assumed to follow an equisolid fisheye model, though the same method can be extended to stereo systems comprised of lenses with any known distortion model. Under the equisolid fisheye model, the relationship between any image point (x_i, y_i) for camera i and its 3D point (X_{oi}, Y_{oi}, Z_{oi}) in the coordinate frame of camera i can be described as follows:

$$r_i = \sqrt{x_i^2 + y_i^2} = 2f \sin\left(\frac{\theta_i}{2}\right), \tag{1}$$

$$\phi_i = \arctan\left(\frac{y_i}{x_i}\right), \tag{2}$$

$$\tan(\theta_i) = \frac{\sqrt{X_{oi}^2 + Y_{oi}^2}}{Z_{oi}}, \tag{3}$$

where f is the focal length; r_i is the distance between (x_i, y_i) and the image center; θ_i is the angle between the Z-axis of camera i and (X_{oi}, Y_{oi}, Z_{oi}); and ϕ_i is the angle between the XZ-plane of camera i and (X_{oi}, Y_{oi}, Z_{oi}), which is the world point (X_o, Y_o, Z_o) in the coordinate frame of camera i. Thus, given a left image point (x_1, y_1), and a known or estimated depth Z_{o1}, one can solve for X_{o1} and Y_{o1}:

$$X_{o1} = \sqrt{\frac{(Z_{o1}\tan(\theta_1))^2}{1 + \tan(\phi_1)^2}}, \quad Y_{o1} = X_{o1}\tan(\phi_1). \tag{4}$$

Both θ_1 and ϕ_1 depend only on the reference point (x_1, y_1) and can be computed from (1) and (2), respectively. After obtaining X_{o1} and Y_{o1} using (4), the world point (X_{o1}, Y_{o1}, Z_{o1}) can be represented in the coordinate frame of camera 2: (X_{o2}, Y_{o2}, Z_{o2}), from which (x_2, y_2), the matching point to (x_1, y_1), can be computed.

We only need to know how to transform between the left and right cameras' coordinate systems, which are defined by $(\mathbf{X}_1, \mathbf{Y}_1, \mathbf{Z}_1)$ and $(\mathbf{X}_2, \mathbf{Y}_2, \mathbf{Z}_2)$, respectively. The translation and rotation matrices used to transform points in either coordinate system into the world coordinate system are:

$$\mathbf{T}_L = \begin{bmatrix} 1 & 0 & 0 & -\frac{B}{2} \\ 0 & 1 & 0 & 0 \\ 0 & 0 & 1 & 0 \\ 0 & 0 & 0 & 1 \end{bmatrix}, \quad \mathbf{T}_R = \begin{bmatrix} 1 & 0 & 0 & \frac{B}{2} \\ 0 & 1 & 0 & 0 \\ 0 & 0 & 1 & 0 \\ 0 & 0 & 0 & 1 \end{bmatrix}, \tag{5}$$

$$\mathbf{R}_{L(R)y} = \begin{bmatrix} \cos\alpha & 0 & \mp\sin\alpha & 0 \\ 0 & 1 & 0 & 0 \\ \pm\sin\alpha & 0 & \cos\alpha & 0 \\ 0 & 0 & 0 & 1 \end{bmatrix}, \tag{6}$$

where baseline B is the distance between the camera centers, and the midpoint along B is the origin of the world coordinate system. We also define a rotation matrix about the X-axis, because each camera is downtilted 20° towards the ground to allow the fisheye cameras to capture enough of the ground to generate complete birdseye-view images:

$$\mathbf{R}_{Lx} = \mathbf{R}_{Rx} = \begin{bmatrix} 1 & 0 & 0 & 0 \\ 0 & \cos 20° & -\sin 20° & 0 \\ 0 & \sin 20° & \cos 20° & 0 \\ 0 & 0 & 0 & 1 \end{bmatrix}. \tag{7}$$

The rigid transformation matrices can then be written as:

$$P_{LW} = T_L R_{Ly} R_{Lx} \quad \text{and} \quad P_{RW} = T_R R_{Ry} R_{Rx}, \tag{8}$$

where P_{LW} and P_{RW} are the transformation matrices to map points in the left and right camera coordinate frames, respectively, into the world coordinate frame. I.e.,

$$\begin{aligned}
\begin{bmatrix} X_o \ Y_o \ Z_o \ 1 \end{bmatrix}^T &= P_{LW} \begin{bmatrix} X_{o1} \ Y_{o1} \ Z_{o1} \ 1 \end{bmatrix}^T \\
&= P_{RW} \begin{bmatrix} X_{o2} \ Y_{o2} \ Z_{o2} \ 1 \end{bmatrix}^T, \tag{9}
\end{aligned}$$

or, more directly,

$$\begin{bmatrix} X_{o2} \ Y_{o2} \ Z_{o2} \ 1 \end{bmatrix}^T = P_{RW}^{-1} P_{LW} \begin{bmatrix} X_{o1} \ Y_{o1} \ Z_{o1} \ 1 \end{bmatrix}^T. \tag{10}$$

Thus, for any point (x_1, y_1) in the left camera, and a known (or estimated) depth Z_{o1}, we can solve for X_{o1} and Y_{o1}. Therefore, a simple sweep over a range of depth values $\{Z_{o1}^k\}, k \in [0, K]$ can yield a set of candidate world points $\{(X_{o1}^k, Y_{o1}^k, Z_{o1}^k)\}$, which, when projected into the right image, comprise the epipolar line segment that serves as our correspondence search space. The objective is then to select the k associated with the best corresponding point.

The parameters in the detailed system are simple due to the known geometry provided by the synthetic construction of the database. In practice, any stereo system can be calibrated to extract intrinsic parameters of the lenses and extrinsic parameters of the stereo camera pair. Image-to-world, camera-to-camera, and camera-to-world relationships can easily be derived and used exactly as one would use Eqs. 1–10 to triangulate depth for any correspondence prediction between images captured by the calibrated cameras.

4.2   Neural Network for Correspondence Prediction

The correspondence prediction network we trained was based on the 2-channel architecture in [25], as shown in Fig. 4, and we followed similar methods for training. We used 200 of the 222 scenes in the database yielding 16, 800 stereo pairs (200 scenes × 21 augmentations/scene × 4 pairs/augmentation). From these pairs, we drew 250, 056 positive (correct) matching patches from feature points and an equal number of negative (incorrect) matches. We used 80% of the data for training, 10% for validation, and 10% for testing. The network inputs are 25 × 25 RGB patches, and we used a regularized hinge loss function, similar to the work in [25], using a manually designed matching cost for patch pairs.

To extract positive matches from the dataset, we used the ground-truth depth maps produced by Blender to compute the 3D location of every pixel in each fisheye image. By projecting all the image points into world coordinates, we matched points projected from fisheye image pairs using a nearest-neighbor search to select the point in the right image closest in 3D space to each point in any given left image. The *cost* of the match was represented by the distance between the matched points. A threshold of 0.1 meters was chosen to be the upper limit on classifying a match as true *positive*. In other words, if the *best* match, i.e., the 3D point from the right image that was closest to the left reference pixel, was more than 0.1 m away, the reference pixel was considered to have *no* stereo match. Thus, all positively matching patches had costs in the range $[0, 0.1]$.

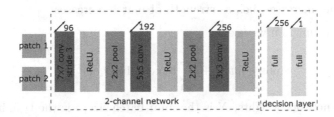

Fig. 4. 2-channel architecture [25]

To extract negative matches, for each positive match, we computed the epipolar curve segment in the right image corresponding to the reference point in the left image. We used a depth threshold of 25 m when computing the epipolar curve segment. In other words, the maximum triangulated depth between the reference point (in the left image) and any point on the epipolar curve segment (in the right image) is 25 m. We then randomly selected a patch from this epipolar curve that was not the true match to be the negative match. To compute the *cost* of the match, we computed the difference between (1) the triangulated depth using the reference point and the positive matching point and (2) the triangulated depth using the reference point and the negative matching point. Throughout this work, depth is computed as a 3D Euclidean distance. Therefore, by computing cost in this way, we still maintain the cost as representing a 3D distance for both positive and negative matches.

In order to achieve balance between the positive and negative matching scores, we flipped and scaled all matching *costs* to obtain *scores* for the positive and negative matches. Specifically, let us represent the distances between positive matches as d_p and the distances between negative patches as d_n:

$$d_p \in [0, 0.1], \qquad d_n \in (0, 25). \tag{11}$$

Note that the endpoints for negative depth distances d_n are non-inclusive, because all depths are non-zero, and the maximum triangulated depth using any point on any epipolar curve segment is 25 m. Therefore, the difference between the true depth and the depth of a negative matching point will be in the range $(0, 25)$. To compute a more balanced set of target match scores, s_p and s_n for the positive and negative matches, respectively:

$$s_p = \frac{0.1 - d_p}{0.1}, \qquad s_n = \frac{-d_n}{25}. \tag{12}$$

By processing the costs for the positive and negative matches in this way, we obtain a set of match scores with the following properties:

- The strongest positive match score is $s_p = 1$, and the strongest negative match score is $s_n = -1$, so all scores lie within $[-1, 1]$.
- All positive matches have positive scores, and all negative matches have negative scores.
- The triangulated 3D distance between matches monotonically decreases as match score increases, i.e., as strength of match increases.

The hinge loss function we use is:

$$\min_{w} \frac{\lambda}{2}||w||_2 + \sum_{i=1}^{N} \max(0, 1 - s_i o_i^{net}), \tag{13}$$

where w are network weights, N is the mini-batch size, i is the training sample index, s_i is the matching score for patch pair i, and o_i^{net} is the network output for patch pair i. We thus train the network to primarily predict patch matches with scores that are indicative of the degree to which patches match, so selecting the patch candidate along an epipolar curve with the *largest predicted magnitude* can then reasonably serve as an effective method of choosing the best match.

To develop a network for real-world data, one could use our synthetically-trained network as a pre-trained model to be further tuned on real data, as past work has done with using virtual data to pre-train a multi-object tracking network to improve the performance of a network trained on real data [4]. To obtain real ground-truth depth, one could follow the work done to construct the KITTI dataset [7], which involved using a calibrated system of camera and lidar sensors to measure depths in stereo scenes.

Training Details. We trained the 2-channel network using mini-batch sizes of 500, a total of 100 epochs, and Adam optimization with learning rate initialized to 0.001. Weights were initialized randomly, and we trained the model from scratch. All pixel values were scaled and centered to fall in the range $[-1, 1]$, i.e., if the original pixel value is v, the processed value is $2\left(\frac{v}{255} - 0.5\right)$, because the images are of type `uint8`. All training examples are shuffled prior to learning.

We monitor the validation loss while training, and every 100 steps, we compute the predicted classification accuracy, among other performance metrics, which we report in Table 1, across the entire validation dataset.

Table 1. Network performance on test data

Acc	TPR	TNR	PPV	NPV	F1
0.860	0.851	0.869	0.872	0.848	0.861

TPR: true positive rate; TNR: true negative rate; PPV: positive predictive value; NPV: negative predictive value; F1 score: harmonic mean between precision and recall.

4.3 Smoothing

The aforementioned network is used as a local predictor, because the 'best' match for any reference patch is that with the highest matching score. Therefore, we also impose an additional smoothness constraint that considers neighboring correspondences when selecting the most appropriate match. Given a set of pixels in a window W_p around any reference image point p, denote the median 'best' depth in W_p as d_m, i.e., the depth which was triangulated from the corresponding patch with the highest matching score. For pixel p, there is also a set of candidate matching pixels C_p in the corresponding image along the epipolar curve, and each candidate pixel $c \in C_p$ has an associated depth d_c, which would result from triangulating depth between p and c. Thus, we penalize match scores for candidate matching pixels based on how far their associated depth is from the median neighboring depth:

$$o'_c = o_c^{net} - \lambda|d_c - d_m|, \tag{14}$$

where o'_c denotes the new matching score, and o_c^{net} is the raw network output for the patch pair comprised of the patch at reference pixel p and candidate matching pixel c. The weight λ determines how heavily to penalize a candidate depth's distance from its neighbors. We used $\lambda = 0.2$, and the neighborhood window was fixed to be 3×3 when computing d_m.

5 Results

The results presented in Table 1 only pertain to the trained network, and we now must evaluate how well the entire pipeline (Fig. 2) performs when we use our trained network to predict point correspondences (Step 4 in Fig. 2). To test how accurately the depth maps are predicted, we use the 22 scenes that were not previously used for training the correspondence prediction network. In other words, we used $1,848$ stereo pairs (22 scenes $\times$ 21 augmentations/scene $\times$ 4 pairs/augmentation), from which no patches were extracted for training, validating, or testing the network for predicting match scores. We used normalized absolute depth error to evaluate the accuracy of predicted depth maps:

$$\text{Error} = \frac{|P - G|}{|G|}, \tag{15}$$

where P is predicted depth, and G is ground-truth depth. We computed the mean normalized absolute depth error over all pixels in each stereo pair and then averaged these values over all $1,848$ pairs.

To establish a benchmark for performance, we first predicted correspondences without the smoothness constraint (Eq. 14) and relied only on the highest match scores (Table 2, 'Network'). We also report the errors when using census similarity, sum of absolute differences (SAD), and sum of squared differences (SSD) to estimate correspondences, because no dense correspondence prediction method exists for native fisheye stereo images, i.e., for fisheye images that are not transformed into the rectilinear space and subsequently rectified. We also compute results when applying the additional smoothing step to our network predictions. Our trained network already performed better than the benchmark similarity metrics, and the error was further reduced when we imposed the smoothness constraint. We show results from our depth-map prediction pipeline and the effectiveness of the additional smoothing step in Fig. 5.

Table 2. Pipeline performance on test scenes

Similarity metric:	Census	SAD	SSD	Network	Network, smoothed
Error (15):	0.93	1.08	1.05	0.75	0.51

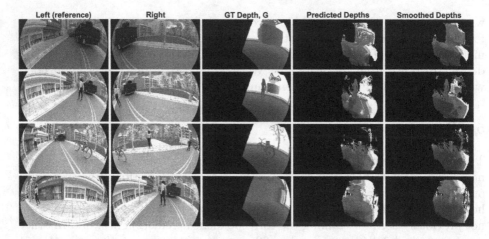

Fig. 5. Example stereo pairs from the 1,848 pairs reserved for testing the depth-map prediction pipeline (Fig. 2). First two columns: left and right fisheye captures; third column: ground-truth depths for pixels visible to both cameras and in the overlapping FoVs; fourth column: predicted depths triangulated using only the network-scored best-match pixel; fifth column: results when the smoothness constraint is imposed.

We compute depths for pixels for which the estimated correspondence has a positive predicted match score. If even the highest match score along an epipolar curve is negative, then the reference point in the left image is considered to not have a predicted matching point in the right image. Additionally, our epipolar search curves were computed only for depths within 25 m (as we discussed in Sect. 4.2), so when we evaluate depth prediction performance, we only consider pixels for which we have a valid depth estimate and can *expect* a valid depth estimate. Therefore, our evaluation results are based on pixels with positive predicted matches and ground-truth depths less than 25 m.

6   Conclusion

We have introduced a new synthetic database for developing and testing fisheye stereo methods and SV algorithms for ADAS, and demonstrated its use in a new depth-map prediction pipeline for fisheye stereo inputs. We trained a neural network to conduct correspondence prediction and, by using the stereo system parameters, showed how it can be used to predict a depth map given a fisheye stereo pair without requiring the removal of fisheye distortions followed by stereo rectification, thus avoiding steps which are both common and necessary when using traditional stereo-matching methods. Despite the difficult conditions, the proposed method is able to produce depth estimates close to the ground-truth.

References

1. Blender: Blender 2.77 Release (2016). http://download.blender.org/release/Blender2.77/
2. Dabral, S., Kamath, S., Appia, V., Mody, M., Zhang, B., Batur, U.: Trends in camera based Automotive Driver Assistance Systems (ADAS). In: IEEE International Midwest Symposium on Circuits and Systems, August 2014
3. Drulea, M., Szakats, I., Vatavu, A., Nedevschi, S.: Omnidirectional stereo vision using fisheye lenses. In: IEEE 10th International Conference on Intelligent Computer Communication and Processing, pp. 251–258, September 2014
4. Gaidon, A., Wang, Q., Cabon, Y., Vig, E.: Virtual worlds as proxy for multi-object tracking analysis. In: CVPR (2016)
5. Gehrig, S., Rabe, C., Krueger, L.: 6D vision goes fisheye for intersection assistance. In: Canadian Conference on Computer and Robot Vision, pp. 34–41, May 2008
6. Gehrig, S.: Large-field-of-view stereo for automotive applications. In: Omnivis 2005, vol. 1 (2005)
7. Geiger, A., Lenz, P., Urtasun, R.: Are we ready for autonomous driving? The KITTI vision benchmark suite. In: IEEE International Conference on Computer Vision and Pattern Recognition (2012)
8. Hamada, K., Hu, Z., Fan, M., Chen, H.: Surround view based parking lot detection and tracking. In: IEEE Intelligent Vehicles Symposium, pp. 1106–1111, June 2015
9. Häne, C., Heng, L., Lee, G.H., Sizov, A., Pollefeys, M.: Real-time direct dense matching on fisheye images using plane-sweeping stereo. In: International Conference on 3D Vision, vol. 1, pp. 57–64, December 2014
10. Hirschmuller, H.: Accurate and efficient stereo processing by semi-global matching and mutual information. In: IEEE International Conference on Computer Vision and Pattern Recognition, vol. 2, pp. 807–814, June 2005
11. Kim, J., Kolmogorov, V., Zabih, R.: Visual correspondence using energy minimization and mutual information. In: IEEE International Conference on Computer Vision and Pattern Recognition, pp. 1033–1040 (2003)
12. Kita, N.: Dense 3D measurement of the near surroundings by fisheye stereo. In: IAPR International Conference on Machine Vision Applications (2011)
13. Kita, N., Kanehiro, F., Morisawa, M., Kaneko, K.: Obstacle detection for a bipedal walking robot by a fisheye stereo. In: Proceedings of the IEEE/SICE International Symposium on System Integration, pp. 119–125, December 2013

14. Luo, W., Schwing, A., Urtasun, R.: Efficient deep learning for stereo matching. In: IEEE International Conference on Computer Vision and Pattern Recognition (2016)
15. Mayer, N., et al.: A large dataset to train convolutional networks for disparity, optical flow, and scene flow estimation. In: IEEE International Conference on Computer Vision and Pattern Recognition (2016)
16. Moreau, J., Ambellouis, S., Ruichek, Y.: 3D reconstruction of urban environments based on fisheye stereovision. In: International Conference on Signal Image Technology and Internet Based Systems, pp. 36–41, November 2012
17. Moreau, J., Ambellouis, S., Ruichek, Y.: Equisolid fisheye stereovision calibration and point cloud computation. In: Conference on Serving Society with Geoinformatics (2013)
18. Nishimoto, T., Yamaguchi, J.: Three dimensional measurement using fisheye stereo vision. In: SICE Annual Conference, pp. 2008–2012, September 2007
19. Okuda, R., Kajiwara, Y., Terashima, K.: A survey of technical trend of ADAS and autonomous driving. In: IEEE International Symposium on VLSI Technology, Systems and Application, pp. 1–4 (2014)
20. Schneider, J., Stachniss, C., Frstner, W.: On the accuracy of dense fisheye stereo. In: IEEE Robotics and Automation Letters, January 2016
21. Thomas, B., Chithambaran, R., Picard, Y., Cougnard, C.: Development of a cost effective bird's eye view parking assistance system. In: IEEE Recent Advances in Intelligent Computational Systems, pp. 461–466 (2011)
22. Yamaguchi, J.: Three dimensional measurement using fisheye stereo vision. In: Advances in Theory and Applications of Stereo Vision (2011)
23. Yu, M., Ma, G.: 360 surround view system with parking guidance. SAE Int. J. Commer. Veh. **7**, 19–24 (2014)
24. Zabih, R., Woodfill, J.: Non-parametric local transforms for computing visual correspondence. In: Eklundh, J.-O. (ed.) ECCV 1994. LNCS, vol. 801, pp. 151–158. Springer, Heidelberg (1994). https://doi.org/10.1007/BFb0028345
25. Zagoruyko, S., Komodakis, N.: Learning to compare image patches via convolutional neural networks. In: IEEE International Conference on Computer Vision and Pattern Recognition, pp. 4353–4361 (2015)
26. Zhang, B., et al.: A surround view camera solution for embedded systems. In: IEEE Conference on Computer Vision and Pattern Recognition Workshop, pp. 662–667 (2014)
27. Zhao, H., Aggarwal, J.K.: 3D reconstruction of an urban scene from synthetic fisheye images. In: IEEE Southwest Symposium on Image Analysis and Interpretation, pp. 219–223 (2000)
28. Ziegler, J., et al.: Making Bertha drive-an autonomous journey on a historic route. IEEE Intell. Transp. Syst. Mag. **6**(2), 8–20 (2014)

Two-Camera Synchronization and Trajectory Reconstruction for a Touch Screen Usability Experiment

Toni Kuronen$^{(\boxtimes)}$, Tuomas Eerola, Lasse Lensu, and Heikki Kälviäinen

Machine Vision and Pattern Recognition Laboratory (MVPR), Department
of Computational and Process Engineering, School of Engineering Science,
Lappeenranta University of Technology, P.O.Box 20, 53851 Lappeenranta, Finland
{toni.kuronen,tuomas.eerola,lasse.lensu,heikki.kalviainen}@lut.fi

Abstract. This paper considers the usability of stereoscopic 3D touch
displays. For this purpose extensive subjective experiments were carried out and the hand movements of test subjects were recorded using
a two-camera setup consisting of a high-speed camera and a standard
RGB video camera with different viewing angles. This produced a large
amount of video data that is very laborious to analyze manually which
motivates the development of automated methods. In this paper, we
propose a method for automatic video synchronization for the two cameras to enable 3D trajectory reconstruction. This together with proper
finger tracking and trajectory processing techniques form a fully automated measurement framework for hand movements. We evaluated the
proposed method with a large amount of hand movement videos and
demonstrated its accuracy on 3D trajectory reconstruction. Finally, we
computed a set of hand trajectory features from the data and show that
certain features, such as the mean and maximum velocity differ statistically significantly between different target object disparity categories.
With small modifications, the framework can be utilized in other similar
HCI studies.

Keywords: Human-computer interaction · Multi-view tracking
3D reconstruction · Stereoscopic touch screen · Image processing
Image analysis

1 Introduction

Advances in gesture interfaces, touch screens, and augmented and virtual real-
ities have brought new usability concerns that need to be studied in a natural
environment and in an unobtrusive way [15]. In this work we focus on the next
generation of user interfaces, combining touch input with stereoscopic 3D (S3D)
content visualization. Stereoscopically rendered views provide additional depth
information that makes depth and structure judgments easier, enhances the ability to detect camouflaged objects as well as increases the ability to recognize

© Springer Nature Switzerland AG 2018
J. Blanc-Talon et al. (Eds.): ACIVS 2018, LNCS 11182, pp. 125–136, 2018.
https://doi.org/10.1007/978-3-030-01449-0_11

the surface material [2,9]. Furthermore, the stereoscopic presentation enhances the accuracy of visually guided touching and grasping movements [14]. Although touch input has already proved its utility and indispensability for various human-computer interaction (HCI) applications interacting with stereoscopically rendered contents is still a challenging task. Usually the touch recognition surface is placed on another plane than the displayed content which being stereoscopically rendered floats freely in front of or behind the monitor. It has been shown that touching an intangible surface, i.e., touching the void leads to confusion and a significant number of overshooting errors [3].

In order to study the usability of the stereoscopic 3D touch screen, it is important to be able to accurately record hand and finger movements of test subjects in 3D. Several robust approaches to hand tracking exist that can measure the hand and finger location with high accuracy, for example, data gloves with electromechanical, infrared, or magnetic sensors [6]. However, such devices affect the natural hand motion and cannot be considered feasible solutions when pursuing natural HCI. Image-based solutions provide an unobtrusive way to study and to track human movement and enable natural interaction with the technology. Commercially available solutions such as Leap Motion[1] and Microsoft Kinect[2] limit the hand movement to a relatively small area, do not allow frame rates high enough to capture all the nuances of rapid hand movements, and are imprecise for accurate finger movement measurements.

This study continues the work done in [11] and [12] where a camera-based hand movement measurement framework for HCI studies was proposed. In order to analyze automatically a large amount of video data, we complement the framework by proposing a video synchronization procedure for a setup consisting of a high-speed camera and a normal-speed camera with different viewing angles. The high-speed camera produces accurate information on hand movements in 2D while the additional normal-speed camera provides the possibility to measure the movements in 3D. The framework is further evaluated with a large scale HCI experiment where the usability of a 3D touch screen is studied with 3D stimuli. Finally, a set of hand trajectory features is computed from the data and they are compared with the different 3D stimuli, i.e., with target objects with different parallaxes.

2   Experiment Setup

The framework was developed for a HCI experiment that uses a S3D touch screen setup. During the trials, test subjects were asked to perform an intentional pointing action towards the observed 3D stimulus. The stereoscopic presentation of the stimuli were done with the NVIDIA 3D Vision kit. The touch screen was placed at a distance of 0.65 m in front of the person. The trigger box contained a button to be pressed to denote the beginning of a single pointing action and was set up 0.25 m away from the screen. The process was recorded

---

[1] Leap motion: https://www.leapmotion.com/product.
[2] Microsoft Kinect: http://www.xbox.com/en-US/kinect.

with two cameras: (i) a Mega Speed MS50K high-speed camera equipped with the Nikon 50 mm F1.4D objective and (ii) a normal-speed Sony HDR-SR12 camera. The high-speed camera was installed on the right side of the touch screen with an approximately 1.25 m gap in-between, while the normal-speed camera was mounted on the top (see Fig. 1). Example frames from both cameras are presented in Fig. 2. The high-speed camera was operated by the trigger resulting a separate video file for each pointing action. The normal-speed recorded the whole session for each test subject into one video file. This resulted in the need of camera synchronization and re-calibration.

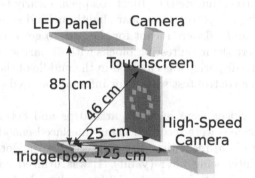

Fig. 1. 3-D touch display experiment.

Similar to earlier pointing action research, e.g., [5], the experiment focused on studying intentional pointing actions. The stimuli were generated by a stereoscopic display with the touch screen to evaluate the effect of different parallaxes, i.e., perceived depth. This arrangement enables study of (potential) conflict between visually perceived and touch-based sensations of depth.

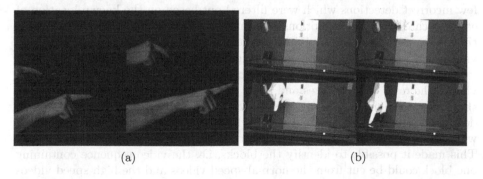

Fig. 2. Example video frames of volunteer interaction with the 3D touch screen display captured with the high-speed camera (a) and normal-speed camera (b).

2.1   Dataset

For the data collection, the pointing action tests were performed by 20 subjects. The pointing actions were divided into nine test blocks based on the interruptions in the high-speed imaging due to limited camera memory. The main test block contained 40 pointing actions per each parallax disparity. Disparity defines the difference in the target object locations between the images seen by the left and right eyes causing the target object to appear in front or behind the screen. Four disparities were considered: (1) 6 pixels causing the object to appear clearly in front of the screen, (2) 2 pixels causing the object to appear slightly in front of the screen, (3) −6 pixels causing the object to appear clearly behind screen, and (4) −2 pixels causing the object to appear slightly behind screen. Blocks 1 and 2 with disparities 6 and −6 were meant for the user to get acquainted with the setup. Blocks 3–6 were the main testing blocks with disparities 6, −6, 2, and −2. In blocks 7 and 8, the disparity was changed in the middle of the pointing action. Finally, block 9 was a control test with color information used as a target for the pointing actions.

The high-speed videos were recorded at 500 fps and 800 × 600 resolution. The normal-speed videos were recorded using interlaced encoding with 50 field rate and 1440 × 1080 (4:3) resolution. For deinterlacing the normal-speed videos the yet another deinterlacing filter (yadif) [1] was utilized with field-to-frame conversion producing double frame rate (50 fps) videos. In total, 2597 pointing actions were recorded with the both cameras.

3   Hand Tracking and Video Synchronization

The hand movements in the high-speed videos were tracked using [8] as proposed in [11]. The tracking window was initialized by a manually placed initial bounding box on the trigger box button image. The normal-speed videos were processed with motion detection near the monitor area. The motion detection was performed using background subtraction (frame differencing). There were few incorrect detections which were filtered out based on the known location of each touch. The detected motions were used to obtain the location of the finger tip which was further used to initialize the tracking window for the normal-speed videos. The tracking was performed using [16] as introduced in [12].

In order to automatically align the normal-speed videos with the high-speed videos, the ratio of framerates and delay are needed. The ratio of framerates of the cameras is known, so only the delay needs to be estimated from data. To do this, first, a coarse alignment was performed using timestamps accompanied with the high-speed videos and the starting time of the normal-speed videos. This made it possible to identify the blocks, i.e. the video sequence containing one block could be cut from the normal-speed videos and the high-speed videos corresponding the same pointing actions can be recognized. Nine blocks were identified based on the longer breaks in pointing actions caused by the limited high-speed camera memory, transfer of the memory contents to the computer and clearing the camera memory. The accurate alignment, i.e. the estimation of the delay was done separately for each block. To do this points of the finger trajectories which can be detected from both videos need to be found. The trigger box had a white button that was visible on the both views, and the point where the trajectory passed the button was used to find a timestamp from both videos. Each block contains several point actions and therefore several timestamps were obtained.

The final alignment was done by searching the delay which maximizes correlation between the timestamp sequences for normal-speed and high-speed videos. As it can be seen from Fig. 3, the timestamp correlations of the corresponding events of passing the white button are periodical so that simply finding the minimal timestamp difference would not work in this case. The event matching was done by binning the trajectory event matches of the timestamp differences. One bin was the length of a single frame (0.02 s) and the maximum for the timestamp correlation was found by summing up 12 frames and finding the largest bin. 12 frames bin size were used because it was the minimal bin size that produced full bin of 20 corresponding time differences of the events for some of the blocks. Figure 3 presents examples of synchronizing blocks of pointing actions. In Fig. 3a all the samples are well correlated within a tight time range (around 3.34 s timestamp difference), but in Fig. 3b it is impossible to say which time difference gives a good result. This is mainly due to the low count of sequences which were tracked correctly in the both videos.

4   Post-processing and 3D Trajectory Reconstruction

The trajectory data retrieved as the result of tracking usually presents an ordered list of object location coordinate points in an image plane. None of the currently available visual trackers achieve immaculate accuracy, and thus, the measures may contain movement noise or completely incorrect position estimation if the tracker lost the target. Moreover, most visual trackers estimate the object location only with a pixel precision and, therefore, the obtained trajectory presents a broken line instead of a desired smooth curve. As noted in [10], the rough-edged transforms between the trajectory points noticeably affect the precision of subsequent calculations. These negative effects can be eliminated by introducing trajectory smoothing and tracking failure detection methods into the processing flow. For the trajectory smoothing, Local Regression (LOESS) [4] was used based on the comparison performed in [11].

To obtain a 3D trajectory, the 2D trajectories estimated using calibrated cameras with a different viewpoint need to be combined. The task of computing a 3D trajectory from multiple 2D trajectories is essentially equivalent to the process of 3D scene reconstruction. For this purpose, we utilized the well-known method [7] sketched in Algorithm 1. The essential matrix is computed with the M-estimator sample consensus (MSAC) algorithm. Finding the best suitable essential matrix was done by minimizing the back-projection errors. The evaluation was performed with confidence levels varying from 90% to 99%, and different Sampson distance [13] thresholds from 5 to 35 pixels.

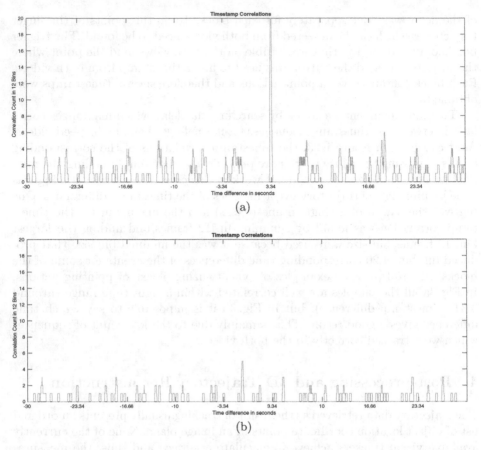

Fig. 3. Example of synchronizing one block of pointing actions from high correlation in one time range (a) and from low correlations within the whole time range (b).

Algorithm 1. The algorithm for three-dimensional scene reconstruction [7]

1. Find point correspondences from the two trajectories.
2. Compute the essential matrix from point correspondences.
3. Compute camera projection matrices from the essential matrix.
4. For each point correspondence, compute the position of a scene point using triangulation.

It was observed that the normal-speed camera did not capture the full trajectories of the hand movement because the touch interface in front of the monitor blocked the view of the finger tip near the monitor. Since the full trajectory was not captured the depth information of the reconstructed 3D trajectories was used to generate the 3D trajectories out of the high-speed video trajectories. The depth information was interpolated by fitting a fourth degree polynomial to all the available sample points, i.e., 3D reconstructed trajectory points which were available from the normal-speed trajectory. The fourth degree polynomial was selected experimentally by examining the fit error and the behavior of the trajectory past its end. The missing parts of the full trajectory depth information at the end of the trajectories were extrapolated by using the last known depth information.

5   Results

5.1   Measuring 3D Trajectories

The success rate of the finger tracking was measured by the proportion of trajectories which reached the predefined end points. For the high-speed videos, the end points were the touch target areas reprojected onto the image plane, and for the normal-speed videos, the defined end point was the triggerbox button. 77% of point actions were tracked correctly from the high-speed videos and 69% from the normal-speed videos. In total, 1161 (58%) of the pointing actions were correctly tracked from the both videos and were synchronized correctly.

To demonstrate the 3D reconstruction of trajectories, 19 pointing actions with 1127 corresponding tracked points were used. Trajectories used for the 3D reconstruction needed to be limited. Otherwise the inlier point selection for the reconstruction task would have been biased to the triggerbox location due to the slow movement speed in the start of the trajectory. This would have resulted in inaccurate reconstruction, and thus the limited set of pointing actions was used for the reconstruction.

The camera calibration was done with a standard checkerboard calibration target with a pattern consisting of 26.5 mm patches. A set of captured calibration images was used to compute the intrinsic camera parameters and distortion coefficients. 1127 finger trajectory points corrected for the lens distortion from the both normal and high-speed videos were used to form the basis of 3D-reconstruction (302 inlier points in the reconstruction). The reconstruction scale ambiguity was eliminated by a fixed setup with known distances between the high-speed camera (HS), normal speed camera (NL), and the monitor. The reconstruction results of all of the 1161 pointing actions are visualized in Fig. 4. The colors in the figure are just for the visualization purpose.

Since there was no ground truth for the 3D trajectories, the 3D reconstruction accuracy was assessed by using the re-projection error measure [7]. The mean re-projection error over all the trajectory points used from 1161 videos in the 3D reconstruction experiment was 31.2 pixels This corresponds to approximately 10 mm in the real world units. The 3D reconstruction results of all of the 1161 pointing actions are visualized in Fig. 4 with the X- (Fig. 4b), Y- (Fig. 4c) and Z-views (Fig. 4d). The test setup is visible in Fig. 4a.

5.2   3D Trajectory Analysis

Pointing actions from the initial 3D reconstruction results with velocity and acceleration curves from one block are visualized in Fig. 5. From the figure it can

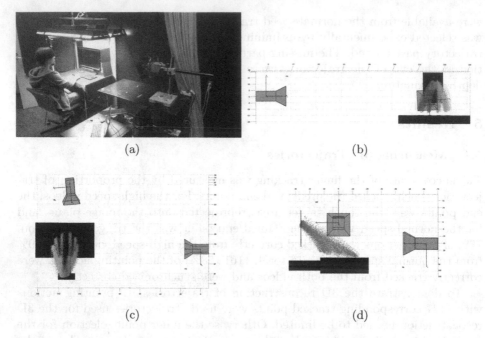

Fig. 4. Visualization of reconstruction results: (a) Setup of the experiments; (b) 3D reconstruction from x-view, and (c) from y-view and (d) from z-view. The trajectories include 1161 captured pointing actions displayed as colored dotted curves. (Color figure online)

be observed that the normal-speed camera did not capture the full trajectories of the hand movement. For the trajectory feature analysis, 3D trajectories reconstructed from the high-speed videos with the extrapolated depth-information were used as described in Sect. 4.

Eleven features were computed from the obtained trajectories: mean velocity, median velocity, maximum velocity, maximum 2nd submovement velocity, maximum 2nd submovement acceleration, mean 2nd submovement velocity, mean 2nd submovement acceleration, the point of the deceleration start, the point where the 2nd submovement acceleration starts for the first time, and the point where the 2nd submovement acceleration starts for the second time. Submovement intervals of the trajectories were detected similarly to [5]. The primary submovement started with the initial acceleration and ended when the acceleration went from negative values to positive values. This was the starting point of the secondary (2nd) submovement of intentional pointing actions where minor adjustments to the trajectory were made and the movement was fixed to the final target position.

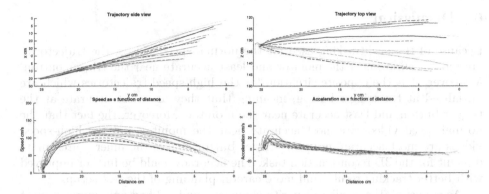

Fig. 5. Trajectory side and top views with velocity and acceleration features extracted from one block of pointing actions. One line style represents one pointing action.

A two sample T-test level was used to analyze the trajectory features. The result h from the test is 1 if the test rejects the null hypothesis at the 5% significance level, and 0 otherwise. It returns a test decision for the null hypothesis that the data in disparity pairs comes from independent random samples from normal distributions with equal means and unequal and unknown variances. Another two sample T-test was used to test the effect of equal, but unknown variances which it did not change the results. These results are visible in Table 1. Using the mean velocities over the whole trajectory in case of the disparity categories −2 and 6, and as well as with 6 and −6 shows that the mean velocities of the disparity categories −2 and 6, and 6 and −6 have unequal means. The null hypothesis is also rejected when using the maximum velocity over the whole trajectory with the disparity categories 2 and 6, as well as with 6 and −6, and with the maximum acceleration over the whole trajectory with the disparity categories 6 and −6. Moreover, using the mean acceleration of the 2nd submovement with the disparity categories 2 and 6, and as well as with −2 and 6 also rejects the null hypothesis.

The 2nd acceleration point of the 2nd submovement showed that the means of the feature in the disparity pairs −2 and 6, and as well as with −2 and −6 are statistically different. Moreover, after the tests with the disparity categories 2 and −2 or 2 and −6, neither could reject the null hypotheses with any of the used features. In Fig. 6 four example cases are shown where statistically significant differences were found in the means of the calculated features with different disparity pairs are shown. It can be observed that the differences are small, but still statistically meaningful.

6 Discussion

Because of the backtracking used for the normal-speed videos, the trajectories are most accurate near the monitor and least accurate near the trigger button. Moreover, it is the opposite situation for the high-speed trajectories which are initialized at the trigger button, meaning that they are most accurate at the trigger button and least accurate near the monitor. Moreover, the fact that the normal-speed videos are most accurate near the monitor and the high-speed videos are most accurate near the trigger button makes the inlier search more difficult for the 3D reconstruction task. The accuracy could be further improved with better tracking results and more careful planning of the test setup.

As expected, the smaller disparity changes 2 and −2 had only minor impact to the hand movements according to the computed features whereas the disparity values 6 and −6 had more significant impact to the movements. Moreover, the large positive disparity 6 (the target object in front of the screen) seemed to have a more prominent effect on the pointing actions than the others. It is shown in Table 1 that there are more features showing the trajectories to be statistically different when using the disparity 6 as one member of the disparity pair than any other disparity category. Furthermore, the velocity features seem to be better than the acceleration features to distinguish the pointing actions towards different disparity values.

Table 1. Rejection of the null hypothesis for the disparity pairs using the selected features.

Features	Disparity pairs						Total
	2 −2	2 6	2 −6	−2 6	−2 −6	6 −6	
Mean velocity	0	0	0	1	0	1	2
Median velocity	0	0	0	0	0	0	0
Maximum velocity	0	1	0	0	0	1	2
Maximum acceleration	0	0	0	0	0	1	1
Maximum 2nd submovement velocity	0	0	0	0	0	0	0
Maximum 2nd submovement acceleration	0	0	0	0	0	0	0
Mean 2nd submovement velocity	0	0	0	0	0	0	0
Mean 2nd submovement acceleration	0	1	0	1	0	0	2
Deceleration start point	0	0	0	0	0	0	0
2nd submovement start 1st	0	0	0	0	0	0	0
2nd submovement start 2nd	0	0	0	1	1	0	2
Total	**0**	**2**	**0**	**3**	**1**	**3**	**9**

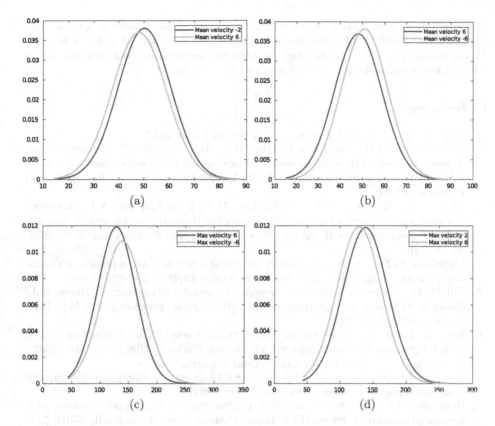

Fig. 6. Normal distribution plot examples where there was statistically significant differences in means of the data. (a) Mean velocity with disparities −2 and 6, (b) Mean velocity with disparities 6 and −6, (c) Maximum velocity with disparities 6 and −6, and (d) Maximum velocity with disparities 2 and 6.

7   Conclusion

In this work, a two-camera framework for tracking finger movements in 3D was evaluated with a large-scale study of human-computer interaction. Moreover, trajectory and video synchronizing processes were introduced and 3D trajectory reconstruction was proposed. The proposed framework was successfully evaluated in the application where stereoscopic touch screen usability was studied with the stereoscopic stimuli.

Overall, the depth information gathered from 3D reconstruction task resulted in full high-speed trajectories with good depth estimation data from the interpolated results of the 3D reconstruction task. Some feature correlation with different parallaxes were already detected, but deeper analysis of the effects of different parallaxes on the trajectories is planned for the future research.

Acknowledgments. The authors would like thank Dr. Jukka Häkkinen and Dr. Jari Takatalo from Institute of Behavioural Sciences from University of Helsinki for producing the data for the research. Their valuable and constructive contributions during the previous steps of the research are very much appreciated.

References

1. FFmpeg (2017). https://ffmpeg.org/. Accessed 1 May 2017
2. van Beurden, M.H., Van Hoey, G., Hatzakis, H., Ijsselsteijn, W.A.: Stereoscopic displays in medical domains: a review of perception and performance effects. In: IS&T/SPIE Electronic Imaging, p. 72400A. International Society for Optics and Photonics (2009)
3. Chan, L.W., Kao, H.S., Chen, M.Y., Lee, M.S., Hsu, J., Hung, Y.P.: Touching the void: direct-touch interaction for intangible displays. In: Proceedings of the SIGCHI Conference on Human Factors in Computing Systems, pp. 2625–2634. ACM (2010)
4. Cleveland, W.S., Devlin, S.J.: Locally weighted regression: an approach to regression analysis by local fitting. J. Am. Stat. Assoc. **83**(403), 596–610 (1988)
5. Elliott, D., Hansen, S., Grierson, L.E.M., Lyons, J., Bennett, S.J., Hayes, S.J.: Goal-directed aiming: two components but multiple processes. Psychol. Bull. **136**(6), 1023–1044 (2010)
6. Erol, A., Bebis, G., Nicolescu, M., Boyle, R.D., Twombly, X.: Vision-based hand pose estimation: a review. Comput. Vis. Image Underst. **108**(12), 52–73 (2007). Special Issue on Vision for Human-Computer Interaction
7. Hartley, R.I., Zisserman, A.: Multiple View Geometry in Computer Vision, 2nd edn. Cambridge University Press, New York (2004)
8. Henriques, J.F., Caseiro, R., Martins, P., Batista, J.: High-speed tracking with kernelized correlation filters. IEEE Trans. Pattern Anal. Mach. Intell. **37**(3), 583–596 (2015)
9. Kooi, F.L., Toet, A.: Visual comfort of binocular and 3D displays. Displays **25**(2), 99–108 (2004)
10. Kuronen, T.: Post-processing and analysis of tracked hand trajectories. Master's thesis, Lappeenranta University of Technology (2014)
11. Kuronen, T., Eerola, T., Lensu, L., Takatalo, J., Häkkinen, J., Kälviäinen, H.: High-speed hand tracking for studying human-computer interaction. In: Paulsen, R.R., Pedersen, K.S. (eds.) SCIA 2015. LNCS, vol. 9127, pp. 130–141. Springer, Cham (2015). https://doi.org/10.1007/978-3-319-19665-7_11
12. Lyubanenko, V., Kuronen, T., Eerola, T., Lensu, L., Kälviäinen, H., Häkkinen, J.: Multi-camera finger tracking and 3D trajectory reconstruction for HCI studies. In: Blanc-Talon, J., Penne, R., Philips, W., Popescu, D., Scheunders, P. (eds.) ACIVS 2017. LNCS, vol. 10617, pp. 63–74. Springer, Cham (2017). https://doi.org/10.1007/978-3-319-70353-4_6
13. Sampson, P.: Fitting conic sections to "very scattered" data: an iterative refinement of the Bookstein algorithm. Comput. Graph. Image Process. **18**(1), 97–108 (1982)
14. Servos, P., Goodale, M.A., Jakobson, L.S.: The role of binocular vision in prehension: a kinematic analysis. Vis. Res. **32**(8), 1513–1521 (1992)
15. Valkov, D., Giesler, A., Hinrichs, K.: Evaluation of depth perception for touch interaction with stereoscopic rendered objects. In: Proceedings of the 2012 ACM International Conference on Interactive Tabletops and Surfaces, ITS 2012, pp. 21–30. ACM, New York (2012)
16. Vojir, T.: Tracking with kernelized correlation filters (2017). https://github.com/vojirt/kcf/. Accessed 1 May 2017

Segmentation and Classification

Comparison of Co-segmentation Methods for Wildlife Photo-identification

Anastasia Popova[1,2], Tuomas Eerola[1(✉)], and Heikki Kälviäinen[1]

[1] Machine Vision and Pattern Recognition Laboratory, Department of
Computational and Process Engineering, School of Engineering Science,
Lappeenranta University of Technology, Lappeenranta, Finland
{anastasia.popova,tuomas.eerola,heikki.kalviainen}@lut.fi
[2] Institute of Mathematics, Mechanics and Computer Science,
Southern Federal University, Rostov-on-Don, Russian Federation

Abstract. Wildlife photo-identification is a commonly used technique
to track animal populations over time. Nowadays, due to large image
data sets, automated photo-identification is an emerging research topic.
To improve the accuracy of identification methods, it is useful to segment the animal from the background. In this paper we evaluate the
suitability of co-segmentation methods for this purpose. The basic idea
in co-segmentation is to detect and to segment the common object in a
set of images despite the different appearance of the object and different backgrounds. Such methods provide a promising approach to process large photo-identification databases for which manual or even semi-
manual approaches are very time-consuming by making it unnecessary
to annotate images to train supervised segmentation methods. We compare existing co-segmentation methods on challenging wildlife photo-
identification images and show that the best methods obtain promising
results on the task.

Keywords: Segmentation · Co-segmentation · Saimaa Ringed Seals
Animal biometrics · Wildlife photo-identification

1 Introduction

Successful conservation of endangered animals requires constant population monitoring. Wildlife photo-identification provides a tool to study and to monitor animal populations over time based on captured images of individuals. It has various
applications in studying the key aspects of the populations such as survival, dis-
persal, site fidelity, reproduction, health, and population size and density. Due
to its non-invasive nature, the photo-identification is a great alternative to more
invasive techniques such as tagging that requires catching the animal and may
cause stress to it, as well as may change its behavior or increase mortality.

Due to the rapid increase in the amount of image data, there is a demand for
automated methods for photo-identification. Various computer vision methods

© Springer Nature Switzerland AG 2018
J. Blanc-Talon et al. (Eds.): ACIVS 2018, LNCS 11182, pp. 139–149, 2018.
https://doi.org/10.1007/978-3-030-01449-0_12

have been developed to identify individual animals, such as ringed seals [3,16], giraffes [6], and humpback whales [10]. Also unified approaches applicable for identification purposes for several animal species have been proposed [5]. A common problem in computer vision based animal individual identification methods is that most of the image data is collected using automatic camera traps [16]. Therefore, the same animal is often captured with the same background. Automatic identification methods that search for similar patterns in images tend to find them in the background instead of animal furs or feathers. This makes the automatic identification considerably more difficult and may cause bias as the supervised methods learn to identify individuals based on the background instead of their characteristics. Therefore, it is necessary to segment the animal from the background before the identification (see Fig. 1).

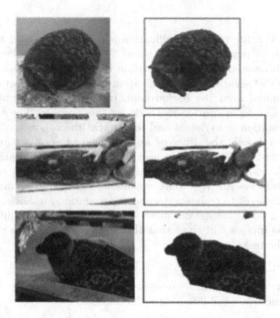

Fig. 1. Examples of animal segmentation: the input images (left) and the segmentation result (right). Modified from [16].

Typically the segmentation is performed by first manually segmenting a set of images to form a training set, and the by using the set to train a segmentation method in supervised manner. The problem with this approach is that the manual annotation of training images is time-consuming and the training has to be performed again every time when a new animal species is studied. Co-segmentation methods provide an attractive alternative segmentation approach. The basic idea of co-segmentation is to detect and to segment the common object in a set of images despite the different appearance of the object and different backgrounds without any supervision. Ideally, the whole collected photo-identification database of one species could be fed to a co-segmentation method that produces a corresponding set of segmented images for the identification method without any manual annotation needed.

Several co-segmentation methods can be found in the literature. However, automatic segmentation of animals is often especially difficult due to the camouflage colors of animals, i.e., the coloration and the patterns are similar to the visual background of the animal. In this study four co-segmentation methods are compared with challenging datasets of animal images. These methods include Discriminative Clustering (DC) [7], Multiple Foreground Co-segmentation (MFC) [8], Multiple Random Walkers (MRW) [11], and Distributed Co-segmentation via Submodular Optimization (CoSand) [9].

The rest of the paper is organized as follows. In Sect. 2 the selected co-segmentation methods are described. In Sect. 3 the datasets, evaluation criteria, and results are presented, and the conclusions are drawn in Sect. 4.

2   Image Co-segmentation

2.1   Discriminative Clustering for Image Co-segmentation

Discriminative clustering (DC) [7] is an approach for co-segmentation based on the combination of bottom-up image segmentation with kernel methods. The algorithm aims to divide all the images into foreground and background pixels in such way that a Support Vector Machine classifier trained with these two classes leads to maximal separation of the object and background classes. The algorithm utlilizes both well-known normalized cuts [13] and kernel methods to formulate a combinatorial optimization problem that is solved via convex relaxation and efficient low-rank optimization. This makes it possible to co-segment tens of images simultaneously.

2.2   Multiple Foreground Co-segmentation

In [8], a method for multiple foreground co-segmentation (MFC) was proposed. The task of MFC is defined as the joint segmentation of K different foregrounds $F = F_1, ..., F_K$ from M input images, each of which contains a different unknown subset of K foregrounds. The proposed approach supports two different scenarios as follows: (1) the unsupervised scenario where a user specifies only the number of foregrounds K and the algorithm automatically distinguish K foregrounds that are most dominant in each image, and (2) the supervised scenario where a user provides bounding-box or pixel-wise annotations for K foregrounds of interest in selected images. Since we aim at a fully automated method, in this study only the unsupervised scenario is considered.

2.3   Distributed Co-segmentation via Submodular Optimization

Distributed co-segmentation via Submodular Optimization (CoSand) [9] is a method for co-segmentation that can cope with a highly variable large-scale image collection. The segmentation task is modeled by temperature maximization on anisotropic heat diffusion. The temperature maximization with K heat sources corresponds to a K-way segmentation that maximizes the segmentation confidence of every pixel in an image.

For a single-image segmentation the image is segmented into superpixels using the TurboPixels method [12]. The next step is to construct an intra-image graph where vertices represent superpixels and edges connect pairs of adjacent superpixels. Weights of the edges are calculated using Gaussian similarity as in [9]. The final step is the agglomerative clustering [9] on the intra-image graph to find out the set of evaluation points where the algorithm greedily selects the largest and most coherent regions.

Co-segmentation is formulated as an optimization problem. The optimization formulation is an extension of the diversity ranking [17] on the intra-image graph where the objective is to maximize the sum of segmentation confidence of every image in the dataset.

2.4 Co-segmentation Using Multiple Random Walkers

Co-segmentation using multiple random walkers (MRW) [11] is another approach for image co-segmentation which constructs a graph-based system to simulate the movements and interactions of multiple random walkers.

To cluster images using random walkers, the graph G is constructed with nodes V corresponding to superpixels which are obtained using simple linear iterative clustering (SLIC) [1]. For the edge set E, the edge connection scheme from [15] is used. Five dissimilarities d_l of node features are proposed in [15], including RGB and LAB super-pixel means, boundary cues, and bag-of-visual-words histograms of RGB and LAB colors [14]. The interaction of random walkers is determined by the reset rule defined in [11]. The algorithm repeats simulating movements of random walkers which form dominant regions until power balance among the walkers is achieved and their distributions converge.

Table 1. Characteristics of the datasets used in the experiments.

	SealVision			iCoseg		
	Easy	Medium	Hard	Alaskan brown bear	Elephants	Goose
Number of images	5	20	239	20	16	32
Multiple animals in one image	−	−	−	+	+	+
Strong obstacles	−	−	+	−	−	-
Minor obstacles	−	+	+	+	+	+
Night time images	−	−	+	−	−	−
Illumination variation	−	+	+	+	+	−

3   Experiments

3.1   Datasets

The comparison of the selected co-segmentation methods was performed using two databases: (1) Saimaa Ringed Seals database [3], and (2) CMU-Cornell iCoseg dataset [2]. From the Saimaa Ringed Seals database three subsets were constructed with increasing complexity. The first subset, called the SealVision easy subset includes five relatively simple images for co-segmentation with clear contrast between the seal and the background. The illumination is similar and

SealVision (easy)

SealVision (medium)

SealVision (hard)

iCoseg (Alaskan brown bear)

iCoseg (goose)

iCoseg (elephants)

Fig. 2. Examples of images from the datasets used in experiments.

there are no large obstacles in the images such as hands, tree branches, or stones. The second subset, called the SealVision medium subset was designed to be more realistic. It includes lighting variations and obstacles like hands and tree branches, but the images still contain clear contrast between the seal and the background. The last subset, called the SealVision hard subset contains all types of seals snapshots, including the most challenging ones with strong obstacles and grayscale night time images where the boundary of an animal is almost invisible. All the images in all three subsets where manually segmented to form a pixcl ground truth for the method evaluation.

The CMU-Cornell iCoseg dataset contains several different types of objects including humans, airplanes, building, and animals. From the dataset images containing animals were selected and three subsets were obtained: (1) iCoseg Alaskan brown bear, (2) iCoseg elephants, (3) iCoseg goose. All these subsets contain strong obstacles, but the illumination variation is relatively small. The pixel ground truth provided with the original dataset was used. The characteristics of all the datasets are summarized in Table 1. Figure 2 shows example images from each subset.

3.2   Implementation

Since the goal is to segment the object from the background the final output of the co-segmentation methods should be a binary maps for all the images. Since this is not case for all the selected co-segmentation methods needed postprocessing steps were performed as discussed next.

The DC approach provides results as a matrix with three types of labels: foreground, background, and unknown pixels. In order to obtain the binary masks all the unknown pixels were set as the background. The selected number of classes was two which means that the separation was done between one foreground class from the background.

CoSand requires the predefined number of clusters k. Preliminary experiments showed that in our dataset $k = 3$ perform better than $k = 2$, since the algorithm tends to segment the brightest object in an image such as blue water

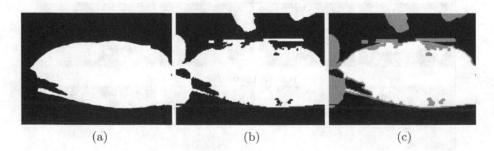

(a)   (b)   (c)

Fig. 3. Visualization of the Jaccard measure calculation: (a) The ground truth (I_{GT}); (b) The segmentation example (I_S); (c) The white area is $I_S \cap I_{GT}$ and the union of the white and gray areas represents $I_S \cup I_{GT}$.

as the foreground. The output of the algorithm is a labeled image with k number of labels. Typically, an animal is presented in the center of images so, in order to obtain the binary map, the cluster which intersects the borders of the image less was selected as the foreground.

The MFC and MRW approaches provide directly the binary masks so the processing of the segmentation results was not needed.

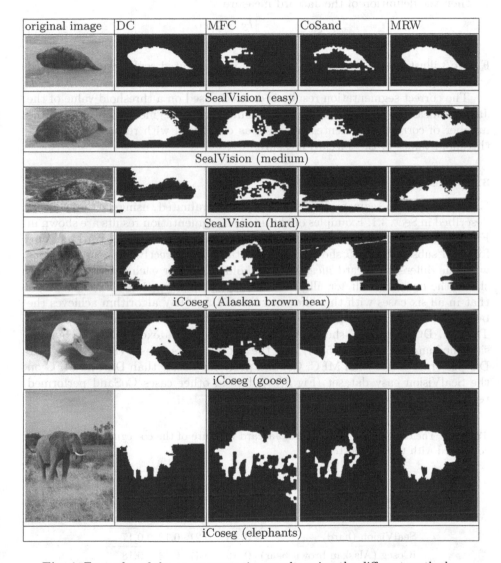

Fig. 4. Examples of the co-segmentation results using the different methods.

3.3 Evaluation Criteria

For the evaluation of the selected co-segmentation methods each obtained binary mask was checked for the similarity with the corresponding ground truth. For measuring similarity the Jaccard measure [4] was chosen which is a common measure for segmentation performance. Suppose that the image I has ground truth I_{GT}, represented as binary mask and I_S is the segmentation result of image I. Then the definition of the Jaccard measure is

$$S_{Jaccard} = \frac{|I_S \cap I_{GT}|}{|I_S \cup I_{GT}|}. \tag{1}$$

Figure 3 illustrates $I_S \cap I_{GT}$ and $I_S \cup I_{GT}$ in segmented seal picture compared with the ground truth. The Jaccard measure for this case is 0.74.

The correct segmentation results is defined based on a threshold value of the Jaccard measure. To evaluate the performance of a co-segmentation method the number of correctly segmented images was calculated with respect to different threshold values of the Jaccard measure.

3.4 Results

Each selected co-segmentation method was evaluated usingall the subsets described in Sect. 3.1. Examples of obtained co-segmentation results are shown in Fig. 4. Table 2 shows the comparison of the mean values of the Jaccard measure for each subset. Figure 5 shows the percentages of correctly segmented images with the different Jaccard measure threshold values for each subset and Fig. 6 shows the overall result for all the subsets combined together. Figure 5 shows that in all six cases with the different subsets the MRW algorithm achieves the best results. Also the mean Jaccard measure is the highest for each dataset (see Table 2). DC and MFC show relatively similar results except for the SealVision medium dataset (Fig. 5(b)) and the iCoseg bears subset (Fig. 5(f)) where DC performs better than MFC. CoSand performs better than DC and MFC on the SealVision easy dataset (Fig. 5(a)), but in other cases CoSand performed relatively similar to MFC and DC.

Table 2. The comparison of the mean Jaccard measure of the co-segmentation results compared with the ground truth.

	MRW	DC	MFC	CoSand
SealVision (easy)	**0.89**	0.59	0.48	0.78
SealVision (medium)	**0.87**	0.66	0.39	0.41
SealVision (hard)	**0.48**	0.21	0.18	0.17
iCoseg (Alaskan brown bear)	**0.56**	0.47	0.13	0.18
iCoseg (elephants)	**0.67**	0.31	0.25	0.21
iCoseg (goose)	**0.69**	0.39	0.36	0.24
All the datasets combined	**0.54**	0.29	0.21	0.20

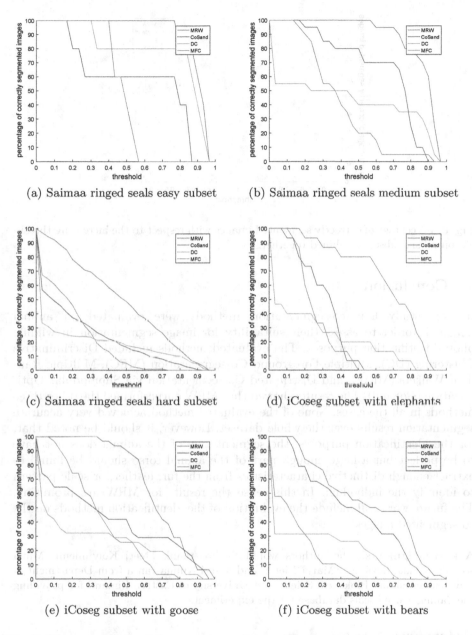

(a) Saimaa ringed seals easy subset

(b) Saimaa ringed seals medium subset

(c) Saimaa ringed seals hard subset

(d) iCoseg subset with elephants

(e) iCoseg subset with goose

(f) iCoseg subset with bears

Fig. 5. The comparison of the percentage of correctly segmented images with respect to the increasing threshold.

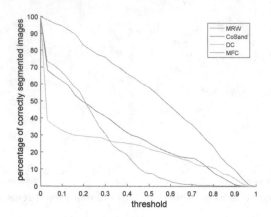

Fig. 6. Percentage of correctly segmented images with respect to the increasing threshold on all six subsets combined together.

4   Conclusion

In this study four co-segmentation methods were evaluated on animal images in order to assess their suitability for image segmentation in wildlife photo-identification purposes. The evaluated methods included Discriminative Clustering (DC), Multiple Foreground Co-segmentation (MFC), Multiple Random Walkers (MRW), and Distributed Co-segmentation via Submodular Optimization (CoSand). The results showed that MRW outperformed the other three methods in all the cases. None of the evaluated method achieved very accurate segmentation results over the whole dataset. However, it should be noted that, for the identification purposes, the segmentation of the animal does not have to be perfect but a large enough part of the animal torso should be found to extract enough distinctive characteristics from the fur, feather, or scale pattern to identify the individual. In this sense the results for MRW are promising. The future work will include the evaluation of the identification methods on the co-segmented images.

Acknowledgments. The authors would like to thank Meeri Koivuniemi, Miina Auttila, Riikka Levä-nen, Marja Niemi, and Mervi Kunnasranta from Department of Environmental and Biological Sciences at University of Eastern Finland for providing the Saimaa ringed seals database for the experiments.

References

1. Achanta, R., Shaji, A., Smith, K., Lucchi, A., Fua, P., Süsstrunk, S.: Slic superpixels compared to state-of-the-art superpixel methods. IEEE Trans. Pattern Anal. Mach. Intell. **34**(11), 2274–2282 (2012)
2. Batra, D., Kowdle, A., Parikh, D., Luo, J., Chen, T.: Interactively co-segmentating topically related images with intelligent scribble guidance. Int. J. Comput. Vis. **93**(3), 273–292 (2011)

3. Chehrsimin, T., et al.: Automatic individual identification of Saimaa ringed seals. IET Comput. Vis. **12**(2), 146–152 (2018)
4. Choi, S.S., Cha, S.H., Tappert, C.C.: A survey of binary similarity and distance measures. J. Syst. Cybern. Inform. **8**(1), 43–48 (2010)
5. Crall, J.P., Stewart, C.V., Berger-Wolf, T.Y., Rubenstein, D.I., Sundaresan, S.R.: Hotspotter - patterned species instance recognition. In: IEEE Workshop on Applications of Computer Vision (WACV), pp. 230–237, January 2013
6. Halloran, K.M., Murdoch, J.D., Becker, M.S.: Applying computer-aided photo-identification to messy datasets: a case study of Thornicroft's giraffe (Giraffa camelopardalis thornicrofti). Afr. J. Ecol. **53**(2), 147–155 (2014)
7. Joulin, A., Bach, F., Ponce, J.: Discriminative clustering for image co-segmentation. In: Proceedings of Conference on Computer Vision and Pattern Recognition (CVPR) (2010)
8. Kim, G., Xing, E.P.: On multiple foreground cosegmentation. In: Proceedings of Conference on Computer Vision and Pattern Recognition (CVPR) (2012)
9. Kim, G., Xing, E.P., Fei-Fei, L., Kanade, T.: Distributed cosegmentation via submodular optimization on anisotropic diffusion. In: International Conference on Computer Vision (ICCV) (2011)
10. Kniest, E., Burns, D., Harrison, P.: Fluke matcher: a computer-aided matching system for humpback whale (Megaptera novaeangliae) flukes. Mar. Mammal Sci. **26**(3), 744–756 (2010)
11. Lee, C., Jang, W.D., Sim, J.Y., Kim, C.S.: Multiple random walkers and their application to image cosegmentation. In: Proceedings of Conference on Computer Vision and Pattern Recognition (CVPR), pp. 3837–3845 (2015)
12. Levinshtein, A., Stere, A., Kutulakos, K.N., Fleet, D.J., Dickinson, S.J., Siddiqi, K.: Turbopixels: fast superpixels using geometric flows. IEEE Trans. Pattern Anal. Mach. Intell. **31**(12), 2290–2297 (2009)
13. Shi, J., Malik, J.: Normalized cuts and image segmentation. IEEE Trans. Pattern Anal. Mach. Intell. **22**(8), 888–905 (2000)
14. Sivic, J., Zisserman, A., et al.: Video Google: a text retrieval approach to object matching in videos. In: International Conference on Computer Vision (ICCV), vol. 2, pp. 1470–1477 (2003)
15. Yang, C., Zhang, L., Lu, H., Ruan, X., Yang, M.H.: Saliency detection via graph-based manifold ranking. In: Proceedings of Conference on Computer Vision and Pattern Recognition (CVPR), pp. 3166–3173 (2013)
16. Zhelezniakov, A., et al.: Segmentation of saimaa ringed seals for identification purposes. In: Bebis, G., et al. (eds.) ISVC 2015. LNCS, vol. 9475, pp. 227–236. Springer, Cham (2015). https://doi.org/10.1007/978-3-319-27863-6_21
17. Zhu, X., Goldberg, A.B., Van Gael, J., Andrzejewski, D.: Improving diversity in ranking using absorbing random walks. In: North American Chapter of the Association for Computational Linguistics: Human Language Technologies, pp. 97–104 (2007)

An Efficient Agglomerative Algorithm Cooperating with Louvain Method for Implementing Image Segmentation

Thanh-Khoa Nguyen[1,2]([✉]), Mickael Coustaty[1], and Jean-Loup Guillaume[1]

[1] L3i laboratory, University of La Rochelle, La Rochelle, France
{thanh_khoa.nguyen,mickael.coustaty,jean-loup.guillaume}@univ-lr.fr
[2] Ca Mau Community College, Ca Mau, Vietnam

Abstract. The idea that brings social networks analysis domain into image segmentation quite satisfies with most authors and harmony in those researches. However, the community detection based image segmentation often produces over-segmented results. To address this problem, we propose an efficient agglomerative homogeneous regions algorithm by considering image histograms which are contributed into bins of the color group properties. Our method is tested on the publicly available Berkeley Segmentation Dataset. And experimental results show that the proposed algorithm produces sizable segmentation and outperforms almost other known image segmentation methods in term of accuracy and comparative PRI scores.

Keywords: Image segmentation · Complex networks · Modularity
Louvain algorithm · Community detection

1 Introduction

Segmentation is one of the most important subjects in image processing applications [12,19,21]. The goal of image segmentation is not only to distinguish them from the background, but also to identify the interesting objects in an image. Particularly, image segmentation could benefit a variety of vision applications, for instance, object recognition, automatic driver assistance, and traffic control systems.

A variety of proposed algorithms has dealt with image segmentation in the literature. These methods can be divided into some main groups according to the underlying approaches, such as feature-based clustering, spatial-based segmentation methods, hybrid techniques and graph-based approaches.

Recently, with the development of complex networks theory, image segmentation techniques based on community detection algorithms have been proposed in the literature [1,10–12,14,21,22]. A community is a group of nodes with dense internal connections and sparse connections with members of other communities. The general idea of those techniques is to highlight the similarity between the

© Springer Nature Switzerland AG 2018
J. Blanc-Talon et al. (Eds.): ACIVS 2018, LNCS 11182, pp. 150–162, 2018.
https://doi.org/10.1007/978-3-030-01449-0_13

modularity criterion in network analysis and the image segmentation process. In fact, the larger the modularity of a network is, the more accurate the detected communities, *i.e.* the objects in the image, are [1,5,14,21]. If the modeling of the image in a graph is well done then we can expect that a good partition in communities corresponds to a good segmentation of the image. The modularity of a partition is a scalar measures the density of links inside communities as compared to links between communities, and its value fall into the interval [−1,1] [4].

Among all the existing community detection algorithms, the Louvain method [4] has received significant attention in the context of image segmentation [5,10,22]. However, it is still facing a problem of over-segmentation. In this paper, we propose a new segmentation approach based on Louvain method that agglomerates homogeneous regions in order to overcome the over-segmented problem. Each sub-segment obtained during the Louvain method phase represents a region. The proposed algorithm then operates by computing similarity distances between two adjacent regions based on the values from the three color channels RGB individually in order to control the aggregation processes.

The rest of this paper is organized as follows. In Sect. 2, we briefly review some well-known image segmentation methods in the literature. In Sect. 3, we introduce complex networks, the concept of community detection and Louvain algorithm to point out how the community detection algorithms can be applied in image segmentation efficiently. Experiments on the publicly Berkeley Segmentation Data Set (BSDS500) are reported in Sect. 4. Finally, our conclusions are presented in Sect. 5.

2 Related Work

In this Section, we briefly review some well-known image segmentation methods.

The *Mean Shift* algorithm [6] considers pixels both from the spatial and the color domains by concatenating the pixel color value and its spatial coordinates into a single vector. Then applying mean shift filtering in this domain yields a convergence point for each pixel. The image is segmented by grouping together all pixels whose convergence points are closer than h_s in the spatial domain and h_r in the range one, where h_s and h_r are parameters. Obviously, the merging process can also manage a constraint on minimum segment size.

The *Watershed* [20] segmentation method approaches image segmentation problem by regarding the gradient magnitude of images as topographic surfaces. The pixels in which water drops will drain into the same local intensity minimum, namely the *catchment basins*. The lines separating the catchment basins are called the watersheds.

The *Normalized Cut* [17] criterion provides a way of integrating global image information into the grouping process. For this work, given an affinity matrix W in which each entry represents the similarity of two pixels, *Normalized Cut* tries to solve for the generalized eigenvectors of the linear system from Eq. (1).

$$(D - W)z = \lambda Dz \tag{1}$$

D is a diagonal matrix with its diagonal entry $D_{ij} = \sum_j W_{ij}$. Then K-means clustering is applied to obtain a segmentation into regions. Cour, et $al.$ approached Normalized Cuts with a variant namely Multiscale Normalized Cuts (NCuts). Sharon, et $al.$ proposed an alternative to enhance the computational efficiency of Normalized Cuts. Beside, many image segmentation methods in this type were mentioned, for instance, Mumford and Shah proposed that the segmentation of an observed image u_0 is given by the minimization of functional:

$$F(u, C) = \int_\Omega (u - u_0)^2 dx + \mu \int_{\Omega \backslash C} |\triangledown(u)|^2 dx + v|C| \qquad (2)$$

where u is piecewise smooth in $\Omega \setminus C$ and μ, v are weighting parameters. In addition, several algorithms have been developed to minimize or simplify the energy by various approach strategies.

Another approach on image segmentation problem comes from the perspective of graph partition. In this approach, the image is regarded as an undirected weighted graph in which each pixel represented as a node in the graph and edge weights measure the similarity distance between nodes. Felzenszwalb and Huttenlocher (Felz-Hutt) [7] attempt to partition image pixels into components. Constructing a graph in which pixels are nodes and edge weights measure dissimilarity between nodes ($e.g.$ color differences), each node is initially placed in its own component. The internal difference of a component $Int(R)$ has been defined as the largest weight in the minimum spanning tree of R. Considering in non-decreasing order by weight of edges, each step of the algorithm merges components R_1 and R_2 connected by the current edge if weight of the edge is less than:

$$min(Int(R_1) + \tau(R_1), Int(R_2) + \tau(R_2)) \qquad (3)$$

where $\tau(R) = k/|R|$, k is a scale parameter that can be used to set a preference for component size.

Recently, the idea of community detection has been applied in image segmentation that offers a new perspective for researchers about image segmentation domain. Li et $al.$ [10] and Mourchid et $al.$ [14] propose using super-pixel and features to solve over-segmented. Both strategies initialize with over-segmented image segmentation in which each subsegment represents as a super-pixel and treat the over segmentation issue by different ways.

Shijie Li, et $al.$ solves the over-segmentation problem by reconstructing the neighborhood system for each region $(super\text{-}pixel)$ and the histogram of states (HoS) texture feature. Then, estimating the distribution of the color feature for each region. Adaptively update the similarity matrix W is computed based on color feature and histogram of states (HoS) texture feature. Youssef Mourchid, et $al.$ approaches the over-segmented problem close the former but compute coefficient for adaptively updating the similarity matrix W based on color feature and histogram of oriented gradients (HOG) texture feature.

3   Our Approach

Complex networks analysis has become one of the most studied topic during last decade [1,14] and in particular many community detection algorithms have been proposed [8]. Applying community detection methods for image segmentation has achieved very good results compared to other techniques [21]. To be inspired, we consider image from the perspective of a complex network and solve the image segmentation based on community detection. The Louvain algorithm has been applied for image segmentation widely [5,21,22]. However, The individual original Louvain method has not overcome the over-segmented problem. Our algorithm can cooperate on the Louvain method to avoid this drawback, and produce accuracy results.

3.1   Complex Networks

A complex network is a graph (network) whose topological structure cannot be trivially described. It comprises properties that emerge as a consequence of global topological organization of the system. Complex network structures describe various systems of high technological and intellectual importance, such as the Internet, World Wide Web, financial, social, neural, and communication networks. One property that has attracted particular attention is the community structure of these networks.

The problem of community detection is usually defined as finding the best partition (or covering) of a network into communities of densely connected nodes, with the nodes belonging to different communities being only sparsely connected. Several algorithms have been proposed to find good partitions in term of a reasonably fast way. These algorithms can be divided into some main types such as, divisive algorithms that detect inter-community links and remove them from the network, agglomerative algorithms that merge similar or close nodes and more generally optimization methods are based on the maximization of an objective function [8]. The quality of partitions resulting from these methods is often measured by the modularity that has been introduced by Newman and Girvan [16]. It is defined as follow:

$$Q = \sum_i (e_{ii} - a_i^2) \tag{4}$$

where e_{ii} denotes the fraction of edges in community i, and a_i if the fraction of ends of edges that belong to i. The value of modularity Q ranges in $[-0.5,1]$ and higher values indicate stronger community structure of the network (Fig. 1).

3.2   Generating a Complex Network from an Image

Complex networks can be generated from images. Each image is represented as an undirected graph $G = (V, E)$, where V is a set of vertices ($V = \{v_1, v_2, ..., v_n\}$) and E is a set of edges $E = \{e_1, e_2, ..., e_k\}$. Each vertex $v_i \in V$ corresponds

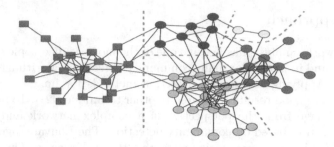

Fig. 1. Community structure in the social network of bottle-nose dolphins population extracted using the algorithm of Girvan and Newman [9]. The squares and circles denote the primary split of the network into two groups and the circles are further subdivided into four smaller group as shown [15].

to an individual pixel and similarity/closeness of pixels are modeled as edges: an edge $e_{ij} \in E$ connects vertices v_i and v_j. A weight on each edge, w_{ij}, is a nonnegative value that measures the affinity between v_i and v_j. The higher affinity is the stronger relation between the pixels. In this paper, each node in the graph represents a pixel and edge weights are defined as:

$$w_{ij} = \begin{cases} 1 & if \quad d_{ij} \leq t \\ nil & otherwise \end{cases} \tag{5}$$

where t is threshold, d_{ij} is a measure of the similarity of pixels i and j intensity for the three channels of color RGB individually. It is defined by $d_{ij} = |I_i - I_j|$ where I_i and I_j represent the intensity of pixel i and j respectively.

For a given pixel, links towards other pixels are created if and only other considered pixels are inside 20 neighboring pixels for rows and columns directions. And if all distances d_{ij} of color channels are lower than t. In this case, the weight is assigned $w_{ij} = 1$. Note that we could have put a weight that reflect the distance (both physical distance and color distance) but this is left for future work.

3.3   Louvain Algorithm

The Louvain algorithm is a well-known method that extracts the community structure of large network [4]. The Louvain method is a hierarchical greedy algorithm that is designed to optimize the modularity on graphs or weighted graphs. Louvain algorithm consists of two phases that are repeated iteratively. Initially, every node is a singleton community. Next, during phase 1, all nodes are considered one by one. Each node is placed in its neighboring community, including its own one, that maximizes the static modularity gain. This process is repeated until no further improvement can be achieved. This first phase stops when the modularity reaches a local maximum. Then, phase 2 consists in building a new graph whose nodes are the communities found during the first phase. To build a new graph, links between nodes of the same community lead to self-loops while the weights of links between new nodes are computed by the sum of the weights of the links between nodes in the corresponding two communities (Fig. 2).

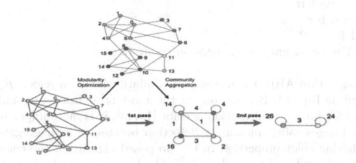

Fig. 2. Process of community detection for Louvain method [4]

3.4   Agglomeration of Homogeneous Regions

Although the Louvain algorithm has been applied for image segmentation widely, it has not still overcome the over-segmented problem. In general, modularity based image segmentation produces many homogeneous regions that could belong to one object in segmented object-level. However, community detection algorithms often deal with the time complexity so their characteristics are heuristics algorithms. Therefore, image segmentation based community detection often lead to over-segmentation problems. In order to solve this problem, homogeneous regions should be combined as its available. Given an over-segmented image that consists of several homogeneous regions represented as a set of regions. The proposed algorithm can merge homogeneous regions in order to generate better segmented image results.

Algorithm AHR
Input: Given an over-segmented image with a set of regions $R = \{R_1, R_2, ..., R_n\}$
01:    **for** each region $R_i \in R$ **do**
02:        **for** each region $R_j \in R$ **do**
03:            **if** R_i and R_j are adjacent regions **then**
04:                **if** R_i or R_j is less than region threshold *(regthres)* **then**
05:                    Merge region R_i and R_j by setting the same label for these pixels of two regions
06:                **else**
07:                    Compute distance of similarity $d(R_i, R_j)$ between R_i and R_j

08:                     **if** $d(R_i, R_j)$ lower than a given threshold **(t) then**
09:                     Merge region R_i and R_j by setting the same label
                         for these pixels of two regions
10:                     **end if**
11:                 **end if**
12:             **end if**
13:         **end for**
14:     **end for**
Output: The set of image segmentation result R.

In the algorithm AHR, the distance of similarity between region R_i and R_j is computed as Eq. (6). Because one of the most straightforward and important feature for image segmentation is color [10,14], it is more important when segmenting images using community detection because of its aggregation communities sharing color properties. In the proposed algorithm, the color images RGB are considered and focused on individual color channels. For each color channel, we compute image histogram (0–255) and distribute it into 16 bins and statistic the percentage of bins. Then, building the 48-dimensional vector consist of the percentages for three color channels of every region. The similarity of two adjacent regions is computed by cosine similarities of two vectors represented to two considering regions, as indicated in Eq. (6).

$$d(R_i, R_j) = Cosine(v_i, v_j) = \frac{v_i^\mathsf{T} v_j}{\|v_i\| . \|v_j\|} \qquad (6)$$

3.5 Noise Removal

In our method, there is pretty important technique that must be point out is noise removal process. As mentioned above, the results, which obtained from Louvain processes consist of over-segmented results, make lower *Probabilistic Rand Index (PRI)* score when it will be evaluated. We consider this issue carefully and attempt some strategies to gain better results and obtain higher *PRI* scores. These strategies are divided into two phases of noise removals: removing noises while performing aggregation region processes, and producing final segmentation results.

The process of removing noise while performing aggregation regions is a crucial part of our algorithm. It supports to merge these so-called small regions neglected considering distance of similarity $d(R_i, R_j)$ between two regions. The criteria for these process performing are existing two adjacent regions and their pixels inside must be less than a threshold defined small regions namely *regthres* in our algorithm. Empirically, we set the threshold for small regions *regthres* = 200 pixels for testing and evaluating on the validation dataset. How to choose an appropriate threshold lead to best results for this noise removal phase is a challenge problem. We attempt to set threshold *regthres* = $\{100, 200, ..., 600\}$ and obtain some potential choices but this will be discussed more detailed on future work.

Although image noises have been reduced by the aggregation regions, the final segmentation images had better be removed noises to produce smooth images. A variety of image noise reduction techniques have been introduced and well-known in the literature. Most of them are supported by Open Source Computer Vision Library such as: Averaging, Gaussian Blurring, Median Blurring and Bilateral Filtering. In our method, Median Blurring technique has been applied to smooth images and obtained efficient smoothing image results.

4   Experimental Evaluation

This section provides experiments that were performed to assess our algorithm. To evaluate the proposed model, we deployed in the *Berkeley Segmentation Data Set 500 (BSDS500)* [13] and evaluated by *Probabilistic Rand Index (PRI)* metric. The qualitative and quantitative evaluation are presented by some figures and comparative results Table 1.

4.1   Berkeley Segmentation Data Set 500

The Berkeley Segmentation Data Set has built with the aim of providing an empirical basis for research on image segmentation and boundary detection. This dataset comprises 500 images, including 200 images for training, 200 images for testing and 100 images for validation. Each image has 481×321 pixels, which yields a graphs of 154401 vertices. BSDS500 also provides ground-truth segmentations that are manually generated by many human subjects. For every image, there are from 5 to 10 ground-truth segmentation maps. Supplying a benchmark for comparing different segmentation and boundary detection algorithms.

4.2   Probabilistic Rand Index

In general, there are three standard evaluation segmentations metrics that have been used by researchers to evaluate their algorithms such as: Variation of Information, Segmentation Covering and Probabilistic Rand Index. Among them, Probabilistic Rand Index is used for evaluating almost algorithms that tested on BSDS500 because of its advantages on multiple ground-truth.

The Probabilistic Rand Index [2,18,19] is a classical evaluation criteria for clusterings. PRI measures the probability that pair of pixels have consistent labels in the set of manual segmentations (ground-truth). Given a set of ground-truth segmentation $\{S_k\}$, the Probabilistic Rand Index is defined as:

$$PRI(S_{test}, \{S_k\}) = \frac{1}{T} \sum_{i<j} [c_{ij} p_{ij} + (1 - c_{ij})(1 - p_{ij})] \qquad (7)$$

where c_{ij} is the event that the algorithm gave the same label to pixels i and j and p_{ij} corresponds to the probability of the pixels i and j having the same label, and is estimated by using sample mean of the corresponding Bernoulli distribution on the ground-truth dataset. T is the total number of pixel pairs. The PRI values range in $[0,1]$ in which a larger value likely indicates a greater similarity between these segmentations.

4.3   Results

For qualitative evaluations, we present some images of the comparison segmentation results as Fig. 3. For these qualitative results, we can see that the proposed algorithm offers good results and produces sizable regions for all selected images. Figure 4 presents some segmentation results of our algorithm on some images. Our algorithm can aggregate homogeneous neighboring regions successfully even if pixels inside each region are dissimilar.

From a quantitative point of view, we evaluated the segmentations results by the *Probabilistic Rand Index* (PRI) by comparing a test segmentation with multiple ground-truth images. We applied this measure on the validation set from the Berkeley segmentation dataset, and obtained the score of 0.80, while the ground-truth (segmentation made by human) got a score of 0.87 [3]. Detailed results are given in Table 1.

The evaluation results reflect the success of our agglomeration process for homogeneous regions. Our method has close performance with gPb-owt-ucm and Fast multi-scale (HOG), and exceeding all the rest algorithms in term of PRI scores.

Table 1. Comparative results using the PRI Index on the Berkeley segmentation dataset [3, 10, 14]

Methods	PRI *(Larger better)*
Human	0.87
gPb-owt-ucm	0.81
Youssef Mourchild's *(Fast multi-scalse (HOG))*	0.81
Youssef Mourchild's *(Modularity optimization based on Danon (HOG))*	0.80
Our method	**0.80**
Shijie Li's method *(L*a*b (HoS))*	0.78
Shijie Li's method *(RGB (HoS))*	0.75
Mean Shift	0.78
Felz-Hutt	0.77
Canny-owt-ucm	0.77
NCuts	0.75

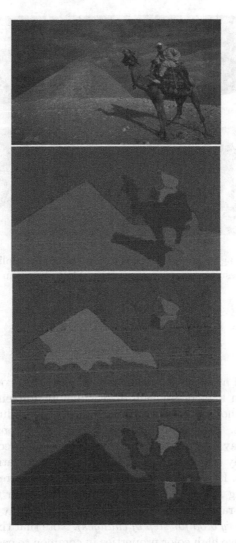

Fig. 3. Visual Segmentation results of some methods: *Top left:* Origin image. *Top right:* Segmentation result of HoS method [10]. *Bottom left:* Segmentation result of HOG method [10]. *Bottom right:* Our segmentation result

Fig. 4. *Left:* Origin images. *Middle:* Segmentation results by applied Louvain method. *Right:* Segmentation results with the cooperative AHR algorithm

5 Conclusion

This paper proposed an efficient agglomerative algorithm cooperating with the community detection namely Louvain algorithms for implementing image segmentation. Our method is significantly accurate and produces efficient image segmentation results. The novelty in this paper is the consideration of color property precisely way in order to build a 48-dimensional vector for each region and proposal to apply cosine similarity distance for aggregation processes using only color properties. Hence, the time complexity has been reduced significantly comparing with using 256 dimensional vector in some other techniques. Empirically, the threshold range for agglomeration process is only taking range from 0.40 to 0.99 *(with every 0.05 for step of changes)*. Note that the regions belong to one segment often have high color properties in common to each other. Extensive experiments have been performed, and the results show that the proposed algorithm can reliably segment the image and avoid over-segmentation in order to produce more accurate objects and enhance computing performance efficiently.

References

1. Abin, A.A., Mahdisoltani, F., Beigy, H.: A new image segmentation algorithm: a community detection approach. In: IICAI (2011)
2. Arbelaez, P., Maire, M., Fowlkes, C., Malik, J.: Contour detection and hierarchical image segmentation. IEEE Trans. Pattern Anal. Mach. Intell. **33**(5), 898–916 (2011). https://doi.org/10.1109/TPAMI.2010.161

3. Arbelaez, P., Maire, M., Fowlkes, C., Malik, J.: From contours to regions: An empirical evaluation. In: 2009 IEEE Conference on Computer Vision and Pattern Recognition, pp. 2294–2301, June 2009
4. Blondel, V.D., Guillaume, J.L., Lambiotte, R., Lefebvre, E.: Fast unfolding of communities in large networks. J. Stat. Mech. Theor. Exp. **10**, 10008 (2008)
5. Browet, A., Absil, P.-A., Van Dooren, P.: Community detection for hierarchical image segmentation. In: Aggarwal, J.K., Barneva, R.P., Brimkov, V.E., Koroutchev, K.N., Korutcheva, E.R. (eds.) IWCIA 2011. LNCS, vol. 6636, pp. 358–371. Springer, Heidelberg (2011). https://doi.org/10.1007/978-3-642-21073-0_32
6. Comaniciu, D., Meer, P.: Mean shift: a robust approach toward feature space analysis. IEEE Trans. Pattern Anal. Mach. Intell. **24**, 603–619 (2002)
7. Felzenszwalb, P.F., Huttenlocher, D.P.: Efficient graph-based image segmentation. Int. J. Comput. Vis. **59**(2), 167–181 (2004). https://doi.org/10.1023/B:VISI.0000022288.19776.77
8. Fortunato, S.: Community detection in graphs. Phy. Rep. **486**(35), 75–174 (2010). http://www.sciencedirect.com/science/article/pii/S0370157309002841
9. Girvan, M., Newman, M.E.J.: Community structure in social and biological networks. Proc. Nat. Acad. Sci. **99**(12), 7821–7826 (2002). https://doi.org/10.1073/pnas.122653799
10. Li, S., Wu, D.O.: Modularity-based image segmentation. IEEE Trans. Circuits Syst. Video Technol. **25**(4), 570–581 (2015)
11. Li, W.: Modularity segmentation. In: Lee, M., Hirose, A., Hou, Z.-G., Kil, R.M. (eds.) ICONIP 2013. LNCS, vol. 8227, pp. 100–107. Springer, Heidelberg (2013). https://doi.org/10.1007/978-3-642-42042-9_13
12. Linares, O.A.C., Botelho, G.M., Rodrigues, F.A., Neto, J.B.: Segmentation of large images based on super-pixels and community detection in graphs. CoRR abs/1612.03705 (2016). http://arxiv.org/abs/1612.03705
13. Martin, D., Fowlkes, C., Tal, D., Malik, J.: A database of human segmented natural images and its application to evaluating segmentation algorithms and measuring ecological statistics. In: Proceedings of 8th International Conference on Computer Vision, vol. 2, pp. 416–423, July 2001
14. Mourchid, Y., El Hassouni, M., Cheri, H.: An image segmentation algorithm based on community detection. In: Cheri, H., Gaito, S., Quattrociocchi, W., Sala, A. (eds.) Complex Networks & Their Applications V. Complex Networks, pp. 821–830. Springer, Cham (2017). https://doi.org/10.1007/978-3-319-50901-3_65
15. Newman, M.E.J.: Detecting community structure in networks. Eur. Phy. J. B **38**(2), 321–330 (2004). https://doi.org/10.1140/epjb/e2004-00124-y
16. Newman, M.E., Girvan, M.: Finding and evaluating community structure in networks. Phy. Rev. E **69**(2), 026113 (2004)
17. Shi, J., Malik, J.: Normalized cuts and image segmentation. IEEE Trans. Pattern Anal. Mach. Intell. **22**(8), 888–905 (2000)
18. Unnikrishnan, R., Pantofaru, C., Hebert, M.: Toward objective evaluation of image segmentation algorithms. IEEE Trans. Pattern Anal. Mach. Intell. **29**(6), 929–944 (2007)
19. Unnikrishnan, R., Hebert, M.: Measures of similarity. In: Seventh IEEE Workshops on Application of Computer Vision 2005. WACV/MOTIONS 2005, vol. 1, pp. 394–394. IEEE (2005)
20. Vincent, L., Soille, P.: Watersheds in digital spaces: an efficient algorithm based on immersion simulations. IEEE Trans. Pattern Anal. Mach. Intell. **13**(6), 583–598 (1991)

21. Mourchid, Y., El Hassouni, M., Cherifi, H.: A new image segmentation approach using community detection algorithms. In: 15th International Conference on Intelligent Systems Design and Applications, Marrakesh, Marocco, December 2015
22. Mourchild, Y., El Hassouni, M., Cherifi, H.: Image segmentation based on community detection approach. Int. J. Comput. Inf. Syst. Ind. Manage. Appl. **8**, 195–204 (2016)

Robust Feature Descriptors for Object Segmentation Using Active Shape Models

Daniela Medley$^{(\boxtimes)}$, Carlos Santiago, and Jacinto C. Nascimento

Institute for Systems and Robotics, Instituto Superior Técnico, Lisbon, Portugal
daniela.medley@tecnico.ulisboa.pt

Abstract. Object segmentation is still an active topic that is highly visited in image processing and computer vision communities. This task is challenging due not only to difficult image conditions (e.g., poor resolution or contrast), but also to objects whose appearance vary significantly. This paper visits the Active Shape Model (ASM) that has become a widely used deformable model for object segmentation in images. Since the success of this model depends on its ability to locate the object, many detectors have been proposed. Here, we propose a new methodology in which the ASM search takes the form of local rectangular regions sampled around each landmark point. These regions are then correlated to variable or fixed texture templates learned over a training set. We compare the performance of the proposed approach against other detectors based on: (i) the classical ASM edge detection; (ii) the Histogram of Oriented Gradients (HOG); and (iii) the Scale-Invariant Feature Transform (SIFT). The evaluation is performed in two different applications: facial fitting and segmentation of the left ventricle (LV) in cardiac magnetic resonance (CMR) images, showing that the proposed method leads to a significant increase in accuracy and outperforms the other approaches.

Keywords: Active shape models · Image segmentation
Texture regions · Histogram of oriented gradients
Scale-invariant feature transform

1 Introduction

Segmenting images containing objects whose appearance vary significantly is a challenging task. Statistical models have become a widely used approach in this context. These models are able to represent large shape and appearance variations of the object of interest. A popular method is the Active Shape Model (ASM). Since its early introduction by Cootes et al. [1], ASM has become a well-recognized powerful tool due to its ability to segment objects with significant shape variability. This method characterizes an object shape by a set of specific points, denoted as landmark points, and models it by a mean shape and its most significant modes of variation learned from a training set. To fit the model to

---

This work was supported by FCT UID/EEA/50009/2013.

J. Blanc-Talon et al. (Eds.): ACIVS 2018, LNCS 11182, pp. 163–174, 2018.
https://doi.org/10.1007/978-3-030-01449-0_14

an object, the ASM searches within the image for candidate positions of the object landmark points, which we call observations. Traditionally, this search consists of finding the strongest edge along a profile line for each landmark point. Although this approach may work in some simple applications, in most real world problems it is a naive approach that might fail to cover all the object features, generating noisy observations (outliers). If some of the observations are outliers, the accuracy of the ASM is severely compromised, resulting in a decreased segmentation performance. Therefore, a crucial component for the success of these models lies in their ability to find the correct position of the object landmark points.

Alternative approaches have since been proposed to overcome this ASM drawback. The Active Appearance Model (AAM), also proposed by Cootes *et al.* [2], is not only able to provide shape information about an object, but also takes into account its variation in appearance, i.e., its textural information. Contrary to ASM, the AAM search method consists of using the texture residual between the learned model and the test image in order to find the best model parameters to match the image. In later work, Cristinacce and Cootes [3] presented the Constrained Local Model (CLM). This method is very similar to AAM, but instead of modelling the whole object texture, it models local templates surrounding each landmark point. It uses Principal Component Analysis (PCA) to learn statistical templates of the appearance from a training set, which are adjustable for each test image.

Powerful feature descriptors have also been combined with these deformable models, in order to improve the observation detection, namely, Histograms of Oriented Gradients (HOG) [4] and Scale-Invariant Feature Transform (SIFT) [5,6]. These models have shown an increased performance in comparison to the standard model approach. Nevertheless, these have an increased complexity and appear to be slower and computational demanding.

This paper proposes two different and efficient observation detection methods that are used within the ASM framework. Both approaches are similar in the sense that they search for observations within a rectangular region around each landmark point and both find the point that maximizes the correlation with a texture template learned from a training data. They differ on how the templates are obtained. In the first method, each template, associated to a landmark point, is computed as the mean texture in the training data, obtaining a *fixed template* that will be the same for the testing stage. The second method uses not only the mean texture, but also the variation modes obtained through PCA. This allows fitting the templates for each test image, i.e., the method uses *variable templates*. This approach resembles the CLM algorithm but uses a different fitting strategy. More specifically, the CLM uses a whole response surface to update the model, whilst we determine the templates around each landmark point and extract the location of best observation point, which is then used to estimate the ASM parameters.

Despite of been easily used for different image interpretations, to show the advantage of the proposed methods we compare them to three other detection approaches in the problem of facial fitting and the segmentation of the left ventricle in cardiac magnetic resonance images: (i) the classical ASM edge detection; (ii) a detector based on HOG features; and (iii) a detector based on SIFT features.

The remaining of this paper is organized as follows: Sect. 2 describes the ASM framework and each observation detection method used in this work. The datasets and evaluation results are shown in Sect. 3. Finally, in Sect. 4 the overall conclusions are presented.

2 Methodology

This section starts by briefly revising the ASM framework used in this work. In Sect. 2.2, we detail all the different detection methods analysed and how they are combined with the ASM methodology.

2.1 Active Shape Model

The ASM algorithm [1] describes the shape of an object by learning its statistics from annotated images in a training set. More specifically, this model uses the mean shape and its main modes of deformation computed using PCA. Formally, any shape $\mathbf{x}$ in the ASM framework can be analytically described as

$$\mathbf{x} \simeq \bar{\mathbf{x}} + \mathbf{Db} \ , \tag{1}$$

where, $\mathbf{x} \in \mathbb{R}^{2N \times 1}$ is a vector representing the mean shape computed from a training set, $\mathbf{D} \in \mathbb{R}^{2N \times K}$ is a matrix containing the first K modes of deformation, and $\mathbf{b} \in \mathbb{R}^{K \times 1}$ contains the deformation coefficients that weight each of the deformation modes. Then, the position of the shape in an image is governed by a similarity transformation, which accounts for the scale, rotation and translation of the shape $\mathbf{x}$.

To fit the model to the object in a new image, the ASM searches for observation points and estimates the parameters that minimize the distance between the landmark points and the corresponding observations. In the next section, we describe five ASM search methods: the classical ASM edge detection, the ASM-HOG, the ASM-SIFT and the proposed fixed and variable texture template methods, denoted as ASM-FTT and ASM-VTT, respectively (see Fig. 1).

2.2 ASM Search Methods

Traditionally, in the ASM framework the observations correspond to the strongest edge along profile lines orthogonal to the contour at each landmark point (see Fig. 1(a)). In most real world problems, this approach generates outliers that misguide the estimation of the ASM parameters. Thus, alternative observation detectors have the potential to improve the ASM performance.

All the four search methods described next share a common framework as they are region-based and use a template descriptor to search for observation

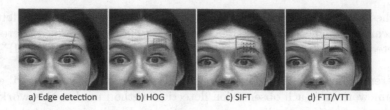

a) Edge detection   b) HOG   c) SIFT   d) FTT/VTT

Fig. 1. Representation of the different search methods for one landmark point (red) of the model (blue), where the search region is represented in yellow: a) the original ASM edge detection; b) the HOG-based detection with the descriptor of the landmark in green; c) the SIFT-based detection with the descriptor of the landmark in green; and d) the proposed search methods based on texture (FTT and VTT). (Color figure online)

points. The first stage of the detectors is to obtain a set of template descriptors and regions in which features are detected. To accomplish this, each training patch is sampled around each landmark point over the training set. A feature extraction function $\mathcal{F}$ is applied to each patch, resulting in a feature vector for each pixel location. Thus, an average descriptor-based template for each landmark can then be built and normalized (e.g. zero mean and unit variance).

Formally, let the training set of images be defined as $\mathcal{D} = \{\mathbf{I}_j\}_{j=1}^{|\mathcal{D}|}$. Assuming the contour is defined as $\mathbf{x} = [\mathbf{x}_i^{\top}]_{i=1,...,N}$ with N landmark points, i.e., $\mathbf{x} \in \mathbb{R}^{2N \times 1}$, we extract N patches for each j-th image $\mathbf{I}_j$, building the following set of patches

$$\mathcal{P}_j = \left\{ \mathbf{P}_j^1, ..., \mathbf{P}_j^N \right\}_{j=1}^{|\mathcal{D}|}, \tag{2}$$

where $\mathbf{P}_j^i$ stands for the i-th patch extracted from the j-th image and is given by $\mathbf{P}_j^i = [\mathbf{I}_j]_{T \times T}^i$. The operator $[.]_{T \times T}^i$ crops the image $\mathbf{I}_j$, centered at the landmark point $\mathbf{x}_i$, with size of $T \times T$ pixels. Now, for each pixel location in each patch $\mathbf{P}_j^i$, we apply a feature extraction function $\mathcal{F}$, obtaining a descriptor-based image $\mathcal{F}(\mathbf{P}_j^i)$.

For each i-th landmark point, the feature based template $\mathcal{T}^i$ of size $T \times T$ can be computed as follows

$$\mathcal{T}^i = \frac{1}{|\mathcal{D}|} \sum_{j=1}^{|\mathcal{D}|} \mathcal{F}(\mathbf{P}_j^i), \quad i = 1, ..., N . \tag{3}$$

When searching in a test image $\mathbf{I}_j$, a larger search region, with $R \times R$ dimension (with $R > T$), is created around each i-th landmark point, *i.e.* $\mathbf{R}_j^i = [\mathbf{I}_j]_{R \times R}^i$. As previously, we extract features from this search region, which we denote as $\mathcal{R}_j^i = \mathcal{F}(\mathbf{R}_j^i)$. By defining a sliding window within the region $\mathcal{R}_j^i$, it is possible to compute the similarity between $\mathcal{T}^i$ and each sub-region of $\mathcal{R}_j^i$ (with $T \times T$ dimension). In the following sections we describe the different methodologies under the above framework.

ASM-HOG. Histogram of oriented gradients [7] is a feature descriptor that characterizes the local appearance of the object based on the gradient orientations in different bins of small portions of an input image (see Fig. 1(b)).

To combine this feature descriptor with an ASM search, we define $\mathcal{F}$ in (3) as a HOG feature extractor applied to each landmark point and use the following approach. Given a training set, we can compute a HOG feature vector for each landmark point and an average template T^i as given in (3). When searching for observation points in the test image, we define the search region $\mathcal{R}_j^i$, sampled around the i-th landmark point, and determine the HOG feature vector for each pixel within this region, through the sliding window. Finally, the resulting HOG features for each pixel in the search region are compared with the corresponding template, T^i. The most similar region will correspond to the new likely location of the object landmark points, *i.e.*, the detected observation. Since HOG feature vectors are histograms, several similarity measures can be applied. Histogram Intersection is one of those metrics that has shown a better performance than the standard Euclidean distance in image applications [8]. For grey images it can be defined as follows [9]

$$\vartheta(h_1, h_2) = \sum_{l=1}^{n} \min(h_1(l), h_2(l)) \ , \tag{4}$$

where h_1 and h_2 are the histograms to be compared of the template T^i and of each sub-region of $\mathcal{R}_j^i$, respectively, with n bins each.

ASM-SIFT. Scale-invariant feature transform [10] is also able to extract distinctive features from images based on the local gradients histograms around each landmark and are invariant to scale, illumination and pose. Contrary to HOGs, these histograms are computed with respect to the dominant orientation of the landmark. It starts by scanning an image to identify potential interest points, known as *keypoints*. Next, the dominant orientation of the keypoint is determined, based on local image gradient direction, and the local descriptor is built.

In order to combine the SIFT descriptor with the ASM algorithm (see Fig. 1(c)), we propose a different strategy for its implementation. For a given training image, we compute a SIFT descriptor by applying a SIFT extraction function $\mathcal{F}$ to each patch (see (3)), forcing the keypoints to be the landmark points, with a gradient orientation relative to the normal vector of the contour at each landmark point. This allows the corresponding feature vector to be used in the ASM search for observations. Thus, an average template patch can also be built and normalized for each landmark point, consisting on SIFT descriptors (see (3)).

Similarly to the ASM-HOG, when analysing a new image, the search process starts by defining a search region $\mathcal{R}_j^i$ around each current landmark point $\mathbf{x}_i$ and determine the SIFT descriptors for each pixel in this region, with a sliding window. Since the SIFT descriptor is also based on histograms, we use the same comparison metric as before, which is defined in (4).

Proposed Search Using Fixed Texture Templates (ASM-FTT). In this method we also follow the procedure addressed in Sect. 2.2. First, for each j-th image in the training set, we extract the set of patches $\mathcal{P}_j$ (see (2)). Then for each landmark point $\mathbf{x}_i$, $i = 1, \ldots, N$, we compute the corresponding $T \times T$ fixed template $\mathcal{T}^i$ following (3), where the feature extraction function $\mathcal{F}$ concerns to the image texture itself (see Fig. 1(d)). Finally, for a given test image, we set the $R \times R$ region $\mathcal{R}_j^i$ in which we compute a (normalized) cross-correlation between the template $\mathcal{T}^i$ and the search region $\mathcal{R}_j^i$ as follows[1] [11]

$$\gamma(u, v) = \frac{\sum_{x,y}[\mathcal{R}(x,y) - \overline{\mathcal{R}}_{u,v}][\mathcal{T}(x - u, y - v) - \overline{\mathcal{T}}]}{\{\sum_{x,y}[\mathcal{R}(x,y) - \overline{\mathcal{R}}_{u,v}]^2 \sum_{x,y}[\mathcal{T}(x - u, y - v) - \overline{\mathcal{T}}]^2\}^{0.5}} \ , \qquad (5)$$

where the notation $\mathcal{R}(x, y)$ stands for the region pixels (x, y) in the region $\mathcal{R}_j^i$; $\overline{\mathcal{T}}$ is the mean of the template $\mathcal{T}$ and $\overline{\mathcal{R}}_{u,v}$ is the mean of the region $\mathcal{R}(x, y)$ under the template.

Therefore, by performing a normalized cross-correlation between the individual training patch $\mathcal{T}^i$ and the search region $\mathcal{R}_j^i$ for each current i-th landmark point, it is possible to determine the new likely location of the object landmark points. This can be determined by analysing the resulting image response, in order to find the strongest match.

Proposed Search Using Variable Texture Templates (ASM-VTT). The search method proposed in this section is similar to the previous one. However, instead of using fixed feature templates, which remain unchangeable during the search process, we describe a method that adapts the texture templates to the test image. More specifically, we apply PCA to the training patches $\mathbf{P}_j^i$, for $j = 1, ..., |\mathcal{D}|$. Similarly to the shape analysis described in Sect. 2.1, we approximate each patch $\mathbf{P}_j^i$ of size $T \times T$ by a vectorized linear combination of K main modes of variation as follows

$$\mathbf{g} \simeq \overline{\mathbf{g}} + \mathbf{D}_g \mathbf{b}_g \ , \qquad (6)$$

where $\overline{\mathbf{g}} \in \mathbb{R}^{T^2 \times 1}$ is the mean normalized texture vector, $\mathbf{D}_g \in \mathbb{R}^{T^2 \times K}$ contains the main modes of variation of the patch, and $\mathbf{b}_g \in \mathbb{R}^{K \times 1}$ is a vector containing the deformation coefficients that weight each variation mode in $\mathbf{D}_g$.

Given a test image, $\mathbf{I}_t$, and an initial guess of the model position, a set of image-specific texture templates can be generated by computing the $\widehat{\mathbf{b}}_{g_t}$ as follows

$$\widehat{\mathbf{b}}_{g_t} = \mathbf{D}_g^{-1}(\mathbf{g}_t - \overline{\mathbf{g}}) \ , \qquad (7)$$

where $\mathbf{g}_t$ is the vectorized texture patch around a specific landmark point of $\mathbf{I}_t$. The vector $\mathbf{b}_{g_t}$ contains the parameters that best match the statistical model

---

[1] In this equation we suppressed the subscripts i and j for the simplicity of the notation.

to the test image. Without additional constraints, $\mathbf{b}_{g_t}$ may correspond to an unrealistic patch. Therefore, an additional step is required to constraint the solution in (7). To achieve this, the Mahalanobis distance $d_{\mathcal{M}}$ is used to measure the acceptability of the generated patches. More concisely, $d_{\mathcal{M}}$ has to be lower than a specific predefined threshold $d_{\max}$

$$d_{\mathcal{M}}^2 = \sum_{l=1}^{L} \frac{\widehat{b}_l^2}{\lambda_l} \leq d_{\max}^2 , \qquad (8)$$

where $\widehat{b}_l$ denotes the l-th component of $\widehat{\mathbf{b}}_{g_t}$ and λ_l is the eigenvalue associated to the l-th deformation mode. If $\widehat{\mathbf{b}}_{g_t}$ does not satisfy (8) the variation mode is rescaled as follows

$$\widehat{\mathbf{b}}_{g_t} = \widehat{\mathbf{b}}_{g_t} \frac{d_{\max}}{d} . \qquad (9)$$

Finally, the normalized cross-correlation can then be applied, as described in (5), between the template and the predefined search region, sampled around each landmark point. By finding the strongest match, the new likely location of the object landmark points can thus be determined.

3   Experimental Setup

This section presents the experimental setup used to evaluate the proposed framework. Two different applications were used to test each feature detection method: facial fitting and segmentation of the left ventricle (LV) in cardiac magnetic resonance (CMR) images.

3.1   Dataset

In the first application, we addressed the problem of fitting an ASM to facial image sequences. For that, we used the publicly available Cohn-Kanade (CK+) database [12,13] of emotion sequences taken from frontal view, where the manual face annotation are available, i.e., the ground truth (GT). Among several emotion sequences, we took the "surprise" sequences, since they contain more challenging lip boundary deformations and large eyebrow displacements. The dataset comprises 56 different sequences, each with 10–20 frames, with a total of 912 images (each with 490×640 size). The leave-one-sequence-out cross validation was used for performance evaluation. For initialization purposes and in order to have an initial guess of the model position, we used the Viola-Jones detector [14], for finding the faces in each image, which results in a rectangle containing the face. The learned mean shape is then aligned to the centre of the rectangle, resulting in a rough initialization of the landmark points.

The second application is the segmentation of the LV in CMR images. For that purpose, we use the publicly available dataset [15], which contains data from 33 different patients. For each patient, the CMR data is a sequence of 20 volumes with 8 to 15 slices, in a total of 7980 images, coupled with the manual LV border annotations, which is used as the ground truth. Each slice is a 256×256 image with an average resolution of 1.4 ± 0.2 mm/pixel, nevertheless, the LV is only present in 4 to 10 of the them (the remaining slices are disregarded). Each volume is segmented independently, whereby, as for the previous application, the leave-one-sequence-out cross validation is also used for each patient volume, i.e., for each one of the 20 volumes of the 33 patients, the statistics are learned from that same volume of the remaining 32 patients. The initialization is only performed for the first slice of each volume of the test sequence and, for that, we use the ground truth. The remaining slices are initialized by successively propagating the previous slice segmentation.

3.2   Error Metric

Segmentations are evaluated by comparing the estimated contour with the true object boundary (the ground truth). Their performance is quantitatively measured using two different metrics: the average distance error [16] for the facial segmentation and the Dice coefficient [17] for the LV segmentation, as detailed next.

Average Distance (d_{AV}). Let us assume $\mathbf{x} = [\mathbf{x}_i^\top]_{i=1,...,N}$ and $\mathbf{y} = [\mathbf{y}_i^\top]_{i=1,...,N}$, with $\mathbf{x}_i$, $\mathbf{y}_i \in \mathbb{R}^2$ being two point vectors representing the estimated and the ground truth of the face, respectively. The AV between $\mathbf{x}$ and $\mathbf{y}$ is defined as the average Euclidean distance, as follows

$$d_{AV}(\mathbf{x}, \mathbf{y}) = \frac{1}{N} \sum_{i=1}^{N} \|\mathbf{y}_i - \mathbf{x}_i\| \ . \tag{10}$$

Dice Coefficient (d_{Dice}). Assuming now $\mathbf{S}_1$ and $\mathbf{S}_2$ as two binary images associated with the estimated contour $\mathbf{s}_1$ and the ground truth $\mathbf{s}_2$, respectively, such that the pixels inside the LV segmentation have value one and the pixels outside have the value zero. We can compute the Dice coefficient as follows

$$d_{Dice}(\mathbf{S}_1, \mathbf{S}_2) = \frac{2\, C(\mathbf{S}_1 \wedge \mathbf{S}_2)}{C(\mathbf{S}_1) + C(\mathbf{S}_2)} \ , \tag{11}$$

where $C(.)$ is a function that counts the number of pixels within the region and the operator $\wedge$ denotes a pixel-wise AND. Note that d_{Dice} is always between 0 and 1, where a value of 1 reflects a perfect match between the two segmentations.

3.3   Results

Facial Segmentation. The performance of the different feature detectors was evaluated and compared first for facial fitting. The same initial guess was used in all the tested methods with $N = 44$ as the number of model landmark points. Figure 2 shows three examples of the segmentation obtained by each method.

ASM   ASM-HOG   ASM-SIFT   ASM-FTT   ASM-VTT

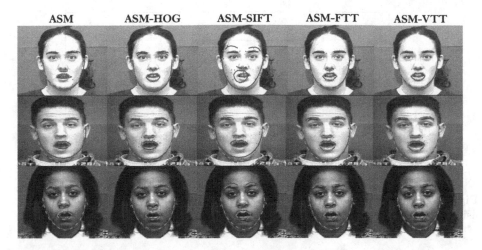

Fig. 2. Examples of the facial segmentation. Each column shows the result of a different method: ASM, ASM-HOG, ASM-SIFT, ASM-FTT and ASM-VTT, respectively The green dashed line shows the ground truth, whereas the blue line shows the estimated segmentation and the red dots represent the observation points in the last iteration. (Color figure online)

These images show the improved accuracy of the proposed methods with fixed (ASM-FTT) and variable (ASM-VTT) templates.

To ascertain the robustness of the proposed approach, in the experimental setup, several patch dimensions were tested, both for the search region and the feature templates. From the extensive experimental evaluation, we found the regions which achieved the best results were in the interval of $\{11 \times 11, 23 \times 23, 35 \times 35\}$ and $\{33 \times 33, 55 \times 55, 67 \times 67\}$ pixels, both for the template and the search region, respectively. It is important to remark that for the ASM implementation the patch size only determines the length of the profile lines, and for larger regions ASM quickly starts to degrade its performance. Figure 3 presents a statistical evaluation of the segmentation accuracy for each of the studied methods. Note that the results shown only concern the statistical performance for a search region of 55×55 pixels and a template of 23×23 pixels, as the results obtained are similar for the different tested values. The results show that the proposed methods performed better than both the standard ASM approach and the alternative ones, namely the ASM-HOG and the ASM-SIFT, which is possible to verify through the decrease of the average distance with the use of the proposed models. Furthermore, both ASM-FTT and ASM-VTT have a similar performance and both lead to a significant improvement over the remaining tested methods.

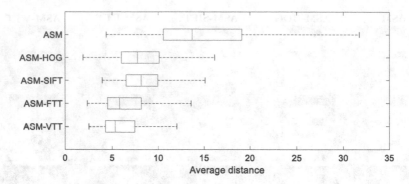

Fig. 3. Comparison of the statistical results of each method for the facial segmentation, using the average distance in pixels, d_{AV}.

Left Ventricle Segmentation. The final application herein presented is the segmentation of the LV. Figure 4 shows three examples of the segmentation obtained with the different methods. These images clearly show the improved accuracy of the proposed methods both with fixed (ASM-FTT) and variable (ASM-VTT) templates.

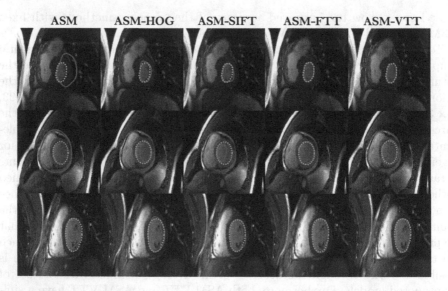

Fig. 4. Examples of LV segmentation with the different methods in three CMR images from different patients. Each column shows the result of a different method: ASM, ASM-HOG, ASM-SIFT, ASM-FTT and ASM-VTT, respectively. The green dashed line shows the ground truth, whereas the blue line represents the estimated segmentation. (Color figure online)

Once more, the parameters were chosen after an extensive evaluation. Note that the images tested here have a reduced size, thus, in accordance to the first application, we find the region of 33×33 pixels as the most suitable for the search region, 15×15 pixels for the template size and $N = 40$ for the model landmark points number. The statistical evaluation of the segmentation accuracy for each of the methods are represented in Fig. 5. The results show that ASM-FTT and ASM-VTT methods have a similar performance and both lead to a significant improvement in accuracy, outperforming the other tested approaches.

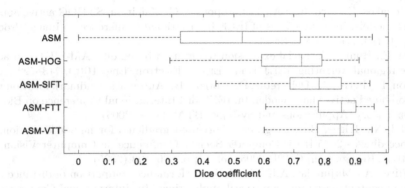

Fig. 5. Comparison of the statistical results of each method for the LV segmentation, using the Dice coefficient, d_{Dice}.

4 Conclusion

This paper proposes two novel ASM-based search methods to detect reliable observation points. The first one is based on a fixed texture template which remains unchangeable during the test phase, the ASM-FTT. The second method is based on statistical texture templates whose variation coefficients change with each test, the ASM-VTT. The two proposed methodologies are compared with: (i) the classical ASM edge detector, (ii) the ASM combined with a HOG detector, and (iii) the ASM combined with a SIFT detector. From the experimental evaluation, we applied these methodologies in two datasets, for the segmentation of faces in image sequences and for the segmentation of the left ventricle in CMR images. The obtained results are relevant and promising, as we have shown that the proposed methodologies lead to a better performance compared to the other three approaches. Further improvements can still be achieved by considering multiple observations points for each landmark point, instead of just one per patch. If the strongest feature in the patch is invalid (i.e., not belonging to the contour), this can jeopardize the segmentation accuracy. This means that more features should be extracted from each patch in the attempt to get the reliable one. Further work will extend the proposed approach to deal with multiple features for each patch.

References

1. Cootes, T.F., Taylor, C.J., Cooper, D.H., Graham, J.: Active shape models - their training and application. Comput. Vis. Image Underst. **61**(1), 38–59 (1995)
2. Cootes, T.F., Edwards, G.J., Taylor, C.J.: Active appearance models. In: Proceedings of European Conference on Computer Vision (ICCV), vol. 2, pp. 484–498 (1998)
3. Cristinacce, D., Cootes, T.F.: Feature detection and tracking with constrained local models. In: Proceedings of the British Machine Vision Conference 2006, pp. 95.1–95.10 (2006)
4. Antonakos, E., Medina, J.A., Tzimiropoulos, G., Zafeiriou, S.: HOG active appearance models. In: Proceedings of IEEE International Conference on Image Processing (ICIP), pp. 224–228 (2014)
5. Ren, F., Huang, Z.: Facial expression recognition based on AAM-SIFT and adaptive regional weighting. IEEJ Trans. Electr. Electron. Eng. **10**(6), 713–722 (2015)
6. Zhou, D., Petrovska-Delacrétaz, D., Dorizzi, B.: Automatic landmark location with a combined active shape model. In: IEEE 3rd International Conference on Biometrics: Theory, Applications and Systems, BTAS 2009 (2009)
7. Dalal, N., Triggs, B.: Histograms of oriented gradients for human detection. In: Proceedings - 2005 IEEE Computer Society Conference on Computer Vision and Pattern Recognition, CVPR 2005, vol. I, pp. 886–893 (2005)
8. Vadivel, A., Majumdar, A.K., Sural, S.: Performance comparison of distance metrics in content-based image retrieval applications. In: International Conference on Information Technology (CIT), September 2003
9. Swain, M.J., Ballard, D.H.: Color indexing. Int. J. Comput. Vis. **7**(1), 11–32 (1991)
10. Lowe, D.G.: Distinctive image features from scale-invariant keypoints. Int. J. Comput. Vis. **60**(2), 91–110 (2004)
11. Lewis, J.P.: Fast normalized cross-correlation. In: Vision Interface: Canadian Image Processing and Pattern Recognition Society, pp. 120–123 (1995)
12. Lucey, P., Cohn, J.F., Kanade, T., Saragih, J., Ambadar, Z., Matthews, I.: The extended Cohn-Kanade dataset (CK+): a complete facial expression dataset for action unit and emotion-specied expression. In: Proceedings of the Third International Workshop on CVPR for Human Communicative Behavior Analysis, pp. 94–101 (2010)
13. Kanade, T., Cohn, J.F., Tian, Y.: Comprehensive database for facial expression analysis. In: Proceedings of the 4th IEEE International Conference on Automatic Face and Gesture Recognition, pp. 46–53 (2000)
14. Viola, P., Jones, M.: Rapid object detection using a boosted cascade of simple features. Comput. Vis. Pattern Recognit. **1**, 511–518 (2001)
15. Andreopoulos, A., Tsotsos, J.K.: Efficient and generalizable statistical models of shape and appearance for analysis of cardiac MRI. Med. Image Anal. **12**(3), 335–357 (2008)
16. Nascimento, J.C., Marques, J.S.: Robust shape tracking with multiple models in ultrasound images. IEEE Trans. Image Process. **17**(3), 392–406 (2008)
17. Dice, L.R.: Measures of the amount of ecologic association between species. J. Neurol. Sci. **26**(3), 297–302 (1945)

Foreground Background Segmentation in Front of Changing Footage on a Video Screen

Gianni Allebosch[1,2(✉)], Maarten Slembrouck[1,2], Sanne Roegiers[1,2], Hiêp Quang Luong[1,2], Peter Veelaert[1,2], and Wilfried Philips[1,2]

[1] TELIN-IPI, Ghent University, St-Pietersnieuwstraat 41, 9000 Gent, Belgium
gianni.allebosch@ugent.be
[2] imec, Kapeldreef 75, 3001 Leuven, Belgium

Abstract. In this paper, a robust approach for detecting foreground objects moving in front of a video screen is presented. The proposed method constructs a background model for every image shown on the screen, assuming these images are known up to an appearance transformation. This transformation is guided by a color mapping function, constructed in the beginning of the sequence. The foreground object is then segmented at runtime by comparing the input from the camera with a color mapped representation of the background image, by analysing both direct color and edge feature differences. The method is tested on challenging sequences, where the background screen displays photo-realistic videos. It is shown that the proposed method is able to produce accurate foreground masks, with obtained F_1-scores ranging from 85.61% to 90.74% on our dataset.

Keywords: Foreground background segmentation
Video screen · Changing background · Color mapping

1 Introduction

Computer guided interactive environments are becoming more and more common. For sports applications e.g. they can be used to provide real time feedback on the execution of certain exercises, which benefits the subject's health or injury recovery. These environments are also found in gaming and industrial applications. As a requirement for most applications, it is desirable to detect the foreground objects (e.g. the people moving around in the scene and the items they carry), e.g. by analysing camera input. The environment can then react to the presence or actions of these objects by modifying what is shown on a screen.

Over the past few years, deep learning has become a mainstream approach to detect or segment foreground objects in video sequences [3,4,6,8] . The general idea behind deep learning is that, when lots of examples are available, an intelligent machine learning system can automatically determine which features are

J. Blanc-Talon et al. (Eds.): ACIVS 2018, LNCS 11182, pp. 175–187, 2018.
https://doi.org/10.1007/978-3-030-01449-0_15

most relevant. Though deep learning based approaches can yield very accurate segmentation results when their networks are properly trained, they do have certain drawbacks. The networks generally require a GPU to be trained and to process input sequences in reasonable time. Furthermore, large interconnected networks are automatically optimised for the available training data, hence a network trained for one specific application can not be directly applied (or modified in a straightforwarded manner) to another.

Another well-known class of algorithms which tackle foreground background segmentation utilize a technique called background subtraction. These methods compare input frames to a model of the background, which is automatically learned from previously captured images. Over the past few years, a plethora of background subtraction based methods has appeared in literature. Modern incarnations, e.g. those described in [2,7,10,11], are able to handle a variety of challenging situations, such as dynamic backgrounds (e.g. waving trees), illumination or weather changes, pan/tilt motions of the camera ... up to a certain degree. However, when the background varies too drastically and unpredictably, these methods struggle to cope with these changes.

One particular case is an indoor environment where the background may consist of video screens on which images are being displayed. At the moment of writing, this specific scenario has attracted little to no interest in foreground detection literature. Many of the issues the aforementioned methods try to handle can be directly avoided, if the following assumptions are made:

(i) Global lighting conditions are not changing throughout the sequence.
(ii) All camera planes and screens are flat and static, such that necessary perspective transformations should remain fixed throughout a given sequence.
(iii) Capturing images from the camera and sending images to screens are all governed by one software process. I.e. it should be possible for the software to compare the input from the camera with what *should* be visible if no person (or object) would be in front of the camera.

The novel segmentation method presented in this paper specifically tackles the problem of detecting foreground objects in front of changing video footage, based on the above assumptions. The main contribution of this paper can be summarised as the combination of efficient color mapping with an informed robust background subtraction method.

Section 2 discusses pixel location transformation between what is projected on the screen to what is recorded by the cameras. Section 3 treats how screen colors are mapped to a captured camera image, by projecting different colored images and analysing how these colors appear to the camera. Section 4 explains how foreground can automatically be detected by comparing a color mapped version of the projected images with the camera input, analysing both color and edge based features. The full algorithm is thoroughly evaluated in Sect. 5.

2   Setup and Scene Calibration

In this paper, the experimental setup is limited to one camera and one projection screen, although the proposed techniques can be straightforwardly extended to multiple cameras and/or screens. A screen may be any (planar) surface on which images are projected or any kind of display. Examples are shown in Fig. 1. Furthermore, the camera and screen images are assumed to be synchronised. Finally, image regions that are not coinciding with the screen can be analysed with any of the foreground detection methods mentioned in Sect. 1, and are thus also omitted from evaluation in this work.

(a)   (b)

Fig. 1. Example lab setups with (a) an LCD screen, (b) a backside illuminated projector screen. The camera's field of view covers the entire screen and there is ample space for people to move around between the camera and the screen in both setups.

Since both the camera and the screen remain static throughout the sequence, determining a pixelwise map between them only once (e.g. at the beginning of a sequence) is sufficient for a given setup. Generally both the camera image plane and the screen are planar. Hence, determining a homography between the image plane and the screen is sufficient to accurately map all camera pixels to their corresponding position on the screen [5].

Let us denote a pixel p_0, coinciding with a certain location on the image that was captured by the camera. This pixel location can be transformed with a homography H, such that the transformed point p corresponds to its location in the image that is sent to the projector or display:

$$p = Hp_0 . \tag{1}$$

H is determined by projecting a predefined checkerboard pattern, such that robust feature points are easy to find (e.g. with the method described in [9]) and match with the checkerboard pattern, as part of the method described in [14]. From these matching points, the 'best fit' homography is determined by using a Direct Linear Transformation [5]. An example can be found in Fig. 2.

Fig. 2. Example of homography estimation with a projected checkerboard pattern. (a) Input image as captured by the camera, with detected pattern denoted, (b) Transformed image.

Fig. 3. Example of the difference between an image that is sent to a projector (a) and that same image as captured by a camera after (backside) projection (b).

For the remainder of this paper, it is assumed that this transformation has been applied to all camera images. Note that transforming pixels in such a fashion generally causes artefacts in the transformed images, which are especially noticeable around image edges. Hence, Lanczos resampling is preferred after applying the homography, due to its desirable properties compared to other resampling techniques [13]. Next, a color map between the camera and the screen can be applied, which will be treated in the following section.

3   Color Mapping

Colors that appear on captured camera images are generally different from those that are sent to the screen (see Fig. 3). This can be attributed to e.g. a limited projector gamut, camera exposure settings, image noise, or external illumination sources in the scene. Explicitly modelling all these potential influences in a given setup is practically an infeasible task. A more suitable approach is to construct a color mapping model that can be estimated directly from a sequence of input images, which is more usable in diverse environments. Note that in theory a model is required for every pixel, since ambient lighting affects different pixel locations in a distinctive manner (e.g. as specular highlights).

An ideal solution would be to project every possible color on the screen once, and then store a value for every combination of red, green and blue (RGB or alternatively another color representation) in a Look-Up-Table (LUT). However, this approach has two main drawbacks. Assume for example all projected color images are represented by 3 bytes (e.g. one for R, G and B) per pixel. Firstly, this means a full LUT would comprise of 256^3 entries per pixel. For a 1 Megapixel image, this would require more than 15 TB of RAM. Secondly, if every color image is projected and captured by the camera once, a total of $256^3 = 16777216$ images need to be captured sequentially, which would take more than 466 hours at a capture rate of 10 frames per second. Hence, this approach would require unrealistic amounts of both memory and time for most applications. In this work, the more realistic approach of obtaining an informed subset of LUT entries is explored. The proposed strategy consists of three parts: pixel dependent contrast stretching, gamma transformation and the creation of a global LUT (one with RGB entries for the entire image).

Firstly, a fully black and a fully white image are projected onto the screen. If the lighting conditions do not change (significantly) throughout the sequence, they represent the darkest and brightest images of the background that can be captured by the camera. Information about the illumination of the scene is also embedded into these images, e.g. in the form of specular highlights. Hence, a pixelwise contrast stretching approach for every pixel can significantly reduce the effect of the scene illumination.

Let $I_{sent,c}(\mathbf{p})$ denote the value for color channel $c \in \{R, G, B\}$ sent to the screen at location $\mathbf{p}$, and $I_{cam,c}(\mathbf{p})$ the corresponding value as captured by the camera (after transformation with H, see Sect. 2). The contrast stretching function that maps the colors of the screen image to the captured image is defined as:

$$I'_{sent,c}(\mathbf{p}) = I_{sent,c}(\mathbf{p}) \frac{I^w_{cam,c}(\mathbf{p}) - I^b_{cam,c}(\mathbf{p})}{255} + I^b_{cam,c}(\mathbf{p}) , \qquad (2)$$

where $I^b_{cam,c}(\mathbf{p})$ and $I^w_{cam,c}(\mathbf{p})$ denote the value for color channel c at $\mathbf{p}$ in the captured black and white images respectively and $\forall c, \mathbf{p} : I_{sent,c}(\mathbf{p}) \in [0, 255]$.

After diminishing the effects of scene illumination, one global LUT can be constructed for the entire image region. Still, 256^3 LUT entries are required. To obtain a close approximation to the desired values in reasonable time, only a subset of the images is projected and the remaining entries in the LUT are filled with trilinearly interpolated values. An ideal solution would yield both high execution speed and interpolation accuracy, but in practise there exists a trade-off between the two. Different subsets are investigated in Sect. 5.

Note that (tri)linear interpolation accuracy is increased when the underlying mapping function behaves more 'linearly'. To this end, before the LUT is created, a gray image where R, G and B are equal to 128, is first projected on the screen. The proposed approach uses this image to determine a gamma transformation (for each color channel individually) to be applied on the contrast stretched images. This image should first be color mapped such that it's average values for R, G and B are equal to those of the captured image. An estimate for the gamma value γ_c for $c \in \{R, G, B\}$ can be obtained as follows:

$$\gamma_c = \frac{\log\left(\frac{\overline{I_{cam,c}^g}}{255}\right)}{\log\frac{128}{255}}, \tag{3}$$

where $\overline{I_{cam,c}^g}$ is the average value of the captured gray image for color channel c.

For the comparison of a captured image with a background image, one can either choose to map the colors of the captured images to the corresponding background image, or the other way around. However, for most applications, it can be assumed that colors of the projected images are mapped to a narrower range than the original one (see Fig. 3). Thus, if the projected (background) image colors are mapped to captured colors, foreground objects of which the colors lie outside this range can very easily be detected.

Note that, for some specific applications, only a limited amount of colors are shown on the screen. In these scenarios it might become feasible to construct a LUT for every color at every pixel location nonetheless. This should yield a very accurate color mapping, leaving only image noise and scene illumination changes as notable sources of error.

4   Foreground Detection

In this section, the proposed foreground detection methodology is discussed. This detection step comprises of two main parts: a color based method and an edge based method, which are then combined through a bitwise AND operation, as was demonstrated in [2].

The color based method is a straightforward approach that can be executed once the color mapping step (see Sect. 3) is complete. Let $I_{cam,c}$ and $I''_{sent,c}$ for $c \in \{R, G, B\}$ be the captured camera image (after applying H) and the projected image (after applying the contrast stretching, gamma transformation and the LUT) respectively. The binary foreground image F_{RGB} is defined as:

$$F_{RGB}(\mathbf{p}) = \begin{cases} 1 & \text{if } \sqrt{\sum_c (I_{cam,c}(\mathbf{p}) - I''_{sent,c}(\mathbf{p}))^2} > T_C \\ 0 & \text{otherwise}. \end{cases} \tag{4}$$

This boils down to a thresholding operation with threshold T_C, based on the L_2 norm in the color space. This method is already capable of detecting foreground by itself. However, small scene illumination changes (e.g. reflections caused by foreground objects) or color mapping errors (e.g. due to the interpolation step) can still introduce notable errors in the foreground masks. Edge features are much less prone to these errors, which is exploited in the edge based method described below. As was originally proposed in [1], foreground edges are derived from a Local Ternary Pattern (LTP) descriptor [12]. For a certain (grayscale) image I, This descriptor is defined as follows:

$$d_s(\mathbf{p}_i) = \begin{cases} 1 & \text{if } I(\mathbf{p}_i) - I(\mathbf{p}) > T_{LTP} \\ -1 & \text{if } I(\mathbf{p}_i) - I(\mathbf{p}) < -T_{LTP} \\ 0 & \text{otherwise} , \end{cases} \tag{5}$$

for $i \in [0, N-1]$, where N represents the number of samples used in the pattern and T_{LTP} a threshold value. These sample points $\mathbf{p}_i$ are selected from a region surrounding the central pixel $\mathbf{p}$ (see Fig. 4). $I(\mathbf{p}_i)$ is the grayscale value for a given sample. When these samples are combined, the descriptor can also be interpreted as a robust edge descriptor, related to the image gradient. Let $\mathbf{e_i}$ be the unit vector pointing from $\mathbf{p}$ towards $\mathbf{p}_i$. The approximated gradient $\mathbf{G}$ can then be found through a sum-of-vectors representation as

$$\mathbf{G}(\mathbf{p}) = \sum_{i=0}^{N-1} d_s(\mathbf{p}_i)\mathbf{e_i} . \tag{6}$$

(a)   (b)

Fig. 4. An example of a Local Ternary Pattern with 8 neighborhood points: (a) Intensity levels in a certain image patch. (b) LTP representation ($T_{LTP} = 3$), with sum-of-vectors representation superimposed.

This LTP based edge descriptor was proven to be very stable, and the resulting foreground masks are generally more robust than standard color based methods, notably in difficult illumination conditions.

In [1] these descriptors are compared to a complex background representation built from multiple background samples. In the proposed application, the background $I''_{sent,c}(\mathbf{p})$ is assumed to be known, thus in this work the edge based foreground mask can be constructed from

$$F_{LTP}(\mathbf{p}) = \begin{cases} 1 & \text{if } \sqrt{\sum_c (G_{cam,c}(\mathbf{p}) - G''_{sent,c}(\mathbf{p}))^2} > T_E \\ 0 & \text{otherwise} , \end{cases} \tag{7}$$

analogously to the color based method described above. $G_{cam,c}$ and $G''_{sent,c}$ are derived from I_{cam} and I''_{sent} respectively (after conversion to grayscale) for

$c \in \{u, v\}$. u an v denote the components of the gradient descriptor along the horizontal and vertical axis respectively and T_E is a threshold. Note that the L_2 norm is preferred, since it is rotationally invariant with regard to the selection of the axes u and v.

The resulting contours are then 'filled' by taking the orthogonal hull of each contour individually, which allows small gaps in the contours. The resulting foreground image is denoted $F_{LTP,\text{filled}}$. One drawback of this method is that the orthogonal hull generally produces foreground silhouettes that are too large when the apparent shapes of the foreground objects contain concavities. However, this enlargement is not present in the color based mask foreground F_{RGB}. Hence, as was first proposed in [2], the edge and color based foreground masks are combined in a bitwise AND operation, yielding a more accurate foreground detection result (see Fig. 5). In the following section, the performance of the color and edge based masks will be validated, both separately and combined.

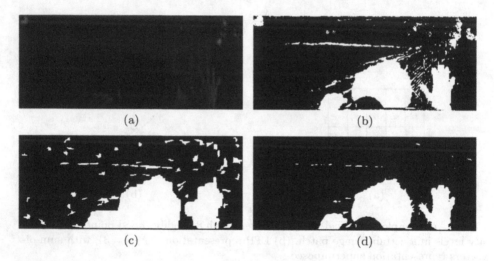

Fig. 5. Combination of different foreground masks. White regions coincide with foreground and black regions with background. (a) input image, (b) color based foreground mask, (c) edge based foreground mask, (d) bitwise AND operation between (b) and (c).

5   Experiments

In this section, the color mapping and foreground detection methods are evaluated in our test setup. Both methods were evaluated on two very different projection objects, one being a LCD screen and the other a backside illuminated projector screen (see Fig. 1).

5.1   Color Mapping

For this part of the evaluation, 4 sequences were recorded. For each screen, one sequence was recorded in a dark environment (where the screen itself was the only illumination source) and another sequence was recorded in a bright environment. No foreground objects were present in these scenarios.

The contrast stretching and gamma transformation are first executed as explained in Sect. 3. Next, a selection of evenly spaced (sub)samples are selected from a 3D space. For each of these samples, an image containing corresponding values for R, G and B is projected on the screen, which is then captured by the camera. Let the spacing between the samples be denoted s_i. The LUT is filled by adding these samples first and than trilinearly interpolating between them at the other entries. The performance of the interpolation method is evaluated by analysing the difference between the observed and interpolated values calculated from the LUT at a second set of samples which are spaced $0.5s_i$ apart. Hence, if the borders (0 and 255) are always included and there are n samples along each dimension in the first subset, there are $2n - 1$ samples along each dimension in the second subset. Except for the coinciding elements, this second subset contains samples that are situated in the middle of the elements in the first subset in 1, 2 or 3 dimensions. Deviations between the observed R, G and B values in I_{cam} and the reconstructed values by the LUT are thus likely close to maximal for a particular interval. Hence, these sample images obtained with step size $0.5s_i$ provide a selection including worst case scenarios for the performance of the LUT. The performance of the color mapping method is evaluated for these images in particular.

Table 1. MSE metrics for the color mapping function, for different scenarios and number of samples n selected along each dimension to construct the LUT, expressed a percentage of the full range (e.g. $[0, 255]$) for R, G and B.

Sequence	n			
	9	5	3	2
LCD screen, dark	0.88	1.20	2.08	3.07
LCD screen, bright	1.06	1.38	2.31	4.05
Backside projection, dark	0.26	0.33	0.55	1.15
Backside projection, bright	0.30	0.38	0.58	1.18

The Mean Squared Error (MSE) between the reconstructed (interpolated) colors and the observed colors for all frames and all pixels was calculated for each of the 4 sequences individually. The results can be found in Table 1. All MSEs are smaller than 1.5% of the full (255 wide) range when $n = 5$ or higher. However it should be noted that the color mapping method performs noticeably better on the backside projection screen. This can be attributed to the narrower range obtained by backside projection compared to an LCD screen, as was clearly noticeable in our experiments. Furthermore, the color mapping is slightly less accurate for bright environments, although this difference is notably less significant.

Note that these mapping results do not take into account effects like blurring of the image or artefacts from the homography transformation, which would be especially noticeable around edges in the image. These effects are not examined explicitly here, but they do implicitly influence the foreground detection results demonstrated in the following subsection.

5.2 Foreground Detection

For evaluating the foreground detection method, 4 sequences with visible foreground were recorded. At the beginning of each sequence, color mapping with $n = 5$ was performed as explained in Sects. 3 and 5.1 . This means 125 images are projected and captured for the LUT. The dataset consists of 2 different movie

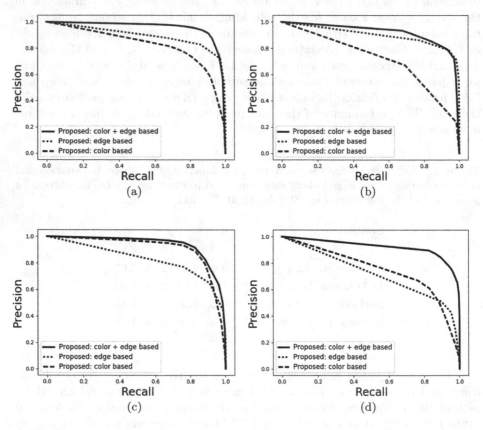

Fig. 6. Precision-recall curves for the foreground detection methods (a) LCD screen, City sequence (b) LCD screen, Havana sequence, (c) Backside projection screen, City sequence (d) Backside projection screen, Havana sequence

trailers shown in the background, each individually projected to both screens in bright circumstances. A test subject was asked to move around in front of the screen, switching between outfits throughout the sequence and carrying different objects. Note that these are very challenging scenarios, since these trailers provide a photo-realistic background. The projected colors are similar to the colors of the test subject, most notably for the LCD screen. This makes the background subtraction problem more difficult.

A total of 54 frames were annotated as ground truth, spread across the 4 sequences. All foreground regions were marked manually. As a means of evaluation, a precision-recall curve was constructed for every sequence, where precision and recall are defined as $\frac{T_P}{T_P+F_P}$ and $\frac{T_P}{T_P+F_N}$ respectively, where T_P, F_P and F_N are the total number of true positive, false positive and false negative pixels in the annotated frames respectively. This process was repeated for the color based, the (filled) edge based and the combined foreground masks per sequence. The results can be found in Fig. 6.

The foreground method that combines the edge and color based foreground masks is overall clearly outperforming the two individual methods. However, there are two sequences where one of the individual methods almost attains the same performance as the combined one. The 'Havana' sequence, when projected on the LCD screen, demonstrates a wide dynamic range of colors over small fragments, causing relatively many false positives in the color based approach. However, the edge based method is capable of dealing with this situation nonetheless. Conversely, the 'City' sequence is much darker. This effect is augmented with the backside illuminated projector screen. Almost no edges are detected in the captured images, and thus the edge based approach has little influence on the end result, as opposed to the color based mask.

From the precision-recall curves, the highest occurring F_1 scores for each sequence can be calculated. This score is defined as the harmonic mean of precision and recall. Hence, high F_1 scores coincide with an accurate method, which also balances false positives with false negatives. It is even argued in [10] that this measure is the best single indicator of the performance of a foreground background segmentation method. For the proposed method which combines edge and color metrics, the highest F_1 scores range between 85.61% to 90.74% on our dataset. The proposed method runs at 2 frames per second for 475 by 200 pixel images on a Core i7 Laptop when written in Python code.

5.3   Future Work

The current foreground method implicitly assumes that noise appears evenly distributed around its central position (e.g. Gaussian). However, this is not necessarily always the case. The construction of more sophisticated noise model can be investigated, which could e.g. lead to a new definition of the used thresholds.

In this work, the efforts were focussed on improving the mapping of images in the RGB color domain. Although the obtained foreground masks are already very robust, color mapping in alternative domains (such as HSV, Lab or YCrCb) is a potentially interesting research step to take in the future.

The proposed method can also be extended to unsynchronised camera/screen combinations. Possibilities include adding a distinctive marker (e.g. a digital watermark) to the projected images, which can be detected at runtime, such that the process can identify the correct image shown at a given point in time.

6   Conclusion

In this paper, an algorithm for detecting foreground objects moving in front of a screen showing changing video footage is proposed. After establishing a homography based pixel transformation between the camera plane and the screen, a color mapping method is used to map what is captured to what is projected. This method analyses different background color images projected in the beginning of a sequence, encompassing contrast stretching, gamma transformation and Look-Up-Table steps. The foreground mask is obtained by comparing color mapped projected images with the captured input images, directly analysing color differences on the one hand and edge based differences on the other. The proposed method is shown to deliver very reliable foreground masks on a self-captured dataset, optimally achieving F_1 scores ranging from 85.61% to 90.74%.

Acknowledgements. The authors acknowledge the financial support from the Flemish Agency for Innovation and Entrepreneurship (Vlaams Agentschap Innoveren en Ondernemen) (imec.ICON project iPlay).

References

1. Allebosch, G., Van Hamme, D., Deboeverie, F., Veelaert, P., Philips, W.: Edge based foreground background estimation with interior/exterior classification. In: Proceedings of the 10th International Conference on Computer Vision Theory and Applications, vol. 3, pp. 369–375. SCITEPRESS (2015)
2. Allebosch, G., Van Hamme, D., Deboeverie, F., Veelaert, P., Philips, W.: C-EFIC: color and edge based foreground background segmentation with interior classification. In: Braz, J., et al. (eds.) VISIGRAPP 2015. CCIS, vol. 598, pp. 433–454. Springer, Cham (2016). https://doi.org/10.1007/978-3-319-29971-6_23
3. Caelles, S., Maninis, K.K., Pont-Tuset, J., Leal-Taixé, L., Cremers, D., Van Gool, L.: One-shot video object segmentation. In: Computer Vision and Pattern Recognition (CVPR) (2017)
4. Cao, Z., Simon, T., Wei, S.E., Sheikh, Y.: Realtime multi-person 2d pose estimation using part affinity fields. In: CVPR (2017)
5. Hartley, R.I., Zisserman, A.: Multiple view geometry in computer vision, 2nd edn. Cambridge University Press, Cambridge (2004) ISBN: 0521540518
6. He, K., Gkioxari, G., Dollr, P., Girshick, R.: Mask R-CNN. In: 2017 IEEE International Conference on Computer Vision (ICCV), pp. 2980–2988, October 2017
7. Hofmann, M., Tiefenbacher, P., Rigoll, G.: Background segmentation with feedback: the pixel-based adaptive segmenter. In: 2012 IEEE Computer Society Conference on Computer Vision and Pattern Recognition Workshops, pp. 38–43, June 2012

8. Redmon, J., Farhadi, A.: Yolo9000: better, faster, stronger. In: 2017 IEEE Conference on Computer Vision and Pattern Recognition (CVPR), pp. 6517–6525, July 2017

9. Shi, J., Tomasi, C.: Good features to track. In: 1994 Proceedings of IEEE Conference on Computer Vision and Pattern Recognition, pp. 593–600, June 1994

10. St-Charles, P.L., Bilodeau, G.A., Bergevin, R.: Subsense: a universal change detection method with local adaptive sensitivity. IEEE Trans. Image Process. **24**(1), 359–373 (2015)

11. St-Charles, P.L., Bilodeau, G.A., Bergevin, R.: A self-adjusting approach to change detection based on background word consensus. In: IEEE Winter Conference on Applications of Computer Vision WACV, pp. 990–997. Waikoloa Beach, Hawaii, January 2015

12. Tan, X., Triggs, B.: Enhanced local texture feature sets for face recognition under difficult lighting conditions. IEEE Trans. Image Process. **19**(6), 1635–1650 (2010)

13. Turkowski, K., Glassner, A.S.: Filters for common resampling tasks. In: Graphics Gems, pp. 147–165. Academic Press, New York (1990)

14. Zhang, Z.: A flexible new technique for camera calibration. IEEE Trans. Pattern Anal. Mach. Intell. **22**(11), 1330–1334 (2000). https://doi.org/10.1109/34.888718

Multi-organ Segmentation of Chest CT Images in Radiation Oncology: Comparison of Standard and Dilated UNet

Umair Javaid[1,2]([envelope]), Damien Dasnoy[1], and John A. Lee[1,2]

[1] ICTEAM, Université Catholique de Louvain, Louvain-la-Neuve, Belgium
{umair.javaid,damien.dasnoy,john.lee}@uclouvain.be
[2] IREC/MIRO, Université Catholique de Louvain, Brussels, Belgium

Abstract. Automatic delineation of organs at risk (OAR) in computed tomography (CT) images is a crucial step for treatment planning in radiation oncology. However, manual delineation of organs is a challenging and time-consuming task subject to inter-observer variabilities. Automatic organ delineation has been relying on non-rigid registrations and atlases. However, lately deep learning appears as a strong competitor with specific architectures dedicated to image segmentation like UNet. In this paper, we first assessed the standard UNet to delineate multiple organs in CT images. Second, we observed the effect of dilated convolutional layers in UNet to better capture the global context from the CT images and effectively learn the anatomy, which results in increased localization of organ delineation. We evaluated the performance of a standard UNet and a *dilated* UNet (with *dilated* convolutional layers) on four chest organs (esophagus, left lung, right lung, and spinal cord) from 29 lung image acquisitions and observed that *dilated* UNet delineates the soft tissues notably esophagus and spinal cord with higher accuracy than the standard UNet. We quantified the segmentation accuracy of both models by computing spatial overlap measures like Dice similarity coefficient, recall & precision, and Hausdorff distance. Compared to the standard UNet, *dilated* UNet yields the best Dice scores for soft organs whereas for lungs, no significant difference in the Dice score was observed: 0.84 ± 0.07 vs 0.71 ± 0.10 for esophagus, 0.99 ± 0.01 vs 0.99 ± 0.01 for left lung, 0.99 ± 0.01 vs 0.99 ± 0.01 for right lung and 0.91 ± 0.05 vs 0.88 ± 0.04 for spinal cord.

Keywords: Multi-organ · Segmentation · Computed tomography

1 Introduction

Radiation oncology treats cancer by irradiating solid tumors with photon or proton beams. Cancerous cells are slightly more sensitive to irradiation than healthy ones and therefore daily targeted irradiation at low dose during several weeks

© Springer Nature Switzerland AG 2018
J. Blanc-Talon et al. (Eds.): ACIVS 2018, LNCS 11182, pp. 188–199, 2018.
https://doi.org/10.1007/978-3-030-01449-0_16

maximizes the probability of local tumor control while minimizing undesired side effects. Medical images are used first as a map to localize the tumor (to be hit) and the organs at risk (to be avoided). Next, images also reveal physical properties of the tissues, which allow simulating deposition of energy by the treatment beams. Most steps of treatment planning are largely automated. However, a noticeable exception is delineation of the tumor and organs at risk (OAR) on the images. It can be seen as a ballistic problem with a trade-off between delivering maximum dosage to the target volume (TV) and minimum dosage to the OAR. Therefore, automatic and accurate delineation of OAR on computed tomography (CT) images is a crucial aspect of treatment planning in radiation oncology. Physicians and radiologists still perform this step manually where OAR and TVs are delineated in a slice-by-slice manner, this requires expertise and is a time-consuming task. Moreover, it is also prone to inter-observer variabilities. Therefore, the need to automate the organ delineation process that can result in efficient and better treatment planning.

Traditionally, automatic organ delineation has relied on non-rigid registrations and atlas-based methods [1–3]. Atlas-based segmentation uses only one or few examples, the atlas(es). Moreover, nonrigid registration considers simplistic deformation models through regularization of the deformation field, in some cases making it difficult to match the atlas with another patient anatomy. Several researchers analyzed hierarchical atlas registration and weighting scheme to generate target specific priors from an atlas database by combining aspects from multi-atlas registration and patch-based segmentation [4]. Others reported increased multi-organ segmentation performance by evaluating organ correlations and prior anatomical knowledge [5]. These approaches need prior information to capture anatomical context for the target organs and therefore, tend to establish anatomical similarities between images from different subjects. Such approaches are mostly handcrafted, therefore limited. They cannot be generalized sufficiently to capture a large number of variabilities of the deformable organs and hence, fail when the regions of interest (OAR or TV) undergo complex physiological deformations.

Recently, deep learning methods have been shown to achieve state-of-the-art for a wider range of tasks in natural image processing such as classification, detection, and segmentation. Deep learning has become prevalent because of its remarkable capabilities to fully learn hierarchical organization of input data thanks to its multi-layered architectures. With convolutional neural networks (CNNs) achieving state-of-the-art performance in medical image segmentation [6], research has been ongoing to achieve better accuracy with higher spatial resolution. Fully convolutional network (FCN) is a variant of CNN that replaces fully connected layers by convolutions [7]. In a typical *supervised* learning setting for image segmentation, a FCN can learn the most suitable data associations between the input images and labels via layers of feature extractors using convolution operations, thus providing a segmentation map per available class in the labels.

In regards to biomedical image segmentation, UNet was introduced. It consists of a contracting path for capturing context and a symmetric expanding path that enables precise localization of target regions in the image [8]. The architecture and the data augmentation of UNet allows efficient learning with superior generalization performance and is one of the most used network for medical image segmentation. Due to its unique architecture, many different approaches have been proposed using the same architectural choice. One study used patch-wise training strategy for bladder segmentation from CT urography in a slice-by-slice manner [9]. Other researchers evaluated several variations of deep ConvNets in the context of hierarchical, coarse-to-fine classification on image patches and regions i.e. superpixels for pancreas segmentation [10]. A third group proposed a hybrid densely connected UNet (H-DenseUNet), which consists of a 2D DenseUNet for efficiently extracting intra-slice features and a 3D counterpart for aggregating volumetric contexts for liver and tumor segmentation [11].

Delineating multiple organs in chest CT images can be formulated as a two-step, challenging task: (1) accurate localization of the organs and (2) delineation of the boundaries between organs and surrounding tissues. Deep learning networks achieve state-of-the-art in object detection and segmentation; their performance is, however, limited in segmenting soft tissues. Soft tissues show high variability in terms of shape, appearance, and relative spatial position with respect to different organs. Therefore they tend to have fuzzy boundaries and potentially same intensity values as the surrounding organs in the CT images that makes it difficult for the network to discriminate.

Even though the networks based upon the notion of UNet are suited for multi-organ segmentation, they are limited to a fixed kernel size i.e. the kernel size of the network does not increase nor decrease during operation. Researchers suggested dilated convolutions that allow multi-scale learning of the context. Dilated convolution preserves the spatial resolution of images i.e. the size of the feature map from one layer to another, which is achieved by increasing the field of view of the kernel [12]. Recently, one study used dilated convolutions to enlarge the network's receptive field size [13] whereas others analyzed à *trous*/dilated convolution for dense prediction tasks [14].

In this paper, we explore the discriminative capabilities of UNet to delineate multiple organs. We also evaluate the effect of dilated convolutions with UNet to capture the global context of the spatial relationship between different organs. We observe that compared to UNet, using dilated convolutions with UNet provides better delineation (*higher localization*) of the soft tissues such as esophagus and spinal cord whereas for lungs, both models performed considerably well and had comparable performance.

The remainder of the paper is organized as follows: Sect. 2 outlines the dataset used and the data pre-processing aspects. We present the different methods and models in Sect. 3. Experimental results are shown in Sect. 4 followed by discussion in Sect. 5. Conclusion and future perspectives are reported in Sect. 6.

2   Image Data and Pre-processing

Dataset. Thirteen patients with histology-proven small cell lung cancer (SCLC, stage 2) were included in this study which is a retrospective use of their data [15]. Each patient was referred for curative intent chemoradiotherapy at UCLouvain University hospital Saint-Luc. Images were acquired at multiple time points during treatment. All scans were acquired on a Gemini TF PET/CT (Philips Medical Systems, Cleveland, OH). Patients were immobilized in treatment position with a thermoplastic thorax mask (Orfit Industries NV, Wijnegem, Belgium).

Each patient had multiple acquisitions where some acquisitions were excluded due to lack of manual annotations. A total of 29 image acquisitions with fully defined TVs and OARs were analyzed, where TVs and OARs were defined by the institutional guidelines and manually delineated by a single experienced radiation oncologist. The chest CT images and their corresponding reference OAR delineations, in particular esophagus, left lung, right lung, and spinal cord were used for this study.

Data Pre-processing. To obtain homogeneous data across all patients, axial slices were selected from the 3D volume along the cranio-caudal direction around lungs followed by geometric re-sampling of the selected slices in terms of uniform pixel spacing (0.97 mm) and slice thickness (2 mm). Image intensity values of all scans were truncated to the range of $[-1000, 3000]$ HU (Hounsfield Unit) to remove possible artifacts in the high HU range. Finally, slices were downsampled from 512×512 to 256×256 and rescaled between 0 and 1 with 8 bits per pixel.

During training, input data was standardized by subtracting mean and dividing by the standard deviation where mean and standard deviation were calculated batch-wise per training iteration. A total of 900 images were considered as input. A test set of 50 slices was created by random selection of the input images, which was only used for evaluation purposes. A 20% validation split was used on the remaining input images, resulting in 680 training and 170 validation images respectively.

Data Augmentation. Data augmentation techniques such as rotation, scaling, and translation were applied to the training images followed by random local elastic deformations. This resulted in local stretch and compression in images [16]. Such data augmentation was applied during training, which provided the network with new transformed examples, hence resulting in more effective learning per epoch. Rotation angle of $10°$ with a shift of 0.1 in height and width was used. Also, the images were zoomed with a factor of 0.01 where the magnitude of distortion for the elastic deformation was drastic i.e., 500.

3   Methods

This study aims at exploiting the discriminative potential of deep learning networks for multiple organ delineation. We first briefly explain about UNet and its use for organ delineation. Then we give an overview about dilated convolutions and their relevance for the delineation.

3.1   Network Architecture and Implementation

We used UNet, which is a multi-scale convolutional network. It consists of a contracting and an expanding path. The *contracting* step extracts features from data where each step of the contracting path consists of two convolutions with a 3 × 3 kernel, each followed by a ReLU activation along with 2 × 2 max-pooling with strides of two in each direction of the domain. On the other hand, the *expanding* path upsamples the feature maps and halves the number of feature channels. Each step in the expanding path uses unpooling, which is concatenated with the features in the contracting path by copy connections (shown in yellow in Fig. 1) during the same step.

The unpooling layers preserve the maximum activation locations which were extracted during the max-pooling step and re-use those locations to place the maximum activations back to their original spatial position. Figure 1 shows the UNet architecture that we used. Images of the size 256 × 256 were used as input whereas the network produces a segmentation mask as output. The output mask is a pixel-by-pixel mask in which each pixel either belongs to one of the organs or the background. For this study, we had five classes: four organs (esophagus, left lung, right lung, and spinal cord) plus the background.

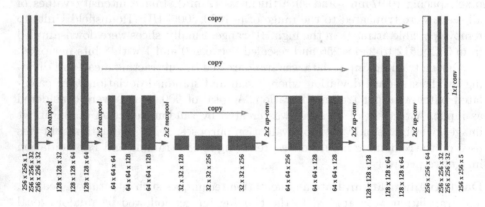

Fig. 1. All convolutions are 3 × 3 kernel size and ReLU as activation function except the last layer that is 1 × 1 and has *softmax* activation, where the output prediction involves five different classes.

3.2   Dilated Convolutions

Dilated convolutions use sparse convolution kernel to enlarge the filters' field of view thus incorporating larger context for dense feature extraction. This is done by deriving a larger filter from the original filter and dilating it by inserting zeros to provide extra context, thereby making it more efficient [17]. For a convolutional $k \times k$ size kernel, the size of the resulting dilated filter is $k_d \times k_d$ where

$k_d = k + (k-1)(r-1)$, r is the dilation rate. Dilation rate introduces spacing between the kernel values. Compared to normal convolution, dilated convolution enables network operation on a coarser scale.

Dilated convolution when used in FCN potentially replaces the max-pooling operation while maintaining the field of view of the corresponding layer [18]. Spatial resolution of the feature map is preserved when the network traverses from one layer to another only by increasing the kernel's field of view. This process is applied iteratively to all layers having downsampling operations, thus the feature map in the output layer can maintain the same resolution as in the input layer. Dilated convolutions have previously been used in various contexts, e.g. signal processing [19], and image segmentation [14].

Precise organ delineation requires knowledge of the spatial relationship between different organs. In the innermost part of UNet, receptive field of the neurons is limited to only a small subset of the image, which limits the information about relative organ positions. We therefore applied the dilated convolutions in the innermost bottleneck of UNet, where we replaced the standard convolutions by dilated convolutions that we call *dilated* UNet. We used dilated convolution rates (2, 4, 8 and 16). Rest of the architecture remains same for the two networks.

3.3 Quantitative Analysis and Evaluation

We use four metrics (three overlap measures and one contour-to-contour distance-based measure) to assess the similarity between predicted delineation per organ resulting from both networks and the ground-truth (manual organ delineation performed by the clinical expert). Dice similarity coefficient (DSC) measures the spatial overlap between ground-truth and predicted segmentations and therefore quantifies the match between two segmentation maps. For two binary masks A and B, DSC is the ratio of their intersection over their average cardinality

$$\text{DSC}(A, B) = \frac{2|A \cap B|}{|A| + |B|}, \tag{1}$$

where $|A|$ counts the number of nonzero pixels in mask A. In the context of binary classification, recall/sensitivity is a measure of how good a classifier is at detecting the positives whereas precision is a measure of how many of the positively classified samples of data are relevant.

$$\text{recall/sensitivity} = \frac{\text{TP}}{\text{TP} + \text{FN}}, \tag{2}$$

$$\text{precision} = \frac{\text{TP}}{\text{TP} + \text{FP}}, \tag{3}$$

where TP, FP and FN stand for True Positive, False Positive, False Negative, and denote the number of pixels relevantly classified, the number of pixels irrelevantly classified, and the number of pixels that are ignored in the ground-truth respectively. The Hausdorff distance (HD) is a measure of maximal error commonly used in clinics to assess contours. It is defined as the maximal distance from a point in the first segmentation (predicted) to a nearest point in manual segmentation (ground-truth). Mathematically, it is given as:

$$HD(A, B) = \max\{\max_{i \in A} \min_{j \in B} d(i, j), \max_{j \in B} \min_{i \in A} d(i, j)\}, \tag{4}$$

where $d(i, j)$ is the Euclidean distance between the centers of nonzero pixels i and j in A and B, respectively. HD is expressed in millimeters (mm). The above-mentioned metrics were computed on the predicted delineation for esophagus, left lung, right lung and spinal cord produced by both networks.

3.4 Network Training

To train the networks, we used the standard pixel-wise unweighted cross-entropy as the loss function, which is given as

$$L(y, \hat{y}) = -\sum_{nc} y_{nc} \log \hat{y}_{nc}, \tag{5}$$

where n refers to pixels and c represents the classes, y_{nc} is the predicted delineation, and $\hat{y}_{nc}$ is the ground-truth delineation.

Networks were trained using the Adam optimizer with learning rate, LR = 0.001 and batch size = 32 for 3.96k iterations. Neither batch normalization nor dropout was used for the simulations. Training each network took almost 3 h where training was done using two Titan X (Pascal) GPUs. Keras [20] was used to implement both networks.

4   Results

We report hereby different results of experiments that were conducted to quantify the performance of both networks in terms of multiple organ delineation. We used spatial overlap measures such as Dice score, precision and recall to evaluate the predicted delineation masks as given in Tables 1 and 2 respectively. It can be seen from the tables that *dilated* UNet yields higher Dice scores for both esophagus and spinal cord, whereas both networks perform almost equally good for delineation of lungs. Bold entries indicate the performance gain between the two networks.

Low recall signifies that the network is not able to identify the whole organ, thus missing some spatial details and resulting in coarse segmentation (*spinal cord with UNet*, Table 1). Low precision means that the network encompasses wrong details as part of the target organ, resulting in fragmented segmentation (*esophagus with UNet*, Table 1). We observe that *dilated* UNet results in higher recall and precision compared to UNet, highlighting more effective learning of the spatial context between organs.

Table 1. Per organ quantitative measures computed on UNet predictions.

Organs	Dice	TPR	FPR	Precision	Recall
Esophagus	**0.71 ± 0.10**	0.78 ± 0.13	0.32 ± 0.15	**0.68 ± 0.15**	**0.78 ± 0.13**
Left lung	0.99 ± 0.01	0.99 ± 0.01	0.02 ± 0.01	0.98 ± 0.01	0.99 ± 0.01
Right lung	0.99 ± 0.01	0.99 ± 0.01	0.02 ± 0.01	0.98 ± 0.01	0.99 ± 0.01
Spinal cord	**0.88 ± 0.04**	0.86 ± 0.09	0.09 ± 0.07	**0.91 ± 0.07**	**0.86 ± 0.09**

Table 2. Per organ quantitative measures computed on *dilated* UNet predictions.

Organs	Dice	TPR	FPR	Precision	Recall
Esophagus	**0.84 ± 0.07**	0.82 ± 0.09	0.15 ± 0.09	**0.85 ± 0.09**	**0.82 ± 0.09**
Left lung	0.99 ± 0.01	0.99 ± 0.00	0.02 ± 0.01	0.98 ± 0.01	0.99 ± 0.00
Right lung	0.99 ± 0.01	0.99 ± 0.00	0.02 ± 0.01	0.98 ± 0.01	0.99 ± 0.00
Spinal cord	**0.91 ± 0.05**	0.92 ± 0.06	0.10 ± 0.09	**0.90 ± 0.09**	**0.92 ± 0.06**

Figure 2 shows the box plots of Dice score per organ evaluated for the predictions from both networks. It can be observed that *dilated* UNet offers higher Dice score than UNet especially for the esophagus and spinal cord.

We checked the Hausdorff distance; (HD) to compute the distance between the predicted delineation and ground-truth. We report the results in Fig. 3. It can be seen that *dilated* UNet offers a significantly better contour consistency compared to standard UNet. As per *dilated* UNet delineation, both lungs lie well below 5 mm whereas esophagus and spinal cord lie below 4 and 3 mm respectively. HD can be considered as finding the worst-case scenario in the delineation. As organs undergo anatomical variations, the mentioned numbers (HD distances per organ) are practical in clinical applications.

Finally, we report the predicted delineations from both networks in Fig. 4. Each image is plotted against the ground-truth delineations where light colors correspond to ground-truth and dark colors correspond to the predicted delineations. It can be observed that both networks capture the context and delineate all the target organs. However, UNet cannot precisely localize the organ boundaries especially for esophagus and spinal cord (low recall and precision scenario). On the other hand, *dilated* UNet incorporates better spatial context and is therefore able to localize the target organs with higher accuracy than UNet.

5   Discussion

Automatic delineation of multiple organs on CT images plays an important role in treatment planning in radiation oncology. As cancerous cells (target volumes - TVs) are treated by irradiation, the purpose of radiotherapy is to provide maximum dosage to the TV and minimum dosage to the OAR. Delineation helps medical staff in the diagnostic process by providing the contours for OAR

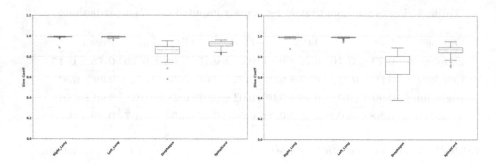

Fig. 2. Box plots showing dice score per organ. Left image shows box plot for dilated UNet whereas right image shows box plot for UNet. It can be seen that there is no significant difference in the Dice score for the lungs however, for esophagus and spinal cord dilated UNet performs better than UNet.

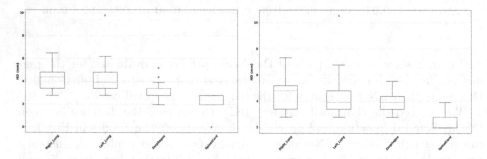

Fig. 3. Box plots showing Hausdorff distance in mm per organ. Left image shows box plot for dilated UNet whereas right image shows box plot for UNet. It can be seen that dilated UNet has lower Hausdorff distance per organ than UNet.

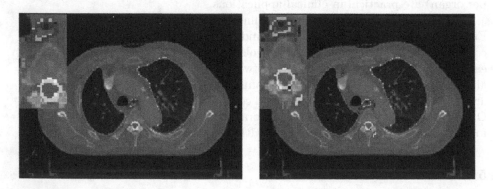

Fig. 4. Organ delineation (Ground-truth, GT vs Predicted Mask, PM), light colors indicate GT whereas dark colors indicate PM. Left image shows the segmentation by dilated UNet whereas right image shows the segmentation by UNet. It can be seen that dilated UNet offers increased localization of organ boundaries (best viewed in color). (Color figure online)

and TV. In this paper, we try to automate the multi-organ delineation step by exploring the discriminative learning capabilities of deep learning networks. In particular, first we use one of the most commonly used network i.e. UNet for delineation and later, we modify the same network by replacing the standard convolutions with dilated convolutions and observe the increased performance in terms of delineation by the modified network that we call *dilated* UNet.

Information about the spatial relationship between organs is needed for accurate delineation of organs. Downsampling operation is a common practice to reduce the spatial frequency information of image to provide a wider receptive field to the neurons. However, downsampling operation is costly as it results in additional network parameters. In this study, instead of using additional downsampling operations, we used dilated convolutions in the innermost bottleneck of UNet to increase the receptive fields of the network. Dilated convolutions allow an exponential growth of receptive field with a linear increase in the number of parameters. We used dilation rates of 2, 4, 8 and 16 to incorporate more spatial context due to wider kernel sizes of the convolutions. We observed increased localization of the organs by *dilated* UNet, thus highlighting the property of dilated convolutions to capture the spatial relationship between organs.

Medical practitioners usually perform manual delineation that is used as a ground-truth for data analysis. There is a risk that the manual annotation might be done in a hurry or lack accuracy. For example, a physician might find some organs less relevant for the radiation treatment planning, therefore such organs might be roughly or not delineated at all. Interpolation of contours across slices also occurs frequently in manual delineation. Therefore, in order to benchmark the deep learning performance versus human-level accuracy, we need an accurately annotated ground-truth that offers effective learning of the anatomical changes and hence, help the doctors to devise better and more precise treatment planning. Relying on retrospective images taken from clinical routine might be suboptimal, compared to images that could be prospectively collected and delineated specifically. While this has been feasible for atlases with only one or few images, the cost in time and money for deep learning, where many examples are necessary, might turn out much too high.

Having a valid ground-truth, we also need to have a standardized data augmentation. As annotated medical data is limited and deep learning networks are data-demanding, data augmentation techniques are a common practice to increase the training dataset. We need to introduce a controlled amount of variability that is consistent with actual variations. For example, flipping a subject does not make sense as the anatomy is not symmetrical and it can degrade the network learning.

There should also be a way to include inter-user variability into the deep learning networks. It is a usual practice to augment the images but we should also try to augment the contours. By doing so, we will potentially end up with multiple contours per organ and hence, multiple ground-truths, which could be a surrogate for the actual inter-observer variabilities.

6   Conclusion and Future Perspectives

In this paper, we assessed UNet for multiple organ delineation in chest CT images. We also explored the effect of dilated convolutional layers in UNet (*dilated* UNet) instead of standard convolutions and observed an increased performance in soft tissue delineation, notably the esophagus and spinal cord. We evaluated an alternative to deep networks as the common notion indicates that for better learning, the network should be deep enough. Here, we benchmark the performance of UNet without increasing its depth. We observed that using dilated convolutions with UNet resulted in improved localization for multiple organs delineation, especially for esophagus and spinal cord.

Having observed the effectiveness of *dilated* UNet to incorporate better global context in the images, we aim to extend our study to 3D and to higher number of organs. We also intend to try our *dilated* UNet on other datasets such as pelvic region and/or head and neck delineation.

Acknowledgments. Umair Javaid is funded by Fonds de la Recherche Scientifique - FNRS, Télévie grant no. 7.4625.16. Damien Dasnoy is a Research Fellow of FNRS. John A. Lee is a Senior Research Associate with the Belgian FNRS. We thank UCLouvain University hospital Saint-Luc for providing the data. We also thank NVIDIA Corporation for providing Titan X (Pascal) GPUs.

References

1. Zhang, T., Chi, Y., Elisa, M., Di, Y.: Automatic delineation of on-line head-and-neck computed tomography images: toward on-line adaptive radiotherapy. Int. J. Radiat. Oncol. Biol. Phys. **68**(2), 522–530 (2007)
2. Gorthi, S., et al.: Segmentation of head and neck lymph node regions for radiotherapy planning using active contour-based atlas registration. IEEE J. Sel. Top. Signal Process. **3**(1), 135–147 (2009)
3. Dolz, J., et al.: Interactive contour delineation of organs at risk in radiotherapy: clinical evaluation on NSCLC patients. Med. Phys. **43**(5), 2569–2580 (2016)
4. Wolz, R., Chu, C., Misawa, K., Fujiwara, M., Mori, K., Rueckert, D.: Automated abdominal multi-organ segmentation with subject-specific atlas generation. IEEE Trans. Med. Imaging **32**(9), 1723–1730 (2013)
5. Okada, T., Linguraru, M.G., Hori, M., Summers, R.M., Tomiyama, N., Sato, Y.: Abdominal multi-organ segmentation from CT images using conditional shape-location and unsupervised intensity priors. Med. Image Anal. **26**(1), 1–18 (2015)
6. Litjens, G., et al.: A survey on deep learning in medical image analysis. Med. Image Anal. **42**, 60–88 (2017)
7. Long, J., Shelhamer, E., Darrell, T.: Fully convolutional networks for semantic segmentation. In: Proceedings of the IEEE Conference on Computer Vision and Pattern Recognition, pp. 3431–3440 (2015)
8. Ronneberger, O., Fischer, P., Brox, T.: U-net: convolutional networks for biomedical image segmentation. In: Navab, N., Hornegger, J., Wells, W.M., Frangi, A.F. (eds.) MICCAI 2015. LNCS, vol. 9351, pp. 234–241. Springer, Cham (2015). https://doi.org/10.1007/978-3-319-24574-4_28

9. Cha, K.H., Hadjiiski, L., Samala, R.K., Chan, H.-P., Caoili, E.M., Cohan, R.H.: Urinary bladder segmentation in CT urography using deep-learning convolutional neural network and level sets. Med. Phys. **43**(4), 1882–1896 (2016)
10. Roth, H.R., et al.: DeepOrgan: multi-level deep convolutional networks for automated pancreas segmentation. In: Navab, N., Hornegger, J., Wells, W.M., Frangi, A.F. (eds.) MICCAI 2015. LNCS, vol. 9349, pp. 556–564. Springer, Cham (2015). https://doi.org/10.1007/978-3-319-24553-9_68
11. Li, X., et al.: H-DenseUNet: hybrid densely connected UNet for liver and liver tumor segmentation from CT volumes. arXiv preprint arXiv:1709.07330 (2017)
12. Yu, F., Koltun, V.: Multi-scale context aggregation by dilated convolutions. arXiv preprint arXiv:1511.07122 (2015)
13. Gibson, E., et al.: Towards image-guided pancreas and biliary endoscopy: automatic multi-organ segmentation on abdominal CT with dense dilated networks. In: Descoteaux, M., Maier-Hein, L., Franz, A., Jannin, P., Collins, D.L., Duchesne, S. (eds.) MICCAI 2017. LNCS, vol. 10433, pp. 728–736. Springer, Cham (2017). https://doi.org/10.1007/978-3-319-66182-7_83
14. Chen, L.-C., George, P., Kokkinos, I., Murphy, K., Yuille, A.L.: Deeplab: semantic image segmentation with deep convolutional nets, atrous convolution, and fully connected CRFs. IEEE Trans. Pattern Anal. Mach. Intell. **40**(4), 834–848 (2018)
15. Di Perri, D., et al.: Evolution of [18f] fluorodeoxyglucose and [18f] fluoroazomycin arabinoside PET uptake distributions in lung tumours during radiation therapy. Acta Oncol. **56**(4), 516–524 (2017)
16. Simard, P.Y., Steinkraus, D., Platt, J.C., et al.: Best practices for convolutional neural networks applied to visual document analysis. ICDAR **3**, 958–962 (2003)
17. Dutilleux, P.: An implementation of the "algorithme à trous" to compute the wavelet transform. In: Combes, J.M, Grossmann, A., Tchamitchian, P. (eds.) Wavelets, pp. 298–304. Springer, Heidelberg (1990). https://doi.org/10.1007/978-3-642-75988-8_29
18. Wang, P., et al.: Understanding convolution for semantic segmentation. arXiv preprint arXiv:1702.08502 (2017)
19. Van Den Oord, A., et al.: Wavenet: a generative model for raw audio. arXiv preprint arXiv:1609.03499 (2016)
20. Chollet, F., et al.: Keras (2015)

Diffuse Low Grade Glioma NMR Assessment for Better Intra-operative Targeting Using Fuzzy Logic

Mathieu Naudin[1,2,3]($\boxtimes$), Benoit Tremblais[1], Carole Guillevin[2],
Rémy Guillevin[2], and Christine Fernandez-Maloigne[1]

[1] ICONES, ASALI, XLIM Laboratory, UMR CNRS 7252, University of Poitiers,
Poitiers, France
mathieu.naudin@univ-poitiers.fr
[2] DACTIM-MIS, LMA, UMR CNRS 7348, University and Hospital of Poitiers,
Poitiers, France
[3] Siemens Healthcare SAS, Saint Denis, France

Abstract. Nowadays, billion images are made each year to discover or follow brain pathologies. The main goal of our tool is to offer an enhanced view of each diffused low grade glioma by using a fuzzy logic targeting. Using this method, we are trying to reproduce the neuroradiologist process to have a better understanding of the images and a better tumor targeting. We can use this method with only one multi-parametric RMN acquisition (anatomical, diffusion, perfusion, spectroscopy). For the diffuse low grade glioma (DLGG), it helps to deal with uncertain bounds of the tumor and helps to target isolated infiltrated tumorous cells. It will help surgeons in their decision process in case of supra-total resection. As results, we obtain color maps which show different parts of the tumor: the main core and spread cells.

1 Introduction

Using the Ultra High Field (UHF) MR scanner, it is now possible to obtain a better understanding of the brain cerebral pathology. So, the MRI became a gold standard for brain studies especially for diagnostic goals. In this paper, we aim to develop a new way to target diffuse low grade gliomas (DLGG) using the pre-operative NMR assessment. Glioma is a brain pathology originating from the support tissue of the brain: the glia. The glioma grading is described within the WHO (World Health Organization) classification from 2007 [8] updated in 2016 [9]. Four classes are currently existing from WHO grade I to WHO grade IV. They are ranked by physiological and histopathological markers. From the French brain tumor databases [21], all gliomas represent 39.2 % of all brain tumors. Gliomas are the most important category from the 57,816 cases referenced in this database. The DLGG (WHO grade I-II) which is a sub-category represents 13% to 16% of all gliomas (Fig. 1). From several brain tumor databases [13,16], the population rate could be estimated at 1 for

© Springer Nature Switzerland AG 2018
J. Blanc-Talon et al. (Eds.): ACIVS 2018, LNCS 11182, pp. 200–210, 2018.
https://doi.org/10.1007/978-3-030-01449-0_17

100,000 person years and a median diagnosis age around 43 years old. Composed by several types (astrocytoma, oligodendroglioma or mixed types), the DLGG are growing relatively fast (approx. 4 mm/year) before and after surgery. When possible the surgeon can remove the tumor and sometimes he can extract more than the tumor itself to remove also some tumorous cells spread around the active core. This method is called supra-total resection. To help surgeons and radiologists, we focus on targeting the tumor and specially the active core of the tumor.

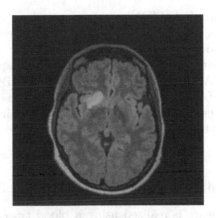

Fig. 1. Example of a low grade glioma (in white, hypersignal) in the insula region of the brain, MRI with a T2 FLAIR sequence

2   Materials and Methods

2.1   Acquisitions

Image acquisitions are planned on a 3T Siemens Magnetom Skyra (Siemens Healthineers, Erlangen, Germany). The machine is already set up for multi-nucleus purposes. A very complete and specific protocol was designed for initial gliomas assessment regarding the international recommendations. The protocol is designed to cover all aspect and estimate as close as possible the tumor grading. All data are acquired with a 64 channels proton coil. The tumor assessment is composed of several parts.

Anatomical

- T2 FLAIR (sagittal, TE = 402.0 ms, TR = 5000,0 ms, $0.5 \times 0.5 \times 0.6$ mm^3)
- T2 SWI (axial, TR = 27.0 ms, TE = 20,0 ms, $0.9 \times 0.9 \times 2.5$ mm^3)
- T1 3D MPRAGE GADO (sagittal, TR = 1900.0 ms, TE = 2,93 ms, $0.5 \times 0.5 \times 1.0$ mm^3).

Anatomical sequences are the standard sequences in MRI. It helps radiologists to visually control tumorous extension. Gadolinium T1 is used to control neo-angiogenesis of the brain. SWI images shows if hemorrhagic phenomena are present. It could help to show the brain blood barrier integrity.

Diffusion

- DTI (axial, 64 dir, TR = 5500.0 ms, TE = 106.0 ms, 2 diffusion ponderations, b0 = 0 s/mm, b1 = 1000 s/mm, $2.0 \times 2.0 \times 2.0$ mm^3).

Diffusion Tensor Imaging (DTI) is a sequence where the water movements are studying. The water diffusion is non-isotropic in different regions of the brain which can give oriented diffusion. This oriented diffusion is used to reconstruct track fibers in the brain and create some parametric maps such as Fractional Anisotropy (FA) and Apparent Diffusion Coefficient (ADC) on which we can respectively bundles of fibers and the diffusion of the free water of the brain. *b*1000 images are also used. This sequence is ponderated in diffusion and shows constrained diffusion when a hypersignal is found.

Perfusion

- Perfusion T2* (axial, TE = 30.0 ms, TR = 1980.0 ms, $1.7 \times 1.7 \times 4.0$ mm^3).

The perfusion sequence is realized using a blood tracer called Gadolinium. High-speed venous injection lead to a bolus effect in the brain. The product diffuses in the whole brain by the blood system. Once the tracer injected, high-speed acquisition is made to follow it through each voxel and finally obtain a curve. Using Arterial Input function (AIF) and deconvolution function, several parameters are extracted from the curve to construct maps. The maps extracted from this sequence are: Cerebral Blood Flow (CBF), Cerebral Blood Volume (CBV), Mean Transit Time (MTT), Time To Peak (TTP).

Metabolic

- Chemical Shift Imaging (CSI) SLASER (Nex = 3, TE = 35.0 and 135.0 ms, TR = 1500.0 ms, $13.8 \times 13.8 \times 20.0$ mm^3, 8 saturation regions)
- Single Voxel Spectroscopy (SVS) PRESS (Nex = 156, TE = 35.0 and 135.0 ms, TR = 1700.0 ms, $15.0 \times 15.0 \times 15.0$ mm^3).

We are tracking the several metabolites as markers of cells function:

- N-Acetyl-Aspartate: neurons health marker
- Choline: membrane marker
- Creatine: cell energy marker
- Lactates: marker anaerobic metabolism of the cell. Metabolite preferred by the glioma as energy.
- Free lipids: marker of cell membrane residue.

2.2   Data Post-processing

Once the acquisition made, data are algorithmically sorted then transformed into NIFTI-1 files [2] using dcm2nii [15] and renamed to be automatically detected by the data pipeline. The full pipeline was developed in language Python 3.5 and launch automatically each process: images sorting, conversion from DICOM to NIFTI (.nii), brain axis computation, parametric maps computation and risks maps evaluation.

To compare images between them, a strong post processing step is needed. Each acquisition which is related to a parameter is registered on the post-gado high resolution 3D T1 MPRAGE.

The T1 MPRAGE is skull striped using FSL-BET [7]. Once this step over, we perform a symmetry analysis of the brain to determine the principal axis. The method will be developed in the parametric map calculation part. The step is important because it established a comparison basis between the two hemispheres.

The T2 3D FLAIR is registered on the T1 and then re-sliced using SimpleITK library [10]. The rigid registration method is based on mutual information [11] with a Powell optimizer. In this case, the rigid registration was used because any pressure modification or gravity modifications will occur during the scan time. Only small head shifting could be noticed easily registered with rigid method. For the low-resolution images, the re-slicing is made using a linear interpolation to generate new slices. This step is important for image comparison.

All images are now with the same voxel size and displayed with a binary mask to represent only the brain. Compared to high resolution imaging, perfusion data are very low resolution to favor time resolution over spatial resolution. Each repetition gives a full head volume each 1.7 s. In return, the resolution is only 64×64 compared to the 256×256 for HR images.

Perfusion Data. The perfusion maps are automatically obtained from the Siemens build-in system (Syngo VE 11C, Siemens Healthineers, Erlangen, Germany). These maps are computed from a Dynamic Susceptibility Contrast (DSC) method using Gadolinium as contrast product.

Data are clean-up using mathematical morphology (geodesic reconstruction) to be registered with less noise than acquired images.

The brain mask is applied on registered and re-sliced images. It eliminates noise generated by superficial blood supply.

Diffusion Data. The DTI sequence is used to compute FA map and color FA map. To post-process the whole set of diffusion data, the Dipy library [5] is used.

Several parameters are extracted from this calculation. They represent the diffusion of the free water in the brain. Data are prepared with filtering and binarization to eliminate artefacts. Then, we used Constrained Spherical Deconvolution method to estimate the Orientation Distribution Function (ODF). To finish, fibers tracks estimation is made and FA is computed.

Spectroscopy 1H. For the spectroscopy, two different sequences could be used in glioma detection or follow-up: SVS or CSI (Single Voxel Spectroscopy and Chemical Shift Imaging). The acquisition VOI (Volume of Interest) is replaced within the brain on anatomical sequence using the information contained into the *.rda* Siemens file system. Then, data are analyzed using JMRUI software [17]. AMARES build-in model is used to quantify metabolites concentrations and ratios. In case of the CSI, several voxels will be used to validate the results map. If the SVS was used, the only spectrum will be used as validation here.

2.3   Parametric Map Calculation

The main purpose is that the brain is basically almost symmetrical in terms of structures and tissues density. In case of tumor, the local structure and the intensity of the observed pixel are modified. Once the registration step completed, a Python framework was developed to automatically compute these parametric maps. They are calculated regarding the neuroradiologist method. First the method is computing the brain axis with manual correction if needed. Then, it projects points over the found axis to compute final maps.

Brain Axis Computation. The brain axis is computed using a 2D Principal Component Analysis (PCA) from the scikit-learn library [14]. In order to get a robust information support, the center slice of the T1 is denoized and binarized, using Otsu thresholding from the scikit-image library [18]. Using the first two eigenvectors of the PCA, we can draw two axes on our data (Fig. 2). From that, a manual correction could be performed in case of wrong result using a dedicated tool. Then, the axis is applied on each slice of the volume.

Projection over Brain Axis. In order to perform a symmetry analysis of the brain, an iteration is made over each hemisphere of the projection. This method is directly inspired from the neuro-radiologists themselves. They always analyze brain images by taking into account the opposite hemisphere in order to find abnormalities.

Once the main axis found, a voxel centered patch is extracted. To qualify the area, mean value of a 7×7 patch (axial plane) is used to take into account the neighborhood of pixels studied. In order to avoid side effects, only patches with half of the voxels in the mask are computed. If it is not the case, the patch is automatically noted as low quality and excluded.

This method could sometimes lead to under or upper-estimation of parameters due to slice effects or in special anatomical area such as insula (Fig. 3) or Willis polygon.

Results Maps. The resulting maps are describing the non-uniformity between the two hemispheres (Fig. 4). It is represented by a color map for each calculation (mean, variance) and each modality. In the first raw, the original acquisition images without any modifications, cleaning or filtering. In the second raw, maps computed from patch mean values and in the third row the corresponding variance maps.

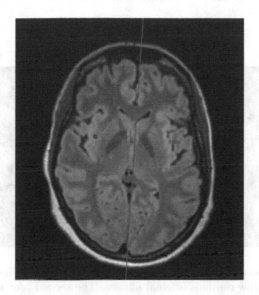

Fig. 2. Calculation of the brain axis (in red) in the middle of the volume, especially in the insula region. It is computed automatically and may be adjusted by hand to fit the anatomy as accurate as possible. In blue, the original point and its patch. In green, the contralateral patch projected over the axis. (Color figure online)

Fig. 3. Projection of the left region from the left hemisphere to the contralateral brain hemisphere

2.4   Fuzzy Logic Rules

In order to determine the classification criteria, we collect some rules to apply them in a fuzzy logic environment. Fuzzy sets are more and more present especially in the medical field [1,12]. Initially theorized by Zadeh *et al.* [20], the fuzzy logic is an extension of the boolean logic. We always want to deal with uncertainty in biomedical engineering. It helps to deal with uncertainty. By giving a set of rules, it is possible to give flexibility to a system or a model. In fuzzy sub-sets, membership functions could be defined from a boolean set to an interval. We can understand medical natural speaking to model it and get values from to compute risks maps. In order to compute the fuzzy sets we used the scikit-fuzzy library [19].

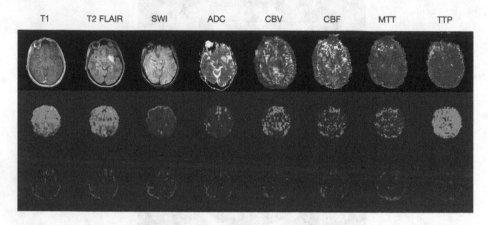

Fig. 4. Parameters extracted from acquisition and computed as maps of similarity in the brain. The first row contains original images. The second row, results mean maps and the third row contains results variance maps.

Anatomical Rules. Regarding the anatomical images, T1 Post gadolinium and T2 FLAIR are used to be highly symmetrical in terms of structure. Another very important rule is related to the image structure. Higher levels of gray, called high intensity in T2 FLAIR, correspond to a signal abnormality which could be a tumor sign. On the T1 post gadolinium, only blood vessels will be colored appears in high intensity due to the gadolinium tracer. The lesion enhancing is not taken into account here because in this case the glioma has already degenerated and reached the WHO Grade III or IV. The following rules are applied:

- **IF** 'Cbv' is 'good' **THEN** 'tumor risk' is 'low'
- **IF** 'Cbf' is 'high' **THEN** 'tumor risk' is 'low'
- **IF** 'Cbv' is 'low' **THEN** 'tumor risk' is 'low'
- **IF** 'Cbf' is 'low' **THEN** 'tumor risk' is 'low'
- **IF** 'Cbv' is 'low' **AND** 'Cbf' is 'low' **THEN** 'tumor risk' is 'high'.

Diffusion Rules. In case of brain tissue alteration and glioma growing, the ADC parameter is decreased in the region of the glioma. It varies in the same way for the Fractional Anisotropy. The glioma will replace fibers by altered cells which appear darker in the FA map.

- **IF** 'FA' is 'high' **THEN** 'tumor risk' is 'low'
- **IF** 'ADC' is 'high' **THEN** 'tumor risk' is 'low'
- **IF** 'B1000' is 'high' **THEN** 'tumor risk' is 'high'
- **IF** 'ADC' is 'high' **AND** 'FA' is 'poor' **THEN** 'tumor risk' is 'high'.

Perfusion Rules. The perfusion rules are highly correlated with the neo-angiogenesis of the glioma. In the first step of the disease, gliomas are hypo perfused. Then, the anaplastic transformation occurs and the neo-angiogenesis starts. As results, the perfusion ratio increases to reach 2.7 as the limit for grading from WHO II grade to WHO III grade. In case for WHO II-I grade, it is most of the time a hypoperfusion of the area.

- **IF** 'Swi' is 'low' **THEN** 'tumor risk' is 'low'
- **IF** 'T2FLAIR' is 'high' **THEN** 'tumor risk' is 'high'
- **IF** 'T1' is 'high' **THEN** 'tumor risk' is 'low'
- **IF** 'T2FLAIR' is 'average' **AND** 'T1' is 'average' **THEN** 'tumor risk' is 'low'
- **IF** 'T2FLAIR' is 'high' **AND** 'T1' is 'average' **THEN** 'tumor risk' is 'high'
- **IF** 'T2FLAIR' is 'high' **AND** 'T1' is 'average' **THEN** 'tumor risk' is 'high'.

2.5 Risks Maps

As final results, we compute risk maps from the three previous maps. First, maps from anatomical rules are computed to give strong tumor boundaries thanks to the T2 sequences. In order to add more information, diffusion map is computed using the dedicated fuzzy logic rules. To finish, perfusion maps are used to verify perfusion of the local area of the tumor. These maps are multiplied between them to obtain a final result (Fig. 5).

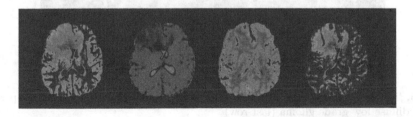

Fig. 5. Result with the contribution of each category (anatomical, diffusion, perfusion) in final result (right)

3 Results

As results, we compute maps from the previous constructed rule system. The defuzzification gives us a value between 0 and 100 for each voxel of the volume. It can be seen as a critical value of the lesion. The closer to 100 the value is, the higher is the danger. We also want to distinguish the different parts of the pathology itself. It can be separated into two different classes [4]. The first one is composed by attached tumor cells to build the main core of the tumor. The other part case is the lonely tumorous cells spread in the adjacent parenchyma.

In case of low diffusion of the glioma, the hypersignal is not very extended (Fig. 6). Then, using our method the DLLG is found and its limits are better defined compared to the simple T2 sequence.

For the diffuse low grade glioma, the glioma is diffuse within the parenchyma. So, we expect a very risky core and an area where the glioma is currently invading new structures.

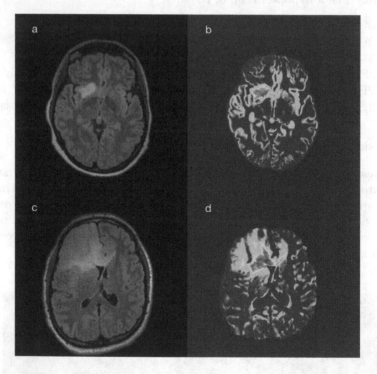

Fig. 6. T2 FLAIR (left) and the Risk map (right) for a low-grade glioma (top row) and a diffuse low grade glioma (last row)

3.1   Results Validation

To validate our new risk maps, we choose to use another NMR sequence type: the spectroscopy. In fact, the spectroscopy allows us to see the concentration of several molecules of the brain. These metabolites bring several markers of brain health. Diamandis et al. [3] show the robustness of such methods in diagnosis environment. In our results, most of the time the SVS is placed in temporal core and the CSI is used to have an overview of the area. Each time, the voxel within the tumor is showing a compatible spectrum with the literature [6] for DLGG core: high levels of choline, low levels of NAA. In the edema, a rise of choline integral value is compatible with a diffuse infiltration of tumor cells and it corresponds to a compatible value of our risk map.

4   Conclusion

To conclude, the fuzzy logic is a great tool to target the low grade glioma and its infiltration. From several MR sequences, we developed a pipeline to sort, converse and post-process brain images. The risk map It allows us to better understand the growth and the impact on brain structures.

This could also be used as help to MRI technicians. Instead of replacing the sequence approximately, he could know used to map to target core and infiltration of the DLGG. The patient will be now following based on the multi-parametric tumor's evolution and anymore not only from the anatomical data.

To better understand the temporal process, we will have to optimize the pipeline by making it fully automatic and more robust to MR artifacts. Then, extends it to 3D acquisitions to process higher volume of data.

References

1. Senén, B., Roque, M.: Fuzzy logic in medicine. Physica **83** (2013)
2. Cox, R.W., et al.: A (sort of) new image data format standard: Nifti-1. Neuroimage **22**, e1440 (2004)
3. Diamandis, E., et al.: MR-spectroscopic imaging of glial tumors in the spotlight of the 2016 WHO classification. J. Neuro-Oncol. **48**(Suppl 4), 1532–10 (2018)
4. Figarella-Branger, D., et al.: Classification histologique et moléculaire des gliomes. Rev. Neurol. **164**(6–7), 505–515 (2008)
5. Garyfallidis, E., et al.: Dipy, a library for the analysis of diffusion MRI data. Front. Neuroinf. **8**, 8 (2014)
6. Guillevin, R., et al.: Proton magnetic resonance spectroscopy predicts proliferative activity in diffuse low-grade gliomas. J. Neuro-Oncol. **87**(2), 181–187 (2008)
7. Jenkinson, M., Beckmann, C.F., Behrens, T.E.J., Woolrich, M.W., Smith, S.M.: Fsl. Neuroimage **62**(2), 782–790 (2012)
8. Louis, D.N., et al.: The 2007 WHO classification of tumours of the central nervous system. Acta Neuropathol. **114**(2), 97–109 (2007)
9. Louis, D.N., et al.: The 2016 world health organization classification of tumors of the central nervous system: a summary. Acta Neuropathol. **131**(6), 803–820 (2016)

10. Lowekamp, B.C., Chen, D.T., Ibáñez, L., Blezek, D.: The design of SimpleITK. Front. Neuroinf. **7**, 45 (2013)
11. Maes, F., Collignon, A., Vandermeulen, D., Marchal, G., Suetens, P.: Multimodality image registration by maximization of mutual information. IEEE Trans. Med. Imaging **16**(2), 187–198 (1997)
12. Mahfouf, M., Abbod, M.F., Linkens, D.A.: A survey of fuzzy logic monitoring and control utilisation in medicine. Artif. Intell. Med. **21**(1–3), 27–42 (2001)
13. Ostrom, Q.T., et al.: Cbtrus statistical report: primary brain and central nervous system tumors diagnosed in the united states in 2008–2012. Neuro-oncology **17**(suppl_4), iv1–iv62 (2015)
14. Pedregosa, F., et al.: Scikit-learn: machine learning in Python. J. Mach. Learn. Res. **12**(Oct), 2825–2830 (2011)
15. Rorden, C.: Dcm2nii dicom 2 nifti conversion (2007)
16. Shibui, S.: Report of brain tumor registry of Japan (2001–2004). Neurologia Medico-chirurgica **54**(Suppl. 1) (2014)
17. Stefan, D.D.C.F., et al.: Quantitation of magnetic resonance spectroscopy signals: the jMRUI software package. Meas. Sci. Technol. **20**(10), 104035 (2009)
18. Van der Walt, S., et al.: scikit-image: image processing in Python. PeerJ **2**, e453 (2014)
19. Warner, J., et al.: Jdwarner-scikit-fuzzy: Scikit-fuzzy 0.3.1 (2017)
20. Zadeh, L.A.: Information and control. Fuzzy Sets **8**(3), 338–353 (1965)
21. Zouaoui, S., et al.: French brain tumor database: general results on 40,000 cases, main current applications and future prospects. Neuro-Chirurgie **58**(1), 4–13 (2012)

Identification of Saimaa Ringed Seal Individuals Using Transfer Learning

Ekaterina Nepovinnykh[1,2], Tuomas Eerola[1(✉)], Heikki Kälviäinen[1],
and Gleb Radchenko[2]

[1] Machine Vision and Pattern Recognition Laboratory, Department
of Computational and Process Engineering, School of Engineering Science,
Lappeenranta University of Technology, Lappeenranta, Finland
{ekaterina.nepovinnykh,tuomas.eerola}@lut.fi
[2] School of Electrical Engineering and Computer Science, South Ural State
University, Chelyabinsk, Russian Federation

Abstract. The conservation efforts of the endangered Saimaa ringed
seal depend on the ability to reliably estimate the population size and to
track individuals. Wildlife photo-identification has been successfully utilized in monitoring for various species. Traditionally, the collected images
have been analyzed by biologists. However, due to the rapid increase in
the amount of image data, there is a demand for automated methods.
Ringed seals have pelage patterns that are unique to each seal enabling
the individual identification. In this work, two methods of Saimaa ringed
seal identification based on transfer learning are proposed. The first
method involves retraining of an existing convolutional neural network
(CNN). The second method uses the CNN trained for image classification to extract features which are then used to train a Support Vector
Machine (SVM) classifier. Both approaches show over 90% identification accuracy on challenging image data, the SVM based method being
slightly better.

Keywords: Animal biometrics · Saimaa ringed seals
Convolutional neural networks · Transfer learning · Identification
Image segmentation

1 Introduction

The Saimaa ringed seal (*Pusa hispida saimensis*) is a subspecies of ringed seal
(*Pusa hispida*) living in Lake Saimaa in Finland (Fig. 1). At present, around
360 seals inhabit the lake, and on the average 60 to 85 pups are born annually.
This small and fragmented population is threatened by various anthropogenic
factors, especially by-catch and climate change [13]. Therefore the long-term and
accurate assessment of the population is needed for conservation purposes.

Successful conservation requires constant population monitoring which is not
easy to do without invasive methods. Traditional population monitoring methods include tagging that requires catching the animal and may cause stress to

© Springer Nature Switzerland AG 2018
J. Blanc-Talon et al. (Eds.): ACIVS 2018, LNCS 11182, pp. 211–222, 2018.
https://doi.org/10.1007/978-3-030-01449-0_18

Fig. 1. Saimaa ringed seal.

it, as well as may change its behavior or increase mortality. This makes non-invasive methods preferable for population monitoring. Wildlife Photo Identification (WPI) is a technology that allows to recognize individuals and to track the movement of animal populations over time. It is based on acquiring images of animals and further identifying individuals.

Recently, camera trapping has been launched as a monitoring tool also for the Saimaa ringed seal [5,12]. The Saimaa ringed seals have a distinctive fur pattern that is never repeated in different individuals and does not significantly change over the course of seal's life [12]. This makes photo identification based on the fur pattern suitable for non-invasive monitoring.

In this work, an automatic photo identification of the Saimaa ringed seals is considered. The proposed method first segments the seal from the background and then uses the fur pattern to identify the individual. The work continues studies presented in [7,20] where the first steps towards automatic individual identification of the Saimaa ringed seal were taken. In this paper, new methods for both the segmentation phase and the identification phase are proposed by utilizing convolutional neural networks (CNNs) and transfer learning.

2   Related Work

A computational approach to the wildlife photo identification is an emergent field that aims to apply formal methods to automate the process of animal biometric identification. There are many advantages over manual identification: traditional methods are time-consuming, highly dependent on the skills of a person who performs identification, and prone to various errors such as observer errors and biases [15]. Moreover, human observers often ignore classification uncertainty, and as such misclassification is often underestimated [9]. Computer methods avoid this problem by utilizing probabilistic methods and often report classification certainty along with other possible classification results. The main advantage of utilizing the animal biometrics system is that it allows researchers to rapidly collect and to robustly analyze the extensive amount of data which ultimately improves research about the seals and their monitoring.

Several approaches for automatic image-based animal identification can be found in the literature. Methods have been developed, for example, for polar bears [3], newts [11], giraffes [10], salamanders [6], and snakes [1], All of these methods use image processing and pattern recognition techniques to identify individuals. Most of the studies limit the individual identification to a certain animal species or species groups.

All the above methods were developed for one species only and as such are not generalizable to the Saimaa ringed seals. In [20], the first steps towards the automatic individual identification of the Saimaa ringed seals were taken. The paper proposes a segmentation method for the Saimaa ringed seals using unsupervised segmentation and texture based superpixel classification. Furthermore, a simple texture based approach for the ringed seal identification was evaluated. In [7], the segmentation method was further developed to decrease its computation time without sacrificing the performance. Moreover, a set of post-processing operations for segmented images was proposed to make the seals easier to identify. Two existing species independent individual identification methods were evaluated to demonstrate the importance of the segmentation and post-processing operations. However, the identification performance of neither of the methods is good enough for most practical applications.

There have been also research efforts towards creating a unified approach applicable for identification purposes for several animal species. For example, in [8], the HotSpotter method to identify individual animals in a labeled database was presented. This algorithm is not species specific and has been applied to Grevy's and plain zebras, giraffes, leopards, and lionfish. HotSpotter uses viewpoint invariant descriptors and a scoring mechanism that emphasizes the most distinctiveness keypoints and descriptors. In [19], a species recognition algorithm based on sparse coding spatial pyramid matching (ScSPM) was proposed. It was shown that the proposed object recognition techniques can be successfully used to identify animals on sequences of images captured using camera traps in nature. One of the problems with the species independent individual identification methods is that they do not provide an automatic method to detect the animals in images. Therefore, either a manual detection or development of a detection method for the studied animal is needed. Furthermore, typically higher identification performance can be obtained by tuning the identification method for one species only.

3   Proposed Method

In this work two Saimaa ringed seal identification methods based on transfer learning are proposed. The goal of the both methods is, given the image of a Saimaa ringed seal, to output the best suitable individual identifier for the specimen.

The both proposed identification algorithms consist of two steps. In the first step, the image is segmented. The segmentation result is an image of a seal without the background or overlapping objects. This is important since most of the image material is obtained using static camera traps. Therefore, the same seal is often captured with the same background increasing the risk that a supervised

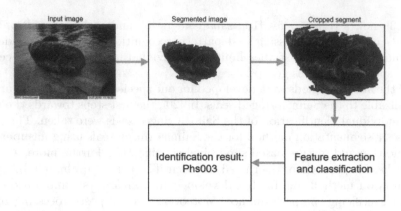

Fig. 2. General seal identification algorithm.

identification algorithm learns to "identify" the background instead of the actual seal if the full image or the bounding box around the seal is used. This may further lead to a system that is not able to identify the seal in a new environment.

The second step is the identification using transfer learning. The first proposed method for the identification is a CNN-based method. It involves retraining a classification CNN by using image extraction layer from another, preexisting convolutional neural network. After the initial experiments, it was concluded that training a CNN from the ground up is too computationally intensive given the constraint. Therefore, it was decided to use a pretrained general purpose CNN from [14] as the source of feature extraction layers. The second method for the identification is a Support Vector Machine (SVM) based method. For this method transfer learning is performed by using the above pretrained CNN for the feature extraction and SVM for the classification. The identification process is visualized in Fig. 2.

3.1 Segmentation

Automatic segmentation of animals is often difficult due to the camouflage colors of animals, i.e., the coloration and patterns are similar to the visual background of the animal. Segmentation results, however, can have a significant impact on identification performance. Segmentation helps to reduce the overfitting by removing the irrelevant background from an image, allowing a standardized object rotation on different images, reducing the dataset bias by only presenting the objects of interest to the training algorithm, and allowing improved color-correction by zeroing out all background colors and only focusing on object colors.

In this work, the segmentation framework proposed in [20] is used. The segmentation pipeline contains the following steps: (1) unsupervised segmentation of an image to produce a set of superpixels, (2) feature extraction from each superpixel, (3) classification of the superpixels to the seal and background classes,

(4) composing of the seal segments into one image, and (5) cropping the resulting image to contain only the seal. Figure 3 shows the segmentation process. For the unsupervised segmentation Multiscale Combinatorial Grouping (MCG) [4] is used. To classify the superpixels, SVM with the feature extraction layers of AlexNet [14] is used, instead of Local Phase Quantization (LPQ) [17] features utilized in [20].

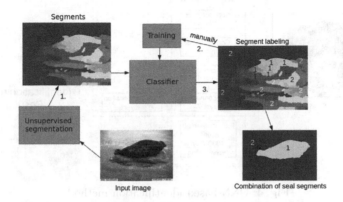

Fig. 3. Segmentation algorithm: (1) unsupervised segmentation, (2) training, (3) classification.

3.2 Identification

This work compares two different ways of building an identification method with transfer learning. The result of the identification is an individual seal id.

The CNN-based identification algorithm is shown in Fig. 4. The method utilizes the well known AlexNet architecture [14]. The classification layer of the original network is removed and replaced with a new classification layer. Instead of the 1000-way classification of the original network, the new classification layer has the number of classes equal to the number of seal individuals. The whole reconstructed neural network is then retrained with cropped and resized input images. The coefficients for training layers 1–7 are intentionally set to be low and for layer 8 high. This is done to reduce the impact of retraining on the feature extraction layers and to focus on the classification layer. The retrained CNN is then used to identify seals.

The SVM-based identification algorithm is shown in Fig. 5. Similarly to the previous method, the AlexNet architecture is utilized. The classification layer is removed after which the output of the neural network is not class probabilities but a 4096-dimensional vector from the fully connected layer 7. This vector is used as a feature vector for the SVM classifier. The 'one against all' strategy [2] is used in generalizing SVM to several classes where there is one binary learner for each class.

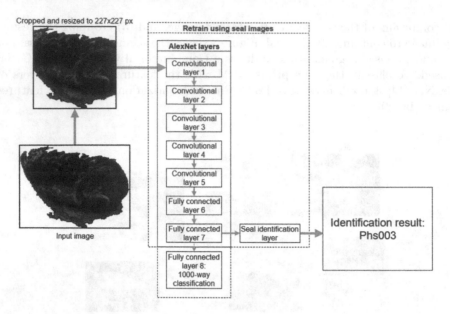

Fig. 4. CNN-based identification method.

4   Experiments

4.1   Datasets

The experiments were performed using two datasets of known seals. The first dataset consisted of four individuals with a large amount of images for each individual. The total number of images was 976 (244 images for each seal). The data was randomly divided into the training set and the test set so that in the test set contained 171 images and the training set 73 images per seal. Figure 6 shows example images from the dataset.

The second dataset consisted of 5585 images of 29 seals. The number of images per individual seal varied significantly from 20 to 860. The dataset was divided into the training and test sets with three different proportions: 30%, 50% or 70% of the images for training and the rest of the images for testing.

4.2   Method Implementation

In order to perform the transfer learning with the neural network, layers that have been trained to extract features from the image have to be selected, but not layers that use these features for classification. The classification layers of AlexNet start with the layer 23 (fully connected layer 8) which is the last fully-connected layer with 1000 outputs. While it is possible to extract the layers 1–22 and to use them as the basis for transfer learning, it is not practical to do so because the layers 21 and 22 are rectifier and dropout layers which are used just for training and do not hold any learned weights themselves. As such, the last

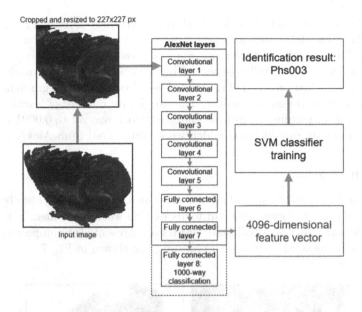

Cropped and resized to 227x227 px

AlexNet layers

Convolutional layer 1

Convolutional layer 2

Convolutional layer 3

Convolutional layer 4

Convolutional layer 5

Fully connected layer 6

Fully connected layer 7

Fully connected layer 8: 1000-way classification

Input image

Identification result: Phs003

SVM classifier training

4096-dimensional feature vector

Fig. 5. SVM-based identification method.

Fig. 6. Examples of seal images.

feature extraction layer is 20 (fully connected layer 7). It is a fully-connected layer with 4096 outputs, meaning that the neural network up to this layer performs dimensionality reduction on input image going from a $227 \times 227 \times 3$ image to a 4096-dimensional vector.

The segmentation procedure was implemented using MATLAB. The threshold value of 0.25 was used to turn the ultrametric contour map obtained using the MCG method [4] into superpixels. The same feature extraction and classifier training procedures as in the SVM-based seal identification was used for the superpixel classification. This technique involves using the feature extraction layers of AlexNet and then training an SVM classifier using these features. In order to collect a training set superpixels were extracted from 100 images of different seals and manually labeled as either belonging to the seal or to the background.

The CNN for the identification was implemented using MATLAB Neural Network Toolbox. The softmax layer and the classification layer were added. Softmax performs normalization of fully connected outputs to a sum of 1 while the classification layer selects a class with the maximum probability and assigns a class label. The following parameters were used during the neural network retraining: The learning rate factors were set low for the early layers of the network and high for the fully connected layer in order to focus training on the new classification layer instead of transfered weights. Stochastic gradient descent with momentum was used as a solver. The maximum number of learning epochs was 20. An initial learning rate of the entire network was set to 0.0001 to further reduce the learning rate of layers which were transferred from AlexNet.

4.3   Segmentation

The proposed segmentation method was used as a preprocessing step before the actual identification. Since the main focus of this work is on identification and not on segmentation, only qualitative results are presented. Examples of the bad and good segmentation results with cropping are shown in Fig. 7.

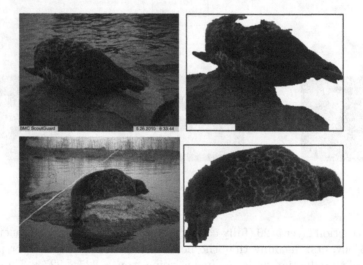

Fig. 7. The example of bad and good segmentation results with cropping.

4.4   Identification

The experiments were performed with the both methods of transfer learning. The both methods were tested in order to determine their ability to identify the Saimaa ringed seals and to measure the identification performance. The SVM classification was significantly faster to train.

The overall identification performance for the first dataset with the CNN-based method was 82.9%, i.e., 82.9% of all testing images were correctly identified. For the SVM based method the overall identification performance was 97.5% and there were no significant confusion between different classes. The confusion matrices are shown in Table 1. More detailed results can be found in [16].

Table 1. Confusion matrices for the CNN and SVM-based method.

Target class

		1	2	3	4	
	1	118 (17.3%)	2 (0.3%)	3 (0.4%)	2 (0.3%)	94.4%
	2	0 (0.0%)	120 (17.5%)	0 (0.0%)	4 (0.6%)	96.8%
Output class (CNN)	3	12 (1.8%)	29 (4.2%)	167 (24.4%)	3 (0.4%)	79.1%
	4	41 (6.0%)	20 (2.9%)	1 (0.1%)	162 (23.7%)	72.3%
		69.0%	70.2%	97.7%	94.7%	**82.9%**
	1	163 (23.8%)	2 (0.3%)	1 (0.1%)	0 (0.0%)	98.2%
	2	2 (0.3%)	168 (24.6%)	0 (0.0%)	4 (0.6%)	96.6%
Output class (SVM)	3	5 (0.7%)	0 (0.0%)	169 (24.7%)	0 (0.0%)	97.1%
	4	1 (0.1%)	1 (0.1%)	1 (0.1%)	167 (24.4%)	98.2%
		95.3%	98.2%	98.8%	97.7%	**97.5%**

With the second dataset, the CNN-based method obtained identification accuracy of 90.5% and the SVM-based method accuracy of 91.2%. These numbers were obtained by using 70% of the dataset for training and the rest 30% of the dataset for testing. Table 2 shows the results for the different ratios of the training and test set sizes.

Table 2. The proportions in the second dataset

Training/testing set ratio	CNN	SVM
30/70%	83.8%	87.3%
50/50%	88.8%	89.5%
70/30%	90.5%	91.2%

While the total accuracy is an important metric, it is not the only way to assess the performance of the identification system. A useful animal identification system should not just present a single identification candidate, but also give researchers a potential choice in ambiguous cases. This is achieved by presenting not just a single top pick of the specimen identifier for each image, but several individuals with the largest posterior probabilities. As such, the rank-based Cumulative Match Score (CMS) was used to assess the identification performance. CMS is commonly used in the face recognition research [18]. It measures how well the identification system ranks the identities in the database with respect to the input image. The Nth bin in the CMS histograms tells the percentage of test images where the correct individual seal was in the set of the N best matches proposed by the identification algorithm.

Figure 8 presents CMS histograms for the both proposed identification methods. CMS for rank 1 is the same as the total accuracy mentioned earlier. It can be seen that the SVM-based method gives better results for all ranks. For example, with SVM-based methods in 99% of the cases the correct seal was within the 5 best matches while CNN-based methods the correct seal was within the 5 best matches in 98% of the cases.

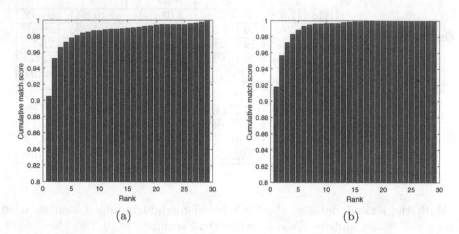

(a)   (b)

Fig. 8. CMS histograms for CNN-based and SVM-based identification methods on the second dataset.

It can be inferred that the classes with more samples have generally higher accuracy and lower relative number of errors. There are 6 seals in the second dataset with more than 300 images. For each of these individuals the identification accuracy was 88.6% or higher with CNN-based method and 92.1% or higher with SVM-based method. With certain individuals with a low number of training images the accuracy was considerably lower. With the CNN-based method the identification accuracy varied from 27.6% to 100.0% with 48% individuals having the identification accuracy of 90% or higher. With the SVM-based method the identification accuracy varied from 51.7% to 100.0% with 41% individuals having the identification accuracy of 90% or higher. More detailed results can be found in [16].

The obtained results are clearly better than the results reported in [20] where only 10% of the seals were correctly identified and in [7] where 44% of seals were correctly identified. However, it should be noted that, in the earlier studies, different datasets were used, and therefore, the results are not directly comparable. The datasets used in the preliminary studies were not suitable for this study as they contained small number of images for each individual seals, making it impossible to train deep CNNs.

The both methods produce very similar results, sharing similar characteristics. In the both cases the least accurately identified individual was the same and the percentage of errors with the individuals with a large number of images followed a similar distribution with the exception that the SVM-based identification method was generally more accurate. The both methods essentially use the same CNN-based feature extraction method and the same dataset which leads to the conclusion that these patterns depend on the feature extraction steps, not on the selection of a classification method.

5   Conclusions

In this paper two methods to identify the Saimaa ringed seal individuals based on the pelage patterns were proposed. The both methods utilize transfer learning. They start with the segmentation step where the seal is extracted from the background. This is done by dividing the image into superpixels and by classifying the superpixels using CNN-based features. The identification step utilizes the well-known AlexNet CNN architecture. The first method is based on retraining the original AlexNet with a new classification layer. The second method uses AlexNet only for the feature extraction and SVM for the classification. The experiments show that both methods provide a good performance on the identification with the challenging image data and the transfer learning contains a great potential on animal biometrics. The SVM-based method produced slightly higher identification accuracy with 91.2% of individuals correctly identified.

Acknowledgments. The authors would like to thank Meeri Koivuniemi, Miina Auttila, Riikka Levä-nen, Marja Niemi, and Mervi Kunnasranta from Department of Environmental and Biological Sciences at University of Eastern Finland for providing the database for the experiments and expert knowledge for identifying the individuals.

References

1. Albu, A.B., Wiebe, G., Govindarajulu, P., Engelstoft, C., Ovatska, K.: Towards automatic modelbased identification of individual sharp-tailed snakes from natural body markings. In: Proceedings of ICPR Workshop on Animal and Insect Behaviour, Tampa, FL, USA (2008)
2. Allwein, E.L., Schapire, R.E., Singer, Y.: Reducing multiclass to binary: a unifying approach for margin classifiers. J. Mach. Learn. Res. 1(Dec), 113–141 (2000)
3. Anderson, C.: Individual identification of polar bears by whisker spot patterns. Ph.D. thesis, University of Central Florida, Orlando, Florida (2007)
4. Arbeláez, P., Pont-Tuset, J., Barron, J., Marques, F., Malik, J.: Multiscale combinatorial grouping. In: Proceedings of the IEEE Conference on Computer Vision and Pattern Recognition (CVPR), pp. 328–335 (2014)
5. Auttila, M., Niemi, M., Skrzypczak, T., Viljanen, M., Kunnasranta, M.: Estimating and mitigating perinatal mortality in the endangered Saimaa ringed seal (phoca hispida saimensis) in a changing climate. Ann. Zool. Fenn. 51(6), 526–534 (2014)
6. Bendik, N.F., Morrison, T.A., Gluesenkamp, A.G., Sanders, M.S., O'Donnell, L.J.: Computer-assisted photo identification outperforms visible implant elastomers in an endangered salamander, *Eurycea tonkawae*. PLoS One 8(3), e59424 (2013)
7. Chehrsimin, T., et al.: Automatic individual identification of saimaa ringed seals. IET Comput. Vis. 12(2), 146–152 (2018)
8. Crall, J., Stewart, C., Berger-Wolf, T., Rubenstein, D., Sundaresan, S.: Hotspotter - patterned species instance recognition. In: IEEE Workshop on Applications of Computer Vision (WACV), pp. 230–237 (2013)

9. Guschanski, K., Vigilant, L., McNeilage, A., Gray, M., Kagoda, E., Robbins, M.M.: Counting elusive animals: comparing field and genetic census of the entire mountain gorilla population of Bwindi Impenetrable National Park, Uganda. Biol. Conserv. **142**(2), 290–300 (2009)
10. Halloran, K.M., Murdoch, J.D., Becker, M.S.: Applying computer-aided photo-identification to messy datasets: a case study of Thornicroft's giraffe (Giraffa camelopardalis thornicrofti). Afr. J. Ecol. **53**(2), 147–155 (2014)
11. Hoque, S., Azhar, M., Deravi, F.: Zoometrics-biometric identification of wildlife using natural body marks. Int. J. Bio-Sci. Bio-Technol. **3**(3), 45–53 (2011)
12. Koivuniemi, M., Auttila, M., Niemi, M., Levänen, R., Kunnasranta, M.: Photo-ID as a tool for studying and monitoring the endangered Saimaa ringed seal. Endanger. Species Res. **30**, 29–36 (2016)
13. Kovacs, K.M., et al.: Global threats to pinnipeds. Mar. Mammal Sci. **28**(2), 414–436 (2012)
14. Krizhevsky, A., Sutskever, I., Hinton, G.E.: Imagenet classification with deep convolutional neural networks. In: Proceedings of Advances in Neural Information Processing Systems (NIPS), pp. 1097–1105 (2012)
15. Kühl, H.S., Burghardt, T.: Animal biometrics: quantifying and detecting phenotypic appearance. Trends Ecol. Evol. **28**(7), 432–441 (2013)
16. Nepovinnykh, E.: Saimaa ringed seal fur pattern extraction for identification purposes. Master's thesis, Lappeenranta University of Technology, Finland (2017)
17. Ojansivu, V., Heikkilä, J.: Blur insensitive texture classification using local phase quantization. In: Elmoataz, A., Lezoray, O., Nouboud, F., Mammass, D. (eds.) ICISP 2008. LNCS, vol. 5099, pp. 236–243. Springer, Heidelberg (2008). https://doi.org/10.1007/978-3-540-69905-7_27
18. Phillips, P.J., Moon, H., Rizvi, S.A., Rauss, P.J.: The FERET evaluation methodology for face recognition algorithms. IEEE Trans. Pattern Anal. Mach. Intell. **22**(10), 1090–1104 (2000)
19. Yu, X., Wang, J., Kays, R., Jansen, P., Wang, T., Huang, T.: Automated identification of animal species in camera trap images. EURASIP J. Image Video Process. **2013**(1), 52 (2013)
20. Zhelezniakov, A., et al.: Segmentation of saimaa ringed seals for identification purposes. In: Bebis, G., et al. (eds.) ISVC 2015. LNCS, vol. 9475, pp. 227–236. Springer, Cham (2015). https://doi.org/10.1007/978-3-319-27863-6_21

Remote Sensing

Enhanced Codebook Model
and Fusion for Object Detection
with Multispectral Images

Rongrong Liu$^{(\boxtimes)}$, Yassine Ruichek, and Mohammed El Bagdouri

LE2i FRE2005, CNRS, Arts et Métiers, Univ. Bourgogne Franche-Comté, UTBM,
90010 Belfort, France
{rongrong.liu,yassine.ruichek,mohammed.el-bagdouri}@utbm.fr

Abstract. The Codebook model is one of the popular real-time models for object detection. In our previous work, we have extended it to multispectral images. In this paper, two methods to impove the previous work are proposed. On one hand, multispectral self-adaptive parameters and new estimation criteria are exploited to enhance codebook model. On the other hand, the approach of fusion is explored to improve the performance on multispectral images by fusing the detection results of the monochromatic bands. For the enhancements of codebook model, the self-adaptive parameter estimation mechanism is developed based on the statistical information of the data themselves, with which, the overall performance has improved, in addition to saving time and effort to search for the appropriate parameters. Besides, the Spectral Information Divergence is used to replace the spectral distotion to evaluate the spectral similarity between two multispectral vectors. Results demonstrate that when the spectral information divergence and brightness criteria are utilized in the self-adaptive codebook method, the performance can be improved slightly even further on average. For the approach of fusion, two strategies, namely pooling and majority vote, are adopted to exploit benefits of each spectral band to obtain better object detection performance.

Keywords: Object detection · Multispectral self-adaptive codebook
Spectral information divergence · Fusion

1 Introduction

1.1 Background Subtraction

Object detection through background subtraction is one of the important research areas in the field of computer vision. It aims at localizing moving foreground targets from images captured by a static camera for further steps like intelligent visual surveillance. Various approaches have been taken in algorithm developments for background subtraction. The work of [1] has provided an

© Springer Nature Switzerland AG 2018
J. Blanc-Talon et al. (Eds.): ACIVS 2018, LNCS 11182, pp. 225–232, 2018.
https://doi.org/10.1007/978-3-030-01449-0_19

overview of all the state-of-the-art background models, like Mixture of Gaussians models, Kernel Density Estimation, cluster models, etc.

As a well-known cluster model, the Codebook technique has attracted many researchers' attention. In the initial Codebook model proposed by [2], a codebook containing several codewords is first constructed from a sequence of images on a pixel-by-pixel basis, then the pixel vector of a new frame is compared with the average vector of the tested codeword in the background model, in order to finally obtain a foreground-background segmentation. During the last decade, many works have been dedicated to improve this model. For example, [3] has adopted two-layer model incorporated with local binary pattern texture measure, to handle dynamic background and illumination variation problems. Other modifications like transferring Red-Green-Blue (RGB) color space to other color space in order to solve the problem of existence of shadows and highlights for foreground detection are also found in [4,5]. In our previous paper [6], the initial Codebook method has been extended to multispectral images for the purpose of investigating the advantages of multispectral images over the conventional RGB images to improve the background subtraction performance.

In this paper, we propose two ways to improve our previous Codebook model: enhancement and fusion. For the enhancement, a multispectral self-adaptive parameters codebook model is developed based on the statistical information extracted from the data themselves; then the Spectral Information Divergence, which is usually adopted in the satellite images, is used to evaluate the spectral distance between the current and reference vectors. For the fusion, two approaches of fusing the monochromatic bands are explored, with an expectation of better foreground detection performance with multispectral images.

Like other parametric methods, the detection result of Codebook is heavily impacted by the parameters. The fashionable way to get these parameters is empirical and experimental. The pioneers of this technique [1] have provided the typical range of these parameters. However, this is still far from adequate, because manual parameters tuning is still required to achieve satisfying results for a specific scene, which is always a really cumbersome and tricky task for researchers. Besides, if the algorithm needs to be run for long periods of time, the parameters should be automatically adjusted according to the environmental changes. Therefore, there is a need for further research with regard to realizing an automatic selection for optimal parameters. Motivated by the work of [7], which has proposed the statistical parameter estimation method for YCbCr color space, we propose a multispectral self-adaptive Codebook method for multispectral images.

As we know, the main idea of Codebook background model construction is that, if the pixel vector of the current image is close enough to the average vector of the tested codeword in the background model, it will be modeled as a perturbation on that codeword, unless, it will seed a new codeword to be associated with that pixel. However, how to measure this closeness, or in another way of comprehension, distance? We should be aware that the combination of the brightness and spectral distortion defined in our previous paper is only one choice among all the possible estimation criteria.

Last but not least, to exploit benefits of each spectral band in multispectral images, a natural and reasonable approach to improve object detection performance is to fuse the detection results of the monochromatic bands.

The remainder of the paper is organized as follows. First, the two techniques are presented separately in Sect. 2. Section 3 demonstrates the experiments results, correspondingly. Concluding remarks are given in Sect. 4.

2   The Proposed Methods

2.1   Enhanced Codebook

2.1.1   Multispectral Self-adaptive Codebook

The codebook model devoted previously [6] is a parametric model with four basic following key parameters: α, β, ε_1 and ε_2. To be specific, α and β are used to obtain the boundaries from the min and max brightness $\check{I}_m$ and $\hat{I}_m$ in a certain codeword, with

$$\begin{cases} I_{low} = \alpha \hat{I}_m \\ I_{high} = \min \left\{ \beta \hat{I}_m, \dfrac{\check{I}_m}{\alpha} \right\} \end{cases} \tag{1}$$

and ε_1 and ε_2 are the spectral distortion thresholds used in the construction and detection phases, respectively. Typically, α is between 0.4 and 0.7, and β is between 1.1 and 1.5 [2].

Motivated by the work of [7], we propose a multispectral self-adaptive method for automatical optimal parameter selection. That is to say, those parameters do not need to be obtained by burdensome experiments, but to be estimated from the data themselves statistically, which can help to save a lot of efforts.

In the self-adaptive method, the definitions of brightness and spectral distortion keep the same with those in our previous work [6], as shown in Eqs. (2) and (3) to remind.

$$I = \sqrt{\sum_{i=1}^{n} X_i^2} \tag{2}$$

where n is the number of bands.

$$spectral\_dist\,(x_t, v_m) = \sqrt{\|x_t\|^2 - p^2} \tag{3}$$

where $p^2 = \|x_t\|^2 \cos^2\theta = \dfrac{\langle x_t, v_m \rangle^2}{\|v_m\|^2}$, in which, v_m is the average vector in the testing codeword.

However, the calculations of the boundaries are changed. During the process of constructing the background model, the statistical information is calculated iteratively for each codeword to automatically estimate both the brightness bounds and spectral distortion threshold. To be specific, the bounds of brightness can be estimated by

$$\begin{cases} I_{low} = \check{I}_m - \sigma_I \\ I_{high} = \hat{I}_m - \sigma_I \end{cases} \tag{4}$$

where σ_I is the standard deviation of brightness in the current codeword till the current frame. And the threshold ε for the spectral distortion is calculated by

$$\varepsilon = max([\sigma_1 ... \sigma_i ... \sigma_n]) \tag{5}$$

where σ_i is the standard deviation of the ith band value in the current codeword till the current frame.

With the self-adaptive mechanism, brightness bounds and spectral distortion threshold are able to adjust themselves with statistical properties of the input images. In the phase of background construction, for each pixel, when a new image arrives, the brightness and spectral distorion are first computed using Eqs. (2) and (3), then the matching process is conducted codeword by codeword. If (1) the brightness of the new pixel lie in the current interval of the brightness bounds and (2) the spectral distortion is smaller than the current threshold of a certain codeword, the new pixel will be modeled as a perturbation on this codeword, whose brightness bounds and spetral threshold will be updated using Eqs. (4) and (5). Unless, it will seed a new codeword. The similiar task is performed in the detection phase. The new pixel is classified as background if an acceptable matching codeword exists and the codeword will be updated using Eqs. (4) and (5) at the same time. Otherwise, the pixel is detected as foreground.

2.1.2 Redefined Codebook Model

In our previous Codebook method [6], two criteria have been adopted to evaluate the distance between two vectors, the brightness (B) and the spectral distortion (SD). Specifically speaking, the brightness is simply the L2-norm of the related bands, and the spectral distortion is measured as a function of the brightness-weighted angle between the current and reference spectral vectors, as illustrated in Eqs. (2) and (3).

Here we adopt another information-theoretic spectral measure, referred to as Spectral Information Divergence (SID) [8], which is applied to determine the spectral closeness or distance between two multispectral vectors. SID models the spectral band-to-band variability as a result of uncertainty caused by randomness, which is based on the Kullback-Leibler divergence to measure the discrepancy of probabilistic behaviors. That is to say, it considers each pixel as a random variable and then defines the desired probability distribution by normalizing its spectral histogram to unity, which is expressed by Eq. (6)

$$\begin{cases} P_x(i) = \dfrac{x_t(i)}{\sum_{i=1}^{n} x_t(i)} \\ P_v(i) = \dfrac{v_m(i)}{\sum_{i=1}^{n} v_m(i)} \end{cases} \tag{6}$$

where n is the number of bands. Then the spectral information divergence d_{SID} between the current spectral vector x_t and the background model v_m can be defined with Eq. (7)

$$d_{SID}(x_t, v_m) = \sum_{i=1}^{n} P_x(i) log \frac{P_x(i)}{P_v(i)} + \sum_{i=1}^{n} P_v(i) log \frac{P_v(i)}{P_x(i)} \tag{7}$$

In the redefined Codebook model, the spectral information divergence is employed to replace the spectral distortion in the previous Codebook model to be the judging criteria together with the brightness condition. In this model, the brightness and spectral information divergence are first computed based on Eqs. (2) and (7) when a new frame arrives. The main constuction and detection procedures keep the same with those in 2.2.1. In the matching process, the same threshold calculating strategy as that for spectral distortion is employed for the SID.

2.2 Fusion

To exploit benefits of each spectral band, a resonable approach on multispectral images to improve object detection performance is to fuse the detection results of the monochromatic bands. One fusion strategy adopted is named pooling: a pixel is detected as foreground if it has been regarded as foreground for more than ρ times in monochromatic images. The second approach is to fuse the detection results of the best K monochromatic bands via majority vote, which has been motivated by the nearest vote in the field of classification.

3 Experiments Results and Analysis

The performance of these techniques have been evaluated with four videos in the public multispectral dataset [9] in terms of F_measure, which is the harmonic mean of recall and precision. The multispectral images have seven bands, six in the visible spectrum and one in the near infrared. RGB sequences are also offered by linear integration of the original mulspectral images. Importantly, this dataset provides ground truth at pixel resolution, which makes detailed quantitative analysis and comparison possible. Figure 1 presents examples of the RGB images and the corresponding ground truth.

3.1 Enhanced Codebook

3.1.1 Multispectral Self-adaptive Codebook
For the Codebook in our previous work [6], the optimal parameters are:

$$\alpha = 0.7, \beta = 1.5, \varepsilon_1 = 0.02, \varepsilon_2 = 0.04 \tag{8}$$

which are empiric values determined experimentally. The experiments have been conducted on the thirty-five different three-band-based combinations for four videos. The best combination is selected and listed in Table 1, together with the results of the corresponding Multispectral self-adaptive (MSA) Codebook.

Fig. 1. Examples of the multispectral dataset

Table 1. F_measure of the best three-band-based combination on 4 videos (previous CB versus MSA CB)

Methods	Video 1	Video 2	Video 3	Video 4	Mean
Previous CB	0.55	**0.60**	0.86	0.69	0.68
MSA CB	**0.70**	0.56	**0.90**	**0.74**	**0.73**

From Table 1, in addition to saving time and effort to search for the appropriate parameters, the overall performance has been improved using the multispectral self-adaptive Codebook method, except for the second video, whose accuracy drops slightly, which may be caused by the tree shadows and occlusion. The accuracy on the other three videos increases to some extent, especially for indoor video 1, which shows a great jump.

3.1.2   Redefined Codebook Model

We then use spectral information divergence to replace the spectral distortion defined in our previous paper. Here we do not need to search the parameters as we have done before. Instead, the multispectral self-adaptive Codebook is adopted directly and the results are illustrated in Table 2.

Table 2. F_measure of the best three-band-based combination with MSA CB (SD versus SID)

Criteria	Video 1	Video 2	Video 3	Video 4	Mean
B+SD	**0.70**	0.56	**0.90**	0.74	0.73
B+SID	0.67	**0.60**	0.88	**0.79**	**0.74**

From the Table 2, when the spectral information divergence is adopted in the multispectral self-adaptive Codebook method, the performance can be improved a little even further on average. This opens a door for other possibilities to seek novel kind of ways to construct the codebook model.

3.2 Fusion

3.2.1 Pooling
Since there are seven bands in the dataset, ρ ranges from 1 to 7, as it is shown in Table 3.

Table 3. F_measure in pooling for monochromatic images

Video	1	2	3	4	5	6	7	RGB
1	**0.67**	0.64	0.51	0.38	0.28	0.20	0.09	0.49
2	0.35	0.52	**0.58**	0.58	0.56	0.51	0.38	0.57
3	0.87	**0.89**	0.88	0.85	0.81	0.71	0.51	0.89
4	0.56	0.84	**0.87**	0.86	0.83	0.72	0.48	0.85

From Table 3, using pooling fusion strategy with the detection results of the monochromatic bands for multispectral images, the detection performance can be better than RGB images.

3.2.2 Majority Vote of the Best K Monochromatic Images
To begin with, we sort the results of seven monochromatic images from largest to smallest for each video. Then a pixel is detected as foreground if the K best results detect the pixel as foreground. Here the parameter K still range from 1 to 7. The results are listed in Table 4.

Table 4. F_measure in majority vote of the best K monochromatic images

Video	1	2	3	4	5	6	7	RGB
1	**0.51**	0.31	0.46	0.35	0.43	0.35	0.38	0.49
2	0.52	0.56	0.56	0.58	**0.60**	0.58	0.58	0.57
3	0.84	0.77	0.85	0.81	0.85	0.83	0.86	**0.89**
4	0.81	0.80	0.84	0.83	0.85	0.84	**0.86**	0.85

From Table 4, with the majority vote of the best K monochromatic images for multispectral images, The F_measures are larger than those of RGB images, except for video 3.

4   Conclusion and Future Work

In this paper, we proposed two techniques to improve our previous Codebook algorithm: enhancement and fusion. For the enhancement, the brightness bound and spectral distortion thresholds are calculated automatically from the image data themselves statistically, which is helpful for researchers to get rid of the cumbersome task of parameters tuning. What's more, the spectral information divergence is adopted to replace the spectral distortion to be the criteria to evaluate the distance between two vectors in the matching process. For the fusion, two approaches of fusing the monochromatic bands are explored to investigate the benefits of multispectral images to obtain better foreground detection performance. In the future, we'd like to induce the texture information, like local binary pattern, in Codebook model to explore the potential of the multispectral object detection.

References

1. Bouwmans, T.: Traditional and recent approaches in background modeling for foreground detection: an overview. Comput. Sci. Rev. **11–12**, 31–66 (2014)
2. Kim, K., Chalidabhongse, T.H., Harwood, D., Davis, L.: Real-time foreground-background segmentation using codebook model. R.-Time Imaging **11**(3), 172–185 (2005)
3. Zhang, Y.-T., Bae, J.-Y., Kim, W.-Y.: Multi-layer multi-feature background subtraction using codebook model framework. World Acad. Sci., Eng. Technol., Int. J. Comput. Inf. Eng. **3**(1) (2016)
4. Huang, J., Jin, W., Zhao, D., Qin, N., Li, Q.: Double-trapezium cylinder codebook model based on YUV color model for foreground detection with shadow and highlight suppression. J. Signal Process. Syst. **85**(2), 221–233 (2016)
5. Krungkaew, R., Kusakunniran, W.: Foreground segmentation in a video by using a novel dynamic codebook. In: 2016 13th International Conference on Electrical Engineering/Electronics, Computer, Telecommunications and Information Technology (ECTI-CON), pp. 1–6. IEEE (2016)
6. Liu, R., Ruichek, Y., El Bagdouri, M.: Background subtraction with multispectral images using codebook algorithm. In: Blanc-Talon, J., Penne, R., Philips, W., Popescu, D., Scheunders, P. (eds.) ACIVS 2017. LNCS, vol. 10617, pp. 581–590. Springer, Cham (2017). https://doi.org/10.1007/978-3-319-70353-4_49
7. Shah, M., Deng, J.D., Woodford, B.J.: A self-adaptive codebook (sacb) model for real-time background subtraction. Image Vis. Comput. **38**, 52–64 (2015)
8. Chang, C.-I.: An information-theoretic approach to spectral variability, similarity, and discrimination for hyperspectral image analysis. IEEE Trans. Inf. Theory **46**(5), 1927–1932 (2000)
9. Benezeth, Y., Sidibé, D., Thomas, J.-B.: Background subtraction with multispectral video sequences. In: IEEE International Conference on Robotics and Automation workshop on Non-classical Cameras, Camera Networks and Omnidirectional Vision (OMNIVIS), pp. 1–6 (2014)

Unsupervised Perception Model for UAVs Landing Target Detection and Recognition

Eric Bazán[1]($\boxtimes$), Petr Dokládal[1], and Eva Dokládalová[2]

[1] PSL Research University - MINES ParisTech, CMM - Center for Mathematical Morphology, Mathematics and Systems, 35, rue St. Honoré, 77305 Fontainebleau Cedex, France
{eric.bazan,petr.dokladal}@mines-paristech.fr
[2] Université Paris-Est, LIGM, UMR 8049, ESIEE Paris, Cité Descartes B.P.99, 93162 Noisy le Grand Cedex, France
eva.dokladalova@esiee.fr

Abstract. Today, unmanned aerial vehicles (UAV) play an interesting role in the so-called Industry 4.0. One of many problems studied by companies and research groups are the sensing of the environment intelligently. In this context, we tackle the problem of autonomous landing, and more precisely, the robust detection and recognition of a unique landing target in an outdoor environment. The challenge is how to deal with images under non-controlled light conditions impacted by shadows, change of scale, perspective, vibrations, noise, blur, among others. In this paper, we introduce a robust unsupervised model allowing to detect and recognize a target, in a perceptual-inspired manner, using the Gestalt principles of non-accidentalness and grouping. Our model extracts the landing target contours as outliers using the RX anomaly detector and computing proximity and a similarity measure. Finally, we show the use of error correction Hamming code to reduce the recognition errors.

Keywords: UAV · Landing target · Perception model
Object detection · Precision landing

1 Introduction

In this paper, we present a novel method for the detection of landing targets for the UAV vision aided landing. We propose to model the landing target by taking into account the principles of the human perception. The methodology presented works in an unsupervised mode, i.e., no need to adjust parameters.

In outdoor environments, many variables affect the vision-based landing target detection. The main problems to face are: the non-controlled light changes that generate shadowing, reflectance and saturation on the surfaces; the perspective and distance of the camera that deforms the objects; the motions and vibrations that blur the images and; the noise generation by a low-quality sensor.

© Springer Nature Switzerland AG 2018
J. Blanc-Talon et al. (Eds.): ACIVS 2018, LNCS 11182, pp. 233–244, 2018.
https://doi.org/10.1007/978-3-030-01449-0_20

The detection of the landing target can be viewed as an image segmentation problem, where there is a wide range of developed methods. The variational framework [13], offers an optimal general method for image segmentation; however, its mathematical complexity and the constant selection of fidelity and a regularization parameters makes its use complex. Also, the number of iterations needed to find the optimal solution avoid having results in real-time. Conversely, thresholding methods have been used for the detection of landing targets [8,9] for its ease of use. However, for a good detection, its use is limited to indoor spaces, where the light conditions are controlled [1].

Recently, convolutional neural networks (CNN) techniques offer the possibility of detecting an object from a large set of classes with a high-reliability [3]. Nevertheless, these methods must have been trained with a database containing the object classes in a wide range of situations and, in case of changes in the object or the scene, the database must be rebuilt [6,19]. Besides, in some cases, the computation is carried out off-board the drone, which implies the need for network infrastructure and limitation of autonomy [10].

Humans can carry out the process of perception in a natural way [14]. We identify meaningful features and exciting events in a scene (such as points, lines, edges, textures, colors, movement) and with the help of our memory and the learning capacity we can recognize and classify objects. The primitives identification is a consequence of their non-accidental apparition, i.e., they are not generated randomly [2]. The Gestalt theory [17] states that we can build a whole (gestalt) through the grouping of non-accidental detected primitives. In this work, we explore the above ideas and propose a novel approach to detect a landing target in the same way as humans do, imitating the human perception process.

The work is organized as follows. In Sect. 2 we develop the perception model. Specifically, Subsect. 2.1 describes how to retrieve image contours as meaningful primitives and Subsect. 2.2 describes how to group the contours to detect a landing target perceptually. Later, in Sect. 3 we present the landing target design and the technique used for the methodology implementation; the results obtained are discussed in Subsect. 3.2. Finally, we present some conclusion and perspectives in Sect. 4.

2   The Unsupervised Perception Model

The proposed algorithm uses the contours as image primitives to obtain information about the scene. For the detection and recognition of landing targets, the algorithm is carried out in three major stages. The first, localize all the image contours and extract meaningful contours using the non-accidentalness principle. The second stage computes some feature contours and, based on them, group the meaningful contours through the Gestalt laws. The last stage performs the target recognition using a decoding technique described in Sect. 3. The diagram in Fig. 1 shows the three major stages, its inputs and outputs, and its subtasks.

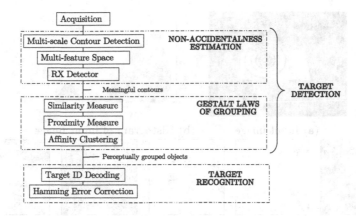

Fig. 1. Diagram of the phases for the landing target detection and recognition

2.1 Non-accidentalness Estimation

We aim to detect object contours in natural images where none, one or more landing targets can be present. Due to its real-time capacity, it is tempting to use a thresholding method to detect the contours of a binary image. We implemented several thresholding methods analyzed in [16], however, given the conditions where a landing target can be found, no method was found robust enough to variations in non-controlled outdoor environments. Figure 2a shows a landing target in an outdoor environment, we also show his histogram to highlight the levels of saturation in the scene. As a comparison, we take one representative method of each class of the taxonomy proposed in [16] to extract the contours of the image; clustering-based (Fig. 2d), entropy-based (Fig. 2e), spacial (Fig. 2f) and local (Fig. 2g) thresholding methods. Namely, there is no guarantee that the contours found by thresholding are present and continuous alongside the object borders.

Contour Detection. Instead of using a thresholding method, we obtain the image without fixing any parameter. The use of the Marr-Hildreth [12] operator guarantees to find continuous and closed contours eliminating the possible noise in the image, while the contours of objects remain unchanged in the presence of shadows. This technique convolves the intensity image f with the 2-D Laplacian of Gaussian operator $\nabla^2 G(x, y, \sigma)$ and generates an image,

$$l_\sigma = \nabla^2 G(\sigma) * f \tag{1}$$

in which we localize the zero-crossings.

The parameter σ in Eq. (1) permits to control the amount of image smoothing, but also acts as scale parameter, that when varies, it generates different scale-space images. Since no single filter can be optimal simultaneously at all scales [12], we use a multi-scale analysis [18] to detect the zero-crossings in l_σ at different scale-spaces to minimize the risk that some contour of interest is not detected.

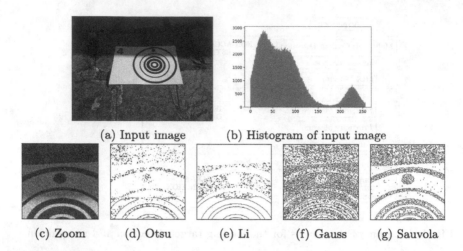

(a) Input image   (b) Histogram of input image

(c) Zoom   (d) Otsu   (e) Li   (f) Gauss   (g) Sauvola

Fig. 2. Landing target under non-controlled illumination conditions and the controus obtained with some thresholding methods

The image l_σ from Eq. (1) contains a set of contours $\mathcal{L}_\sigma = \{L_i^\sigma, \ i = 0, 1, \ldots, N\}$ for a given scale σ. Then,

$$\mathcal{L} = \bigcup_\sigma L_\sigma \tag{2}$$

represents all the contours of an image obtained at different scale-spaces. Figure 3d shows the set of contours $\mathcal{L}$ found for $\sigma = [1, 2, 3]$. Besides, it is also appreciated that at a fine scale (Fig. 3a) we can see more characteristics of the objects, i.e., there are more contours. Conversely, in coarse scales (Fig. 3c), due to the smoothing, there is a spatial distortion, and fewer contours appear. However, those contours that had already appeared at a coarse scale, will not disappear. Then, exist the probability that those contours that spatially coincide on two or more scales belong to a change of intensity generated by the border of an object.

Multi-feature Space. The Helmholtz principle states that meaningful characteristics appear as large deviations from randomness and that is how the human perception automatically works to identify an object [2]. The a contrario model proposed in [4], formulates this principle statistically by setting the number of false alarms (NFA) below some acceptable level; however, this method cannot be easily extended to more complex shapes. Instead of setting the NFA, we use the RX detector [15] to detect outliers. Initially called the constant false alarms rate detection algorithm (CFAR) it can detect the presence of a know signal pattern in several signal-plus-noise channels. For that, it uses a $N \times Q$ multi-variable space $Z = [Z_1, \ldots, Z_Q]$ with Q observation vectors of dimension N. In our approach, the primitive is a closed contour. We build the multi-variable space with observations based on internal (geometrical features, e.g., circularity, roundness,

(a) $\mathcal{L}_\sigma$ for $\sigma = 1$ (b) $\mathcal{L}_\sigma$ for $\sigma = 2$

(c) $\mathcal{L}_\sigma$ for $\sigma = 3$ (d) Set $\mathcal{L}$ for $\sigma = [1, 2, 3]$

Fig. 3. The image contours found at three different scales joined in the set $\mathcal{L}$

area, perimeter) and external (e.g., mean gradient intensity, intensity inner area) properties of the contours.

Let $L_i \in \mathcal{L}$ be a contour, A_i its area and P_i its perimeter; we compute the circularity Eq. (3) and the mean gradient intensity Eq. (4) to build the multivariable space $Z = [Z_1, Z_2]$.

$$Z_1 = \left[\frac{4\pi A_i}{P_i^2}, \quad i = 0, \ldots, N \right]^T, \quad N = card(\mathcal{L}) \tag{3}$$

$$Z_2 = \left[\frac{1}{P_i} \sum_{x \in L_i} | \nabla f(x) |, \quad L_i \in \mathcal{L} \right]^T \tag{4}$$

RX Detector. The RX anomaly detector [15] is commonly used to detect outliers on such data. The space Z models the set of contours $\mathcal{L}$ with $Q = 2$ feature vectors describing the circularity Eq. (3) and the mean gradient intensity Eq. (4). The RX detector gives an anomaly score to each contour taking into account the mean of the distribution and covariance between the Q-features through the Mahalanobis distance,

$$y_i = (z_i - \mu_Z)^T \Sigma_Z^{-1} (z_i - \mu_Z) \tag{5}$$

where $\mu_Z = [\mathrm{E}[z_1], \ldots, \mathrm{E}[z_N]]^T$ is the observations mean vector and Σ_Z^{-1} the $N \times Q$ covariance matrix of the data. If the data have normal random distribution, then the score vector $Y = [y_i, \ldots, y_N]$ follows a chi-square distribution $\chi_Q^2(\varphi)$ with Q degrees of freedom, where φ is a confidence level [11]. The value of $\chi_Q^2(\varphi)$ with a confidence value $\varphi = 99.9\%$ operates as a threshold to identify all contours that behave as outliers in the multi-variable distribution. In our case, the contours belonging to a landing target appear as outliers in the vast majority of random contours belonging to the background.

With the previous strategy we preserve the anomalous contours having a value of mean gradient and circularity deviating from the principal mode of the distribution in the set $\widetilde{\mathcal{L}} = \{L_i \mid y_i > \chi_Q^2(\varphi)\}$. $\chi_Q^2(\varphi)$ is the value of the cumulative distribution at the confidence level φ and $\widetilde{\mathcal{L}} \subset \mathcal{L}$.

In the set $\widetilde{\mathcal{L}}$ some contours make not part of a landing target. For example, in the Fig. 4, we can see that the paper sheet contours remain because they have a high value of circularity. The same occurs with the contours of those objects with an important value of mean gradient, as the number 4 at the top-left of the sheet or the rock textures of the background.

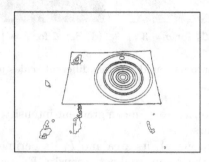

Fig. 4. The contours from Fig. 3d that behave as outliers in the multi-feature space Z with a confidence value of $\varphi = 99.9\%$

2.2   Gestalt Laws of Grouping

We use the Gestalt theory [17] to group the meaningful contours $L_i \in \widetilde{\mathcal{L}}$ and detect landing targets.

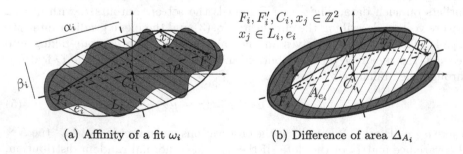

(a) Affinity of a fit ω_i (b) Difference of area Δ_{A_i}

Fig. 5. Visual description of affinity of ellipse and difference of area

Goodness of Shape. Since the landing targets have only circular contours, we evaluate the resemblance with an ellipse (to deal with the perspective deformation) of all contours. Considering an ellipse e_i that fits one gray contour L_i in Fig. 5a, we recover the centroid C_i, the rotational angle ρ, the semi-major axis α_i, the semi-minor axis β_i and the coordinates F_i and F_i' of the ellipse's foci. Since the sum of the distances from any point of the ellipse $x_j \in e_i$ to the foci is $\overline{x_j F_i} + \overline{x_j F_i'} = 2\alpha_i$; if the contour L_i is an ellipse, the value $d_i = \left| (\overline{x_j F_i} + \overline{x_j F_i'}) - 2\alpha_i \right|$ must be zero or negligible $\forall x_j \in L_i$.

Based on the form of the landing target we estimate the similarity using two measures,

$$\omega_i = \exp^{-\frac{d_i^2}{2\sigma^2}} \quad \text{the affinity of the fit and,} \tag{6}$$

$$\Delta_{A_i} = 1 - \frac{|A_{e_i} - A_i|}{\max(A_{e_i}, A_i)} \quad \text{the difference of area.} \tag{7}$$

The affinity $\omega_i \to 1$ for contours closed to an ellipsoidal shape. However, if the contour L_i is a croissant shape (as in Fig. 5b) then, the Eq. (6) also has a high value (near to 1) but the contour is from being an ellipse. The variable in Eq. (7) complements the affinity ω_i taking into account the area of the ellipse A_{e_i} and the area of the contour A_i. To calculate the similarity to an ellipse, we use the harmonic mean of both.

$$\kappa_i = \mathcal{H}(\omega_i, \Delta_{A_i}), \quad \kappa_i \in (0, 1) \tag{8}$$

where $\kappa_i \to 1$ for contours ressembling to an ellipse and $\kappa_i \to 0$ otherwise. $\mathcal{H}$ denotes the harmonic mean $\mathcal{H} = N \left(\sum_{i=1}^{N} \xi_i^{-1} \right)^{-1}$.

Proximity Measure. The Gestalt law of proximity states that we group those meaningful elements if they are spatially close to each other. In the case of contours, we take the coordinates of their centers C_i to measure their spatial proximity.

Affinity Clustering. The normalized coordinates of the centroid C_i and the ellipse similarity κ_i map the contour $L_i \in \tilde{\mathcal{L}}$ into the 3-D space $(0, 1) \in \mathbb{R}^3$. We use the affinity propagation clustering method [5] to group the contours using the matrix $X = [C_i, \kappa_i]$. This technique yields a set of clusters $\mathcal{C}_K \in \mathcal{C}(X)$. Because the landing target has ten different contours (see Sect. 3), the clusters with $card(\mathcal{C}_K) \geqslant 10$ and an important similarity value $\mathcal{H}(\kappa_i) \geqslant 0.8$, represent the candidate contours of a landing target.

The affinity propagation technique groups in $K = 12$ clusters the image contours from Fig. 4. In a 3D plot (Fig. 6a), we see the influence of κ_i at clustering process. Projecting the clusters in a 2-D plane (Fig. 6b), we notice that even if the contours are nearby, it can form a new cluster if there is a distant κ. A clear

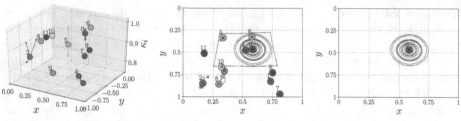

(a) Clusters obtained by affinity propagation

(b) Clusters projected on the image domain

(c) Target candidate cluster

Fig. 6. Clusters of contours from Fig. 4

example is the clusters 0 and 4 (blue and purple, respectively) that correspond to the contour centers of the landing target and the center of the sheet of paper, they are close to each other but the similarity not. Applying the threshold values $card(\mathcal{C}_K) \geqslant 10$ and $\mathcal{H}(\kappa_i) \geqslant 0.8$ we obtain the candidate clusters to form a landing target (see Fig. 6c).

Heretofore, we have built a model based on perceptual characteristics for the landing target detection. However, there could be false detections if there are circular objects with concentric borders in the image. We code an ID number in the target design to differentiate a landing target from an object with concentric circular edges. The coding of information allows discriminating between several landing targets and circular objects. The following section describes the landing target design as well as the coding and decoding technique.

3   Implementation

3.1   Landing Target Description

The landing target is formed by a set of black and white circles (see Fig. 7) that generate contours when stacked. Two of the circles ($\varnothing_9$ and $\varnothing_{10}$) have a constant diameter and serve as a reference to determine the nominal size of the target in the image. The black circle ($\varnothing_{11}$) is an orientation reference and has the same diameter as the smallest circle, $\varnothing_{11} = \varnothing_1$. The other circles $\varnothing_1, \ldots, \varnothing_8$ are coding circles.

Landing Target ID Encoding. Let $\varnothing = (\varnothing_1, \varnothing_2, \ldots, \varnothing_n)$ denote the nominal diameters of the coding circles. We can set the nominal diameters, e.g., $\varnothing_i = \frac{i}{n}\varnothing_n$ for a target without the encoding capability. To encode a number in the target form, we modify the nominal diameters $\varnothing$ to obtain $\varnothing' = (\varnothing'_1, \varnothing'_2, \ldots, \varnothing'_n)$ by adding/subtracting a positive constant Δh

$$\varnothing'_i = \begin{cases} \varnothing_i + \Delta h, & \text{if } w_i = 1 \\ \varnothing_i - \Delta h, & \text{otherwise} \end{cases} \tag{9}$$

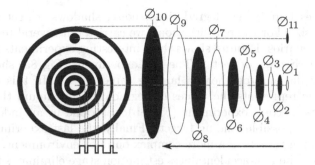

Fig. 7. Landing target design and description

and obtain a binary message $W = [w_1, \ldots, w_n]$. The message W is protected from errors by Hamming error-correction code [7]. It provides a set of different codewords $W = D \times M$ of size $n = k+m$, where D is useful data, $M = [I_k \mid 1-I_k]$ the generator matrix and I_k is the $k \times k$ the identity matrix. The data vector D comes from the decimal to binary conversion of the landing target ID[1] number. In our representation, we have experimented with $n = 8$ coding circles allowing to have four rings and $n = 8$ contours $\varnothing_1, \ldots, \varnothing_8$. This allows us to use the extended $[n, k]$ Hamming code with $k = 4$ data bits and $m = 4$ parity bits to generate $2^4 = 16$ landing targets.

Landing Target ID Decoding. After the clustering stage of Sect. 2.2, we rank by size the ellipses' major axes α_i by size and normalize them w.r.t. the largest value α_{10} to obtain $\boldsymbol{\alpha} = \frac{\varnothing_{10}}{\alpha_{10}}(\alpha_1, \ldots, \alpha_{10})$.

We compare the received and normalized axis $\boldsymbol{\alpha}$ with the nominal diameters of the coding circles $\varnothing$ and transform them into a binary vector $\widehat{W}$;

$$\widehat{W} = \begin{cases} 1, & \text{if } \alpha_i - \varnothing_i > 0 \\ 0, & \text{otherwise} \end{cases} \quad \forall i = 1, \ldots, n \qquad (10)$$

The Hamming syndrome vector $S = \widehat{W} \times H^T$ (with $H = [1 - I_k \mid I_k]$ as the parity-check matrix) indicates whether an error has occurred. The syndrome is a null vector $S = 0$ when no error has occurred, otherwise, $S \neq 0$ and $\widehat{W} = W + E$. The element $e_i = 0$ of the error vector $E = H^T - S$ indicates an error at the position i. The $[n, k]$ Hamming code with $n = 8$ and $k = 4$ can find up to two erroneous bits and correct one. Once the algorithm corrects the error (if there is), the vector $\widehat{W}$ is decoded by using the modulo 2 of the product $\widehat{D} = \widehat{W} \times M^T$.

3.2 Validation and Tests

The presented strategy was validated on landing target images under simulated and real situations. We tested the algorithm in a synthetic image database

---

[1] Identification number

which simulates four image degradations: noise, shadows, target deformation and change of size. For the real situations, we carried out several tests in indoor and outdoor scenarios. Figure 8 shows three interesting experiments and the output image of each stage of Sect. 2. The first experiment (Fig. 8a) shows the four synthetic degradations together on landing target ID 14. In this context, the synthetic image represents the values of degradation maximum that the algorithm supports. Second experiment (Fig. 8b) was done in an indoor space to show the sixteen possible landing targets. Finally, the last experiment (Fig. 8c) shows five landing targets in a more complex outdoor environment. In the three experiments, (i) the non-accidentalness estimation stage eliminates the contours generated by noise with low circularity and mean gradient values; (ii) the grouping stage filters random contours generated by intensity changes like shadows to keep contours with an important value of similarity and proximity.

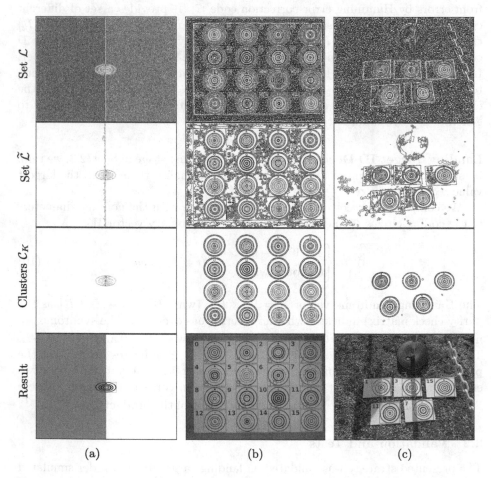

Fig. 8. Algorithm validation: (a) Target under simulated degradations, (b) The 16 targets in an indoor environment, (c) Five targets in an outdoor scenario under non-controlled image degradations

A compilation of the experiments carried out under real conditions can be seen in https://youtu.be/igsQc7VEF2c. In there, notice the presence of other objects, different background textures, irregular shadows, perspective deformations and the change of scale of the landing targets. Despite of that, we always detect all the targets. Conversely, as the recognition and decoding stage depends on the good identification of the reference diameters, only a part of the errors is solved by the Hamming codes. We work on a better decoding phase.

4   Conclusion and Future Extensions

We have described the procedure for the landing target detection and recognition based on a perception model. The detection part is based on the Helmholtz non-accidentalness principle and the Gestalt theory. The non-accidentalness estimation is performed in a multi-feature object space built from the image contours at different scales. This approach allows us to obtain scene information avoiding the loss of information because of the objects' change of size or the presence of shadows and noise. We have used the similarity and proximity Gestalt laws to group the contours and to build a perceptual object of high level; the landing target. Additionally, we perform a recognition stage taking advantage of the target design and the Hamming error codes. The experiments show that the proposed methodology for the detection of landing targets is robust to uncontrolled light conditions and other images degradations existing in complex environments.

For this particular work, the objective was to detect a landing target, a circular object (basic geometric shape) with concentric borders. However, the presented methodology has a wide extension capacity. At the moment we use only the image contours to measure the circularity and the mean intensity gradient. A future extension could include the use of other image primitives, such as points, regions, texture or color, and other object features. Similarly with the grouping laws, we explored only the proximity and similarity laws. It is possible to use other laws such as alignment, symmetry or continuity. These ideas will allow creating new descriptive feature spaces of scenes and objects in an image. That gives the possibility of make combinations between grouping laws, features and image primitives to detect some particular object.

Finally, our contribution proposes a perception model that utilizes the a contrario theory but avoids the mathematical complexity of setting the NFA to adjust threshold values or parameters for specific situations.

Acknowledgments. This research is partially supported by the Mexican National Council for Science and Technology (CONACYT).

References

1. Araar, O., Aouf, N., Vitanov, I.: Vision based autonomous landing of multirotor UAV on moving platform. J. Intell. Robot. Syst. **85**(2), 369–384 (2017)
2. Attneave, F.: Some informational aspects of visual perception. Psychol. Rev. **61**(3), 183–193 (1954)
3. Carrio, A., Sampedro, C., Rodriguez-Ramos, A., Cervera, P.C.: A review of deep learning methods and applications for unmanned aerial vehicles. J. Sensors **2017**, 3296874:1–3296874:13 (2017)
4. Desolneux, A., Moisan, L., Morel, J.M.: From Gestalt Theory to Image Analysis: A Probabilistic Approach. Interdisciplinary Applied Mathematics. Springer-Verlag, New York (2008). https://doi.org/10.1007/978-0-387-74378-3
5. Frey, B.J., Dueck, D.: Clustering by passing messages between data points. Science (New York, N.Y.) **315**(5814), 972–976 (2017)
6. Furukawa, H.: Deep Learning for End-to-End Automatic Target Recognition from Synthetic Aperture Radar Imagery. arXiv:1801.08558 [cs], January 2018
7. Hamming, R.W.: Error detecting and error correcting codes. Bell Syst. Tech. J. **29**(2), 147–160 (1950)
8. Lacroix, S., Caballero, F.: Autonomous detection of safe landing areas for an UAV from monocular images. In: IEEE/RSJ International Conference on Intelligent Robots and Systems (2006)
9. Lange, S., Sünderhauf, N., Protzel, P.: Autonomous landing for a multirotor UAV using vision. In: SIMPAR 2008 International Conference on Simulation, Modeling and Programming for Autonomous Robots, pp. 482–491 (2008)
10. Lee, J., Wang, J., Crandall, D., Šabanović, S., Fox, G.: Real-time, cloud-based object detection for unmanned aerial vehicles. In: 2017 First IEEE International Conference on Robotic Computing (IRC), pp. 36–43, April 2017
11. Lu, C.T., Chen, D., Kou, Y.: Multivariate spatial outlier detection. Int. J. Artif. Intell. Tools **13**(04), 801–811 (2004)
12. Marr, D., Hildreth, E.: Theory of edge detection. Proc. R. Soc. Lond. B **207**(1167), 187–217 (1980)
13. Mumford, D., Shah, J.: Optimal approximations by piecewise smooth functions and associated variational problems. Commun. Pure Appl. Math. **42**(5), 577–685 (1989)
14. Petitot, J.: Neurogéométrie de la vision: modèles mathématiques et physiques des architectures fonctionnelles. Editions Ecole Polytechnique (2008)
15. Reed, I.S., Yu, X.: Adaptive multiple-band CFAR detection of an optical pattern with unknown spectral distribution. IEEE Trans. Acoust. Speech Signal Process. **38**(10), 1760–1770 (1990)
16. Sezgin, M., Sankur, B.: Survey over image thresholding techniques and quantitative performance evaluation. J. Electron. Imaging **13**(1), 146–168 (2010)
17. Wertheimer, M.: FormsUntersuchungen zur Lehre von der Gestalt II. Psycologische Forsch. **4**, 301–350 (1923)
18. Witkin, A.: Scale-space filtering: a new approach to multi-scale description. In: ICASSP 1984. IEEE International Conference on Acoustics, Speech, and Signal Processing, vol. 9, pp. 150–153, March 1984
19. Yao, H., Yu, Q., Xing, X., He, F., Ma, J.: Deep-learning-based moving target detection for unmanned air vehicles. In: 2017 36th Chinese Control Conference (CCC), pp. 11459–11463, July 2017

Parallel and Distributed Local Fisher Discriminant Analysis to Reduce Hyperspectral Images on Cloud Computing Architectures

Rania Zaatour$^{(\boxtimes)}$, Sonia Bouzidi, and Ezzeddine Zagrouba

Université de Tunis El Manar, Institut Supérieur d'Informatique El Manar,
LR16ES06 Laboratoire de recherche en Informatique, Modélisation et Traitement de
l'Information et de la Connaissance (LIMTIC), 2 Rue Abou Raihane Bayrouni,
2080 l'Ariana, Tunisie
rania.zaatour@fst.utm.tn, sonia.bouzidi@insat.rnu.tn,
ezzeddine.zagrouba@uvt.tn

Abstract. Hyperspectral images are data cubes that offer very rich spectral and spatial resolutions. These images are so highly dimensioned that we generally reduce them in a pre-processing step in order to process them efficiently. In this context, Local Fisher Discriminant Analysis (LFDA) is a feature extraction technique that proved better than several commonly used dimensionality reduction techniques. However, this method suffers from memory problems and long execution times on commodity hardware. In this paper, to solve these problems, we first added an optimization step to LFDA to make it executable on commodity hardware and to make it suitable for parallel and distributed computing, then, we implemented it in a parallel and distributed way using Apache Spark. We tested our implementation on Amazon Web Services (AWS)'s Elastic MapReduce (EMR) clusters, using different hyperspectral images with different sizes. This proved higher performances with a speedup of up to 70x.

Keywords: Local Fisher Discriminant Analysis (LFDA)
Cloud computing · Dimensionality reduction · Hyperspectral images
Amazon Web Services (AWS) · Elastic MapReduce (EMR)

1 Introduction

Through the recent years, hyperspectral sensors have evolved so fast to give new hyperspectral images (HSIs) that offer higher spectral and spatial resolutions, and, hence, larger volumes. For example, the space-borne sensor Hyperion can collect up to 71.9 GB of hyperspectral data per hour [1].

Although these HSIs offer richer amounts of information to interpret the captured scene, their higher correlations, redundancy and mainly dimensionality pose important challenges for analysis and processing tasks.

© Springer Nature Switzerland AG 2018
J. Blanc-Talon et al. (Eds.): ACIVS 2018, LNCS 11182, pp. 245–257, 2018.
https://doi.org/10.1007/978-3-030-01449-0_21

Traditionally, instead of working with raw HSIs, we reduce them to their most representative features in order to make the processing task more efficient. However, when dealing with higher-dimensioned HSIs, even this pre-processing step may cause a computational burden and fail to proceed in reasonable amounts of time, using commodity hardware.

Given these facts, the HSI's processing community opted for the cloud computing in order to distribute the existent techniques that work efficiently on small data but fail to scale. To name a few, this was adopted to distribute PCA for HSIs' dimensionality reduction [1], to distribute K-means for HSIs' clustering [2], to define an architecture for HSIs' classification [3], and to restore HSIs [4].

In this paper, we are interested in distributing the Local Fisher Discriminant Analysis (LFDA) [5] a dimensionality reduction technique that proved to outperform many commonly-used methods to reduce various types of data [6–11] including HSIs [12–14], but fails to process some regular lower-dimensioned HSIs (in comparison with Hyperion HSIs) on commodity hardware.

In fact, aiming at maximizing the between-class variance and at minimizing the within-class one, LFDA is computed in a class-wise fashion where we consider one class at a time to compute the final transformation matrix. Not only does this sequential treatment take a long time, but, when reducing large HSIs with large classes on a commodity hardware, the greedy LFDA will fail since it requests memory that is beyond the machine's limited available RAM.

One intuitive solution is to use more performing machines. This may solve the problem of one lower-dimensioned data, but will fail again with larger HSIs as the Hyperion ones. Besides, even if a very large HSI does not contain very large classes and therefore do not need much memory, processing the classes sequentially will take a huge amount of time which is inconvenient.

That is why, we hereby propose a parallel and distributed implementation of LFDA on a cloud architecture using Apache Spark [15] as a computing engine. This latter, extending the MapReduce model, provides parallel distributed processing, fault tolerance and scalability while being fast, general-purpose and easy to use thanks to its online high-level APIs.

To distribute LFDA, we dispatch the work such that, at a time, every worker processes a class. This reduces the time consumption problem but still can fail since our workers are commodity hardware and larger classes will cause the same memory issues as in LFDA's sequential version. To avoid this, we propose to divide larger classes into smaller sub-classes.

Compared to the original result, this will give an altered result but not a wrong one. Actually, assuming that LFDA's function is to maximize the between-class variance and minimize the within-class one, the only difference we will get is the distances between the samples of the same large class that we divided into smaller handleable sub-classes.

In our experiments, we used Amazon's Web Services (AWS) Elastic MapReduce (EMR) clusters and Simple Storage Service (S3) to read/write files from/to the cloud. The efficiency of our proposed approach was measured in terms of the time consumption and the speedup according to the variation of several variables.

The rest of this paper is organized as follows. Section 2 reminds the basics of LFDA. Section 3 details our optimization for the distributed implementation and introduces the parallel and distributed implementation. Section 4 assesses the efficiency of the distributed LFDA. Finally, Sect. 5 concludes this work and gives some future work lines.

2   Local Fisher Discriminant Analysis - LFDA

In [5], Sugiyama introduced the Local Fisher Discriminant Analysis (LFDA) as a combination of the supervised Fisher Discriminant Analysis (FDA) [16], and the unsupervised Locality Preserving Projection (LPP) [17].

LFDA reduces a given data by projecting it in a lower-dimensioned space, where the between-class variance is maximized and the within-class variance is minimized, without altering the samples' locality. That is to say, samples of the same class would not be inevitably made close in the embedding space.

Consider $X = \{x_i | x_i \in \mathbb{R}^d, i \in [1, ..., n]\}$ a data made of n d-dimensioned samples representing c classes, and $Y = \{y_i | y_i \in [1, ..., c], i \in [1, ..., n]\}$ its class-label vector. Let n_l be the number of samples of class l.

In order to reduce X's n-dimensions to $r < n$, LFDA defines its transformation matrix $T_{LFDA} \in \mathbb{R}^{d \times r}$ by maximizing an objective function that uses the local between-class scatter matrix S^{lb} and the local within-class scatter matrix S^{lw}, respectively defined in (2) and (3), as shown in (1).

$$T_{LFDA} = \arg\max_{T} [tr((T^{\top} S^{lw} T)^{-1} T^{\top} S^{lb} T)]. \tag{1}$$

$$S^{lb} = \frac{1}{2} \sum_{i,j=1}^{n} W_{i,j}^{lb} (x_i - x_j)(x_i - x_j)^{\top} \tag{2}$$

$$S^{lw} = \frac{1}{2} \sum_{i,j=1}^{n} W_{i,j}^{lw} (x_i - x_j)(x_i - x_j)^{\top} \tag{3}$$

Where W^{lb} and W^{lw} are respectively defined in (4) and (5), using LPP's affinity matrix A.

$$W_{i,j}^{lb} = \begin{cases} A_{i,j}(\dfrac{1}{n} - \dfrac{1}{n_l}) & \text{if } y_i = y_j = l, \\ \dfrac{1}{n} & \text{if } y_i \neq y_j, \end{cases} \tag{4}$$

$$W_{i,j}^{lw} = \begin{cases} \dfrac{A_{i,j}}{n_l} & \text{if } y_i = y_j = l, \\ 0 & \text{if } y_i \neq y_j. \end{cases} \tag{5}$$

Sugiyama provided a MATLAB implementation of LFDA[1] [5] that takes in two (.mat) files of the data and its labels vector, and the expected reduced

---

[1] Available online at www.ms.k.u-tokyo.ac.jp/software.html#LFDA.

dimension r. As output, it gives the reduced data and T_{LFDA} which is the result of solving a generalized eigenvalue problem that uses S^{lb} and S^{lw}. These initialized-to-zeros matrices are updated, in a class-wise manner, using the $n_l \times n_l$ matrix A that accounts for the distances between the samples of class $l \in [1, c]$.

3   Parallel and Distributed LFDA

3.1   Optimized LFDA for Parallel and Distributed Implementation

If we are to reduce LFDA to three main steps, it would be (a) processing each class to update S^{lb} and S^{lw}, (b) solving the generalized eigenvalue problem based on these matrices in order to define T_{LFDA}, and (c) computing the reduced data. The first step consumes most of the time and memory.

Since realized in a class-wise fashion, we can, intuitively, distribute this first step based on the classes in the HSI: every executor extracts the information from a given class and updates the shared variables S^{lb} and S^{lw}. That is to say, unlike the by-default principle of according a random set of samples to every worker to process, in our case, every worker will be given a class to process.

One major problem of this idea is that when the HSI contains very large classes, these latter require huge amounts of memory to be processed, which can not always be handled by the executors, being commodity hardware, and which may lead to a `Memory error`.

In order to avoid this problem, we propose to consider very large classes as sets of smaller handleable sub-classes. To do so, we start by defining a value of maximum number of samples per class, which we are calling n_{max}. This latter has to satisfy the following constraints: (a) it must define sub-classes that can be processed by the used machines, and (b) since LFDA looks for the k nearest neighbors of every sample, with k being an argument set by default to 7, the smallest sub-class n_{max} will define must contain at least $k + 1$ samples.

Once we set n_{max}, we limit the HSI's large classes, containing more than n_{max} samples, to their first n_{max} samples, and affect new class-labels to the rest of the samples, while assuring that every newly-added class contains at most n_{max} samples. Using this optimizing step, which only changes the labels vector of the HSI, LFDA will be able to run on a commodity hardware and will give a different result that is not a wrong one.

In fact, since LFDA's main purpose is to maximize the between-class variance and to minimize the within-class one while preserving the samples' locality, considering a large class whose samples would be made close using LFDA, with this optimizing step, some of these samples will be made closer according to the defined sub-classes, while having the same locality and still being far from the other classes. Therefore, we can consider the obtained altered result accurate.

3.2   Parallel and Distributed Implementation of LFDA on Spark

Our parallel and distributed LFDA is implemented, following the steps illustrated in Fig. 1, using Python, an in-vogue Spark-supported programming language with a very large and active community.

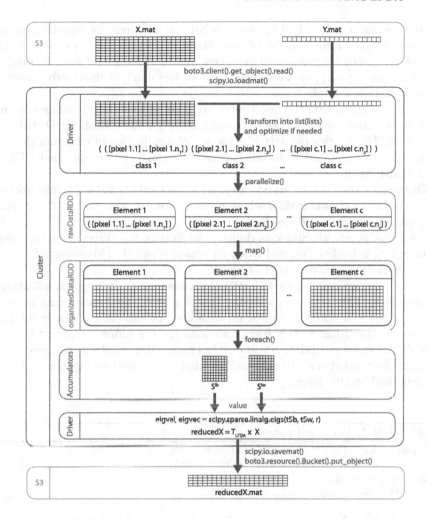

Fig. 1. Overview of the distributed and parallel implementation of LFDA

As shown in Fig. 1, the input HSI, its labels vector and the resulting reduced HSI are stored on Amazon Web Services[2] (AWS)'s Simple Storage Service (S3)[3], in (.mat) format to follow the original MATLAB implementation. Hence, we use Boto 3, the AWS SDK for Python[4], to read/write from/to S3, and Scipy [18] to load/write data from/to (.mat) files.

Our implementation takes in the url of the S3 bucket, two file-names of the HSI and its labels vector, the expected reduced dimension r, and n_{max} for in case the data's large classes need to be divided into smaller sub-classes.

---

[2] Amazon Web Services (AWS) - Cloud Computing Services (aws.amazon.com).

[3] Amazon S3 aws.amazon.com/s3.

[4] Available online at www.github.com/boto/boto3.

First, in one pass, the HSI loaded in a Numpy [19] array, is transformed into a list whose every item is a list of one class's pixels. By pixel, we refer to the array of the spectral signatures of one pixel, and by class, we refer to the original classes containing n_{max} or less pixels, and to the newly-defined sub-classes, if the optimization is applied.

Using `SparkContext`, the main entry point to Spark's functionalities, we use the transformation `parallelize` to transform the list into a Resilient Distributed Dataset (RDD), Spark's primary abstraction of a collection of elements that are distributed and operated in parallel across the cluster. The obtained RDD is called `rawDataRDD`.

Then, using the transformation `map`, we transform `rawDataRDD` into another RDD called `organizedDataRDD` whose every element is a matrix of the spectral signatures of a given class' pixels.

Until this point, all Spark is doing is to update its Direct Acyclic Graph (DAG) which is lazily evaluated when an action is called. This eventually guarantees the fault tolerance aspect of Spark.

In our case, we call `foreach(PerClassProcessing)` on `organizedDataRDD` to evaluate it. This action applies the function `PerClassProcessing` on each element of the RDD, in other words, on each class or sub-class of the input HSI. Detailed in Algorithm 1, this function computes the affinity matrix A of the given class in order to update S^{lb} and S^{lw}. These latter are $d \times d$ matrices stored in Spark's `Accumulators`, shared variables that can be updated by the executors and only read by the driver.

Algorithm 1. PerClassProcessing

Input : $X_c = \{x_i\}_{i=1}^{n_c}$, samples of class c
n, total number of samples in the HSI

1 **for** $i \leftarrow 1$ **to** n_c **do**
2     $x_i^{(7)} \leftarrow$ 7th nearest neighbor of x_i among X_c
3     $\sigma_i \leftarrow \left\| x_i - x_i^{(7)} \right\|$

4 **for** $i, j \leftarrow 1$ **to** n_c **do**
5     $A_{i,j} \leftarrow exp(-\left\| x_i - x_j \right\|^2 / (\sigma_i \sigma_j))$
6 $G \leftarrow X_c \, diag(A1_{n_c}) X_c^\intercal - X_c A X_c^\top$
7 $S^{lb} \leftarrow S^{lb} + G/n + (1 - n_c/n) X_c X_c^\top + X_c 1_{n_c} (X_c 1_{n_c})^\top / n$
8 $S^{lw} \leftarrow S^{lw} + G/n_c$

Now that the most consuming step is realized, and that S^{lb} and S^{lw} are updated by every class' information, the driver reads their `value` and proceed with solving the generalized eigenvalue problem in order to define T_{LFDA}.

4 Experimental Results

4.1 Experimental Setup

In order to assess the efficiency of our parallel and distributed implementation of LFDA, we used AWS's Elastic MapReduce (EMR)[5] clusters.

Our cluster is made of one driver (master) of type `m2.2xlarge` with 4 vCPUs and 34.2 GiB of RAM, and 8 workers (slaves) of type `m4.xlarge` with 4 vCPUs and 16 GiB of RAM each. Each worker's vCPU is a hyperthread of an Intel Xeon E5-2676 v3 [6]. All nodes have Spark 2.3.0, Numpy, Scipy and Boto3 installed.

Our cluster runs, as all EMR clusters, on YARN[7]. This latter is configured to `maximizeResourceAllocation`, to set `spark.rpc.message.maxSize` to 2048, and to set `spark.executor.memoryOverhead` to 2048.

We have submitted our spark applications in client mode while setting the `driver-memory`, the `executor-memory` and the `executor-cores` to 20 GB, 10 GB and 6, respectively.

For comparison, we used a serial Python version which follows *by the book* the MATLAB implementation and adds the optimization step of using n_{max}. This version is tested on a desktop PC equipped with an Intel Core i7-6700HQ (at 2.6 GHz) and 16 GB of RAM.

For our experiments, we used the well-known Pavia Center scene captured by the Reflective Optics System Imaging Spectrometer (ROSIS) sensor over Pavia, northern Italy. It is composed of 1096×715 pixels and of 102 bands. It represents 9 classes of soil covers excluding the background. It is of size 116.5 MB.

We generated 7 other larger datasets by mosaicking Pavia Center, hereinafter noted PaviaC, as follows: PaviaC3, sized 457.4 MB and made of 3 instances of PaviaC; PaviaC5, sized 762.3 MB and made of 5 instances of PaviaC; PaviaC7, sized 1 GB and made of 7 instances of PaviaC; PaviaC10, sized 1.5 GB and made of 10 instances of PaviaC; PaviaC15, sized 2.2 GB and made of 15 instances of PaviaC; PaviaC20, sized 3 GB and made of 20 instances of PaviaC; PaviaC25, sized 3.7 GB and made of 25 instances of PaviaC.

4.2 Accuracy Assessment

Before assessing the performance of our parallel and distributed LFDA, we first assess its accuracy.

Table 1 compares the three highest and lowest eigenvalues we obtain when applying the MATLAB and Python implementations of LFDA to the AVIRIS Indian Pines, a well-known small HSI which LFDA can process on our machine.

As shown in Table 1, both implementations give exactly the same eigenvalues, except one slight difference.

---

[5] Amazon EMR aws.amazon.com/emr.

[6] Amazon EC2 Instance Types (aws.amazon.com/ec2/instance-types).

[7] Apache Hadoop YARN - Yet Another Resource Negotiator hadoop.apache.org/docs/stable/hadoop-yarn/hadoop-yarn-site/YARN.html.

Table 1. Comparison of the obtained eigenvalues when applying MATLAB and Python implementations of LFDA to AVIRIS Indian Pines, with $r = d = 200$

	MATLAB implementation	Python implementation
First highest eigenvalues	14420.87342645	14420.87342645
	3583.75113002	3583.75113000
	1167.22369795	1167.22369795
Last lowest eigenvalues	41.06061845	41.06061845
	44.93439216	44.93439216
	45.61566629	45.61566629

Now that we are sure that MATLAB and Python implementations work the same, we compare the sequential Python implementation to the parallel and distributed one on Spark. As shown in Table 2, both implementations give exactly the same eigenvalues, and hence the same result.

Table 2. Comparison of the obtained eigenvalues when applying sequential and parallel and distributed LFDA to ROSIS Pavia Center, with $r = d = 102$ and $n_{max} = 6530$

	Sequential implementation	Parallel and distributed implementation
First highest eigenvalues	104076.39140465	104076.39140454
	49828.26250469	49828.26250469
	7325.26753262	7325.26753262
Last lowest eigenvalues	212.40586054	212.40586054
	226.32011271	226.32011271
	237.95159471	237.95159471

4.3   Performance Assessment

In this subsection, we computationally assess the performance of our proposed parallel and distributed LFDA.

In order to set the value of n_{max}, we consider $avg = n/c$, the average of the number of samples n in the HSI representing c classes, and set n_{max} to one of its divisors that satisfies the constraints we earlier defined in Sect. 3.1. In our case, the highest handleable divisor of $(1096 \times 715/10)$ is 6530.

Traditionally, in cases where executors treat sets of randomly chosen samples, partition's size has a big impact on the execution time and hence on the speedup. In our case, we can not specify partitions' size since at a time, an executor processes a given predefined class. However, since changing the value of n_{max} affects the sizes of the classes, we can consider it as an equivalent parameter to the partitions' size. That is why we will vary n_{max}'s value in our experiments.

In our first experiment, we studied how the number of executors in the cluster affects the time consumption and the speedup of our parallel and distributed LFDA applied to reduce PaviaC to 3 dimensions, using different values of n_{max}. Figure 2 illustrates the different obtained results.

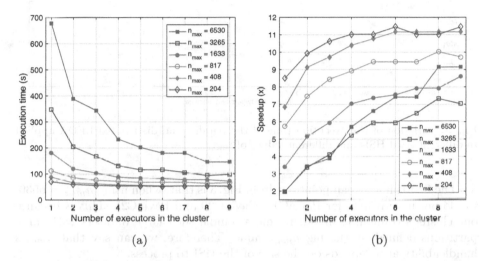

(a)   (b)

Fig. 2. Variation of the performance of the parallel and distributed LFDA applied to ROSIS Pavia Center measured in terms of (a) execution time and (b) speedup, according to different values of n_{max}

As we can see from Fig. 2, the more executors we use, the shorter the execution time gets and the higher the speedup we obtain. This is explained by the fact that the more workers we use, the more classes are processed at a time. Thus, the parallelization will reduce the execution time and increase the speedup.

Besides, when using bigger values of n_{max}, as we use more executors, we observe a more significant decreasing in the execution time and a more important increasing in the speedup, than when using smaller values of n_{max}. In this latter case, adding executors do not seem to add much of a difference to the performance since we are considering smaller classes that do not need much time to be processed. Therefore, we are distributing many light workloads and adding the communication costs that will offset the gain in execution time and speedup.

It is worth mentioning that, in the last experiment, we reduced the lowest-dimensioned of our datasets. In the next experiment, we are assessing the scalability of our parallel and distributed LFDA by trying to reduce the different higher-dimensioned datasets we generated, using 9 executors in the cluster and several values of n_{max}. Figure 3 illustrates the obtained execution times.

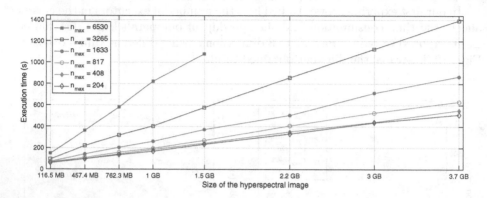

Fig. 3. Variation of the execution time of the parallel and distributed LFDA applied to different sized HSIs and different values of n_{max}

The first thing we can point out from Fig. 3 is that, when using $n_{max} = 6530$, we do not have results for PaviaC15, PaviaC20 and PaviaC25. This is because our cluster's configuration can no more handle such large datasets with large partitions defined by the big n_{max} value. Therefore, we can say that n_{max}'s handleability also depends on the size of the HSI to process.

If we consider the other values of n_{max}, we can conclude that the higher-dimensioned the HSI is, the longer it takes to reduce it. Besides, our proposed parallel and distributed implementation of LFDA scales very efficiently since it gives results below the linearly expected ones. For example, Considering $n_{max} = 3265$, PaviaC is reduced in 94 s and PaviaC25 is reduced in 1394 s, which is lower than the expected 94 s$\times$ 25 = 2350 s.

Aside from this, we can also conclude that $n_{max} = \{408, 204\}$ almost give exactly the same results. This is because, as we mentioned earlier, the communication time covers the time we gain in the execution. Therefore, we can say that under a certain value, it is useless to further reduce the value of n_{max}.

In our final experiment, we studied the variation of the speedup according to the variation of the HSI's size and to several values of n_{max} as shown in Fig. 4.

As we can see from Fig. 4, the larger the data is, the more significant the speedup is. Besides, we can confirm that our distributed LFDA scales well, mainly when using smaller values of n_{max}, where the parallel and distributed implementation of LFDA can run more than 70 times faster.

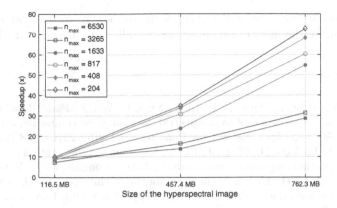

Fig. 4. Variation of the speedup of the parallel and distributed LFDA applied to different sized HSIs and different values of n_{max}

It is worth noting that we do not have speedups for HSIs larger than PaviaC3. This is because our commodity hardware has limited resources that can not handle these data.

5   Conclusion and Future Work

In this paper, we proposed a parallel and distributed implementation of the Local Fisher Discriminant Analysis (LFDA) using Python and Apache Spark.

LFDA is a feature extraction that tries to maximize the between-class variance and to minimize the within-class variance while accounting for the locality of the samples. It proved to be efficient in reducing hyperspectral images (HSIs) but it still suffers from being greedy and unable to process higher-dimensioned HSIs on commodity hardware.

In this work, we first optimized LFDA in order to make it able to run on a commodity hardware and therefore to make it suitable for parallel and distributed computing. To do so, we considered a HSI's large classes as collections of smaller sub-classes that can be handled by the limited resources of a commodity hardware, or of a worker in a cluster. To decide whether a class is large enough to be divided or not, we used a value called n_{max}. Classes that contain more than n_{max} samples are divided into smaller sub-classes with at most n_{max} samples each.

Once LFDA is ready for parallel and distributed computing, we distributed its class-wisely computed part across the cluster so that it still runs in a class-wise fashion, but while many classes get processed at the same time.

Using Amazon Web Services (AWS)'s Elastic MapReduce (EMR) clusters, we tested our proposed parallel and distributed implementation of LFDA using a cluster of one master and 8 workers. The obtained results confirmed the efficiency of our implementation, and mainly guaranteed its scalability property.

In our future works, we will further focus on how to set the values of n_{max} in order to make the most of our proposed parallel and distributed LFDA.

References

1. Wu, Z., Yonglong, L., Antonio, P., Li, J., Xiao, F., Wei, Z.: Parallel and distributed dimensionality reduction of hyperspectral data on cloud computing architectures. IEEE J. Sel. Top. Appl. Earth Obs. Remote. Sens. **9**(6), 2270–2278 (2016)
2. Haut, J.M., Paoletti, M., Plaza, J., Plaza, A.: Cloud implementation of the k-means algorithm for hyperspectral image analysis. J. Supercomput. **73**(1), 514–529 (2017)
3. Quirita, V.A.A., et al.: A new cloud computing architecture for the classification of remote sensing data. IEEE J. Sel. Top. Appl. Earth Obs. Remote. Sens. **10**(2), 409–416 (2017)
4. Yin, X., Wu, Z., Liao, W., Wei, Z., Tan, C.: Cloud implementation of hyperspectral image restoration with PCA and total variation based on spark. In: 2017 IEEE International Geoscience and Remote Sensing Symposium (IGARSS), pp. 3405–3408. IEEE (2017)
5. Sugiyama, M.: Dimensionality reduction of multimodal labeled data by local fisher discriminant analysis. J. Mach. Learn. Res. **8**, 1027–1061 (2007)
6. Shen, P., Lu, X., Liu, L., Kawai, H.: Local fisher discriminant analysis for spoken language identification. In: 2016 IEEE International Conference on Acoustics, Speech and Signal Processing (ICASSP), pp. 5825–5829. IEEE (2016)
7. Chen, H.-L., Liu, D.-Y., Yang, B., Liu, J., Wang, G.: A new hybrid method based on local fisher discriminant analysis and support vector machines for hepatitis disease diagnosis. Expert. Syst. Appl. **38**(9), 11796–11803 (2011)
8. Pedagadi, S., Orwell, J., Velastin, S., Boghossian, B.: Local fisher discriminant analysis for pedestrian re-identification. In: Proceedings of the IEEE Conference on Computer Vision and Pattern Recognition, pp. 3318–3325 (2013)
9. Zhang, S., Lei, B., Chen, A., Chen, C., Chen, Y.: Spoken emotion recognition using local fisher discriminant analysis. In: 2010 IEEE 10th International Conference on Signal Processing (ICSP), pp. 538–540. IEEE (2010)
10. Rahulamathavan, Y., Phan, R.C.-W., Chambers, J.A., Parish, D.J.: Facial expression recognition in the encrypted domain based on local fisher discriminant analysis. IEEE Trans. Affect. Comput. **4**(1), 83–92 (2013)
11. Guo, J., Chen, H., Li, Y.: Palmprint recognition based on local fisher discriminant analysis. JSW **9**(2), 287–292 (2014)
12. Li, W., Prasad, S., Fowler, J.E., Bruce, L.M.: Locality-preserving dimensionality reduction and classification for hyperspectral image analysis. IEEE Trans. Geosci. Remote. Sens. **50**(4), 1185–1198 (2012)
13. Li, W., Prasad, S., Fowler, J.E.: Hyperspectral image classification using gaussian mixture models and markov random fields. IEEE Geosci. Remote. Sens. Lett. **11**(1), 153–157 (2014)
14. Zaatour, R., Bouzidi, S., Zagrouba, E.: Impact of feature extraction and feature selection techniques on extended attribute profile-based hyperspectral image classification. In: Proceedings of the 12th International Joint Conference on Computer Vision, Imaging and Computer Graphics Theory and Applications, vol.4, VISAPP, (VISIGRAPP 2017), pp. 579–586. INSTICC, ScitePress (2017). ISBN 978-989-758-225-7, https://doi.org/10.5220/0006171305790586

15. Zaharia, M.: Fast and interactive analytics over hadoop data with spark. Usenix Login **37**(4), 45–51 (2012)
16. Fisher, R.A.: The use of multiple measurements in taxonomic problems. Ann. Eugen. **7**(2), 179–188 (1936)
17. Niyogi, X.: Locality preserving projections. In: Neural Information Processing Systems, vol. 16, p. 153. MIT (2004)
18. Oliphant, T.E.: Scipy: open source scientific tools for python. Comput. Sci. Eng. **9**, 10–20 (2007)
19. Oliphant, Travis E.: A Guide to NumPy, vol. 1. Trelgol Publishing USA (2006)

Bayesian Vehicle Detection Using Optical Remote Sensing Images

Walma Gharbi[1,2]([envelope]), Lotfi Chaari[2,3], and Amel Benazza-Benyahia[1]

[1] COSIM laboratory, University of Carthage, Tunis, Tunisia
benazza.amel@supcom.rnu.tn
[2] Digital Research Center of Sfax, University of Sfax, Sfax, Tunisia
gharbi.walma@gmail.com, lotfi.chaari@enseeiht.fr
[3] IRIT-ENSEEIHT, University of Toulouse, Toulouse, France

Abstract. Automatic object detection is a widely investigated problem in different fields such as military and urban surveillance. The availability of Very High Resolution (VHR) optical remotely sensed data, has motivated the design of new object detection methods that allow recognizing small objects like ships, buildings and vehicles. However, the challenge always remains in increasing the accuracy and speed of these object detection methods. This can be difficult due to the complex background. Therefore, the development of robust and flexible models that analyze remotely sensed data for vehicle detection is needed. We propose in this paper a hierarchical Bayesian model for automatic vehicle detection. Experiments performed using real data indicate the benefit that can be drawn from our approach.

Keywords: Image processing · Object detection · Vehicle detection
Bayesian methods · Markov Chain Monte Carlo (MCMC)

1 Introduction

Automatic object detection using optical remotely sensed data is of a paramount importance in many application fields such as environmental monitoring [1] and military surveillance [2]. It involves detecting and localizing objects within a studied area using aerial or satellite images. The significant growth of satellite sensors such as Quickbird, GeoEye, Wordldview and IKONOS, has enabled the acquisition Very High Resolution (VHR) optical images in terms of spectral and spatial resolutions. The spatial resolution is very often less than one meter in the panchromatic band and few meters in the spectral bands (approximately less than 5 m). Therefore, the improvement of the spatial resolution has motivated the need to design new object detection methods that allow the recognition of small objects like ships [3], buildings [4] and vehicles [5,6].

During the last decade, thanks to the witnessed progress of machine learning techniques, many approaches have been developed for object detection and more precisely vehicle detection. This category of techniques often involves learning

© Springer Nature Switzerland AG 2018
J. Blanc-Talon et al. (Eds.): ACIVS 2018, LNCS 11182, pp. 258–269, 2018.
https://doi.org/10.1007/978-3-030-01449-0_22

the classifier using a training dataset. Indeed, the classifier receives as input a region that can be either a sliding window [6] or region proposal [7] along with their feature representation (e.g. bounding boxes) and outputs a predicted label (e.g. vehicle or non-vehicle). Feature extraction and classifier training are highly important components in object detection performance. In [8], Canny detector was applied for road location which is an edge detector and it is practically used to isolate the disposable edges and keep the vehicle edges. A prior stating that vehicles are on a road was used to make their extraction easier. In [5], Haar and Histograms of Oriented Gradients (HOG) features were combined as an attempt to detect vehicles in multi-orientations which was successful in classifying vehicle and non vehicle regions. Feature fusion can be valuable for object detection performance since it takes into consideration the different variations of an object and allows a more faithful feature representation. In this context, the authors of [9] have integrated HOG, Haar wavelets, and local binary patterns [10] using an AdaBoost classifier. After feature extraction, a classifier can be trained using different learning approaches such as AdaBoost [9], Support Vector Machine (SVM) [11], conditional random field [12] and Artificial Neural Network (ANN) [6]. In [11], a vehicle detection approach was proposed using Unmanned Aerial Vehicles (UAV) and applied the SVM classifier along with scale invariant feature transform to extract features. A contextual assistance was applied to reduce false positives caused by the very high pixel resolution of UAV that may lead to a duplicate detection of a vehicle.

ANN are able to learn complex patterns that are difficult to analyze by the conventional approaches. Convolutional Neural Network (CNN) are one of neural network architectures that have been successfully applied in vehicle detection. In [6], a CNN classifier was applied and have achieved a successful rate of accuracy which is always important for any object detection problem. This work [6] adopted a sliding window technique that allows to focus on a specific object of interest such as vehicle to guarantee the performance of vehicle detection. Unfortunately, this technique can be difficult to generalize and also time consuming. As an attempt to increase the speed of classification using CNN while maintaining the rate of accuracy, the authors of [7] employed the binary normed gradient as a preliminary step to capture region proposals or candidate vehicles. The training phase involves no less than 3,000 samples to obtain satisfactory accuracy rates. The region proposals is tested using the trained CNN samples. Further success have been achieved using Region-based CNN (R-CNN) methods [13], unfortunately they suffer from a high computational cost. Fast R-CNN [14] and faster R-CNN [15] were introduced as extensions of R-CNN to improve both speed and accuracy. More precisely, faster R-CNN [15] was proposed to deal with the limitations of computing region proposals using Edge Boxes [16] or Selective Search [17] techniques independently from CNN training. In fact, faster R-CNN implements region proposal mechanism using the same CNN designed for training and thereby regional proposal would be a part of CNN which allows to increase the speed. The aforementioned deep learning techniques allow an automatic vehicle detection with a highly satisfying accuracy rates. However,

VHR images are submitted to different disturbance factors and generally have a complex background that should be taken into consideration in the training process. This requires a lot of training samples (e.g. objects and non-objects samples) to learn which leads to a high computational cost. Since deep learning techniques are considered highly dependent on training, low accuracy rates would be obtained if not enough samples are available.

Object detection performance or accuracy rate is highly related to the robustness of the adopted method that should faithfully represent the data and built-in the detection and prediction upon the object of interest. In this context, we can mention Bayesian methods that are known to be highly flexible when handling complex problems in the signal and image processing literature [18,19]. In fact, Bayesian inference is performed to estimate the model parameters and hyperparameters directly from the data based on reliable priors without needing to train the model on a ground truth dataset.

We propose a fully automatic hierarchical Bayesian algorithm for vehicle detection. Our method reformulates vehicle detection as an inverse problem for which the inference is performed in a Bayesian framework. The model relies on the definition of a difference image between the target image and a smoothed one. This difference image is modeled using a Bernoulli-Laplace mixture model involving a Bernoulli distribution to capture zero coefficients, while the non-zero coefficients are modeled by the Laplace distribution. A Markov Chain Monte Carlo (MCMC) [18] sampling scheme is designed to build the inference and derive estimators. The remainder of this paper is organized as follows. Section 2 introduces the proposed Bayesian method for vehicle detection and inference scheme. Experimental results were performed in Sect. 3. Conclusions and perspectives are outlined in Sect. 4.

2 Proposed Bayesian Method for Vehicle Detection

2.1 Problem Formulation

Let X_1 in $\mathbb{R}^{M \times N}$ be the gray scale image. Let also X_2 be a low-pass filtered version of X_1. Denoting x_1 and x_2 as the vectorized version of X_1 and X_2, we can build the difference vector d as

$$d = x_1 - x_2. \tag{1}$$

Vehicle detection is performed by analyzing the difference between X_1 and X_2. In the difference vector d, only significant fluctuations of the image X_1 will be captured since d is supposed to reflect the high frequency behavior of the image X_1. The areas of the scene containing vehicles will have a large value, since they correspond to high frequency content due to the edges of the vehicles while other areas are expected to provide values close to zero. Vehicle detection is formulated as a denoising problem where the observation is a perturbation of the clean version of d. We aim at detecting vehicle objects based on the observation model denoted as

$$y = d + n, \tag{2}$$

where $y \in \mathbb{R}^{M \times N}$ and $d \in \mathbb{R}^{M \times N}$ are the noisy and clean difference vectors, respectively, and $n \in \mathbb{R}^{M \times N}$ is an additive i.i.d. Gaussian noise vector with a covariance matrix $\sigma_n^2 I$. This problem then amounts to derive $\hat{d}$, an estimated version of d from the observation y, where pixels corresponding to vehicles will have a significant difference value.

2.2 Hierarchical Bayesian Model

This section introduces the proposed hierarchical Bayesian model. A vectorized notation of all the variables is adopted. Moreover, we assume that y and d are realizations of random vectors Y and D. In the following, the used likelihood and the different prior distributions are detailed.

Likelihood: Assuming the variance of noise σ_n^2, is an additive Gaussian distribution, thus the likelihood can be denoted as

$$f(y|d, \sigma_n^2) = \left(\frac{1}{2\pi\sigma_n^2}\right)^{\frac{M \times N}{2}} \exp\left(-\frac{\|y - d\|^2}{2\sigma_n^2}\right) \tag{3}$$

where $\| \cdot \|$ denotes the Euclidean norm.

Priors: Our method aims at estimating the unknown parameter vector $\theta = \{d, \sigma_n^2\}$ To this end, we use the prior distributions that conveniently describe these two parameters.

Prior for d: The choice of the Bernoulli-Laplace as a prior distribution for each pixel can be explained by the importance of promoting the sparsity of the target signal in the original space. This facilitates the isolation of non-vehicle pixels that will be captured by the Bernoulli distribution and set to zero.

$$f(d_i|\omega, \lambda) = (1 - \omega)\delta(d_i) + \frac{\omega}{2\lambda} \exp\left(-\frac{|d_i|}{\lambda}\right) \tag{4}$$

where d_i is the ith component of the d vector corresponding to the pixel i in the difference image, ω is a weight belonging to $[0, 1]$, $\lambda \geq 0$ is the parameter of the Laplace distribution and $\delta(.)$ is the Dirac delta function. Assuming the independence between pixels, the prior distribution of the difference vector d writes

$$f(d|\omega, \lambda) = \prod_{i=1}^{M \times N} f(d_i|\omega, \lambda). \tag{5}$$

Prior for σ_n^2: An inverse gamma (IG) prior distribution is employed to model the parameter σ_n^2, since it is a real positive scalar

$$f(\sigma_n^2|\alpha, \beta) = IG(\sigma_n^2|\alpha, \beta) = \frac{\beta^\alpha}{\Gamma(\alpha)}\sigma_n^{-2(\alpha+1)} \exp\left(-\frac{\beta}{\sigma_n^2}\right) \tag{6}$$

where $\Gamma(.)$ stands for the standard gamma function and where the positive reals α and β are the shape and scale hyperparameters. The choice of the inverse gamma distribution for the unknown variance of the normal distribution is common as a non-informative prior [20]. A non-informative prior distribution is chosen to reflect uncertainty when no or little prior information on σ_n^2 is available. The hyperparameters α and β can be either estimated or manually fixed.

Hyperparameter Priors: This work defines a hierarchical Bayesian model that allows to estimate the hyperparameters $\boldsymbol{\Phi} = \{\lambda, \omega\}$ from the observed data.

Hyperprior for ω: Since ω denotes a weight reflecting the rate of non-zero coefficients (sparsity factor), we use a uniform distribution on $[0, 1]$.

Hyperprior for λ: Since $\lambda \in [0, +\infty[$, we assume that $\lambda \sim \mathcal{IG}(\lambda | a, b)$. In order to keep a non-informative prior, the hyperparameters a and b are adjusted to 10^{-3}. Assuming that the individual hyperparameters are statistically independent, the full hyperparameter prior distribution for $\boldsymbol{\Phi} = \{\lambda, \omega\}$ writes

$$f(\boldsymbol{\Phi} | a, b) = f(\lambda | a, b) f(\omega). \tag{7}$$

2.3 Bayesian Inference Scheme

Based on the likelihood and the priors defined in the previous subsections, our goal is to derive an estimation of the unknown parameter vector of $\boldsymbol{\theta}$ as well as the hyperparameter vector $\boldsymbol{\Phi}$. The joint posterior distribution of the parameter vector $\{\boldsymbol{\theta}, \boldsymbol{\Phi}\}$ can be expressed as

$$f(\boldsymbol{\theta}, \boldsymbol{\Phi} | \boldsymbol{y}, \alpha, \beta, a, b) \propto f(\boldsymbol{y} | \boldsymbol{\theta}) f(\boldsymbol{\theta} | \boldsymbol{\Phi}) f(\boldsymbol{\Phi} | a, b). \tag{8}$$

According to the model priors and hyperpriors, the joint posterior distribution can be reformulated as

$$f(\boldsymbol{\theta}, \boldsymbol{\Phi} | \boldsymbol{y}^b, \alpha, \beta, a, b) \propto \left(\frac{1}{2\pi\sigma_n^2} \right)^{\frac{M \times N}{2}} \exp\left(-\frac{\|\boldsymbol{y}^b - \boldsymbol{d}^b\|^2}{2\sigma_n^2} \right)$$

$$\times \prod_{i=1}^{M \times N} \left[(1 - \omega)\delta(d_i^b) + \frac{\omega}{2\lambda} \exp\left(-\frac{|d_i^b|}{\lambda} \right) \right] \times \mathcal{U}_{[0,1]}$$

$$\times \frac{\beta^\alpha}{\Gamma(\alpha)} \sigma_n^{-2(\alpha+1)} \exp\left(-\frac{\beta}{\sigma_n^2} \right) \times \frac{b^a}{\Gamma(a)} \lambda^{-a-1} \exp\left(-\frac{b}{\lambda} \right).$$

Due to the complexity of the posterior distribution in (2.3), it is difficult to analytically derive a simple closed-form expression of the estimators related to $\boldsymbol{\theta}$ and $\boldsymbol{\Phi}$. For this reason, resorting to MCMC sampling techniques is a common strategy to obtain a numerical approximation of the target posterior distribution [18]. To this regard, the Gibbs sampler (GS) [21] allows sampling from the joint posterior distribution in order to build the inference. More precisely, an iterative procedure is used, where at each iteration, Markov Chain samples are generated for each variable by sampling from its conditional distribution with the remaining variables fixed to their current values until convergence. Generally, the convergence is attained when the Markov chain reaches a steady-state that asymptotically characterizes the posterior distribution. After applying the GS algorithm, the obtained samples will not all be considered. For this purpose, the burn-in period is of great significance, since the samples generated during this period have to be discarded because they are not yet asymptotically sampled according to the target distribution. Consequently, the rest of the samples is used to calculate the target estimators. The iterative procedure is used to generate samples, at each iteration, according to the following conditional posteriors: $f(\boldsymbol{d}|\boldsymbol{y}, \omega, \lambda, \sigma_n^2)$, $f(\sigma_n^2|\boldsymbol{y}, \boldsymbol{d}, \alpha, \beta)$, $f(\lambda|\boldsymbol{d}, a, b)$ and $f(\omega|\boldsymbol{d})$. The main steps of the proposed sampling algorithm are summarized in Algorithm 1. where S

Algorithm 1. Gibbs Sampler (GS).

- Initialize with some $\boldsymbol{d}^0$.
  for $s = 1 \ldots S$ **do**
      Sample σ_n^2 according to Eq.(9);
      Sample λ according to Eq.(10);
      Sample ω according to Eq.(11);
      for $i = 1$ *to* $M \times N$ **do**
         | Sample d_i according to Eq.(12);
      end
  end

---

indicates the number of iterations fixed properly, to be large enough, in order to ensure reaching the convergence. As a preliminary stage, we fixed the number of iterations based on a couple of experiments while gradually varying its value. Inference was drawn using the Maximum A Posteriori (MAP) estimator $\widehat{\boldsymbol{d}}$ that seeks to maximize the posterior distribution. For the proposed model, this estimator operates in two steps where the first step introduces a discrete variable γ_i to determine if each coefficient d_i is equal to zero. The second step allows to estimate a non-zero value of each d_i if this coefficient belongs to the non-zero class of coefficients. The hierarchical MAP allows to properly recover zero coefficients promoted by the Bernoulli distribution, which enforces the sparsity of the target image. As regards the estimators $\widehat{\sigma_n^2}$, $\widehat{\lambda}$ and $\widehat{\omega}$, they are obtained by minimizing the expected error. Hence, we estimate the variance of noise and the hyperparameters according to the Minimum Mean Square Error (MMSE) principle. To this end, the estimators $\widehat{\sigma_n^2}$, $\widehat{\lambda}$ and $\widehat{\omega}$ can be computed by averaging the samples obtained after discarding those generated during the burn-in period. The conditional distributions used in the GS (namely Algorithm 1) are detailed in the following.

Sampling According to $f(\sigma_n^2|\boldsymbol{y}, \boldsymbol{d}, \alpha, \beta)$: The following posterior is computed by combining the likelihood and the prior distribution of σ_n^2. The target distribution writes

$$\sigma_n^2|\boldsymbol{d}, \boldsymbol{y}, \alpha, \beta \sim \mathcal{IG}\left(\sigma_n^2|\alpha + \frac{M \times N}{2}, \beta + \frac{\|\boldsymbol{y} - \boldsymbol{d}\|^2}{2}\right). \tag{9}$$

Sampling According to $f(\lambda|\boldsymbol{d})$: Calculations lead to the following expression of this distribution

$$\lambda|\boldsymbol{d} \sim \mathcal{IG}\left(\lambda|a + \|\boldsymbol{d}\|_0, b + \|\boldsymbol{d}\|_1\right) \tag{10}$$

where the number of non-zero coefficients is computed using the l_0 pseudo-norm denoted as $\|.\|_0$, while $\|.\|_1$ denotes the l_1 norm defined as $\|\boldsymbol{d}\|_1 = \sum_{i=1}^{M \times N} |d_i|$.

Sampling According to $f(\omega|\boldsymbol{d})$: Straightforward calculations similar to [22] show that the posterior of ω is a beta distribution

$$\omega|\boldsymbol{d} \sim \mathcal{B}(1 + \|\boldsymbol{d}\|_0, 1 + M \times N - \|\boldsymbol{d}\|_0). \tag{11}$$

according to which it is easy to sample.

Sampling According to $f(\boldsymbol{d}|\boldsymbol{y}, \omega, \lambda, \sigma_n^2)$: Under the assumption of independence between signal coefficients, the conditional distribution can be expressed as

$$f(d_i|\boldsymbol{y}, \omega, \lambda, \sigma_n^2) = \omega_{1,i}\delta(d_i) + \omega_{2,i}\mathcal{N}^+(\mu_{i+}, \sigma_n^2) + \omega_{3,i}\mathcal{N}^-(\mu_{i-}, \sigma_n^2). \tag{12}$$

where $\mathcal{N}^+$ and $\mathcal{N}^-$ denote the truncated Gaussian distribution on $\mathbb{R}_+$ and $\mathbb{R}_-$, respectively. The weights $(\omega_{l,i})_{1 \leq l \leq 3}$ in (12) are further detailed in [23] and can be expressed as $\omega_{l,i} = \frac{\mu_{l,i}}{\sum_{l=1}^{3} \mu_{l,i}}$. To sample from (12), a two-step procedure is requested. We first introduce a discrete variable γ_i that will be equal to 0 if the candidate coefficient has to be sampled according to the Dirac function (equal to zero). If $\gamma_i = 1$, an additional discrete variable κ_i has to be introduced to determine, using the weights $\omega_{2,i}$ and $\omega_{3,i}$, whether the candidate coefficient has to be sampled according to $\mathcal{N}^+(\mu_{i+}, \sigma_n^2)$ or $\mathcal{N}^-(\mu_{i-}, \sigma_n^2)$.

3   Experimental Results

In this section, two experiments are conducted. The test were performed on an Intel Core i5CPU @ 1.60 GHz using a MATLAB implementation. The first experiment is a validation of the proposed model over a real dataset while the second experience allows to evaluate the performance of the proposed model by comparing it to another state of the art vehicle detection method.

3.1   First Experiment: Validation

A Google Earth image of the region of Benguerdane, Tunisia is employed. Figure 1 depicts the acquired satellite image of interest of size 500×900 with 26 m spatial resolution. Two images of size 140×140 have been extracted from the original one to represent two different contexts. The first context involves the analysis of two vehicles that have a similar shape. However, the second context introduces two vehicles of different shapes that overlap. The three images are illustrated in Fig. 2. A 2-D Gaussian smoothing kernel with standard deviation of 2 was applied to X_1 in Fig. 2(a), in order to get the low frequency image X_2. After running the proposed model for 200 iterations (823 s) and a burn-in period of 100, the obtained binary detection map is depicted in Fig. 2(b). The overlay of the binary detection map vehicle boundaries over X_1 is displayed in Fig. 2(c). It is clear that the detected edges match the vehicles in the scene. The second context contains two vehicles of different shapes: a truck and a car. Figure 3(a) depicts the image X_1. Based on the proposed model, the estimation of the binary detection map is illustrated in Fig. 3(b). Figure 3(c) depicts the vehicle boundaries overlay. In both contexts, we have succeeded to detect two vehicles in different shapes even when they are overlapping.

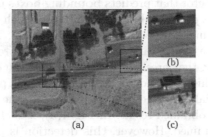

(a)   (b)   (c)

Fig. 1. Studied area over the region of Benguerdane, Tunisia (a), images of the first (b) and second context (c).

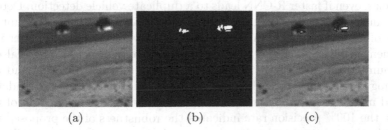

(a)   (b)   (c)

Fig. 2. Image X_1 (a), the binary detection map (b) and vehicle boundaries overlay (c).

(a)   (b)   (c)

Fig. 3. Image X_1 (a), the binary detection map (b) and vehicle boundaries overlay (c).

3.2   Second Experiment: Comparison

The performance and effectiveness of the proposed model can be further evaluated by conducting a comparison with another vehicle detection method. Thus, a comparison with a recent deep learning technique such as Faster R-CNN implemented in MATLAB is performed. This Faster R-CNN procedure[1] is (available for public) initially trained with a 295 images and inspired from the work in [15]. The authors proposed a deep fully convolutional network that proposes regions and a Fast R-CNN detector that predicts bounding boxes using hand-picked priors. The classification is performed in the final layer which is based on a softmax function [24] that computes a probability between 0 and 1. This exponential function will increase the maximum value probability of the previous layer compared to other values. Finally, the threshold is applied over the computed probability to separate between vehicle and non-vehicle objects. The pre-trained detector of faster R-CNN was initially applied over the images in the previous experiment. Figure 4 illustrates the obtained result. In Fig. 4(a), it is clear that no successful vehicle detection is obtained. However, this detection is partial in Fig. 4(b). A second trial was performed using another optical remotely sensed dataset which consists of an aerial imagery with 15 cm spatial resolution[2]. The network of the faster R-CNN was trained using 263 similar labeled images. Figure 5 illustrates the obtained results for our proposed hierarchical Bayesian approach and faster R-CNN. It is clear that both methods perform similarly and are unable to detect all vehicles, even if faster R-CNN leads to a duplicate vehicle detection. Detection performance is evaluated using the displayed criteria in the Table 1. Count is the number of detected vehicles in the studied scene where the true number in this experiment is 16. TP, FP and FN are true positive, false positive and false negative counts. F-measure is the weighted harmonic mean of precision and recall. According to the obtained values, we can note that the proposed model have detected more vehicles than faster R-CNN and obtained a better rate of recall. Besides, the 100% precision rate indicates the robustness of the proposed model to complex background and noise. As a conclusion, this comparison allowed us

---

[1] https://www.mathworks.com/help/vision/examples/object-detection-using-faster-r-cnn-deep-learning.html.

[2] http://gdo-datasci.ucllnl.org/cowc/.

to position along another recent state of the art method proposed for object detection and more precisely vehicle detection. Although deep learning methods allow high accuracy rates, their performance is highly dependent on the training. Since a lot of training samples (e.g. objects and non-objects samples) is needed for learning-based methods (approximately in the order of 1000 examples per class for a valid detection), the computational time can reach days; in our case 6 hours of training for only 263 images. However, the acquisition of objects and non-objects samples can be difficult due to the unavailability of reference data which limits the ability to add semantic relevance. In contrary, our method do not require any expert action. Although, the average response rate for an area of 500×500 pixels is approximately 5 h, our method is data-independent and can be applied over images with different spatial resolutions while maintaining a valid detection. In addition, it minimizes the need for conventional field investigation to acquire reference samples in real applications and thereby reduce the operational cost. Moreover, our proposed Bayesian method is more flexible than the faster R-CNN. In fact, many parameters were chosen empirically such as the size of filter and the threshold. However, analyzing remotely sensed data requires a certain flexibility for handling complex problems.

Fig. 4. Faster R-CNN obtained results over the first (a) and second context (b).

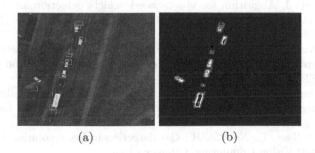

Fig. 5. Results of the Faster R-CNN detector (a) and the proposed model (b).

Table 1. Vehicle detection accuracy measures comparison.

Methods	Count	TP	FP	FN	Precision (%)	Recall (%)	F-measure (%)
Faster R-CNN detector	10	6	4	10	60	37.5	46
The proposed model	8	8	0	8	100	50	66.67

4 Conclusion

In this work, we addressed the topic of object detection and more precisely vehicle detection using optical remotely sensed images. We proposed a flexible hierarchical Bayesian model that enabled the achievement of promising results. In contrary to learning-based methods, the proposed method allows a fully automatic vehicle detection without any user intervention. For future works, we will focus on applying our proposed method on larger datasets with higher resolutions.

References

1. Chen, Y.-L., et al.: Intelligent urban video surveillance system for automatic vehicle detection and tracking in clouds. In: 27th International Conference on (AINA), pp. 814–821. IEEE (2013)
2. Lemelson, J.-H., Pedersen, R.-D.: GPS vehicle collision avoidance warning and control system and method, November 9, US Patent 5,983,161 (1999)
3. Zhu, C., Zhou, H., Wang, R., Guo, J.: A novel hierarchical method of ship detection from spaceborne optical image based on shape and texture features. TGRS **48**(9), 3446–3456 (2010)
4. Stankov, K., He, D.-C.: Detection of buildings in multispectral very high spatial resolution images using the percentage occupancy hit-or-miss transform. IEEE J. Sel. Top. Appl. Earth Obs. Remote. Sens. **7**(10), 4069–4080 (2014)
5. Hu, Y., et al.: L Algorithm for vision-based vehicle detection and classification. In: IEEE International Conference on Robotics and Biomimetics (ROBIO), pp. 568–572. IEEE (2013)
6. Chen, X., Xiang, S., Liu, C.-L., Pan, C-H.: Vehicle detection in satellite images by parallel deep convolutional neural networks. In: Pattern Recognition (ACPR), pp. 181–185. IEEE (2013)
7. Qu, S., Wang, Y., Meng, G., Pan, C.: Vehicle Detection in Satellite images by incorporating objectness and convolutional neural network. J. Ind. Intell. Inf. **4**(2) (2016)
8. Ram, T.-Z., Zhao, T., Nevatia, R.: Car detection in low resolution aerial images. In: Image and Vision Computing. Citeseer (2001)
9. Grabner, H., Nguyen, T.-T., Gruber, B., Bischof, H.: On-line boosting-based car detection from aerial images. ISPRS J. Photogramm. Remote. Sens. **63**(3), 382–396 (2008)

10. Ojala, T., Pietikainen, M., Maenpaa, T.: Multiresolution gray-scale and rotation invariant texture classification with local binary patterns. IEEE Trans. Pattern Anal. Mach. Intell. **24**(7), 971–987 (2002)
11. Moranduzzo, T., Melgani, F.: Automatic car counting method for unmanned aerial vehicle images. TGRS **52**(3), 1635–1647 (2014)
12. Zhong, P., Wang, R.: A multiple conditional random fields ensemble model for urban area detection in remote sensing optical images. TGRS **45**(12), 3978–3988 (2007)
13. Girshick, R., Donahue, J., Darrell, T., Malik, J.: Rich feature hierarchies for accurate object detection and semantic segmentation. In: Proceedings of the IEEE conference on computer vision and pattern recognition, pp. 580–587 (2014)
14. Gkioxari, G., Girshick, R., Malik, J.: Contextual action recognition with R-CNN, in: Proceedings of the ICCV, pp. 1080–1088 (2015)
15. Ren, S., et al.: Faster R-CNN: Towards real-time object detection with region proposal networks. In: Advances in NIPS, pp. 91–99 (2015)
16. Zitnick, C.L., Dollár, P.: Edge boxes: locating object proposals from edges. In: Fleet, D., Pajdla, T., Schiele, B., Tuytelaars, T. (eds.) ECCV 2014. LNCS, vol. 8693, pp. 391–405. Springer, Cham (2014). https://doi.org/10.1007/978-3-319-10602-1_26
17. Uijlings, J.: Selective search for object recognition. Int. J. Comput. Vis. **104**(2), 154–171 (2013)
18. Robert, C., Castella, G.: Monte Carlo Statistical Methods. Springer, New York (2004). https://doi.org/10.1007/978-1-4757-4145-2
19. Huerta, G.: Multivariate bayes wavelet shrinkage and applications. J. Appl. Stat. **32**(5), 529–542 (2005)
20. Tiao, G.-C., Tan, W.Y.: Bayesian analysis of random effect models in the analysis of variance. I. Posterior distribution of variance-components. Biometrika **52**(1/2), 37–53 (1965)
21. Robert, C.-P.: Monte carlo methods, Wiley Online Library (2004)
22. Dobigeon, N., Hero, A.-O., Tourneret, J.-Y.: Hierarchical bayesian sparse image reconstruction with application to MRFM. IEEE Trans. Image Process. **18**(9), 2059–2070 (2009)
23. Tourneret, J.-Y., Chaari, L., Batatia, H.: Sparse bayesian regularization using bernoulli-laplacian priors. In: (EUSIPCO), Marrkech, Morocco, September, 9–13 (2013)
24. Shen, W., Wang, X., Wang, Y., Bai, X., Zhang, Z.: Deepcontour: A deep convolutional feature learned by positive-sharing loss for contour detection. In: Proceedings of the IEEE Conference on Computer Vision and Pattern Recognition, pp. 3982–3991 (2015)

Integrating UAV in IoT for RoI Classification in Remote Images

Loretta Ichim and Dan Popescu$^{(\boxtimes)}$

University Politehnica of Bucharest, Bucharest, Romania
{loretta.ichim,dan.popescu}@upb.ro

Abstract. The paper presents a cheap and efficient solution for remote processing of images taken by a team of UAVs (Unmanned Aerial Vehicles). The work objective was to implement an integrated system for detection and classification of regions of interest (RoIs), in the case of flood events. The UAVs are considered as objects of the internet. This means the integration of UAVs in IoT (Internet of Things) as intelligent objects. The investigated RoIs are: water, grass, forests, buildings, and roads. A multi UAV – multi GCS (Ground Control Station) solution is proposed. Due to this integration, land segmentation by image processing can be efficiently made in real time. For RoI detection and evaluation a multi CNN structure is used as a multi classifier structure. Particularly, a CNN classifier is implemented for each type of RoI and all the CNNs work in parallel. The orthophotoplan obtained from remote acquired images are successively decomposed in adjacent images of size 6000 × 4000 and next in overlapping patches of size 65 × 65 pixels for classifier learning or for testing. Finally, the images are segmented in RoIs by a multi-mask technique and the percentage of each RoI is calculated. The accuracy of segmentation and the processing time, evaluated from 10 real images, was better than in other reported cases.

Keywords: UAV surveillance · CNN based classification · Remote images
RoI segmentation · IoT integration

1 Introduction

The integration of UAS (Unmanned Aerial System) in IoT (Internet of Things) is very useful in various fields like: critical infrastructure surveillance, search and rescue missions, border surveillance and aerial photogrammetry [1]. This is because a system based on multiple GCS (Ground Control Station), GDT (Ground Data Terminal) and UAV (Unmanned Aerial Vehicle) make these missions more efficient. Also it can be observed that a good channel for transferring data between local GCSs and central GCSs is using internet as a communication layer [2]. In order to manage more sophisticated tasks such as large area surveillance and monitoring or multiple target tracking, teams of multiple UAVs should be deployed. This requires more complex control, coordination and cooperation mechanisms. These mechanisms have already been studied and developed, first by simulation for virtual UAVs as a part of multi-agent systems. In [3] the authors present solutions to the problems that appear in the

© Springer Nature Switzerland AG 2018
J. Blanc-Talon et al. (Eds.): ACIVS 2018, LNCS 11182, pp. 270–282, 2018.
https://doi.org/10.1007/978-3-030-01449-0_23

process of integrating two hardware UAVs into an existing multi-agent simulation system with additional virtual UAVs.

Flat trajectory generation for way-points relaxations and obstacle avoidance is used in [4] to solve the photogrammetry problem where the trajectory has to efficiently cover an a priori given region (according with some cost, in our case taken as the total path length). Otherwise, this particular topic is the subject of major interest in the literature [5, 6].

There are many studies based on the application of cloud computing to the rapid processing of remote sensing images, especially in many large-scale real-time monitoring, such as meteorological monitoring and natural disaster warning [1]. Cloud computing contributes to remote sensing applications for maximizing imagery resources, disseminating on-demand insight to end users globally, and improving efficiency while reducing cost and risk. Some researchers approach IoT for monitoring air pollutants with an UAV in a smart city one of the biggest problems nowadays. Air pollution causes serious problems which can affect people's health and not only. The IoT concept allows UAVs to become an integral part of IoT infrastructure and is also, capable of measuring anything anywhere, and capable of flying in a controlled airspace with a high degree of autonomy.

Recently, there are a number of papers to detect RoIs from large aerial imagery [7–11]. Saito et al. [7], have developed a system based on CNN, for automatically detecting buildings and roads directly from aerial imagery. Also, a deep learning approach to semantic segmentation of very high resolution (aerial) image is developed in [10, 13]. The classified categories were: impervious surfaces, buildings, low vegetation, trees, and cars. In [8] the authors proposed a system to train discriminatively CRF models (Conditional Random Field) for semantic segmentation of urban classes from high resolution satellite/aerial imagery. Eight different classes were considered in the application proposed by the authors. Also, another study [9] addresses the multiclass semantic segmentation of urban areas in high-resolution aerial images such as, satellite with class specific object priors for buildings and roads.

The main goal of this paper is the detection, classification, and evaluation of RoIs from UAV high resolution images, based on a multi CNN classifier structure. For a flexible, fast processing and storage solution an integrated UAV-IoT configuration is proposed.

2 Proposed Method

The images acquired by the UAV team during the flight mission are primary processed in a local processing unit (UAV and local GCS) and next transmitted via internet to a central processing and control unit. UAV control is also remotely made via the internet [2]. The images are integrated into an orthophotoplan and then, the cropped images are decomposed in patches to be classified as different RoIs: flood, grass, trees, agricultural crops, buildings, and roads. For each of them a CNN (Convolutional Neural Network) based classifier is implemented.

2.1   Image Acquisition by UAV Surveillance

For image acquisition a team of UAVs is used (Fig. 1). The functional model of the system composed from two fixed–wing, electrically powered type UAVs connected with a remote central unit via internet is presented. The abbreviations are detailed in Table 1.

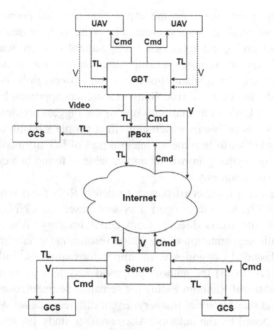

Fig. 1. Experimental model UAV- IoT

Table 1. Abbreviations in Fig. 1.

Abbreviations	Significance
UAV	Unmanned Aerial Vehicle
Cmd	Commands
TL	Telemetry data
V	Video frames or Images
GDT	Ground Data Terminal
GCS	Ground Control Station

There are two groups of components: aerial components (UAVs) and ground system components (GCS, GDT, IPBox). Two GCS categories contribute to UAV control: local GCS and remote GCS, connected by internet. The hardware developed in MUROS project [14] used to develop GCS is similar to a PC, with some differences: (a) Proprietary shell (design and manufacturing in our facility); (b) 120 GB or 240 GB

SSD; (c) High frequency CPU (>3 GHz) with 4 or more cores; (d) Compatible motherboard (it is mandatory the presence of the Serial Port for communication with our control chip for GCS controllers (Buttons, joystick, etc.); (e) Industrial monitor (1024 × 768 resolution); and (f) Joystick, buttons and a custom-made chip for controlling them.

The advantage of multiple UAVs surveillance consists in better area coverage per unit time. The path generation for image acquisition depends on the imposed mission: photogrammetry or isolated images. In Fig. 2 the trajectory for an orthophotoplan generation is presented in the case of the three UAVs. The images are successively taken such that the images are partially overlapped.

Fig. 2. Dividing the photogrammetry route between multiple UAVs

2.2   Image Processing Subsystem

The basic image processing methodology consists on image decomposition in patches and their classification in different classes corresponding to RoIs. The images acquired by UAV are transmitted via internet to the image processing subsystem (IPS) located in the Server. The IPS contains two main modules: Image Primary Processing (IPP), and Image Segmentation (IS). In IPP module (Fig. 3) the overlapped images taken by UAV are first cleaned by noise (Buffer/Noise filter) and then assembled to create the orthophotoplan. Adjacent images of size 6000 × 4000 are cropped and sent to IS (Fig. 4) to be segmented in five RoIs - classes: water (W), forest (F), grass (G), buildings (B), and roads (R). Here, the images are first decomposed in color components and then they are partitioned in patches of size 65 × 65 pixels. The IPS has two phases: learning and operating phase. Each class (W, F, G, B, and R) has a CNN (Convolutional Neural Network) as assigned classifier (Figs. 4 and 5): CNN W, CNN F, CNN G, CNN B, and CNN R. This ensemble constitutes a network of parallel CNNs named NCNN. If the output of a particular CNN is positive (meaning that it indicates the patch contains the corresponding RoI) for an analyzed patch, then this is marked in the initial image (segmentation step – Segm.) and its area is summed for RoI evaluation (evaluation step – Eval.). If the analyzed patch does not belong to any classes, then it is considered as non RoI (belonging to the complementary class NROI).

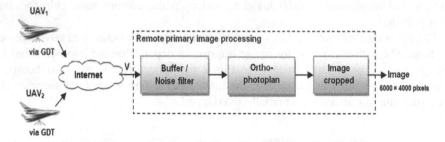

Fig. 3. The IPP module

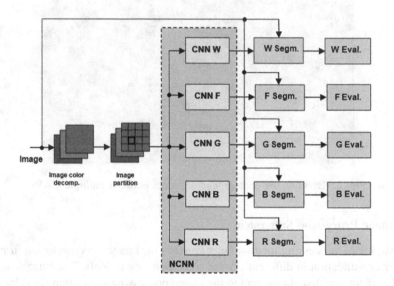

Fig. 4. The IS module

2.3   CNN for RoI Classifier

Each ROI has dedicated a CNN as classifier. The CNN is comprised by five convolutional layers followed by five pooling layers, one fully connected layer, one ReLu activation layer, and a Softmax layer (Fig. 5). As it was mentioned before, the input data for the proposed CNN structure are color patches of 65×65 pixels. These are passed through the entire network to get their classification into five pairs of classes: WATER (W) and NON-WATER (NW), FOREST (F) and NON-FOREST (NF), GRASS (G) and NON-GRASS (NG), BUILDINGS (B) and NON-BUILDINGS (NB), and ROAD and NON-ROAD (NR). The CNN architecture (CNN W) for water patch classification is presented in Fig. 5 as an example.

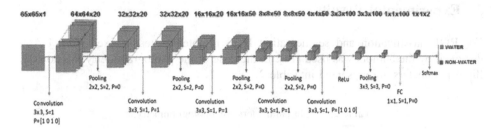

Fig. 5. The CNN for defining the classes WATER (W) and NON-WATER (NW)

All the primary processing steps applied on the learning set will be performed on the testing set too. In order to provide a large dataset, patches of 65 × 65 pixels were extracted in the learning phase, from different images. The patch dimension is experimentally chosen taken into account the image resolution and the RoI types. Each patch is labeled using the manual segmentation provided. Each positive patch has four associated learning patches: the initial patch and patches rotated with 90, 180, and 270°.

The CNN receives preprocessed images of the selected color components (red, green, and blue) from the Sliding Box Decomposition module. The first layer is a convolutional layer with the convolution window of 3 × 3 pixels, the stride 1 and the padding equal to [1010]. This means that the results are boxes with dimension of 64 × 64 pixels. Thus, next successive divisions with 2 and pooling operations without losses are permitted. The layer contains 20 filters considered as neurons. They are initialized with random numbers from a Gaussian distribution. The second layer in the CNN structure is a pooling layer which reduces the space dimension by half with a sliding box of 2 × 2 pixels. The second convolutional layer has its parameters similar to those of the first convolutional layer, with the difference that the classical padding (P = 1) is made in order to preserve the box size. The corresponding pooling layer is the same as the precedent, reducing to half the size of the feature mapping from the previous layer. The third convolutional layer does not change the size of the feature maps, but through it the deep is increased to 50 neurons. Using similar structures, after a ReLu layer (rectified linear unit), boxes of dimension 3 × 3 and deep of 100 filters are obtained. The last convolutional layer is a fully connected layer which reduces the spatial resolution to a spatial resolution of 1 × 1 and a depth of 2 (corresponding to labels WATER and NON-WATER). The last layer of this neural network architecture is a layer called Softmax. This is a "loss" layer because a loss function measures how badly the network is doing for the input.

In the classification process (operating phase) the CNN receives an input box, with the same size like the training box (65 × 65 pixels), and returns, for the central pixel, a weight for each class. For each pixel, we consider the class with the higher weight. The pixels classified by CNN are lastly considered for the segmented image reconstruction. In this module the pixels are integrated in the segmented image. Now, morphological operations (erosion and dilation) are made in order to eliminate possible noises due to segmentation process.

3 Experimental Results

For ROI classification and segmentation from remote images, a fixed-wing UAV, designed by the authors was used [14]. In the case of the photogrammetry mission the flight parameters are presented in Table 2.

Table 2. Flight parameters for photogrammetry.

Parameters/Characteristics	Values
Eight from ground	300 m
Ground resolution	3.8×3.8 cm^2/pixel
Width (ground)	154 m
Length (ground)	230 m
Image resolution	6000×4000 pixels
Longitudinal coverage	80%
Cross coverage	65%
Total area coverage for experiment	1.5×1 km^2
Payload type	Gyro-stabilized
Navigation	GIS-based
Camera characteristics	Objective 50 mm, 24.3 megapixels, 10 fps

For the first step of learning phase (CNN training) 100 images of size 6000×4000 are arbitrary cropped in 2500 patches of dimension 65×65 pixels manually classified in the five classes (500 for each class). The patches are rotated with 90, 180, and 270° – totally 2000 patches per class for the learning phase. Examples of patches for different classes (positive samples) and also example of negative samples for W (NW), F (NF), G (NG), B (NB), and R (NR) are given in Fig. 6. The negative samples contain images complementary to the corresponding class. For the negative images of each class, 2000 patches from complementary classes are considered. For example, NW is learning with 500 patches from each complementary class (F, G, B, and R); similarly for NF, NG, NB, and NR. A patch is considered as belonging to NRoI class if the conditions (1) are simultaneously fulfilled.

$$P \in NW, \quad P \in NF, \quad P \in NG, \quad P \in NB, \quad and \quad P \in NR \tag{1}$$

In order to represent a patch in MatConvNet soft [15] for CNN, each patch is transformed from the matrix form to a vector form by concatenating the lines. Each line is composed by concatenating the lines from a box and also contains the box label. Thus, a training file is a structure with two fields: data and labels. Each line from a training file contains a vector of 4225 elements, in the data field, which represents a positive or a negative box (65×65 pixels), and also the corresponding label, in the label field. We trained the proposed CNN for 50 epochs (experimentally chosen). We can consider that the network mostly completed their training within 50 epochs. The learning rate varied between 0.0005 and 0.005 depending of epoch number. Note that

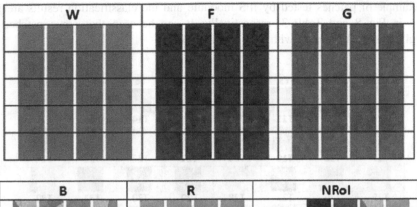

Fig. 6. The patches for defining the classes *W, F, G, B, R*, and *NRoI* (learning phase)

the Fig. 7 represents in the left graphic the training error (blue) and the validation error (green) across the epochs. The kernel sizes and the number of filters per layer (the deep) were also experimentally and lead to the best road segmentation accuracy.

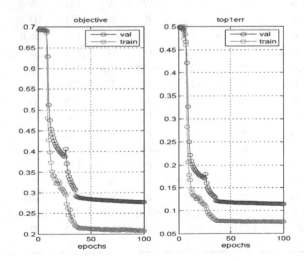

Fig. 7. Objective function and error versus epochs (Color figure online)

Example of patches tested by IPS moodule and the classification results are presented in Fig. 8 and Table 3, respectivelly. It can be seen that some patches are wrongly classified (marked with grey in Table 3).

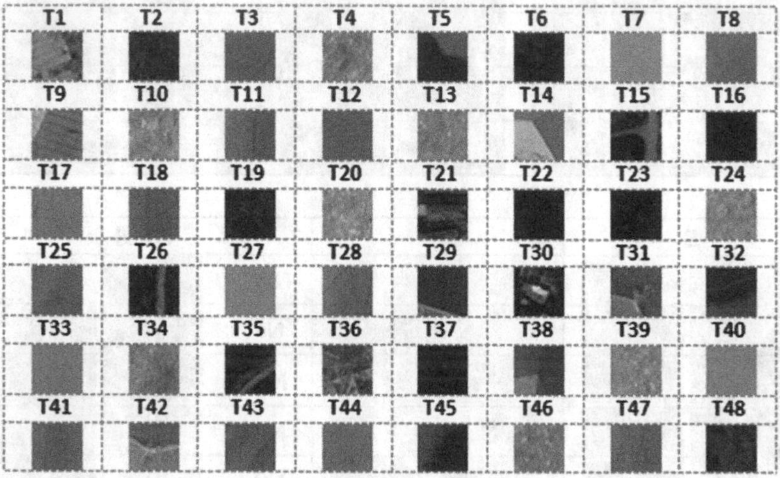

Fig. 8. Examples of patches for system testing

Table 3. True class and CNN classification results for examples in Fig. 8.

Patch	Real class	CNN classif.	Patch	Real class	CNN classif.	Patch	Real class	CNN classif.
T1	B	B	T17	W	W	T33	W	W
T2	F	F	T18	R	R	T34	G	G
T3	W	W	T19	F	F	T35	NROI	NROI
T4	G	G	T20	G	G	T36	B	B
T5	NROI	NROI	T21	NROI	NROI	T37	F	F
T6	F	F	T22	F	F	T38	B	B
T7	W	W	T23	F	F	T39	G	G
T8	R	R	T24	G	G	T40	W	W
T9	B	R	T25	R	R	T41	R	R
T10	G	G	T26	NROI	F	T42	B	B
T11	R	R	T27	W	W	T43	R	R
T12	W	W	T28	R	R	T44	W	W
T13	G	G	T29	B	B	T45	NROI	F
T14	B	R	T30	NROI	B	T46	G	G
T15	NROI	NROI	T31	B	B	T47	R	R
T16	F	F	T32	NROI	NROI	T48	F	F

For each CNN we have established weights that will be considered in the final classification process. This operation represents the second step in the learning phase. For weight assignment 500 patches different from first step were used in a supervised technique (100 patches for each RoI). The weights were calculated taking into account the confusion matrices in the learning phase (examples in Table 4). These weights were calculated on all color channels, and for each class, the channel which gives the highest weight was chosen (Table 5). If a patch is recognized at positive by two classes it is considered that the patch belongs to the class with the highest weight. As it was indicate in (1) the patch belongs in NRoI if it is negative for all classifiers.

Table 4. The confusion matrices and the resulting weights for the RoI classifiers.

Classifier	CNN W	CNN F	CNN G	CNN B	CNN R
Confusion matrix (learning phase)	49 1 2 48	46 4 3 47	50 0 4 46	44 6 4 46	46 4 2 48
Weights	$w_1 = 0.97$	$w_2 = 0.93$	$w_3 = 0.96$	$w_4 = 0.90$	$w_5 = 0.94$
Color	G	R	B	G	G

Table 5. Weights for color channel selection of RoI classification (Orange – selected weights).

Color channel	CNN W	CNN F	CNN G	CNN B	CNN R
R	0.95	0.93	0.93	0.85	0.92
G	0.97	0.89	0.95	0.90	0.94
B	0.90	0.91	0.96	0.87	0.89

Based on the patch classification, the RoIs were segmented and different colored masks were created above the initial images (Fig. 9: water segmentation - Im1, asphaltic road segmentation - Im2, and forest segmentation - Im3). Finally, the segmented RoI is evaluated in size as percent from the entire image. For examples presented in Fig. 10, the percents of RoI occupation are the following: water in Im1 – 10%, asphaltic road in Im2 – 4.5%, and forest in Im3 – 43%.

Taken into account the performance statistics, the precision - *PPV* (Positive Predictive Value) and the accuracy - *ACC* - (2) were calculated from a set of 1000 patches from 10 images, where *TP* is true positive, *TN* is true negative, *FP* is false positive, and *FN* is false negative.

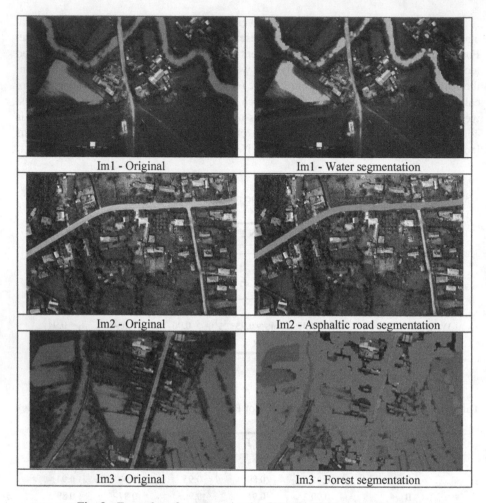

Fig. 9. Examples of segmentation by CNN W, CNN R and CNN F

$$PPV = \frac{TP}{TP + FP}, \quad ACC = \frac{TP + TN}{TP + TN + FP + FN} \tag{2}$$

The performances of RoI detection and segmentation depend on the altitude of flight (low or mid-altitude), image resolution, RoI type, patch dimension and also on the CNN structure. From Table 6 it can be observed that the performances of our method are similar, or better than other works.

Table 6. Comparison with performance metrics of other works.

Indicator	W	F	G	B	R
Precision [%] [7]	–	–	–	92.30	88.66
Precision [%] [9]	–	86.00	65.00	86.00	89.00
Accuracy [%] [8]	94.31	95.10	86.92	87.04	84.02
Accuracy [%] [10]	–	92.30	81.80	89.40	–
Our method PPV/ACC	94.80/95.70	90.10/94.80	78.20/87.10	91.30/92.40	92.40/93.50

4 Conclusions

Unmanned aerial systems integrated in internet offer great opportunity concerning flexibility, speed of data processing and storage capacity. The segmentation method of aerial images is based on a network of CNNs, each of them assigned to a RoI (class). The learning process is a supervised one and, during it, the color channels and also the weights associated with the classifiers are chosen. Because the classes are not very different, a distance-based classifier is not effective. The image processing system operates in a parallel manner and the classification of small patches of dimension 65×65 pixels. The ground RoI segmentation, which is based on this method, gives similar or better results than the separate criteria.

Acknowledgements. This work has been funded by Romanian projects: CAMIA GEX 25/2017, MAARS 185/2017, MUWI, 122/1/2018.

References

1. Wang, P., Wang, J., Chen, Y., Ni, G.: Rapid processing of remote sensing images based on cloud computing. Futur. Gener. Comput. Syst. **29**, 1963–1968 (2013)
2. Popescu, D., Ichim, L., Stoican, F: Unmanned aerial vehicle systems for remote estimation of flooded areas based on complex image processing. Sensors **17**(3), 446, 1–24 (2017)
3. Selecky, M., Meiser, T.: Integration of autonomous UAVs into multi-agent simulation. Acta Polytech. **52**(5), 93–99 (2012)
4. Stoican, F., Prodan, I., Popescu, D.: Flat trajectory generation for way-points relaxations and obstacle avoidance. In: 23th Mediterranean Conference on Control and Automation, Athens, Greece, pp. 695–700 (2015)
5. Ragi, S., Chong, E.: Decentralized control of unmanned aerial vehicles for multitarget tracking. Proceedings of the International Conference on Unmanned Aircraft Systems, pp. 260–268 (2013)
6. Liang, Y., Wu, J., Jia, Y., Du, J.: UAV path planning for passive multi-target following in complex environment. In: Proceedings of the 32-nd Chinese Control Conference, Xi'an, China, pp. 7863–7868 (2013)
7. Saito, S., Aoki, Y.: Building and road detection from large aerial imagery. In: Proceedings of SPIE - The International Society for Optical Engineering, vol. 9405, pp. 1–12 (2015)

8. Volpi, M., Ferrari, V.: Semantic segmentation of urban scenes by learning local class interactions. IEEE Conference on Computer Vision and Pattern Recognition, Boston, MA, USA, pp. 1–9 (2015)

9. Montoya-Zegarra, J.A., Wegner, J.D., Ladicky, L., Schindler, K.: Semantic segmentation of aerial images in urban areas with class-specific higher-order cliques. https://www.inf.ethz.ch/personal/ladickyl/roads_pia15.pdf

10. Marmanis, D., Wegner, J.D., Gallian, S., Schindler, K., Datcu, M., Stilla, U.: Semantic segmentation of aerial images with an ensemble of CNNS, ISPRS_Congress_2016, Prague, pp. 1–8 (2016)

11. Popescu, D., Ichim, L., Caramihale, T.: Flood areas detection based on UAV surveillance system. In: 19th International Conference on System Theory, Control and Computing, Sinaia, Romania, pp. 1–6 (2015)

12. Wei, Y., Wang, Z., Xu, M.: Road structure refined CNN for road extraction in aerial image. IEEE Geosci. Remote Sens. Lett. **14**(5), 709–713 (2017)

13. Ayoul, T., Buckley, T., Crevier, F.: UAV navigation above roads using convolutional neural networks (2017). http://cs231n.stanford.edu/reports/2017/pdfs/553.pdf

14. MUROS. https://trimis.ec.europa.eu/project/multisensory-robotic-system-aerial-monitoring-critical-infrastructure-systems

15. Vedaldi, A., Lenc, K., Gupta, A.: MatConvNet Convolutional Neural Networks for MATLAB (2018). https://arxiv.org/pdf/1412.4564.pdf

Biometrics

Enhanced Line Local Binary Patterns (EL-LBP): An Efficient Image Representation for Face Recognition

Hung Phuoc Truong[1,2] and Yong-Guk Kim[1(✉)]

[1] Department of Computer Engineering, Sejong University, Seoul, Korea
tphung@sju.ac.kr, ykim@sejong.ac.kr
[2] Department of Computer Science, University of Science, VNU-HCM,
Ho Chi Minh City, Vietnam

Abstract. Local Binary Patterns (LBP) is one of the efficient approaches for image representation, especially in the face recognition field. The motivation of the present study is to find a compact descriptor which captures texture information and yet is robust against several visual challenges such as illumination variation, facial expressions and head pose variation. The proposed approach, called it Enhance Line Local Binary Patterns (EL-LBP), is an improvement of 1D-Local Binary Patterns (1D-LBP) by reducing the dimension of feature vectors within 1D-LBP histogram and it leads to decrease the time cost during the matching stage. Experiments using ORL, Yale and AR datasets show that EL-LBP outperforms previous LBP methods in terms of recognition accuracy with much lower time cost, suggesting that this new representation scheme would be more powerful in the embedded vision systems where the computational cost is critical.

Keywords: Local binary patterns · Line local binary patterns
Face recognition · Short bins in histogram · YALE · ORL · AR

1 Introduction

Face recognition plays an important role in the field of computer vision and pattern recognition. Nowadays, many face recognition methods have been proposed and implemented for the real-world applications. And yet, face recognition has many challenges such as variations of facial expressions, head pose, illumination as well as background complexity. The key for successful face recognition would be how one can extract valuable and crucial facial features. The aim of feature extraction, of course, is to reduce the dimensionality of data while preserving their discriminative information for the classification task. There are two main approaches: the first is the holistic-based and the second is the component-based one. For the holistic case, there are two well-known techniques: Principal Component Analysis (PCA) and Fisher Discriminant Analysis (FDA) [8,10,11,13].

© Springer Nature Switzerland AG 2018
J. Blanc-Talon et al. (Eds.): ACIVS 2018, LNCS 11182, pp. 285–296, 2018.
https://doi.org/10.1007/978-3-030-01449-0_24

The basic idea of this approach is to utilize all pixels on the whole face image, and yet it encounters some limitations including several unconstrained cases such as illumination, head pose or facial expressions. The component-based approach, which is also known as feature-based method, utilizes the local facial feature for image representation. It considers the geometry structure of each patch within the given image. Then, the features obtained with this arrangement can cover more useful information than the holistic-based case. Among many, LBP approach [1,2,14,15] is one of the outstanding methods, by which surface texture is captured and the extracted features are used for the classification task.

Inspired from the success of LBP for face recognition [16] and representation of X-ray bone images [3], we propose a novel descriptor, called it EL-LBP, for the face image presentation. The proposed method aims to fuse the 1D-Local Binary Patterns (1D-LBP) with a short bins strategy in counting 1D-LBP patterns to produce a set of effective and compact features. Figure 1 shows our proposed framework: first, the original face images are preprocessed by some denoising techniques and partitioned into smaller blocks; second, each block is projected into two orthogonal directions of 1D-LBP to obtain new representation, which is robust against variation of illumination and scale; third, a short bins strategy is applied to 1D-LBP histogram to produce powerful descriptors while reducing the number of bins within the histogram; fourth, each descriptor corresponding to each block is concatenated into a single feature vector and then these vectors are trained using several classifiers for the face recognition task. The evaluation results suggest that the proposed EL-LBP outperforms previous LBP approaches in terms of recognition accuracy as well computational cost.

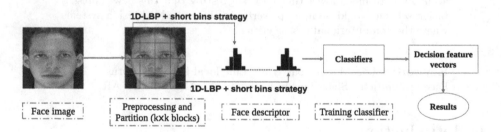

Fig. 1. Architecture of face recognition system based on our EL-LBP

The paper is organized as follows: Sect. 2 describes briefly principles of LBP and a feature selection strategy for image representation because they are the motivation of this study; Sect. 3 presents our algorithm; our experiments on some datasets are shown in Sect. 4; conclusion and references are given in the rest of the paper.

2   Related Works

2.1   Local Binary Patterns (LBP)

The basic LBP operator encodes the relationship between the central pixel and its 3×3 pixels surrounding neighborhood to generate a binary bit string. This bit string can be easily converted to its corresponding decimal value given the LBP code. The process of encoding LBP bit string is illustrated in Fig. 2.

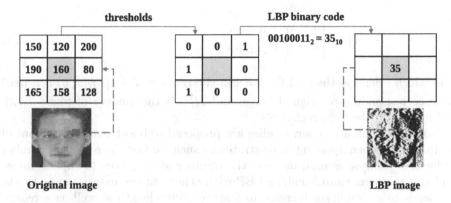

Fig. 2. The basic LBP operator with P = 8 and R = 1 denoted by $LBP_{8,1}$

Then, the LBP operator was extended [14] to utilize different sizes of the neighborhood to capture dominant features of large scale structures. Equation (1) describes the extended version of LBP, where (P, R) denotes a neighborhood of P sampling points on a circle of radius R at a given central pixel (x_c, y_c):

$$LBP_{P,R}(x_c, y_c) = \sum_{p=0}^{P} 2^p s(g_p - g_c), \quad \text{where } s(x) = \begin{cases} 1, & x \geq 0 \\ 0, & x < 0 \end{cases} \quad (1)$$

The LBP operator produces 2^P different output values, corresponding to 2^P different binary patterns formed by P pixels in the neighborhood. However, there is a small of subset of patterns, called "uniform patterns", which occur frequently and have the discriminative power. These special patterns contain at most two bitwise transitions from 0 to 1 or vice versa when the corresponding bit string is considered in a circular shape. From this definition, the number of different uniform output labels for mapping patterns of P bits in $LBP_{P,R}^{u2}$ is $P(P-1) + 3$. Adopting uniform patterns instead of all patterns leads to the significant lower number of possible LBP labels and makes a reliable estimation of their distribution with fewer samples, which was mentioned in [14].

2.2   LBP Feature Selection

In LBP-based face recognition, a face image is decomposed into m small regions $S^{(j)}$ ($j = 0..m - 1$ and $m = k^2$) having the same size and non-overlapping regions where LBP descriptors can be extracted and then concatenated into a single feature vector to represent the texture of a face image locally and globally. A histogram for j^{th} local region $S^{(j)}$ is computed as follows:

$$H_i^{(j)} = \sum_{(x,y) \in S^{(j)}} T\{f_l(x,y) = i\}, \quad i = 0, ..., L - 1 \tag{2}$$

$$T\{A\} = \begin{cases} 1, & \text{if A is true} \\ 0, & \text{if A is false} \end{cases} \tag{3}$$

in which i denotes the ith LBP label in the range $[0, L-1]$, $f_l(x,y)$ is a LBP image corresponding to original image and $H_i^{(j)}$ is the number of pixels (with LBP label i) in the local region $S^{(j)}$.

From this approach, many studies are proposed to boost the discriminant of feature vector when applying some strategies such as: first, increase the number of blocks of images; second, increase the number of neighbors to capture more dominant textures; third, combine LBP with other features extraction methods. This leads to a significant increase in feature vector length as well as a redundant representation including unnecessary information. Thus, the problem of LBP feature selection has recently been studied in many works. According to a recent survey [7], LBP feature selection techniques are classified into two main categories:

- The first is LBP subspace learning which maps the dataset from a high dimensional space to a lower dimensional space [4,17]. This approach tries to boost LBP features by combining many individual LBP features or LBP features with other holistic/ local features to take advantages of each feature type without caring the dimensionality. Then, several well-known dimensionality reduction methods such as PCA, FDA and Gabor wavelet are applied to obtain a low-dimensional compact representation.
- The second is to reduce the dimension of feature vectors inside the algorithm of LBP based on a rule strategy. This strategy is divided into two sub-strategies: first, reducing the number of LBP labels such as Local Line Binary Patterns (LLBP) [16] or Center-symmetric Local Binary Patterns (CS-LBP) [5] to produce shorter histograms while considering the large neighborhood; second, selecting a subset of bins available instead of the whole set of patterns [9,18]. Our present approach belongs to this strategy whereby we aim to boost the discriminant of these compact descriptors.

3 Enhanced Line Local Binary Patterns-Based Face Descriptor

3.1 1D-Local Binary Patterns

The motivation of 1D-LBP comes from the idea of Petpon [16] and Houam [6] where they consider only the pixel neighbors of a given central pixel in the horizontal or the vertical direction instead of the grid shape of neighborhood. Our basic idea of 1D-LBP is illustrated in Fig. 3 and it is similar to the original LBP but there are some differences:

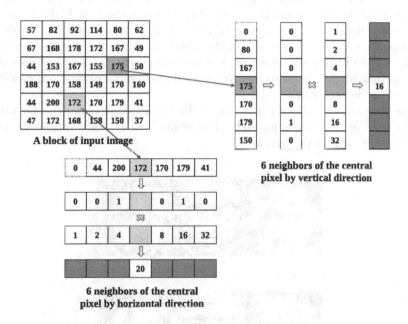

Fig. 3. An example of 1D-LBP operator with 6 neighbors corresponding to a 6-bits string considered by two directions

1. The neighborhood shape has a straight line in one direction: either the horizontal or the vertical direction.
2. The number of neighbors of a central pixel is the length of a bit string. With a given central pixel, the number of neighbors of left and right part (with horizontal direction) as well as the number of neighbors of upper and lower (with vertical direction) part are the same. If the number of one part is less than the other, we use a 0-padding mask to balance the bit string. The total number, called it N, has to be an even number.
3. A bit string comes from comparing all neighbors to the central value: the binary bit is 1 if the neighbor pixel is greater or equal to the current element and 0 otherwise. Then, each element in the bit string is multiplied by a weight

according to its position (described in Figs. 3, (4) and (5)) and the current pixel is replaced by the sum of all values.

$$1D - LBP_h(N, c) = \sum_{n=0}^{N-1} 2^n s(h_n - g_c) \qquad (4)$$

$$1D - LBP_v(N, c) = \sum_{n=0}^{N-1} 2^n s(v_n - g_c) \qquad (5)$$

where N is the number of neighbors, c is a given central pixel, g_c is the value of center c, h_n is neighborhood pixel on the horizontal line and v_c is the neighborhood pixel on the vertical line, and s function is same as in LBP. Then, this descriptor is defined by the histogram of 1D-LBP patterns and the number of bins depends on the size of the neighborhood. For instance, 8 neighbors yield 256 bins in the histogram. However, because the number of bins by this way is too long, then we propose a strategy to reduce the number of bins and described its detail in the next section.

The algorithm of 1D-LBP calculates the binary code along the horizontal and the vertical direction separately. The example of 1D-LBP images for these directions is shown in the Fig. 4.

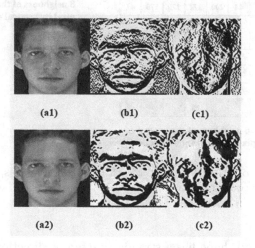

Fig. 4. Example of face image procced by 1D-LBP operator in two cases, the first row: (a1) is an original image, (b1) and (c1) are 1D-LBP images in vertical and horizontal direction of (a1), respectively; the second row: (a2) is a preprocessed image from original image, and the remaining images are processed as the first row case.

3.2   A Short Bins Strategy in Counting the 1D-LBP Patterns

Motivated by the second category that reduces the dimension of feature vectors inside the algorithm of LBP based on a rule strategy (described in Sect. 2.2), we propose a novel strategy for compact descriptors of 1D-LBP operator, called a short bins strategy, described with 3 steps as below:

1. Step 1: Let $A^{short}_{1D-LBP} = \{a'_{i,j}\}$ denote the result matrix after doing 1D-LBP (described in (4) or (5)). All the values $a_{i,j}$ of matrix A_{1D-LBP} are converted into a new range of values by (6) and Fig. 5.

$$A^{short}_{1D-LBP} = f(a_{i,j}|\forall a_{i,j} \in A_{1D-LBP}) = \begin{cases} a_{i,j} \bmod 256, a_{i,j} > 255 \\ a_{i,j}, \quad 0 \le a_{i,j} \le 255 \end{cases} \quad (6)$$

2. Step 2: Find all uniform patterns corresponding to the number of neighbors and convert all of them into decimal numbers as shown in Table 1. Here, #BITS denotes the number of bits in a bit string (the number of neighbors), whereas #BINS denotes the number of bins in the histogram when counting 1D-LBP patterns.
3. Step 3: Select values in Table 1 corresponding to the number of bits as a subset of 1D-LBP patterns, and divide the entire range of values into series of intervals which are consecutive and non-overlapping intervals of values, such as $[0, 1), [1, 2), [2, 3), [3, 4), [4, 6), [6, 7), [7, 8), [8, 9), [9, 11), [11, 12), [12, 13), [13, 14), [14, 15]$ (in case #BITS = 4). Then, we count the occurrence frequency of the value in matrix A^{short}_{1D-LBP} to obtain a compact histogram.

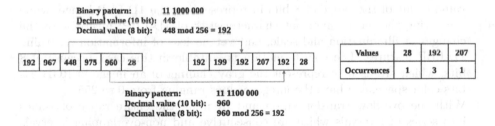

Fig. 5. Converting values 10-bits string to decimal values in a new range of values

3.3 Summary on EL-LBP

Since the proposed scheme was inspired from 1D-LBP and a short bins strategy, it has some unique characteristics. This section describes why EL-LBP can be used for an efficient representation in the face recognition task.

1. The process of face image data acquisition is often affected by a few factors, such as the angle of exposure, the source of incoming light or facial expressions. By considering micro-patterns in line, 1D-LBP is robust against changes of illumination and scale. One limitation of 1D-LBP is that it does not guarantee the rotation-invariant while traditional LBP features are invariant to

Table 1. All values of uniform patterns in different neighborhoods

#BITS	The range of values between bins in the histogram (All decimal values of uniform patterns)	#BINS
2	0, 1, 2, 3	3
4	0, 1, 2, 3, 4, 6, 7, 8, 9, 11, 12, 13, 14, 15	13
6	0, 1, 2, 3, 4, 6, 7, 8, 12, 14, 15, 16, 24, 28, 30, 31, 32, 33, 35, 39, 47, 48, 49, 51, 55, 56, 57, 59, 60, 61, 63	30
8	0, 1, 2, 3, 4, 6, 7, 8, 12, 14, 15, 16, 24, 28, 30, 31, 32, 48, 56, 60, 62, 63, 64, 96, 112, 120, 124, 126, 127, 128, 129, 131, 135, 143, 159, 191, 192, 193, 195, 199, 207, 223, 224, 225, 227, 231, 239, 240, 241, 243, 247, 248, 249, 251, 252, 253, 254, 255	57

rotation because it is done in a circular or square neighborhood. Nevertheless, this characteristic is not required for face recognition, but anisotropic information is more important. In fact, if the direction of incoming lights lies in vertical direction or illuminate one part of a face (left or right), extracted features of 1D-LBP$_v$ are done better than 1D-LBP$_h$ and otherwise.

2. The image quality produced by a transformation is "good" if that image may be described in terms of how well it approximates the principal components of original image. Equation (6) is actually an overflow transformation if the value is out of the range of 8 bits in representing the 1D-LBP image, while preserving the most important characteristics of 1D-LBP: invariant to the changes of illumination and scale, and yet no loss of information regarding 1D-LBP features. The transformation is suitable in this case, because of 8 bits, which are used to represent the gray channel of an image in RGB 24-bits color space, can have the integer values ranging from 0 to 255.

3. With the overflow transformation and dividing the entire range of values into series of intervals which are consecutive and non-overlapping intervals of values (in step 3 of Sect. 3.2), the number of bins in histogram are always in the range 3, 13, 30 and 57, despite the fact that the number of neighbors can be greater than 8. The number of bins by the present strategy, compared to 1D-LBP [3,6] or LBP uniform, is smaller. Therefore, the computational time cost of counting 1D-LBP patterns becomes a constant and leads to the algorithmic complexity O(1).

4   Experimental Results

4.1   Datasets and Configuration of Experiments

Three standard datasets used for evaluating face recognition performance are ORL, AR [12] and YALE. The ORL dataset contains 10 images (92 × 112 gray image) of 40 individuals in different facial expressions (open/ close eyes, smile/

not smile), several conditions of illumination, facial details (hairstyles or with/ without beard, and with glasses/ without glasses), diverse head poses (the angle of tilting from 0° up to 20°). The YALE dataset contains 11 grayscale images, having 243×320 pixels, of 15 individuals. The AR dataset collects 26 images of 116 individuals (63 men and 53 women) in the RGB color format (768×576 pixels). Figure 6 illustrates some original images and their corresponding 1D-LBP images on three datasets. The images in YALE and AR datasets contain some challenges for different facial expressions (happy, sad, neutral, sleepy, anger, surprised and wink) or diverse configurations (with/ without glasses, left-light, right-light or all side lights).

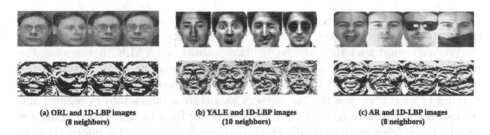

| (a) ORL and 1D-LBP images | (b) YALE and 1D-LBP images | (c) AR and 1D-LBP images |
| (8 neighbors) | (10 neighbors) | (8 neighbors) |

Fig. 6. Some original and their 1D-LBP images respectively on three datasets

To evaluate the face recognition performance, we used K-fold cross validation method with $K = 2$ on each dataset. It means that half of the samples are randomly selected for training and the remaining half for testing, and this evaluation is repeated for 100 times. Therefore, recognition results are the average values for 100 random permutations. Moreover, each original image was preprocessed by applying denoising techniques for ORL and by applying histogram equalization for YALE and AR after converting images into gray levels, respectively. The k-NN and SVM classifiers with several kernels were used to produce the recognition results. The result suggests that Linear SVM is the best classifier among many.

4.2 Results

Our proposed scheme (EL-LBP) is compared to some existing LBP methods. The best performance (the average of recognition rate in %) of these methods is shown in Table 2. We can see that EL-LBP (vertical) shows the highest performance on three datasets and the reason for this result was explained in Sect. 3.3.

As described in Sect. 2.2, for an effective descriptor of the face recognition, each image is partitioned into m ($m = k^2$) regions, thus the dimension of feature descriptor is very huge $m \times q$ (q is the number of bins in the histogram). With $m = 49$, the feature descriptor for 1D-LBP [6] (8 neighbors) requires 12544 ($= 49 \times 256$) bins. While the best neighborhood configurations of EL-LBP for three cases, ORL (4), YALE (10), and AR (8), require only 637 ($= 49 \times 13$) or

Table 2. The recognition rate of several methods on three datasets

Methods	ORL	YALE	AR
$LBP^{u2}(8,1)$	82.5	80.1	87.5
$LBP^{u2}(16,2)$	86.1	83.4	89.2
LBP(8,1) + PCA	91.4	89.2	93
1D-LBP (vertical)	95.41	81.75	92.6
EL-LBP (horizontal)	93.62	93.65	94.35
EL-LBP (vertical)	**97.12**	**95.27**	**98.27**

$2793 (= 49 \times 57)$ bins, suggesting that EL-LBP outperforms 1D-LBP because the feature descriptor length is reduced by 4.5 times.

We also compared the face recognition results on YALE in the case the number of neighbors is greater than 8, simply because EL-LBP obtained the best performance with 10 neighbors. The result, shown in Table 3 (N is the number of neighbors, k is the parameters for dividing image into m regions and #BINS is the dimensionality of obtained descriptor from each method), indicates clearly the efficiency of EL-LBP which can preserve the useful information in the descriptor as well as speed up the matching process for a large dataset when applying a short bins strategy. For instance, with $N \geq 8$ and $k = 7$, the proposed EL-LBP descriptors only have a constant of the number of bins (2793 bins) while the number of bins of 1D-LBP increases dramatically.

Table 3. The comparison of two methods (1D-LBP [7] and EL-LBP) on YALE data

	N = 8		N = 10		N = 12	
	1D-LBP	EL-ELBP	1D-LBP	EL-ELBP	1D-LBP	EL-ELBP
k = 3	81.37	83.64	81.67	84.25	81.60	84.71
#BINS	2304	513	9216	513	36864	513
k = 5	89.29	91.80	89.68	91.81	90.14	91.85
#BINS	6400	1425	25600	1425	102400	1425
k = 7	**91.50**	**94.79**	**91.61**	**95.27**	**91.11**	**95.07**
#BINS	12544	2793	50176	2793	200704	2793

5 Conclusion

In this paper, a novel LBP scheme is proposed for face image representation and face recognition. The proposed face representation shows the higher face recognition rate than the traditional face recognition by fusing 1D-LBP with a short bins strategy, called it EL-LBP. It makes a compact descriptor with tolerance for several information regarding variations of illumination, head poses, scale and facial expressions. Thus, our proposed method has several strengths because it can describe within-class changes while differentiating between-class; EL-LBP features can be easily extracted from raw images; the dimensionality of EL-LBP features are smaller than other LBP to avoid high computational cost of classifiers. We expect that the present representation scheme would make a better sense in the embedded vison system where computational resources are critical.

Acknowledgement. This work was supported by Institute for information & communications Technology Promotion (IITP) grant funded by the Korea government (MSIT) (No. 2017-0-00731, Personalized Advertisement Platform based on Viewers Attention and Emotion using Deep-Learning Method).

References

1. Ahonen, T., Hadid, A., Pietikainen, M.: Face description with local binary patterns: application to face recognition. IEEE Trans. Pattern Anal. Mach. Intell. **28**(12), 2037–2041 (2006)
2. Ahonen, T., Hadid, A., Pietikäinen, M.: Face recognition with local binary patterns. In: Pajdla, T., Matas, J. (eds.) ECCV 2004. LNCS, vol. 3021, pp. 469–481. Springer, Heidelberg (2004). https://doi.org/10.1007/978-3-540-24670-1_36
3. Benzaoui, A., Boukrouche, A., Doghmane, H., Bourouba, H.: Face recognition using 1DLBP, DWT and SVM. In: 2015 3rd International Conference on Control, Engineering Information Technology (CEIT), pp. 1–6, May 2015
4. Chan, C.-H., Kittler, J., Messer, K.: Multi-scale local binary pattern histograms for face recognition. In: Lee, S.-W., Li, S.Z. (eds.) ICB 2007. LNCS, vol. 4642, pp. 809–818. Springer, Heidelberg (2007). https://doi.org/10.1007/978-3-540-74549-5_85
5. Heikkil, M., Pietikinen, M., Schmid, C.: Description of interest regions with local binary patterns. Pattern Recogn. **42**(3), 425–436 (2009)
6. Houam, L., Hafiane, A., Jennane, R., Boukrouche, A., Lespessailles, E.: Trabecular bone anisotropy characterization using 1D local binary patterns. In: Blanc-Talon, J., Bone, D., Philips, W., Popescu, D., Scheunders, P. (eds.) ACIVS 2010. LNCS, vol. 6474, pp. 105–113. Springer, Heidelberg (2010). https://doi.org/10.1007/978-3-642-17688-3_11
7. Huang, D., Shan, C., Ardabilian, M., Wang, Y., Chen, L.: Local binary patterns and its application to facial image analysis: a survey. IEEE Trans. Syst., Man, Cybern., Part C (Appl. Rev.) **41**(6), 765–781 (2011)
8. Huang, P., Gao, G., Qian, C., Yang, G., Yang, Z.: Fuzzy linear regression discriminant projection for face recognition. IEEE Access **5**, 4340–4349 (2017)
9. Lahdenoja, O., Laiho, M., Paasio, A.: Reducing the feature vector length in local binary pattern based face recognition. In: IEEE International Conference on Image Processing 2005, vol. 2, pp. 914–917, September 2005
10. Le, D.T., Truong, H.P., Le, T.H.: Facial expression recognition using statistical subspace. In: 2014 IEEE International Conference on Image Processing (ICIP), pp. 5981–5985, October 2014

11. Le, T.H., Truong, H.P., Do, H.T.T., Vo, D.M.: On approaching 2D-FPCA technique to improve image representation in frequency domain. In: Proceedings of the Fourth Symposium on Information and Communication Technology, SoICT 2013, pp. 172–180. ACM, New York (2013). https://doi.org/10.1145/2542050.2542061
12. Martinez, A., Benavente, R.: The AR face database. CVC Technical report, 24 (1998)
13. Mi, J., Liu, T.: Multi-step linear representation-based classification for face recognition. IET Comput. Vis. **10**(8), 836–841 (2016)
14. Ojala, T., Pietikainen, M., Maenpaa, T.: Multiresolution gray-scale and rotation invariant texture classification with local binary patterns. IEEE Trans. Pattern Anal. Mach. Intell. **24**(7), 971–987 (2002)
15. Ojala, T., Pietikinen, M., Harwood, D.: A comparative study of texture measures with classification based on featured distributions. Pattern Recogn. **29**(1), 51–59 (1996)
16. Petpon, A., Srisuk, S.: Face recognition with local line binary pattern. In: 2009 Fifth International Conference on Image and Graphics, pp. 533–539, September 2009
17. Tan, X., Triggs, B.: Fusing gabor and LBP feature sets for kernel-based face recognition. In: Zhou, S.K., Zhao, W., Tang, X., Gong, S. (eds.) AMFG 2007. LNCS, vol. 4778, pp. 235–249. Springer, Heidelberg (2007). https://doi.org/10.1007/978-3-540-75690-3_18
18. Topi, M., Timo, O., Matti, P., Maricor, S.: Robust texture classification by subsets of local binary patterns. In: Proceedings 15th International Conference on Pattern Recognition, ICPR-2000, vol. 3, pp. 935–938, September 2000

Single Sample Face Recognition by Sparse Recovery of Deep-Learned LDA Features

Matteo Bodini[1], Alessandro D'Amelio[1], Giuliano Grossi[1],
Raffaella Lanzarotti[1(✉)], and Jianyi Lin[2]

[1] PHuSe Lab - Dipartimento di Informatica, Università degli Studi di Milano,
via Celoria 18, 20133 Milano, Italy
{bodini,damelio,grossi,lanzarotti}@di.unimi.it
[2] Department of Mathematics, Khalifa University of Science and Technology,
Al Saada Street, PO Box 127788, Abu Dhabi, United Arab Emirates
jianyi.lin@ku.ac.ae

Abstract. Single Sample Per Person (SSPP) Face Recognition is receiving a significant attention due to the challenges it opens especially when conceived for real applications under unconstrained environments. In this paper we propose a solution combining the effectiveness of deep convolutional neural networks (DCNN) feature characterization, the discriminative capability of linear discriminant analysis (LDA), and the efficacy of a sparsity based classifier built on the k-LiMapS algorithm. Experiments on the public LFW dataset prove the method robustness to solve the SSPP problem, outperforming several state-of-the-art methods.

1 Introduction

Although the overwhelming adoption and achievements of deep neural networks for face recognition systems, some challenging problems are still open, mainly concerning the quality and quantity of usable data. For instance, actual recognition techniques suffer when the images are captured in uncontrolled conditions, entailing variations in illumination, pose, expression and occlusion. Moreover, the recognition task becomes very hard when only one image per subject is available in the gallery/train construction, tackling the so called single sample per person (SSPP) problem [8]. SSPP is an extreme case of small sample size (SSS) problem, and grounds its motivations in applicative scenarios such as e-passport control, law enforcement, surveillance, human-computer interaction, to name but a few. Hence, designing effective automatic recognition systems for SSPP referring to images acquired in uncontrolled conditions is still a major challenge in computer vision, whose success may have a considerable impact on the real world.

In recent years, some methods extending state-of-the-art systems have been proposed to tackle the SSPP problem, achieving promising but not yet satisfactory results [8]. Without expecting to be exhaustive, in Sect. 1.1 we recall some recent works that have given a great impulse in this research field.

© Springer Nature Switzerland AG 2018
J. Blanc-Talon et al. (Eds.): ACIVS 2018, LNCS 11182, pp. 297–308, 2018.
https://doi.org/10.1007/978-3-030-01449-0_25

In this paper (cfr Sect. 2) we propose a new effective technique[1] facing the SSPP problem by coupling deep-learned features with the sparse representation paradigm. The adoption of DCNN features was already proven discriminative with small galleries [5]. In our proposal, the sparse representation paradigm relies on k-LiMapS algorithm [2], accomplishing the task of recovering a succinct coding of a probe image. The method takes as input a feature and activates an optimization process exploiting a target objective function that relaxes the pseudo-norm ℓ_0. This approach, although not using a convex cost function, is fast and does not suffer from bottlenecks of high computational times, as typically happens for simplex-based solver using ℓ_1-norm, such as the method SRC (Sparse Representation Classifier), proposed in [17].

The effectiveness of the method is proven in Sect. 3 where evaluations and comparisons are computed on the very challenge dataset LFW. Finally, Sect. 4 provides the conclusions and draws potential future directions in this domain.

1.1   Related Works

To a first approximation, the methods tackling the SSPP problem can be divided into three categories [8]: (1) generic learning methods, that pre-train some feature extractors or part of the model on a generic unlabeled image set, distinct from both gallery and test sets, (2) virtual image generation models, which produce new synthetic face images in order to enrich the gallery, and (3) methods of decomposition into local regions, that form a larger and more discriminative training set.

Among the generic learning approaches, the authors of [4] propose the idea of mapping images of distinct subjects into points that are equidistant in an embedding space, maximizing the inter-subject margin of each prototype. Such mapping is realized through a computationally optimized incremental algorithm of linear regression analysis on LBP and Gabor feature descriptors. The paper [18] proposes JCR-ACF, a joint collaborative representation based classifier operating on features learned by a DCNN from local patches that are rather discriminative of facial characteristics. They first train a variated DeepID [15] network on highly-discriminative local patches of a large dataset, thus obtaining an adaptive convolutional feature (ACF) extractor network. The ACF vectors resulting from the gallery and the generic dataset respectively form the gallery dictionary and the intra-class variation dictionary: the first contains prototype atoms for different subjects, while the latter accounts for facial characteristic variations in local regions within a subject. The joint dictionary is then used in an ℓ_2-regularized CRC model with a functional term accounting for within-class scattering. The work [5] introduces S^3RC, a sparse representation based classification method dealing with small number of samples per subject, and even single sample in the extreme case. Assuming a sparse linear representation by gallery and variation dictionary, the method models the residual error of a sample as a Gaussian

---

[1] A demo code is available on the website: https://github.com/phuselab/SSPP-face_recognition.

Mixture Model noise, whose parameters (centroid and covariance matrix) are estimated in maximum likelihood sense by an EM algorithm initialized with the empirical distribution as class prior. Hence, the estimated centroids form the gallery dictionary, while the variation dictionary is obtained from single samples or from standard within-class centering of labeled samples.

An example of recent methods proposing virtual image generation, [20] utilizes an auto-encoder neural network for the nonlinear estimation of the distinct manifolds encapsulating pose variation and individual's information; such separation allows to generate virtual samples of each person synthesized in different poses. Subsequently the augmented dataset is fed to another auto-encoder for training the classifier.

Within image decomposition methods, the authors of [9] propose a multi-phase ℓ_2-regularized CRC based on local structure of decomposed sample image. A complete structured dictionary is then formed by separate subdictionaries, each built by the local patches in a large region across all sample images. A test image, preprocessed by the same local structure scheme, undergoes a CRC classification, producing a vector of votes for the classes, and then some minority classes with their atoms are removed to promote vote distribution's sparsity. The phases of classification and minority pruning are repeated until reaching a satisfactory majority voting outcome. The experiments indicate an improvement compared to state-of-the-art alternatives.

More recently, the authors of [14] have studied a very particular subproblem of SSPP, namely the problem of classifying the identity of a whole batch of "probe" images for each subject, instead of a single image at a time. In light of these premises, they use the probe images of each subject to learn a customized intra-class variation dictionary, later exploited in an extended joint sparsity model (CD-EJSR) that accounts for the gallery (represented as additional dictionary) containing single samples. Notably, they analytically derive a closed-form solution of such two-phase quasi-quadratic model through the ADMM optimization, demonstrating efficient computational times. However, their method does not take advantage of any informative feature extraction, but rather uses suitably cropped-resized grayscale images.

2   Proposed Method

To address the SSPP problem, in this section we detail a new effective technique that combines deep-learned features with the sparse representation paradigm. More precisely, as shown in the block diagram of Fig. 1, in a first stage new images are yielded by applying simple transformations, i.e. flipping, zooming and shifting, commonly used in the CNN domain to augment the training set. Next, deep neural network features are extracted by applying the VGG-face net [11] to the so obtained images. Given the high dimensionality of the learned features, we resort to the linear discriminant analysis (LDA) to provide highly discriminative and reduced features. Lastly, arranging the discriminative features of the gallery images in a number of dictionaries and leveraging on ℓ_0-norm minimization, the

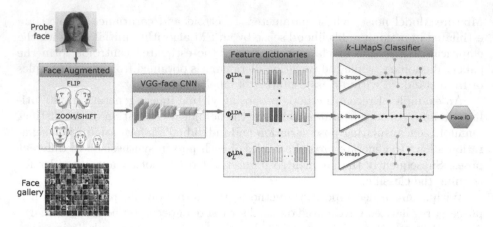

Fig. 1. Block diagram of the classification process. First, image augmentation, second, feature characterization via DCNN, third, data organization in dictionaries and LDA transformation, finally k-LiMapS classifier to produce the face identity.

k-LiMapS algorithm [2] is used to sparsely represent the features of the probe images against the dictionaries, aiming at seeking the corresponding subject identity.

2.1 Augmented Features

The SSPP problem can refer either to the unique target images available [16], or try to enrich the gallery as stated in Sect. 1.1. In this work, as we use a sparsity based classifier, multiple instances per subject are required. To attain them, several techniques could be adopted [12]. Here we refer to the traditional transformation of shifting, zooming in/out, and flipping, deputizing the feature extraction step to capture more complex data information.

Specifically, for each face image I belonging to the image space $\mathcal{I}$, after the face localization [3], a pool of transformations on $\mathcal{I}$ is applied: flips F_1, F_2, shifts $S_1, \ldots, S_a$, and zooms $Z_1, \ldots, Z_b$. These provide the set of new transformed images

$$A_I = \bigcup_{(f,s,z)\in\{1,2\}\times\{1,\ldots,a\}\times\{1,\ldots,b\}} \left\{ Z_z(S_s(F_f(I))) \right\}$$

of cardinality $d = 2\,a\,b$, together with the associated index set $D = \{1, \ldots, d\}$, collecting all kind of transformations applied, and isomorphic to the Cartesian product giving rise all triples (f, s, z) in previous union.

To be successful in classification, each face image $I_j \in A_I$, has to be represented in a proper feature space. Recently, Gao et al. [5] have shown that coupling even simple classifiers with deep-learned features significantly improve the system performances. Supported by this evidence, here we derive the feature characterization adopting the VGG-face net [11], a public deep convolutional neural network (DCNN) trained to recognize faces, and referring to the last full connected output. In particular, for each augmented image I_j^i of subject i, with $j \in D$, we compute the feature $\varphi_j^i = \text{VGG-face}(I_j^i)$, which is a p-dimensional sparse vectors, with $p = 4096$. The feature set associated to the image I is then

$$\mathcal{F}_I = \{\varphi_1^i, \dots, \varphi_d^i\} \subseteq \mathbb{R}^p.$$

2.2   Feature Projection onto LDA Space

Suppose we are given q subjects $\mathcal{C} = \{1, \dots, q\}$ with a single image per subject for the gallery construction, and a target image of a subject $s \in \mathcal{C}$. As described in Sect. 2.1, both gallery and probe images undergo the face augmentation module, followed by the extraction of features by the VGG-face net. As shown in the block diagram of Fig. 1, gallery features are then arranged in L dictionaries, $\Phi_1, \dots, \Phi_L$, as columns.

In principle, we could admit arbitrary collections of features in each dictionary so as to increase the covariance within each one. Here the simplest case where the features are arranged k by k is taken into account. In this way, a specific dictionary matrix Φ_l results in an group of $k \le d$ features independently picked from the set $\mathcal{F}$, i.e.,

$$\Phi_l = \left[\varphi_{l_1}^1 \mid \dots \mid \varphi_{l_k}^1 \mid \dots \dots \mid \varphi_{l_1}^q \mid \dots \mid \varphi_{l_k}^q \right] \tag{1}$$

where $\{l_1, \dots, l_k\} \subseteq D$ specify the features chosen for the l-th dictionary and $1, \dots, q$ the subject's IDs in the gallery. Naturally, the same transformations are addressed for the probe image I^s, thus producing the features φ_j^s for all $j \in D$.

In order to gain a more discriminatory capacity, we apply an additional transformation to the features by means of the Fisher's linear discriminant analysis (LDA) [13], a widely used technique for pattern classification problems. It employs Fisher's ratio, which is the ratio of the between-class scatter matrix and the within-class scatter matrix allowing to derive a set of feature vectors by projecting high-dimensional data onto a low-dimensional data space while maximizing class separability. Very briefly, given a dictionary of features Φ_l corresponding to the classes in $\mathcal{C}$, let

$$\mu = \frac{1}{qk} \sum_{i=1}^{q} \sum_{j=1}^{k} \varphi_{l_j}^i \quad \text{and} \quad \mu_i = \frac{1}{k} \sum_{j=1}^{k} \varphi_{l_j}^i$$

be the global mean and the mean of class i respectively,

$$S_W = \sum_{i=1}^{q} \sum_{j=1}^{k} (\varphi_{l_j}^i - \mu_i)(\varphi_{l_j}^i - \mu_i)^T$$

the between-class scatter matrix, and

$$S_B = k \sum_{i=1}^{q} (\mu_i - \mu)(\mu_i - \mu)^T$$

within-class scatter matrix The Fisher's discriminant analysis finds a weight matrix $W \in \mathbb{R}^{p \times (q-1)}$ that projects all samples $\varphi_j \in \Phi_l$ in the low-dimensional feature space $\mathbb{R}^{q-1}$ to achieve the best possible class separability. The linear transformation W_l is optimal in the sense of Raleigh quotient which is the ratio of the between-class scatter to the within-class scatter, and is achieved by optimizing the functional

$$J(W) = \frac{|W_l^T S_B W_l|}{|W_l^T S_W W_l|},$$

in terms of the generalized eigenvalue problem. In previous formula, $|\cdot|$ stands for the determinant of a matrix.

Downstream of the LDA analysis, we obtain both new dictionaries and new probe features actually used in the identity recovery via sparse representation, as explained in Sect. 2.3. Hereafter, the new train and probe features lying in the LDA space will be denoted with the superscript LDA and computed as follows:

$$\Phi_l^{\text{LDA}} = W_l \Phi_l, \quad \text{for all} \quad l = 1, \dots, L \tag{2}$$

for the dictionaries and

$$\varphi_{l_f}^{\text{LDA},i} = W_l \varphi_{l_f}^i, \quad \text{for all} \quad f \in \{1, \dots, k\} \tag{3}$$

for the features of the probe identity i.

2.3 Sparse Representation by ℓ_0-norm Minimization

The general framework of sparse representation relies on a linear combination of few atoms, i.e., representative samples, to approximate a probe sample at hand. To calculate the solution, that is the representation coefficients of atoms, many optimizers could be used, being each one characterized by different norm minimization incorporating the sparsity constraints [19]. In this work we will concerns with ℓ_0-norm[2] minimization that is undertaken by iterative algorithms working on feature space both for dictionary and probe images.

Let $x_1, \cdots, x_m \in \mathbb{R}^n$ be the known samples arranged on a matrix $X \in \mathbb{R}^{n \times m}$ (with $n < m$) as columns defining an overcomplete dictionary. Under this setting, for a given vector $y \in \mathbb{R}^n$, the goal is to find a solution to the underdetermined linear system $y = X\alpha$. From the side of linear algebra, the problem is ill-posed since it will never have a unique solution. To overcome this ambiguity, it is feasible to impose suitable regularization constraints to the solution α such as those aimed at recovering the sparsest solution. A way to find the sparsest representation solution is to solve the aforementioned linear system with the ℓ_0-norm minimization constraint. More precisely, the problem can be recast as

$$\underset{\alpha \in \mathbb{R}^m}{\arg\min} \|\alpha\|_0 \quad \text{subject to} \quad X\alpha = y. \tag{P_0}$$

---

[2] Strictly speaking the ℓ_0-norm is not actually a norm, it is a cardinality function counting the number of nonzero elements in a vector.

It is worth emphasizing that problem P_0 is a well-known NP-hard optimization problem [10], thus forcing to seek out plausible variants of the problem, provided they can be efficiently solvable.

Because real data, as well as the features representing images, are generally noisy, the original exact model $y = X\alpha$ could be realistically replaced by the noisy model $y = X\alpha + \eta$, where data contains a small or controlled amount of noise $\eta \in \mathbb{R}^n$, having bounded energy or variance $\|\eta\| \leq \varepsilon$. With the presence of noise, the sparse solutions of problems P_0 can be approximately obtained by resolving the following optimization problem

$$\underset{\alpha \in \mathbb{R}^m}{\operatorname{argmin}} \|\alpha\|_0 \quad \text{subject to} \quad \|X\alpha - y\|_2^2 \leq \varepsilon. \qquad (P_\varepsilon)$$

Furthermore, in many contexts it is very common to introduce a strong approximation variant of the exact problem P_0 by emphasizing the required sparsity level k, thus rewriting the problem in the form

$$\underset{\alpha \in \mathbb{R}^m}{\operatorname{argmin}} \|X\alpha - y\|_2^2 \quad \text{subject to} \quad \|\alpha\|_0 \leq k. \qquad (P_a)$$

As a source of inspiration, to tackle the latter problem, we resort to the sparsity solver called k-LiMapS [2], which is an algorithm able to solve the sparse approximation problem P_a over redundant dictionaries, where the input signal is expressed as a linear combination of k atoms or fewer from the dictionary.

Roughly speaking, the basic strategy of the method rests on a family of non-linear mappings which results to be contractive in a interval close to zero. By iterating contractions and projections, the method is able to extract the most significant components also for noisy signals which subsume ideal underlying signals having sufficiently sparse representation. For reasonable error level, the fixed point solution of such a iterative scheme provides a sparse approximation containing only the nonzero terms characterizing the unique sparsest representation of the ideal noiseless sparse signal. Algorithm k-LiMapS adopts the ℓ_0-norm optimization, and is based on a suitable parametric family of Lipschitzian type mappings providing an easy and fast iterative scheme. In [1,6] we have already proven its effectiveness tackling the FR problem. Here we show how to apply it to the SSPP problem posed at the origin of this work.

2.4 Identity Recovery via k-LiMapS Sparsity Promotion

As argued in the previous section, the problem of recognizing the identity of a probe image among a number of subjects can be casted into the problem of finding k-sparse solutions to a linear system collecting k features per subject. The latter statement represents the rationale behind our recognition strategy, which is better detailed below:

1. split the feature set D into L pools each of cardinality k, $\{l_1, \ldots, l_k\}$ so that
$$D = \{ \underbrace{1, \ldots, k,}_{\text{feats in dict 1}} \mid \underbrace{k+1, \ldots 2k,}_{\cdots} \mid \cdots \mid \underbrace{(l-1)k+1, \ldots, lk,}_{\text{feats in dict } l:\, l_1, \ldots, l_k} \mid \ldots, \underbrace{Lk}_{d\text{-th feat}} \}.$$

2. for each pool $\{l_1, \ldots, l_k\} \subseteq D$ according to (2) build the dictionary $\Phi_l^{LDA} \in \mathbb{R}^{(q-1) \times m}$, where q is the amount of subjects in the gallery and $m = k\,q$,
3. for a probe image of generic identity $i \in \mathcal{C}$, according to (3), compute the features $\varphi_{l_f}^{LDA,i}$ for each $f \in \{1, \ldots, k\}$, and for each $l \in \{1, \ldots, L\}$,
4. for each pair dictionary/feature, i.e., for all $(l, f) \in \{1, \ldots, L\} \times \{1, \ldots, k\}$, solve the problem P_a, meaning to find the k-sparse solution $\hat{\alpha}_{l,f}$ satisfying

$$\hat{\alpha}_{l,f} = \operatorname*{argmin}_{\alpha \in \mathbb{R}^m} \|\Phi_l^{LDA}\alpha - \varphi_{l_f}^{LDA,i}\| \quad \text{subject to} \quad \|\alpha\|_0 \leq k. \tag{4}$$

Notice that the leading idea boils down to the very essence of covariance between atoms in a dictionary. Indeed, fixing at k the sparsity level of the approximate solutions to the linear systems at hand, we hope to intercept solely the atoms corresponding to the k features of Φ_l^{LDA} associated to subject i, as highlighted below

$$\Phi_l^{LDA} = \left[\varphi_{l_1}^1 \mid \cdots \mid \varphi_{l_k}^1 \mid \cdots \mid \underbrace{\varphi_{l_1}^i \mid \cdots \mid \varphi_{l_k}^i}_{\text{subject } i} \mid \cdots \mid \varphi_{l_1}^q \mid \cdots \mid \varphi_{l_k}^q\right].$$

In other words, the probe features $\varphi_{l_1}^{LDA,i}, \ldots, \varphi_{l_k}^{LDA,i}$ and those highlighted above in the dictionary should be highly correlated and therefore the latter widely preferred to the others in the dictionary.

A singular aspect of this process is the fact that the support[3] of sparse solution to the linear system (4) is more informative than the linear combination of atoms multiplied by the representation coefficients. Indeed, we empirically shown that this classification approach improves the global recognition rates and certainly overcomes the weakness of widely used methods based on residual measures (e.g., least square minimization) when dealing with noisy images.

On the basis of these considerations, at the end of previous steps yielding the pool $A = \{\hat{\alpha}_1, \ldots, \hat{\alpha}_d\}$ of d[4] sparse solutions, we take the support of each vector and use the following very easy approach for the classification step.

1. Define the mapping $\mathcal{L} : \{1, \ldots, d\} \to \mathcal{C}$ from the column-index i of Φ_l^{LDA} to the corresponding subject $\mathcal{L}(i) \in \mathcal{C}$,
2. using previous mapping, define

$$V_j = \{\mathcal{L}(i) \in \mathcal{C} : i \in \operatorname{supp}(\hat{\alpha}_j), \hat{\alpha}_j \in A\}$$

as the multiset of votes collected from each subject in $\mathcal{C}$,
3. put together the votes in the set $V = \bigcup_{j=1,\ldots,d} V_j$ and choose the final identity according to the mode of V,
4. in case of multiple winners, apply the canonical least square criterion between the probe features of the winners and the linear combination of their dictionary atoms, to achieve a ranking among them.

---

[3] For a given of vector α, the support $\operatorname{supp}(\alpha)$ is the index pool of nonzero entries of α.

[4] Here we have simplified the notation to refer to the sparse solutions $\alpha_{l,f}$ to α_j, knowing that the couple set (l, f) has cardinality d.

3   Experiments on LFW

In this section, we evaluate the effectiveness of our method on the LFW funneled database [7] that contains more then 13000 images of 5749 different individuals acquired in unconstrained environments and already centered. The variances in pose, illumination, expression, and the presence of partial occlusions make SSPP problem extremely challenging. Tuning and analysis are performed on the subset of the LFW dataset corresponding to the 158 individuals which includes no less than 10 samples per subject (in the following LFW-158), while the evaluation refers to several sub-galleries obtained from the LFW-158, plus two tests referring to the subset of 1680 subjects for which at least two images are present in the dataset (in the following LFW-1680).

The first step of our method concerns with image augmentation, aiming at generating augmented images of size 224×224, as required by the VGG-face DCNN, for the entire LFW dataset. In this regards, we provide either the original or flipped images, varying the zooming according to the values $\{1.5, 1.75, 2, 2.25\}$, and applying horizontal and vertical shifts, each within the set $\{-10, 0, +10\}$. Thus, for each image I we attain $d = 2 \times 4 \times 9 = 72$ augmented images I_j and consequently 72 features φ_j.

In order to perform the tuning and analysis, we split randomly the images in the LFW-158 into three sets: gallery, validation and test sets, and we run 10 trials to assess the method. The gallery contains one image per subject, while the remaining images are randomly split, putting 4 images per subject in the validation set, and the remaining images in the test set.

Given that the feature set cardinality is d, and organizing the features k by k in L dictionaries so that $k = d/L$, we are interested in investigating the system behaviour varying the number of dictionaries L. In Fig. 2 we plot both the performances and the computational costs obtained on the validation set, showing that they both take advantage from the subdivision of features in sub-dictionaries, arriving to a plateau around 12 dictionaries. The performance improvement happens because, the subdivision increment restricts the search domain for each feature φ_j helping in reducing false matching. Concerning the computational time, it decreases when L is augmented, implying a reduction of the dictionary sizes, making the k-LiMapS faster.

Furthermore, we access the method reliability by comparing the distribution of the votes when dealing with error or hit cases. Specifically, for each validation image, we refer to the collection of votes V produced by the classifier, and we compute the difference between the number of votes obtained by the two most frequently voted IDs. We then collect and arrange these differences separately according to correctness/incorrectness of the classification. In Fig. 3 the two empirical PDFs are depicted.

The distributions suggest that when the model classifies correctly, than the vast majority of the atoms will vote for the correct ID, with typically large differences with respect to the second most voted ID, thus denoting high confidence in the choice. On the other hand, when a misclassification occurs, the distribution of the differences shows higher probabilities around smaller ones, suggesting

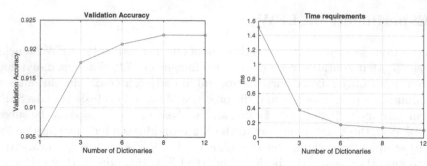

Fig. 2. Accuracy *(left)* and computational costs *(right)* arranging the 72 features varying the number of dictionaries.

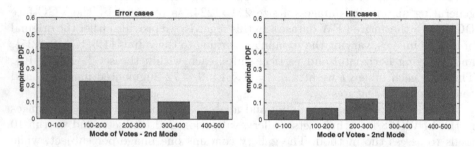

Fig. 3. Distribution of the differences between the two most voted IDs, in case of misclassification (**left**) or hit cases (**right**).

the presence of (at least) two close modes in the empirical distribution of the votes. This analysis suggests that it would be worth exploiting the empirical distribution just discussed for better discriminating misclassification.

Finally, to assess the method performances and to compare them with the state-of-the-art in SSPP problem, we set $d = 72$, $L = 12$, and consequently $k = 6$, and $q = \{50, 100, 158, 793, 1680\}$. The experiments with $q \leq 158$ are attained referring to the LFW-158, while the others referring to the LFW-1680. In Table 1 the average results over 10 trials are reported together with comparisons when available, proving the effectiveness of our method.

Table 1. Face recognition rates (%) on several subsets of the LFW database identified by the number of subjects (sbj) in the gallery. Our results report also the standard deviation. For comparison, the state-of-the-art on the LFW, referring to different gallery cardinalities are reported.

Method	50 sbj	100 sbj	158 sbj	793 sbj	1680 sbj
Our	95.93 ± 0.02	94.20 ± 0.01	92.48 ± 0.01	87.63 ± 0.01	84.11 ± 0.01
Others	86 [18]	92.7 [5]	37.9 [9], 50 [14]	65.3 [18]	-

4   Conclusions and Future Works

In this paper we proposed a method coupling the strength of deep features with the effectiveness of our sparse representation classifier based on the k-LiMapS algorithm. Concerning the evaluation, we have shown our approach outperforms the state-of-the-art concerning the SSPP problem, and it is robust even on a large gallery of 1680 individuals. These preliminary results encourage further investigations, deepening several aspects arisen in the analysis. One concerns the role of the augmented images: would a larger number of augmented images or virtual images obtained from generative models (e.g. the GAN, generative adversarial networks) improve further the performances? Also the feature organization in dictionaries would deserve further analysis, investigating other partitions besides the k by k adopted in this paper. Finally the possibility to introduce a criterion for discriminating correct classification from false positives would deserve an in-depth analysis. These questions are planned to be faced in our future work.

References

1. Adamo, A., Grossi, G., Lanzarotti, R.: Local features and sparse representation for face recognition with partial occlusions. In: 2013 IEEE International Conference on Image Processing, pp. 3008–3012 (2013)
2. Adamo, A., Grossi, G., Lanzarotti, R., Lin, J.: Sparse decomposition by iterating lipschitzian-type mappings. Theor. Comput. Sci. **664**, 12–28 (2017)
3. Cuculo, V., Lanzarotti, R., Boccignone, G.: Using sparse coding for landmark localization in facial expressions. In: 5th European Workshop on Visual Information Processing (EUVIP), pp. 1–6, December 2014
4. Deng, W., Hu, J., Zhou, X., Guo, J.: Equidistant prototypes embedding for single sample based face recognition with generic learning and incremental learning. Pattern Recogn. **47**(12), 3738–3749 (2014)
5. Gao, Y., Ma, J., Yuille, A.L.: Semi-supervised sparse representation based classification for face recognition with insufficient labeled samples. IEEE Trans. Image Process. **26**(5), 2545–2560 (2017)
6. Grossi, G., Lanzarotti, R., Lin, J.: Robust face recognition providing the identity and its reliability degree combining sparse representation and multiple features. Int. J. Pattern Recognit. Artif. Intell. **30**(10) (2016)
7. Huang, G.B., Ramesh, M., Berg, T., Learned-Miller, E.: Labeled faces in the wild: a database for studying face recognition in unconstrained environments. Technical report 07–49, University of Massachusetts, Amherst (2007)
8. Lahasan, B., Lutfi, S.L., San-Segundo, R.: A survey on techniques to handle face recognition challenges: occlusion, single sample per subject and expression. Artif. Intell. Rev., 1–31 (2017)
9. Liu, F., Tang, J., Song, Y., Bi, Y., Yang, S.: Local structure based multi-phase collaborative representation for face recognition with single sample per person. Inf. Sci. (Ny) **346–347**, 198–215 (2016)
10. Natarajan, B.K.: Sparse approximate solutions to linear systems. SIAM J. Comput. **24**, 227–234 (1995)

11. Parkhi, O.M., Vedaldi, A., Zisserman, A.: Deep face recognition. In: British Machine Vision, pp. 1–12 (2015)
12. Perez, L., Wang, J.: The effectiveness of data augmentation in image classification using deep learning. CoRR abs/1712.04621 (2017)
13. Rao, C.R.: The utilization of multiple measurements in problems of biological classification. J. R. Stat. Soc.-Ser. B **10**(2), 159–203 (1948)
14. Shang, K., Huang, Z.H., Liu, W., Li, Z.M.: A single gallery-based face recognition using extended joint sparse representation. Appl. Math. Comput. **320**, 99–115 (2018)
15. Sun, Y., Wang, X., Tang, X.: Deep learning face representation from predicting 10,000 classes. In: 2014 IEEE Conference on Computer Vision and Pattern Recognition, pp. 1891–1898. IEEE (2014)
16. Wiskott, L., Fellous, J., Kruger, N., von der Malsburg, C.: Face recognition by elastic bunch graph matching. In: Intelligent Biometric Techniques in Fingerprints and Face Recognition, pp. 355–396 (1999)
17. Wright, J., Yang, A.Y., Ganesh, A., Sastry, S.S., Ma, Y.: Robust face recognition via sparse representation. IEEE Trans. Pattern Anal. Mach. Intell. **31**(2), 210–27 (2008)
18. Yang, M., Wang, X., Zeng, G., Shen, L.: Joint and collaborative representation with local adaptive convolution feature for face recognition with single sample per person. Pattern Recogn. **66**, 117–128 (2017)
19. Zhang, Z., Xu, Y., Yang, J., Li, X., Zhang, D.: A survey of sparse representation: algorithms and applications. IEEE Access **3**, 490–530 (2015)
20. Zhuo, T.: Face recognition from a single image per person using deep architecture neural networks. Clust. Comput. **19**(1), 73–77 (2016)

Recursive Chaining of Reversible Image-to-Image Translators for Face Aging

Ari Heljakka[1,2]($\boxtimes$), Arno Solin[2], and Juho Kannala[2]

[1] GenMind Ltd., Espoo, Finland
{ari.heljakk,arno.solin,juho.kannala}@aalto.fi
[2] Department of Computer Science, Aalto University, Espoo, Finland

Abstract. This paper addresses the modeling and simulation of progressive changes over time, such as human face aging. By treating the age phases as a sequence of image domains, we construct a chain of transformers that map images from one age domain to the next. Leveraging recent adversarial image translation methods, our approach requires no training samples of the same individual at different ages. Here, the model must be flexible enough to translate a child face to a young adult, and all the way through the adulthood to old age. We find that some transformers in the chain can be recursively applied on their own output to cover multiple phases, compressing the chain. The structure of the chain also unearths information about the underlying physical process. We demonstrate the performance of our method with precise and intuitive metrics, and visually match with the face aging state-of-the-art.

Keywords: Deep learning · Transfer learning · GAN · Face synthesis
Face aging

1 Introduction

Generative Adversarial Network (GAN) [8] variants have been successful for various image generation and transformation tasks. For image-to-image translation (such as mapping sketches to photographs), they have achieved state-of-the-art results, with paired training data [9] and without it [13,17,27,28].

This paper generalizes image-to-image mapping to a sequential setting. Previous works have not focused on recursive application of the models on their own outputs, and there have been no extensions to apply the method for a sequence of domains, even though, *e.g.*, [5] allow applying several *different kinds* of domain transformations to the same image. We propose a recursive adversarial domain adaptation approach that is capable of producing step-wise transformations for human aging, as visualized in the examples in Fig. 1. We use the *reversible* image translation approach of [13,27,28].

© Springer Nature Switzerland AG 2018
J. Blanc-Talon et al. (Eds.): ACIVS 2018, LNCS 11182, pp. 309–320, 2018.
https://doi.org/10.1007/978-3-030-01449-0_26

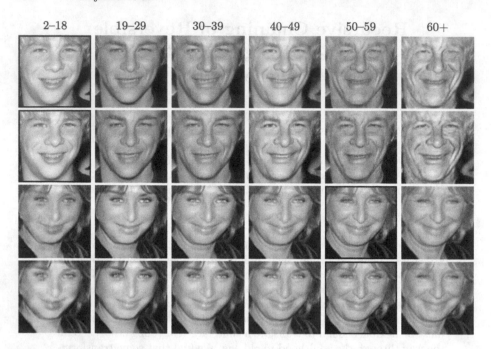

2–18   19–29   30–39   40–49   50–59   60+

Fig. 1. Examples of face aging transformations. Row 1: Non-recursive transformation from 15-year-old original to 65-year-old synthetic. Row 2: Partially recursive transformation by re-using the transformer of 25 → 35 for also 35 → 45. Row 3: Non-recursive transform of an approximately 55-year-old original to older (65-year-old) and younger (towards 15-year-old). Row 4: Partially recursive transformation by re-using the transformer of 35 → 25 for also 45 → 35.

Previously, Antipov *et al.* [3] applied a conditional GAN to simulating human face aging with good initial results. Their model distinguishes itself by explicitly enforcing the preservation of identity in face transformations. However, in reversible transformers such as ours, the preservation of identity requires no explicit measures. By learning a reversible mapping between two faces, the transformer is inclined to preserve the identity information. Face aging has been tackled before in approaches such as [7,12,16,25]. One can also find adjustable parameters to modulate the prevalence of general attributes in an image [15]. While learning adjustable knobs allows for fine-grained control, our approach is suitable for robust learning of pronounced sequential changes.

The rationale of this paper follows from modeling progressive changes. Consider the development of an entity that can be approximated as a closed system (*e.g.*, an aging human face, a deteriorating surface, a growing tree, or a changing city outline). Visual representation of such development can be approximated as a succession of snapshots taken over time, so that one snapshot is translated to the next one, then to the next one, *etc.* Such a translation requires Markovian development, so that each image carries sufficient information to enable translating it to the next stage. This allows for *e.g.* face aging, but not for modeling the day–night cycle.

Utilizing unpaired image-to-image translation methods, we can learn these translation steps. Suppose we have training samples from an entity class P such that we can at least roughly assign each one to a bin P_i that corresponds to its developmental phase i. Knowing the ordering of the phases, the model can learn to transform an image from stage P_i to stage P_{i+1}. The proposed method can simulate one such possible line of development, with the hope that the real underlying process has sufficient regularity to it. In this paper, the application of interest is human face aging, which is both a challenging modeling problem and the results are easy to validate by human readers (as humans are highly specialized in face perception).

The contributions of this paper are as follows.

- We show how a chain framework of unsupervised reversible GAN transformers can be constructed to convert a human face into a desired age, along with the necessary image pre-processing steps.
- We show that using an auxiliary scoring method (*e.g.* a face age estimator) to rank the transformers between training stages, we can leverage transfer learning with a simple meta-training scheme that re-uses the best transformers and compresses the chain with no performance cost. The auxiliary score provides no learning gradient for an individual transformer. However, we show that the score does consistently reflect the training progress.
- Consequently, our ability to compress the transformer chain can be used to separate the high-level linear (*e.g.* increasing formation of wrinkles) and non-linear stages of development, producing a compact human-interpretable development descriptor.

2   Proposed Method

2.1   Sequential Transformations

In reversible GAN transformers, such as [28], there are two generator networks and two discriminator networks. One generator transforms images from domain A to domain B and the other from B to A. One discriminator separates between whether an image originates from the training set of B or from the first generator. The other discriminator does the same for A and the second generator. The resulting generators will have learnt to map image features between the domains.

We can treat this kind of a network as a single building block of a chain of domain transformers. For paired domains such as summer–winter or horses–zebras, the notion of such chaining does not apply. However, it does apply for classes of closed systems that evolve over time along a path of certain regularity. If we can disentangle the relevant patterns of development, a transition between two stages may occur by repeated or one-time application of those patterns.

A sequential transformation can be described as a composition of operators, with fundamental relations such as reflexivity and transitivity. Given a mapping

F from domain A to domain B, what is the intended meaning of $F(F(a))$, $a \in A$? In the context of regular non-sequential transformations such as [28], a reasonable option would be to require either reflexivity as $F(F(a)) \in A$ or anti-reflexivity as $F(F(a)) \notin A$ or $F(F(a)) \in B$. These constraints could be added to existing image-to-image translators.

The more relevant case here is transitivity. For domains A_t, $t = \{1, 2, 3\}$, and $a_i \in A_i$, we would like to have $F(a_1) \in A_2$, and also $F(F(a_1)) \in A_3$. This allows us to enforce the transformer's ability to perform multiple transformations, turning it into an extra loss term. One may also vary the 'domain thickness' with respect to F, *i.e.* the measure of how many times F should be applied in order to move from domain i to domain $i + 1$ (*e.g.* 2, 1, 1/2, *etc.*). With enough domains, thickness of 1 suffices.

2.2   Meta-Training of Transformer Models

We train a chain of transformers to cover each developmental stage, as given in Algorithm 1. The algorithm requires a way to measure model performance at meta-level, i.e. between the actual training runs. For GANs, traditional likelihood measures are problematic [24,26] and in some cases inversely correlated with image quality [6]. GANs are often prone to collapse into a specific mode of the distribution, missing much of the diversity of the training data. Trainable auxiliary evaluators like [6,18] do help, but for many application domains, we already have automated deterministic scoring tools for images. For human face aging, such auxiliary evaluators exist. We chose [2].

The algorithm starts by training a separate transformer network Φ_1 between the first two consecutive datasets, *e.g.* 15 and 25-year-old face images, for N steps (N determined empirically). Transformers trained in this way are called *baseline*. We use CycleGAN [28], but the algorithm is applicable for any image-to-image transformer.

Then, we train another model Φ_2 in the exact same way for the next stage pair, *e.g.* mapping from 25 to 35-year-olds. But now, we also re-train a copy of the model Φ_1 recycled from the previous stage with additional $N/2$ steps on its earlier data and $N/2$ on the next stage data. For example, Φ_1, originally trained for $15 \rightarrow 25$ conversion, is now re-trained also on $25 \rightarrow 35$ data. In order to benchmark Φ_1 and Φ_2, we measure the perceived age of the faces in the images they produce, with the estimator [2]. The transformations should age by 10 years, so we can score each transformer simply by the error in the mean age of its output distribution, normalized by its standard deviation (alg. lines 8–9).

If the re-trained transformer is not significantly worse on this new transformation than the baseline transformer, we discard the baseline transformer and replace it with the re-trained transformer. Otherwise we discard the re-trained one. The remaining transformer will then be tried out on the next stage, and continuing until the end of the transformer chain. By virtue of reversible transformers, we could also run the same algorithm backwards, finding the best models that make an old face young again (rows 2 and 4 in Fig. 1).

Algorithm 1. Greedy forward-mode recursive transformer chain with two-step backward-compatibility. In the experiments, we used $\mu_i^{\text{target}} = [15, 25, \ldots, 65]$, $N_D = 6$, $S = 600,000$, $\varepsilon = 0.1$.

1: **Require:** Number of stages N_D, data sets D_i and target mean age μ_i^{target} with $i \in [1, N_D]$, trainable models $\Phi_j : \mathbb{R}^{256 \times 256} \to \mathbb{R}^{256 \times 256}$ with $j \in [1, N_D - 1]$, number of steps S, auxiliary age estimator $\Gamma : \mathbb{R}^{256 \times 256} \to \mathbb{R}$

2: **Initialize:** $a \leftarrow 1$ ▷ Denote the index of the Φ model we try to re-use

3:     $\Phi_1 \leftarrow \text{train}(\Phi_1, [(D_1, D_2)], S)$

4: **for** $i = 2, \ldots, (N_D - 1)$ **do**

5:   $\Phi_i \leftarrow \text{train}(\Phi_i, [(D_i, D_{i+1})], S)$

6:   $\Phi_a' \leftarrow \text{copy}(\Phi_a)$

7:   $\Phi_a' \leftarrow \text{train}(\Phi_a', [(D_{i-1}, D_i), (D_i, D_{i+1})], S)$

8:   $E' \leftarrow \left| \mathbb{E}_{d \sim D_i} [\|\Gamma(\Phi_a'(d)) - \mu_{i+1}^{\text{target}}\|_1] / \sigma(\Gamma(\Phi_a'(d))) \right|$

9:   $E \leftarrow \left| \mathbb{E}_{d \sim D_i} [\|\Gamma(\Phi_i(d)) - \mu_{i+1}^{\text{target}}\|_1] / \sigma(\Gamma(\Phi_i(d))) \right|$

10:   **if** $|E - E'| < \varepsilon$ **then**   ▷ Recycled model wins

11:    release(Φ_i)

12:    release(Φ_a)

13:    $\Phi_a \leftarrow \Phi_i \leftarrow \Phi_a'$ ▷ Upgrade Φ_a and try re-using again

14:   **else**

15:    release(Φ_a')

16:    $a \leftarrow i$ ▷ Next, try re-using the most recent base model

The auxiliary scoring method is only used for a binary choice between two competing transformers during stage transitions. The score, therefore, provides no gradient for the training. Furthermore, our scoring model has been trained on an altogether different dataset and model architecture. Retrospectively, we calculate what the scores would have been during training (Fig. 2b).

The algorithm uses a simple greedy search. By enforcing backward compatibility only to the previous transformation stage, we will only prevent forgetting in short sequences. For the six domains, we found the incurring performance penalty to be minor (Fig. 2b). To scale up, we could simply cap the maximum number of re-uses of a model or the allowed error from forgetting, or add equal parts of even earlier domains to the training set of the re-used model.

3 Experiments

3.1 Dataset and Auxiliary Age Estimator

For training, we use the Cross-Age Celebrity Dataset (CACD, [4]), with a large number of age-annotated images. Our chosen auxiliary age estimator [2], pre-trained on the IMDB-Wiki dataset [21], was used as-is (the estimator has not seen any of our training data). We found it necessary to improve the CACD data in two ways. First, we used the off-the-shelf face alignment utility of [22] to crop and align the faces based on landmarks. Second, despite the existing age annotations of CACD, they are not accurate enough for age-estimation [1]. We ran the auxiliary age estimator on the data and found major discrepancy between the annotations and the results of the estimator. We also confirmed visually that the estimator was closer to the ground truth. Using the estimator as ground-truth, we re-annotated the whole CACD data accordingly. Our model, however, has access neither to the age information nor the estimator. It only knows that the pictures come from different domains, at 10-year accuracy. For validation, we use a small subset of the IMDB-Wiki dataset.

We divided the data into slots 2–18, 19–29, 30–39, 40–49, 50–59, and 60–78 (six domains, with five direct transformation paths). This enables direct comparison with [3]. We removed enough samples so that the mean age in the sets is 10 years apart—15, 25, 35, 45, 55, and 65, respectively. Since the numbers of samples may differ between domains, we measure training progress by the number of steps, not epochs, so as to maintain commensurability between stages.

3.2 Architecture and Training

The architecture of each transformer module follows [28] (which re-uses structures from [9,10]), with two generators and two discriminators per module. The generators utilize convolution layers for encoding, 9 ResNet blocks for transformation, and deconvolution layers for decoding (original source code adopted from [19]). Our image preprocessing reduces the real face resolution to 132×132, which is then upscaled to 256×256 for computational efficiency using off-the-shelf Mitchell–Netravali filter [20]. The full chain is composed of five independently trained successive modules, so that one module feeds its output to the next. The results of each transformation stage can be evaluated independently, and external input can be fed in at any stage of the chain. For recursion, one can co-train with 3 stages (e.g. $15 \rightarrow 25$ and $25 \rightarrow 35$), or apply the model twice ($15 \rightarrow 35$).

We trained with ADAM ([14], learning rate 0.0002, $\beta_1 = 0.5$, $\beta_2 = 0.999$) with a batch size one. As in [23,28], we update the discriminators using a buffer of 50 recently generated images rather than only the most recently produced ones. The training time on an NVIDIA P100 workstation was 540 h (300 h for the baseline, 60 h for each re-training session).

3.3 Human Face Age Progression

Our solution is based on a pipeline of successive transformations. In order to train a single transformation, say $25 \rightarrow 35$, we train a single transformer network. By the reversible architecture, we simultaneously train the network to carry out the full inverse transformation, $35 \rightarrow 25$.

We trained and evaluated each transformer network according to Algorithm 1. For human face aging, our hypothesis was that the network trained for transforming faces from $15 \rightarrow 25$ would not generalize to the other stages, whereas the network that transforms from $25 \rightarrow 35$ would generalize to multiple later stages as well. We confirm this both during training-time (Fig. 2a) and validation (Fig. 2b). The algorithm drops the baseline transformer $35 \rightarrow 45$ off the chain ($\epsilon > 0.35$ would drop the next one, too). From this result, we can read off the development descriptor of the form $F(\text{age} = 65) = F_{55 \rightarrow 65}(F_{45 \rightarrow 55}(F^2_{25 \rightarrow 35}(F_{15 \rightarrow 25}(\text{age} = 15))))$. We expected that if we can reuse some transformers, this provides us with high-level information about the underflying development pattern. While our result alone only gives tentative support to this idea, we expect this approach to be also applicable with more fine-grained precision.

3.4 Comparison

In Fig. 1, we show sample transformation paths, with a 15-year-old successively transformed to 65, and a 55-year-old transformed back to 15 and forward to 65. The results are visually at least at the level of [3]. Additional results are shown in Figs. 3 and 4.

As most methods are concerned with singular transformations, our most relevant points of comparison are [3,15]. However, both show results where several successive stages of development show effectively no changes at all, and no age measurements. Our results show greater variation between each stage, with not too many artifacts even at relatively low resolution. The improvements are presumably due to both the differences between reversible GANs and conditional GANs, and to the use of the transformer chain. The chain, of course, comes with additional computational cost.

In average, our compressed chain produced the target age in the validation set with an error less than 4.5% (maximum of 3.2 years) and standard deviations between 2.4–5.6 in different age groups. The performance drop from the baseline due to compression was negligible. [3] reports an estimation accuracy 17% lower for synthetic images than natural ones. This may reflect the improvements in visual accuracy and variation in our model.

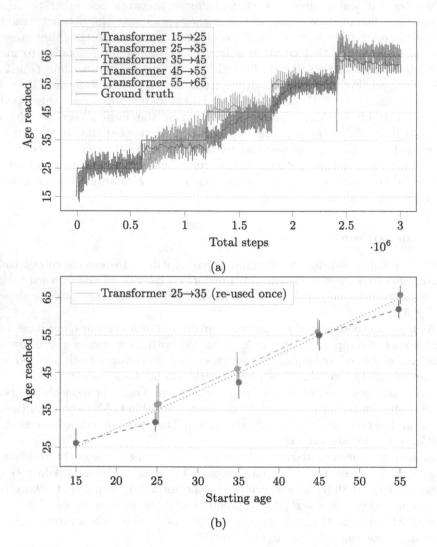

(a)

(b)

Fig. 2. (a) Performance of modules as a function of training steps (with stage-specific transformation target), greedily selecting the best-performing models for re-use. The error bars indicate the standard deviation of the ages of generated test samples. Some of the variance is by design, as we intentionally challenge the model with a wide *range* of ages for robustness and generalization in training data (for clarity of visualization, the ground truth variance not shown). (b) Resulting age after transformation in validation set, as a function of the input age. During training, the model learned that $25 \rightarrow 35$ transformer can be recycled for $35 \rightarrow 45$ to transform better than the freshly trained $35 \rightarrow 45$ transformer. The $35 \rightarrow 45$ transformer was thus discarded (see Algorithm 1). $15 \rightarrow 25$ and $45 \rightarrow 55$ transformers succeeded on their baseline but were not re-usable.

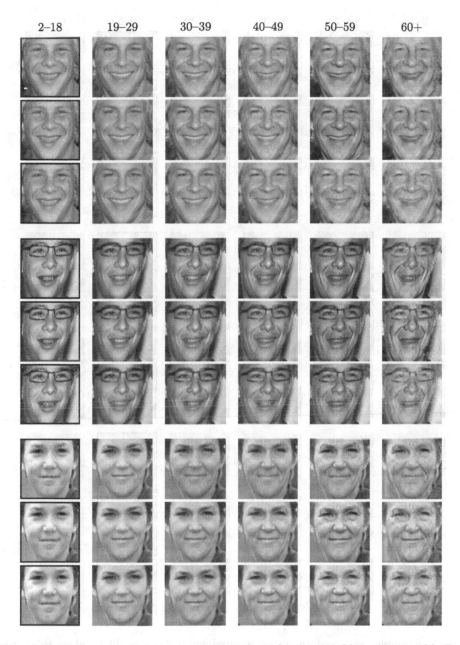

Fig. 3. Examples of transforming an approximately 15-year-old to 65-year-old. For each identity, row 1 shows the non-recursive transformation (applying the baseline transformer on each stage). Row 2 shows the partially-recursive transformation, with the double-trained $25 \rightarrow 35$ transformer applied also to $35 \rightarrow 45$ (the best chain, according to Algorithm 1). Row 3 shows the transformation with the most recursive steps, with the triple-trained $25 \rightarrow 35$ transformer applied also to both $35 \rightarrow 45$ and $45 \rightarrow 55$ (the most compact chain, picked manually).

Fig. 4. Examples of transforming an approximately 55-year-old to older (65-year-old) and younger (towards 15-year-old). For each identity, row 1 shows the non-recursive transformation (applying the baseline transformer on each stage). Row 2 shows the partially-recursive transformation, with the double-trained 35 → 25 transformer applied also to 45 → 35 (the best chain, according to Algorithm 1). Row 3 shows the transformation with the most recursive steps, with the triple-trained 35 → 25 transformer applied also to both 45 → 35 and 55 → 45 (the most compact chain, picked manually).

4   Discussion and Conclusion

In this paper, we showed that a transformer chain composed of reversible GAN image transformer modules can learn a complex multi-stage face transformation task. The domain-agnostic base algorithm is expected to generalize to other kinds of temporal progression problems. Notably, we found that a single transformer can carry out the face transformations $25 \rightarrow 35$ and $35 \rightarrow 45$ (and, with minor loss of accuracy, also $45 \rightarrow 55$). One might suspect this to be because not many visible changes happen during this time. However, the auxiliary age estimator can still discern the ages 25, 35, and 45 easily. As the estimator and transformer chains are independent, this indicates the presence of real changes that the networks capture in a systematic way, even when they are hard for humans to discern.

On some problem domains, the full chain may be relatively long and slow to train. Also, there is currently little control over the transformation paths, and the chain may not yield a range of varied outcomes in contexts where the developmental paths could diverge in different directions. Recent methods for improving the resolution of GAN-generated images [11] could be combined with our method. Follow-up work should evaluate the extent to which the layers of separate transformer modules can be shared, so as to reduce the total training time. More comprehensive evaluation of the models in both directions would likely find more compact chains. In semi-supervised setting, using a small number of paired examples would likely improve our results.

The code for replicating the results is available online: https://github.com/ AaltoVision/img-transformer-chain.

Acknowledgments. This research was supported by GenMind Ltd. and the Academy of Finland grants 308640, 277685, and 295081. We acknowledge the computational resources provided by the Aalto Science-IT project.

References

1. Cross-age reference coding for age-invariant face recognition and retrieval. http:// bcsiriuschen.github.io/CARC/. Accessed 18 May 2018
2. Antipov, G., Baccouche, M., Berrani, S., Dugelay, J.: Apparent age estimation from face images combining general and children-specialized deep learning models. In: CVPR Workshops (2016)
3. Antipov, G., Baccouche, M., Dugelay, J.L.: Face aging with conditional generative adversarial networks. In: ICIP (2017)
4. Chen, B.-C., Chen, C.-S., Hsu, W.H.: Cross-age reference coding for age-invariant face recognition and retrieval. In: Fleet, D., Pajdla, T., Schiele, B., Tuytelaars, T. (eds.) ECCV 2014. LNCS, vol. 8694, pp. 768–783. Springer, Cham (2014). https:// doi.org/10.1007/978-3-319-10599-4_49
5. Choi, Y., Choi, M., Kim, M., Ha, J.W., Kim, S., Choo, J.: StarGAN: unified generative adversarial networks for multi-domain image-to-image translation. arXiv preprint arXiv:1711.09020 (2017)

6. Danihelka, I., Lakshminarayanan, B., Uria, B., Wierstra, D., Dayan, P.: Comparison of maximum likelihood and GAN-based training of real NVPs. arXiv preprint arXiv:1705.05263 (2017). https://arxiv.org/abs/1705.05263

7. Duong, C., Quach, K., Luu, K., Savvides, M.: Temporal non-volume preserving approach to facial age-progression and age-invariant face recognition. In: ICCV (2017)

8. Goodfellow, I.J., et al.: Generative adversarial networks. In: NIPS (2014)

9. Isola, P., Zhu, J.Y., Zhou, T., Efros, A.A.: Image-to-image translation with conditional adversarial networks. In: CVPR (2017)

10. Johnson, J., Alahi, A., Li, F.: Perceptual losses for real-time style transfer and super-resolution. In: ECCV (2016)

11. Karras, T., Aila, T., Laine, S., Lehtinen, J.: Progressive growing of GANs for improved quality, stability, and variation. arXiv preprint arXiv:1710.10196 (2017)

12. Kemelmacher-Shlizerman, I., Suwajanakorn, S., Seitz, S.: Illumination-aware age progression. In: CVPR (2014)

13. Kim, T., Cha, M., Kim, H., Lee, J.K., Kim, J.: Learning to discover cross-domain relations with generative adversarial networks. In: ICML. PMLR, vol. 70, pp. 1857–1865 (2017)

14. Kingma, D.P., Ba, J.: Adam: a method for stochastic optimization. In: ICLR (2015). https://arxiv.org/abs/1412.6980

15. Lample, G., Zeghidour, N., Usunier, N., Bordes, A., Denoyer, L., Ranzato, M.: Fader networks: Manipulating images by sliding attributes. In: NIPS (2017)

16. Lee, M., Seok, J.: Controllable generative adversarial network. arXiv preprint arXiv:1708.00598 (2017)

17. Liu, M.Y., Breuel, T., Kautz, J.: Unsupervised image-to-image translation networks. In: NIPS (2017)

18. Lopez-Paz, D., Oquab, M.: Revisiting classifier two-sample tests. In: ICLR (2017). https://arxiv.org/abs/1610.06545

19. LynnHo (GitHub user): CycleGAN Tensorflow PyTorch. https://github.com/LynnHo/CycleGAN-Tensorflow-PyTorch-Simple, gitHub repository

20. Mitchell, D.P., Netravali, A.N.: Reconstruction filters in computer-graphics. ACM Siggraph Comput. Graph. **22**(4), 221–228 (1988)

21. Rothe, R., Timofte, R., Gool, L.V.: DEX: Deep EXpectation of apparent age from a single image. In: ICCV, Looking at People Workshop (2015)

22. Schroff, F., Kalenichenko, D., Philbin, J.: FaceNet: a unified embedding for face recognition and clustering. In: CVPR (2015)

23. Shrivastava, A., Pfister, T., Tuzel, O., Susskind, J., Wang, W., Webb, R.: Learning from simulated and unsupervised images through adversarial training. In: CVPR (2017). https://arxiv.org/abs/1612.07828

24. Theis, L., van den Oord, A., Bethge, M.: A note on the evaluation of generative models. In: ICLR (2016). https://arxiv.org/abs/1511.01844

25. Tiddeman, B., Burt, M., Perrett, D.: Prototyping and transforming facial textures for perception research. IEEE Comput. Graph. Appl. **21**(5), 42–50 (2001)

26. Wu, Y., Burda, Y., Salakhutdinov, R., Grosse, R.: On the quantitative analysis of decoder-based generative models. In: ICLR (2017). https://arxiv.org/abs/1611.04273

27. Yi, Z., Zhang, H., Tan, P., Gong, M.: DualGAN: unsupervised dual learning for image-to-image translation. In: ICCV (2017)

28. Zhu, J.Y., Park, T., Isola, P., Efros, A.A.: Unpaired image-to-image translation using cycle-consistent adversarial networks. In: ICCV (2017)

Automatically Selecting the Best Pictures for an Individualized Child Photo Album

Floris De Feyter[✉], Kristof Van Beeck, and Toon Goedemé

KU Leuven — EAVISE, Leuven, Belgium
floris.defeyter@kuleuven.be

Abstract. In this paper we investigate the best way to automatically compose a photo album for an individual child from a large collection of photographs taken during a school year. For this, we efficiently combine state-of-the-art identification algorithms to select relevant photos, with an aesthetics estimation algorithm to only keep the best images. For the identification task, we achieved 86% precision for 86% recall on a real-life dataset containing lots of specific challenges of this application. Indeed, playing children appear in non-standard poses and facial expressions, can be dressed up or have their faces painted etc. In a top-1 sense, our system was able to correctly identify 89.2% of the faces in close-up. Apart from facial recognition, we discuss and evaluate extending the identification system with person re-identification. To select out the *best-looking* photos from the identified child photos to fill the album with, we propose an automatic assessment technique that takes into account the aesthetic photo quality as well as the emotions in the photos. Our experiments show that this measure correlates well with a manually labeled general appreciation score.

Keywords: Face recognition · Person re-identification
Aesthetics analysis · Emotion classification · Child identification

1 Introduction

Many teachers in preschool and elementary school take a large amount of photographs during the school year. These photos form a tangible memory for both child and parent. In many current schools, at the end of the school year, parents receive gigabytes of photographs of all the pictures that were taken that year in class. Mostly, however, parents are only interested in images containing their own child. This forces them to painstakingly go through the enormous amount of images and select only those of interest. To help parents collect photographs of their child without the inconvenience of manually going through thousands of files, we propose an innovative system that automates this whole process. Our system selects pictures that are well-suited candidates for a photo book with a specific child as the *main character*. For this, we focus on two main aspects: identifying children on a photograph and selecting photographs which are generally perceived as beautiful.

© Springer Nature Switzerland AG 2018
J. Blanc-Talon et al. (Eds.): ACIVS 2018, LNCS 11182, pp. 321–332, 2018.
https://doi.org/10.1007/978-3-030-01449-0_27

Both the identification and the aesthetic selection demand a careful procedure to deal with their intrinsic challenges. The identification of children in an environment of school activities, includes identifying children when they are playing, running, dressed up etc. Figure 1 illustrates this with naturally occurring situations where identification is challenging. We propose an approach based on state-of-the-art facial recognition using 8 reference descriptors per identity. Additionally, we developed and employed experiments with a hybrid system that combines facial recognition and state-of-the-art re-ID to handle the highly challenging task of identifying persons of whom the faces are undetectable.

Fig. 1. Examples of challenges in automatic facial recognition for children: dressed up, difficult poses, extreme facial expressions and painted faces.

Due to the underlying subjectivity, selecting the *best-looking* photographs is another delicate task. We correlated the emotions on photographs and scores on multiple aesthetics criteria with appreciation scores expressed by a panel of human annotators. These correlations suggest a general preference towards positive emotions and compositional aspects like using the rule of thirds. We show that it is—to a certain extent—possible to predict the general appreciation for a photograph by combining emotion classifications with aesthetics criteria scores.

2 Related Work

Identifying persons in photo albums has mainly been done in the context of family photo albums [4,10,12]. Family photos impose less challenges than pictures taken in a school context. Although the amount of photos might be similar, there are less identities and therefore the amount of photos per identity is higher, increasing the chances for a correct hit. These papers do not take image quality into account and simply focus on the identification aspect, making use of obsolete techniques. Zhao et al. [12], for example, use 2D Hidden Markov Models (HMMs) [2] achieving an Equal Error Rate (EER) of about 40% precision and recall. Chen et al. [4] propose a block wavelet approach, but only focus on *semi-automatic* person recognition, i.e. their system suggests a handful identities for a user to select from. Previous literature on image selection based on quality, focuses only on aesthetics criteria [3,8]. Although important, we are convinced that not only the aesthetic quality of the image must be taken into account, but also the expressed emotion of the faces in the photograph. The contribution of our paper is twofold: (i) apply state-of-the-art person identification techniques to the problem of composing a personalized photo album; (ii) use current advances in aesthetics analysis and emotion detection to create an appreciation prediction for selecting the best images for the photo album.

3 Approach

Automatically selecting the best photos for an individualized photo album involves two tasks. Firstly, one has to recognize the identities in the photos in order to select those having a specific child as the *main character* (Sect. 3.1). From the set of selected photos, in a second step only the most beautiful pictures have to be selected for the album. For this, we propose a photo appreciation scoring approach (Sect. 3.2), combining both aesthetic and emotional criteria.

3.1 Identification

Figure 2 shows a schematic overview of how our proposed system identifies children on a given photo.[1] It combines the results of two identification techniques: facial recognition and re-ID. The facial recognition branch consists of three steps. First, a face detector detects the faces that are present in the image. Then the image is categorized into one of four shot types: (i) *ambiance photo*: a picture without any faces; (ii) *group photo*: a photo with more than four faces; (iii) *close-up*: a picture containing at least one face that covers more than 10% of the total picture area; (iv) *non-close-up*: a photo that is neither of the previous types. Only for close-up and non-close-up photos, the recognition step is actually performed. We chose to use the Dlib framework [5] because of its state-of-the-art accuracy of 99.38% on the LFIW dataset. [6] The Dlib face recognition algorithm represents each detected face with a 128-dimensional descriptor vector. By comparing these descriptors with a gallery of reference descriptors of which the identity is known, we estimate the query's identity. The gallery of reference photos is given beforehand and is labeled by, for example, teachers or parents.

Our use case demands the identification of children during everyday activities. This includes identifying children that are playing, running etc. without actually posing in front of the camera. Hence, it is likely that many of the images will not contain the entire face of the child that was photographed. This requires an additional identification approach, apart from facial recognition. Therefore, we added a re-ID branch to our identification system. Person re-identification is the task of re-identifying a person in different images by making use of features from over the *entire body*, taking into account e.g. the colours of their clothing. In the re-ID branch shown in orange in Fig. 2, a person detector first delimits regions where persons are present, i.e. regions containing a head, a body etc. Next, the detected persons are compared with images of persons whose identities were already estimated. This is done by representing both the identified

---

[1] The names are, of course, fictitious.

persons (gallery) and the unidentified persons (query) with a descriptor vector and comparing the unidentified descriptors with the identified descriptors. We trained a ResNet-34 [1,6] network on the CUHK03 [9] dataset using the Open-ReID [11] framework which extracts these descriptors. It is important to note that the re-ID comparison is done *within the same date*, i.e. a given query image will be compared with gallery images that were taken on the same date. That is because the general appearance of a person's body might drastically change every day, mainly due to different clothing.

The facial recognition branch and the re-ID branch are combined to obtain a final estimation of the identities on the image. Once these identities are obtained, the image is added to the re-ID collection of identified photos taken on the respective date. Hence, the re-ID gallery will be gradually extended with potential hits for a given person query.

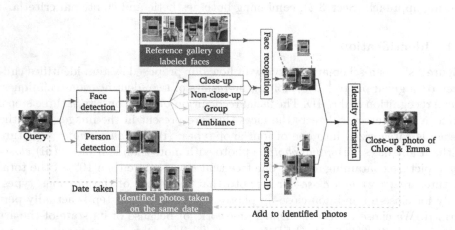

Fig. 2. Overview of the identification part.

3.2 Estimating Appreciation

For an individualized child photo album to be satisfactory, it not only should be actually individualized, but it should also contain photographs that are generally perceived as beautiful. Because parents evidently prefer photos in which their children are smiling rather than crying, we combined aesthetics analysis *and* emotion detection. Figure 3 shows a schematic overview of the appreciation estimation component. The first branch in the diagram contains an emotion classifier. This classifier [1] detects faces in a given photograph and assigns a score to the face for each of seven emotions: happy, surprise, angry, disgust, fear, sad, neutral. The second branch makes use of the Deep Image Aesthetics Analysis (DIAA) [7] framework which consists of a Convolutional Neural Network (CNN) that was trained on a dataset with photos rated by professional photographers. Each photo received a rating on multiple criteria (rule of thirds, depth of field etc.) and a general aesthetics score ("Aesthetics" in the diagram). All aesthetics and emotion features are linearly combined to obtain a single appreciation estimation.

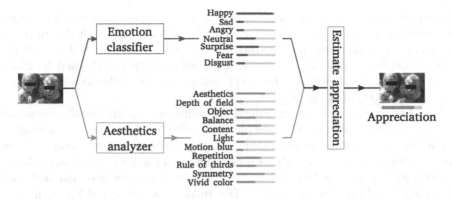

Fig. 3. Overview of the appreciation estimation part.

4 Results

For our experiments, we used a dataset of 2086 photos taken by the teacher of a class during the entire school year. It consists of images from children both inside the classroom (during arts and crafts activities, birthdays etc.) and outside the classroom (during a library visit, an ice skating activity etc.). Figure 1 shows some examples of images in the dataset. We annotated the dataset for both the faces and bodies of 33 identities. As can be seen from Fig. 4, there are great differences in the amount of images for each shot type, with specifically a large amount of non-close-up images. To thoroughly evaluate our proposed system, we performed experiments for each specific component of the pipeline. This provides a more detailed analysis than simply evaluating the system as a whole.

4.1 Face Recognition

As described in Sect. 3.1, the facial recognition algorithm yields a descriptor for each detected face. To evaluate the performance of the facial recognition algorithm, we (i) extracted a descriptor for each face in the labeled dataset; (ii) assigned the corresponding identity to each descriptor using the manual annotations; (iii) randomly split up the set of descriptors into a *gallery* set and

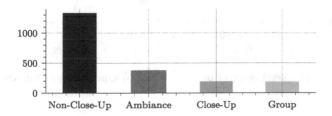

Fig. 4. The distribution of shot types for our dataset.

a *query* set such that each identity had $N \geq 1$ descriptors in the gallery which were used as reference descriptors for that identity; (iv) for each descriptor in the query set, looked for the descriptor in the gallery set to which it had the smallest Euclidean distance (i.e. its nearest neighbour); (v) checked to which extent the labeled identity of the query images matched the identity of their nearest neighbour in the gallery set.

In a first experiment, we examined the influence on the facial recognition accuracy of the amount of reference photos on the one hand, and the averaging of reference descriptors on the other. For the full dataset, we compared the mean Average Precision (mAP) with the amount of gallery descriptors chosen per identity (N in step (iii) above). Each query descriptor was matched with its nearest neighbour in the gallery. The mAP was calculated for each $N \in \{1, 2, \ldots, 15\}$. This experiment was performed for two types of galleries: one where the N descriptors were averaged to always obtain a single descriptor per identity (Fig. 5b) and one where the N descriptors per identity were left untouched (Fig. 5a). The vertical bars in Fig. 5 mark the standard deviations for 5 experiments. As we can see, the mAP only increases for the case where the gallery descriptors are averaged. This is mainly because the average of an inlier and an outlier reference descriptor will be less of an outlier. By averaging over multiple reference descriptor, the average will get closer and closer to a reliable representation. When the reference descriptors are not averaged, N has no influence on the accuracy because outliers will be present, no matter the value of N. From Fig. 5, we see that the optimal choice is to average the descriptors with $N = 8$, since the mAP hardly increases after $N = 8$ with mAP $\approx 83\%$.

Our second facial recognition experiment (Fig. 6), compares the Precision-Recall (PR) curves for non-averaged and averaged galleries with $N = 8$. We split up the dataset in the shot types described in Sect. 3.1 to be able to get more detailed insights. *Ambiance* photos are, of course, not relevant here since they do not contain any faces. It can be seen that the overall performance when using averaged reference descriptors is indeed significantly better with an EER

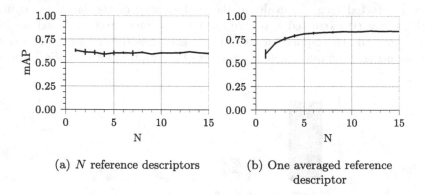

(a) N reference descriptors   (b) One averaged reference descriptor

Fig. 5. The mAP per gallery size when using facial recognition with (a) all reference descriptors per identity; (b) one averaged reference descriptor.

of 86% precision and 86% recall for close-up photos. There is, however, a big difference between the shot types. Evidently, the visibility of a face determines the performance of the facial recognition algorithm. In group photos, the faces are a lot smaller and the chances of occlusion are higher compared to close-up photos.

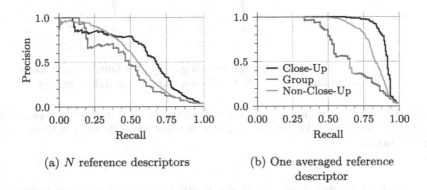

(a) N reference descriptors

(b) One averaged reference descriptor

Fig. 6. PR curves when using facial recognition ($N = 8$) using (a) all N reference descriptors per identity; (b) a single averaged reference descriptor per identity.

4.2 Person Re-identification

In order to be able to identify challenging images where no faces could be detected, we propose to use re-ID (see Fig. 2). The re-ID framework we integrated [11], returns a 128-dimensional descriptor for a given image of a person.

To evaluate the potential of re-ID for our use case, we (i) extracted a descriptor for each person image in the labeled dataset; (ii) assigned the corresponding identity to each descriptor using manual annotations; (iii) added all the descriptors to both the query and the gallery set; (iv) for each descriptor in the query set, looked for the descriptor in the gallery set to which it had the smallest Euclidean distance (i.e. its nearest neighbour), excluding itself; (v) checked to which extent the labeled identity of the query images matched the identity of their nearest neighbour in the gallery set.

As a first experiment, we investigated how the performance increases when we impose that gallery and query must be of the same date. We created a CMC curve for the case that a query and its matched gallery item can be of different dates (Fig. 7a), and one for the case that they must be of the same date (Fig. 7b). A CMC curve shows the percentage of queries for which the correct identity is in the top-k nearest gallery descriptors for $k \in \{1, 2, \ldots, k_{max}\}$. We see that the top-1 accuracy is about 20% higher for close-ups when we impose gallery and query to be of the same date. It is remarkable, however, that the CMC curve does not seem to converge to 100% in the same-date case. Indeed, when we plotted this curve for $k_{max} = \#$gallery items we saw that it instead converged to a value of about 60%. This means that it was *impossible* for about 40% of the queries

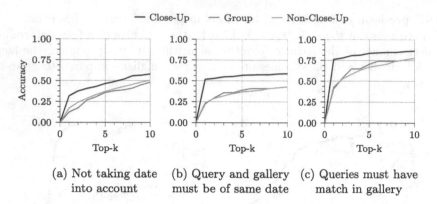

(a) Not taking date into account　　(b) Query and gallery must be of same date　　(c) Queries must have match in gallery

Fig. 7. CMC curves when using re-ID (a) not taking the date into account (b) taking the date into account (c) taking the date into account and filtering out queries with no gallery match.

to match with a correct gallery item of the same date, because there were none. In other words, there is a considerable amount of photos for which the person on the photo was only photographed once that day. Hence, in our use case, re-ID is intrinsically restricted in accuracy due to its need for same-date images. Therefore, in Fig. 7c, we show a similar plot as in Fig. 7b, but now queries for which there were no possible gallery matches were left out. This figure shows that for about 80% of the (close-up) queries, the nearest neighbour in the gallery is a true match, given two restrictions: (i) a query image can only be matched with a gallery image that was taken on the same day; (ii) each person that is photographed on a certain day must have a corresponding gallery image of that same day.

In a second experiment, we analyzed the complementarity of facial recognition and re-ID. The hypothesis was that we would get a better accuracy by combining re-ID with facial recognition. We applied re-ID wherever the face detector did not detect a face while there was a person present in the image, according to manual annotations. Figure 8a shows the CMC curve of facial recognition by itself (using $N = 8$ averaged gallery descriptors per identity). Figure 8b shows the CMC curve of the combination of facial recognition and re-ID. The addition of a state-of-the-art re-ID algorithm did not seem to improve the accuracy of the identification system. It is important to note, however, that there are significant differences between the CUHK03 [9] dataset on which the re-ID model was trained, and our own dataset. CUHK03 consists mainly of pedestrians with, more or less, their entire bodies present in the image. Images where a child's body is entirely present, are rather uncommon in our dataset. Future research is needed to examine the possibilities of developing a re-ID model based on datasets that incorporate similar challenges as our own dataset.

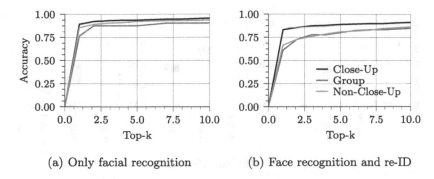

(a) Only facial recognition   (b) Face recognition and re-ID

Fig. 8. CMC curves for (a) facial recognition ($N = 8$, averaged); (b) a combination of facial recognition and re-ID.

4.3   Estimating Appreciation

To gather validation data and get an idea about how beautiful a picture is generally perceived, we created a website on which human annotators could label photos as either good or bad. These annotators were selected to be representative for the target audience of parents with children in preschool or elementary school. For each picture, we calculated the average amount of positive votes and used that average as a single appreciation score for the picture. When performing our experiments, each picture was viewed by an average of 9 annotators. As we discussed in Sect. 3.2, the aesthetics analysis framework returns scores for a set of aesthetics criteria and the emotion detection framework returns—per face—a score for each of seven emotions. For images with multiple faces (and therefore, multiple sets of emotion scores), the emotion scores were averaged by weighing them with the area of the corresponding face detection.

In a first experiment, we calculated the correlations between the features (aesthetics and emotion) and the average appreciation score given by the human annotators. Figure 9 shows the correlations with the largest absolute values, for emotions and aesthetics criteria respectively. Bars that are oriented upward, correlate positively with the average score given by annotators, while bars that are oriented downward, correlate negatively. On Figs. 9a–c, we see that positive emotions (happy) tend to correlate positively with the opinion of the human annotators while negative emotions (angry, sad) tend to correlate negatively. This means that, in general, the annotators preferred pictures with positive emotions over pictures with negative emotions, as could be expected beforehand. Figures 9d–f show that the annotators mostly appreciated a shallow depth of field and a clear focus on a single object (person) in close-up and non-close-up photos. For group photos, the general aesthetics score seems to be an important factor for the appreciation.

As explained before, we propose to linearly combine these different scores into one appreciation score. By employing a ridge regression, we derive that linear combination of the aesthetics scores and the emotion scores. We trained 10 linear

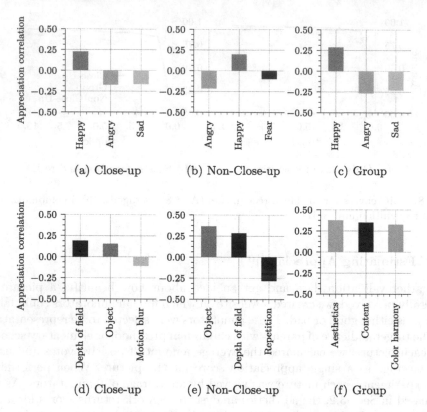

Fig. 9. Highest correlations (in absolute value) between (a)–(c): annotators appreciation and emotion; (d)–(f): annotators appreciation and aesthetic features.

regression models for each shot type and selected those with the highest R2-score. Next, these models were evaluated on a test set. Figure 10 shows the highest correlations (for the test set) between a set of features—aesthetics, emotion and the trained linear combination of these—and the average appreciation from the annotators. The correlation between the trained linear feature combination and the average annotator appreciation is indicated with "Prediction" in Fig. 10. A clear advantage is found for non-close-up images where the prediction is the most correlated feature in the test set. For close-up and group images, however, the prediction correlates for 0.04 and −0.27, respectively; which is less than the 6 most correlated features shown in Figs. 10a and c. It is possible, however, that due to the small amount of close-up and group images present in the dataset (see also Fig. 4), the test sets for these shot types were not representative. This can also be seen when comparing Figs. 9 and 10. For the non-close-up test set in Fig. 10b, the correlations are similar to those in Figs. 9b and e. For the correlations of the features in the close-up and group test set, however, this is less the case.

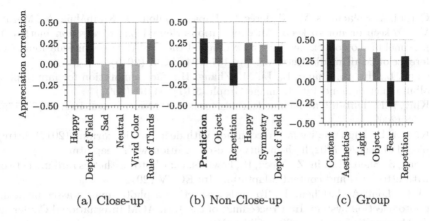

Fig. 10. Correlations with human annotators appreciation and linear regression prediction test set.

5 Conclusion

In this paper, we proposed a system to automate the individualized selection of child photos taken during a school year. Our approach consisted of two parts: correctly identifying children on photographs and assigning an appreciation score to each image.

We showed that facial recognition is very suitable for our use case, especially when taking the average of about 8 reference descriptors to obtain a single reference face descriptor per identity. We obtained 86% precision for 86% recall. We also presented that the facial recognition technique by itself is able to achieve a top-1 accuracy of 89.2%. Furthermore, we evaluated the accuracy gain of adding re-ID to the identification system. This, however, resulted in a *decrease* in accuracy.

We proposed a method of estimating the appreciation generally expressed by our target audience with respect to given images. For non-close-up photos, the prediction model combined the emotion and aesthetics features in a way to obtain a better correlation with the scores given by the annotators.

Future research needs to point out whether a prediction model can be obtained that works for all shot types by using (i) a more representative appreciation questionnaire with, for example, at least 30 annotations per image; (ii) other aesthetics analysis or emotion detection frameworks, eventually more aimed at child photos; (iii) a complexer prediction model (e.g. a neural network).

References

1. Arriaga, O., Valdenegro-Toro, M., Plöger, P.: Real-time convolutional neural networks for emotion and gender classification. arXiv preprint arXiv:1710.07557 (2017)
2. Cardinaux, F., Sanderson, C., Bengio, S.: Face verification using adapted generative models. In: Sixth IEEE International Conference on Automatic Face and Gesture Recognition, Proceedings, pp. 825–830. IEEE (2004)

3. Ceroni, A., Solachidis, V., Niederée, C., Papadopoulou, O., Kanhabua, N., Mezaris, V.: To keep or not to keep: An expectation-oriented photo selection method for personal photo collections. In: Proceedings of the 5th ACM on International Conference on Multimedia Retrieval, pp. 187–194. ACM (2015)
4. Chen, L., Hu, B., Zhang, L., Li, M., Zhang, H.: Face annotation for family photo album management. Int. J. Image Graph. **3**(01), 81–94 (2003)
5. King, D.E.: Dlib-ml: a machine learning toolkit. J. Mach. Learn. Res. **10**(7), 1755–1758 (2009)
6. King, D.E.: High quality face recognition with deep metric learning (2017). http://blog.dlib.net/2017/02/high-quality-face-recognition-with-deep.html
7. Kong, S., Shen, X., Lin, Z., Mech, R., Fowlkes, C.: Photo aesthetics ranking network with attributes and content adaptation. In: ECCV (2016)
8. Li, C., Loui, A.C., Chen, T.: Towards aesthetics: a photo quality assessment and photo selection system. In: Proceedings of the 18th ACM International Conference on Multimedia, pp. 827–830. ACM (2010)
9. Li, W., Zhao, R., Xiao, T., Wang, X.: DeepReID: deep filter pairing neural network for person re-identification. In: CVPR (2014)
10. O'Hare, N., Smeaton, A.F.: Context-aware person identification in personal photo collections. IEEE Trans. Multimed. **11**(2), 220–228 (2009)
11. Xiao, T.: Open-ReID (2017). https://github.com/Cysu/open-reid
12. Zhao, M., Teo, Y.W., Liu, S., Chua, T.-S., Jain, R.: Automatic person annotation of family photo album. In: Sundaram, H., Naphade, M., Smith, J.R., Rui, Y. (eds.) CIVR 2006. LNCS, vol. 4071, pp. 163–172. Springer, Heidelberg (2006). https://doi.org/10.1007/11788034_17

Face Detection in Painting Using Deep Convolutional Neural Networks

Olfa Mzoughi[1]([⊠]), André Bigand[2], and Christophe Renaud[2]

[1] Prince Sattam Bin Abdulaziz University, Al-Kharj, Kingdom of Saudi Arabia
o.mzoughi@psau.edu.sa
[2] LISIC, ULCO, Calais Cedex, France
{bigand,renaud}@univ-littoral.fr

Abstract. The artistic style of paintings constitutes an important information about the painter's technique. It can provide a rich description of this technique using image processing tools, and particularly using image features. In this paper, we investigate automatic face detection in the Tenebrism style, a particular painting style that is characterized by the use of extreme contrast between the light and dark. We show that convolutional neural network along with an adapted learning base makes it possible to detect faces with a maximum accuracy in this style. This result is particularly interesting since it can be the basis of an illuminant study in the Tenebrism style.

Keywords: Face detection · Realist art
Illumination comprehension · Deep learning

1 Introduction

Automatic face detection is a well studied problem in literature. A great success has been realized especially for photographs with a frontal-posed faces in a non-complex background. Recent researches is dealing mainly with unconstrained environment where faces may be close or cluttered and with various poses and contrasts. Nowadays, thanks to the development and massive use of smartphones and fast mobile networks in the world, that the number of photos has been exploded. The obtained photos are mainly used in social networks (Facebook, ...), for family purpose or even for other new applications such as sentiment analysis or trend detection. These millions of photos make it possible to use deep convolutional neural networks (CNN) to develop efficient and successful face detection algorithms [6] in photographs with unconstrained environment.

Such success has opened up new fields of Humanity researches. One of these is the study of art painting, which has been one of the basic tools that can be used to discover and understand the history. In these last past years, deep learning has made it possible to automatically recognize art styles in paintings [2]. However, other painting characteristics are still not well explained. One of these characteristics was the study of the illumination distribution of paintings,

© Springer Nature Switzerland AG 2018
J. Blanc-Talon et al. (Eds.): ACIVS 2018, LNCS 11182, pp. 333–341, 2018.
https://doi.org/10.1007/978-3-030-01449-0_28

especially in the Tenebrism style. Stork [3] has focused on this issue and has shown that the illumination technique used by the painter is mainly linked to the face viewpoint of people appearing in the painting.

For that, in this paper, we propose a CNN-based approach to detect faces in Tenebrism paintings. The Tenebrism art is one of the realistic style that was developed in the seventeenth century. It is characterized by deep shadows and a violent contrast between light and dark areas. The interpretation of such paintings, with such beautiful representation of suffering and such mysterious strong contrasts between light and darkness, has been always interested many historians. Non-perceptual characteristics such that estimating the behaviour of the external source light (the "spotlight") that is used to make contrast between light and dark or, in general, estimating the difference between real and drawn scenes may be interesting to understand the intention of the painter.

As previously mentioned, the first step toward modeling the illumination in such paintings was the automatic detection of painted characters' faces. Such task is particularly difficult because of the small number of Tenebrism paintings examples as well as the complex composition between light and dark. In this paper, we investigate the use of a transfer learning based on an adapted data augmentation technique to handle this problem.

The paper is organized as follows, we first present related works considering face detection and the challenge of face detection in painting. Then we describe our approach in Sect. 3, and we present experimental results in Sect. 4. Finally we conclude with some comments and future directions.

2 Related Works

2.1 Face Detection

There is a widely held belief that computer vision, in general, and face detection, in particular, are to a large extent solved problems. A lot of works have been presented on the subject. Here, we give a brief summary on the evolution of face detection methods w.r.t constraints considered: from frontal face without background images... and the different employed techniques. We first remind the classic and the frequently used technique of face detection: the Viola-Jones face detector technique [8], then, we discuss more recent deep learning techniques.

2.2 Basic Face Detection Approaches

Zang and Zang [10] proposed a complete survey about the main approaches of face detection. They give an overview about the main features as well as the learning algorithms that were used for face/object detection. We now review one of the basic face detection approaches: the Viola-Jones face detector technique [8]).

The Viola-Jones technique is a well-known and efficient technique, with very low computational complexity. Nevertheless, directly applying this technique on a classic Tenebrism painting does not give good result. We can see in Fig. 1 the bad obtained result on a classic Tenebrism painting and the color histogram (R, G, B) that is representative of the painting and can also explain this bad result. Indeed the "darkness" of such paintings need image pre-processing (histogram equalization, image enhancement) before applying face-detection. Moreover, Viola-Jones techniques can accurately find visible up-right faces but they often fail to detect faces from different angles (side view, partially occluded faces) that occur in these paintings.

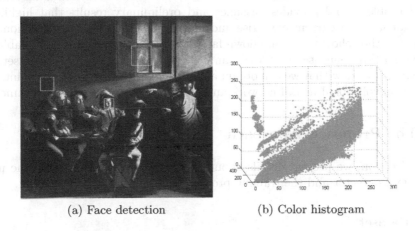

(a) Face detection   (b) Color histogram

Fig. 1. Results with the Viola-Jones algorithm.

Thus these classic techniques give good results in the general situations but they are very sensitive to varying poses, varying illumination, ... and a lot of variants have been proposed.

2.3   Neural Network Based Approaches

More recently, Zafeiriou *et al.* [5] completed the previous survey and classified face detection techniques into two general frames: rigid templates using boosting or deep learning, and deformable methods using constellation of parts. Deep learning, and particularly Convolutional Neural Networks (CNNs), is well adapted to recognize objects in non-artistic photographs as shown by Kryzhevsky *et al.* [1]. Particularly, convolutional Neural Networks were used for face detection in different ways. In [4], Haoxiang-Li *et al.* have proposed a Cascade of Convolutional Neural Network face detectors of increasing resolutions: 12, 24 and 48. A calibration nets is also proposed in order to enhance the quality of bounding boxes. In [7], Farfade *et al.* have proposed to fine-tune AlexNet model *et al.* (firstly defined for the Imagenet challenge) in the context of face detection. Training examples are examples are extracted from the training examples from the AFLW dataset.

For simplicity reasons and in order to investigate the impact of convolutional nets in paintings, we propose, in this paper, to extend the use this last model to the context of Tenerism paintings.

2.4   Face Detection in Paintings

Recently, Wechsler and Toor presented an interesting paper about modern art face detection challenges [9]. This paper is devoted to face authentication using examples from modern art that significantly confound face detection. In that paper, the challenges are made concrete using a modern art face detection composed mostly of modern art examples that cover much diversity in style and artists. The authors show that singleton and crowd face detection is an almost solved problem, and provides baselines and preliminary results that highlight the inadequacy of current expertise and methods to address face detection. In particular, they show that well-known face detection algorithms are only able to achieve an F-measure score of less than 35% overall across the new dataset. In this paper, we show that we can obtain a good accuracy for Tenebrism paintings using a convolutional neural network and an appropriate fine-tuning technique.

3   The Proposed Approach

In this section, we present details about our particular dataset, then, the proposed face detector and the training process, appropriate for this dataset.

3.1   Dataset

We used a database of 65 Tenebrism style paintings (Fig. 2). This dataset holds several challenges: The first challenge was the presence of violent contrasts of light and dark, which characterizes the Tenebrism style. The second challenge was the significant difference in appearance and dress of painted characters as

Fig. 2. Some Dataset samples, Tenebrism style.

well as in the scene in general, compared to real-scene photographs. The last challenge was the low number of dataset images. To handle these challenges, we propose to apply a fine-tuning method using an appropriate data augmentation process (see Subsect. 3.3).

3.2 Network Architecture

The AlexNet architecture is now well-known. The input is a $227 \times 227 \times 3$ matrix. Its structure is made of five convolutional layers and three fully connected layers. Dropout is thus implemented after those fully-connected layers.

3.3 Data Augmentation and Fine-Tuning

In order to increase the number of learning examples as well as to handle different challenges of our dataset (light variation and multiple face views), we propose to increase our dataset following the steps below:

- We first vary the contrast in positive example using a basic contrast stretching approach. This consists in modifying image intensity range from $[mean - k * \sigma, mean + k * \sigma]$ to $[0\ 255]$ for different values of k and for the three RGB color channels. Some issued samples are illustrated in Fig. 3a.
- In order to ensure invariance of the model to multiple views of faces, we also randomly flipped training examples (see Fig. 3b).
- Negative examples are composed from randomly cropping the background of images (the remained part without face) to small pictures with different sizes. Some samples, obtained by this process, are illustrated in the Fig. 4

This resulted in a total number of positive $40K$ images and $13K$ negative training examples.

These examples were then resized to 227×227 and used to finetune the AlexNet model (finetuned previously on face photographs following [7]). This complementary learning dataset makes it possible face detection whatever illumination and pose of faces in paintings.

For re-tuning, we used 10K iterations and learning rate $= 0.001$. All quantitative evaluations of the proposed method were computed on https://www-calculco.univ-littoral.fr CALCULCO, University of Littoral's computing server.

4   Experiments and Results

In this section, we assess the performance of the proposed method for the problem of face detection in Tenebrism style.

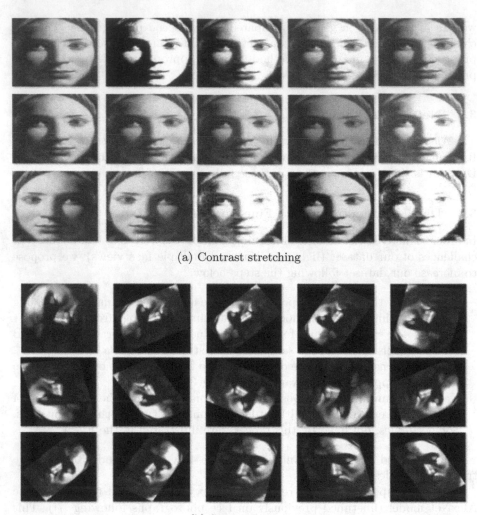

(a) Contrast stretching

(b) Some rotations

Fig. 3. Some positive examples used in fine-tuning.

We first illustrate in Fig. 5 some results obtained by the proposed method. From this figure we can clearly see that our method can accurately detect faces in different orientations and with variable illumination.

Fig. 4. Samples of some negative examples.

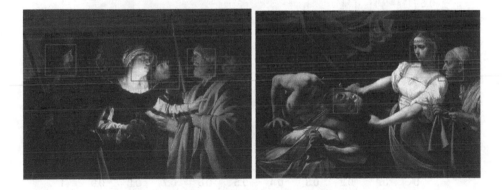

Fig. 5. Some results obtained by our algorithm.

4.1   Comparison with Viola-Jones Method

We now discuss the performance of the proposed face detector with regard to the Viola-Jones algorithm to show how style and face representations may impact these algorithms. For the evaluation, we use the average precision metric. The following Table 1 shows the good obtained results considering average precision

Table 1. Comparison of the Average precision (Ap) between Viola-Jones face detector algorithm and the proposed method.

Viola-Jones face detector	0.081
Our algorithm	0.2589

4.2   Impact of Fine-Tuning

Finally, we compare the AlexNet network which was proposed in [7] with the model trained with our complementary training dataset.

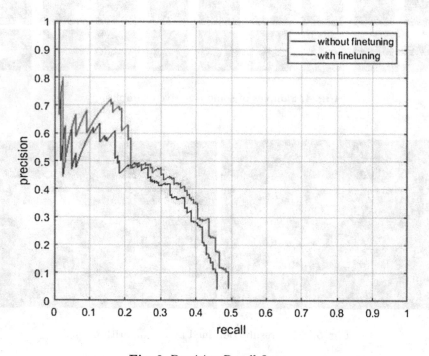

Fig. 6. Precision-Recall figure.

The precision-recall curve for the fine-tuning dataset is shown in Fig. 6. It can be seen that the proposed fine-tuning has a very good recall at 99% precision and justify the good results of the dataset.

5   Conclusion

This paper addresses the problem of face detection on paintings of past centuries. Towards that goal we describe a fine-tuning of interest for such purposes. The proposed method gives good results for Tenebrism painting style. The expected impact and outcomes from ancient art challenge to face detection should aim art historians to better understand illumination techniques. We also plan to deploy this technique for other parts of body in the paintings towards further progress of art technique understanding by art historians. Thus this work open interesting perspectives for the future.

References

1. Krizhevsky, A., H., Hinton, G.E.: Imagenet classification with deep CNN. In: Advances in Neural Information Processing Systems, pp. 1097–1105 (2012)
2. Lecoutre, A., B., Yger, F.: Recognizing art style automatically in painting with deep learning. In: Proceedings of the 9th Asian Conference on Machine Learning, ACML, vol. 3, pp. 327–342 (2017)
3. Stork, D.G.: Computer vision, image analysis, and master art: part 1, 2, 3. Artful Media **24**, 16–173 (2016)
4. Li, H., Lin, Z., Shen, X., Brandt, J., Hua, G.: A convolutional neural network cascade for face detection. In: The IEEE Conference on Computer Vision and Pattern Recognition (CVPR), June 2015
5. Zafeiriou, S., Zhang, C., Zhang, Z.: A survey on face detection in the wild: past, present and future. Comput. Vis. Image Underst. **138**, 1–24 (2015)
6. Farfade, S.S., M.S., Li, L.J.: Multi-view face detection using deep convolutional neural networks. In: Proceedings of the 5th ACM on International Conference on Multimedia Retrieval, vol. 2, pp. 643–650 (2015)
7. Farfade, S.S., M., Li, L.J.: Multi-view face detection using deep convolutional neural networks, vol. 24, pp. 118–173. ACM (2015)
8. Viola, P., Jones, M.: Rapid object detection using a boosted cascade of simple features. In: Proceedings of CVPR 2001, vol. 2, pp. 643–650 (2001)
9. Wechsler, H., Toor, A.S.: Modern art challenges face detection. Pattern Recogn. Lett. (2018, in Press)
10. Zhang, C., Zhang, Z.: A survey of recent advances in face detection (2010)

Robust Geodesic Skeleton Estimation from Body Single Depth

Jaehwan Kim[✉] and Howon Kim

Electronics and Telecommunications Research Institute, Daejeon, Republic of Korea
jh.kim@etri.re.kr

Abstract. In this paper, we introduce a novel and robust body pose estimation method with single depth image, whereby it is possible to provide the skeletal configuration of the body with significant accuracy even in the condition of severe body deformations. In order for the precise identification, we propose a novel feature descriptor based on a geodesic path over the body surface by accumulating sequence of characters correspond to the path vectors along body deformations, which is referred to as GPS (Geodesic Path Sequence). We also incorporate the length of each GPS into a joint entropy-based objective function representing both class and structural information, instead of the typical objective considering only class labels in training the random forest classifier. Furthermore, we exploit a skeleton matching method based on the geodesic extrema of the body, which enhances more robustness to joints misidentification. The proposed solutions yield more spatially accurate predictions for the body parts and skeletal joints. Numerical and visual experiments with our generated data confirm the usefulness of the method.

Keywords: 3D body parts classification · Joints identification
Skeleton estimation · Random forest · Geodesic descriptor
Joint entropy · Dynamic time warping

1 Introduction

3D body pose estimation, whose goal is to recover the poses of body parts and joints with naturally articulated movements, plays a key role and is a well investigated problem in variety of areas such as computational vision, human-computer interface, and computer animations, and so on. Especially, in the works of Shotton et al. [1,2], random forest algorithm proposed by [3] is employed to predict body poses from single depth image. The random forest is an ensemble learning method, which has proven fast and effective multi-class classifiers for various works such as image classification, object tracking, facial expression recognition, pose estimation and so on [4–6]. The solution proposed by Shotton et al. [1] is embedded within the commercial product 'Microsoft Kinect sensor^{TM}', which is readily available off-the-shelf gaming system. Moreover, the depth comparison proposed by [1] is popularly used in many works [2,7] as learning features

© Springer Nature Switzerland AG 2018
J. Blanc-Talon et al. (Eds.): ACIVS 2018, LNCS 11182, pp. 342–353, 2018.
https://doi.org/10.1007/978-3-030-01449-0_29

for the random forest classifier. Within the framework of body pose estimation based on the classified body parts [1], the accuracy and reliability of the body parts classification are important because they might influence the consequent learning process to infer the positions of 3D body joints. Furthermore, although the depth comparison features proposed by [1] are easy to compute and efficient in characterizing the change in body parts, the features themselves encode the only local information for the body parts not a global information such as the deformed whole body or the skeletal structure of the body joints. The depth comparison features are insufficient to empower the discriminative ability of the classifier.

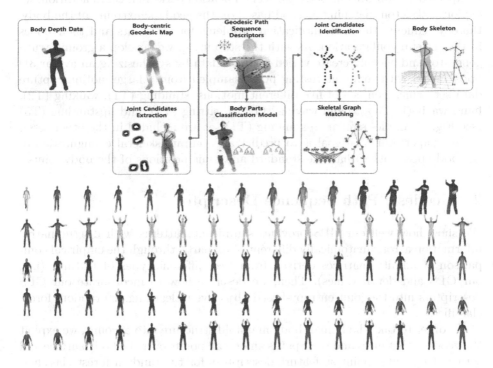

Fig. 1. Systematic overview of our system and our ground-truth samples (normalized to the depth [0,1]): from the top row, forward walking (T2), hand waving1 (T3.a), hand waving2 (T3.b), sitting (T4), and upstanding (T5) motions.

Our approach for 3D body pose estimation from a single depth-map is related to the previous works from [8,9] as they exploit a geodesic distance graph of the body depth image to localize the skeletal joints of the body.

In works [10], a variety of objective functions with the geodesic distance transforms based features for identifying interest objects in the semantic image segmentation with random forest. Moreover, in the context of a decision forest, a joint objective function for pixel classification and shape regression is introduced in [11], which yields more spatially consistent predictions than results from the typical objective function only considering the data labels.

Motivated by existing works [1,8,11], we propose a new feature descriptor based on a geodesic path over the body surface, referred to as GPS (Geodesic Path Sequence), which is derived by concatenating sequence of characters correspond to the vectors along deformation paths. In order for the body parts classification, we also incorporate the length of each GPS descriptor into the joint entropy-based objective function involving both the body parts labels themselves and their geodesic structural information, leading to more accurate predictions. The geodesic descriptors reflect a geometry of body surface well, which is expected to improve our body parts classification performance. In addition, we exploit a skeleton matching method based on the geodesic extrema of the body, thereby reducing the misidentification problems for the joints and their bones in the skeletal configuration. As with the step in [1], we develop a ground-truth generator and cheaply create varied realistic data by synthesizing an avatar 3D body model with some interesting poses sampled from a large motion capture data set, which consists of five different motions: standing (T1), walking (T2), hand waving1 (T3.a), hand waving2 (T3.b), sitting (T4), and upstanding (T5) (see Fig. 1, samples similar to standing (T1) set are included in the other sets). In this paper, our final goal is to predict an accurate skeletal configuration of the body pose rather than the standard anatomic positions of the body joints.

2   Geodesic Path Sequence Descriptor

We show how well our GPS provides significant patterns with discriminative information across anatomically different body parts, through the empirical comparison of affinity matrices derived from two different types of features (i.e., our GPS and depth values). Then, we describe how to incorporate our GPS descriptors into the joint entropy-based objective in learning the random forest classifier.

In order to take the human body manifold structure into account, we exploit the geodesic distances and their paths among all points over the body surface and a their barycenter point as feature descriptors for the random forest classifier. At first, we construct an undirected weighted graph $\mathcal{G} = (\mathcal{V}, \mathcal{E})$ from the body points set $\{p_x\} \subseteq \mathcal{V}$, where $\mathcal{V}$ and $\mathcal{E}$ denote a set of vertices and a set of edges with pairwise distances being assigned as edge weights, and each p_{x_i} is a 3D position vector consisting of a 2D coordinate x_i and its depth $d_D(x_i)$ in the body depth image. The set of edges are defined as:

$$\mathcal{E} = \{d_E(p_{x_i}, p_{x_j}) \in \mathcal{V} \times \mathcal{V} \mid (\|p_{x_i} - p_{x_j}\|_2 < \delta)$$
$$\wedge \ (\|x_i - x_j\|_\infty \leq 1)\}, \tag{1}$$

Each edge $d_E(p_{x_i}, p_{x_j}) \in \mathcal{E}$ is stored as a weight $w(d_E)$, where a 3D Euclidean distance of less than δ. The Dijkstra geodesic distance d_G is then computed along the shortest path $\mathcal{P}$ between $\boldsymbol{p}_{\boldsymbol{x}_p}$ and $\boldsymbol{p}_{\boldsymbol{x}_q}$, which is defined as:

$$d_G(\boldsymbol{p}_{\boldsymbol{x}_p}, \boldsymbol{p}_{\boldsymbol{x}_q}) = \sum_{d_E \in \mathcal{P}(\boldsymbol{p}_{\boldsymbol{x}_p}, \boldsymbol{p}_{\boldsymbol{x}_q})} w(d_E) \tag{2}$$

The graph based geodesic descriptors are invariant to large motion deformations and geometric transforms as long as the local connection relationships remain, which well reflect the local body structure [8,12]. We then generate a body-centric geodesic map by measuring the Dijkstra geodesic distances for all N points on the body, $\{d_G(\boldsymbol{p}_{\boldsymbol{x}_i}, \boldsymbol{p}_{\boldsymbol{x}_0})\}_{i=1}^N$. Each Dijkstra geodesic distance, $d_G(\boldsymbol{p}_{\boldsymbol{x}_i}, \boldsymbol{p}_{\boldsymbol{x}_0})$, is associated with the sum of edge weights along a shortest path between a point, $\boldsymbol{x}_i$, over the body surface and a barycenter of the body, $\boldsymbol{x}_0$, under an assumption that points on anatomically similar body parts maintain a nearly constant geodesic distance. From the body-centric geodesic map, we finally define a descriptor based on the geodesic path which is represented by accumulating sequences of characters correspond to the vectors along the body's deformation path. The GPS for a point $\boldsymbol{x}_i$ is defined as:

$$d_g(\boldsymbol{x}_i) = [c_1, c_2, \cdots, c_i], \tag{3}$$

where c_i is a character indicating the direction of between $\boldsymbol{p}_{\boldsymbol{x}_{i-1}}$ and $\boldsymbol{p}_{\boldsymbol{x}_i}$.

Dynamic time warping (DTW) is a powerful algorithm for measuring similarity between two time series by finding an optimal alignment. In here, we employ the fast DTW algorithm [13] in order to compute the similarity between two GPS descriptors in the binary test function of random forest within linear time. Fig. 2 shows that affinities between the inter- and intra- body parts for data aligned in the parts. All distance values for the affinities are normalized between 0 and 1. The more the affinity matrix has well-formed block diagonal structure, the better the partitioning of different parts. As shown in Fig. 2, the simple depth comparison features empirically do not provide enough discriminative power in learning the classifier. In case of two points having similar depth values, but located at different parts of the body, the features likely lead to erroneous predictions in the classification problem. Meanwhile, our proposed GPS is robust to large motion deformation, and it is effectively discriminative for different body parts.

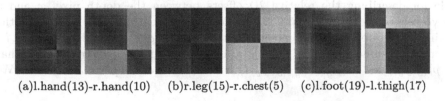

(a)l.hand(13)-r.hand(10)   (b)r.leg(15)-r.chest(5)   (c)l.foot(19)-l.thigh(17)

Fig. 2. From the left, each pair of affinity matrices are depth-based and GPS-based similarities between two different body parts for data in Fig. 1's overview.

3   Joint Entropy-Based Body Parts Classification

For formulation of the body parts classification from single depth image, we assume that a set of N training samples $Q = \{(f_{\theta i}, l_i)\}_{i=1}^{N}$ is given. The input variable $f_{\theta i}$ corresponds to a feature for an individual pixel x_i. The output variable is a discrete label $l_i \in C$, where C is a finite set of body labels.

In a given pixel x in depth image D, we propose a GPS comparison feature similar to the existing depth comparison feature [1], which is defined as:

$$f_\theta(D, x) = d_W \left(d_g \left(x + \frac{i}{|d_g(x)|} \right), d_g \left(x + \frac{j}{|d_g(x)|} \right) \right), \quad (4)$$

where $d_W(d_g(x_{\theta i}), d_g(x_{\theta j}))$ is a warp path distance between $d_g(x_{\theta i})$ and $d_g(x_{\theta j})$ descriptors. $\theta = (i, j)$ is a pair of offsets to the pixel x, and the scale invariance of depth is considered through the normalized by the length of $d_g(x)$. Each node in tree is trained over a set of splitting candidates $\phi = \{(\theta, \tau)\}$, where feature parameter θ and partition threshold τ. The split candidates ϕ are randomly sampled from uniform distribution. For each ϕ ($m = |\phi|$), the subsets Q_L and Q_R partitioned from the original set of data Q are evaluated with our various energy functions at the current node. The partitioning is performed as follows:

$$\begin{aligned} Q_L(\phi) &= \{(D, x) \mid f_\theta(D, x) < \tau\} \\ Q_R(\phi) &= Q \setminus Q_L(\phi) \end{aligned} \quad (5)$$

For the forest training procedure, the goal is to find optimal splitting parameters of each node and build partitioning binary tree which minimizes the objective function J defined as follows:

$$\phi^* = \arg\min_\phi J(Q, \phi). \quad (6)$$

An optimal criteria $\phi^* = \{\theta^*, \tau^*\}$ is defined as the split parameters of the node, and later used for prediction of new input data. The entropy is the expected value of the information contained in each message. The Shannon's entropy is generally used for training forests. Our goal is now to learn the joint probability $p_t(l, g | f_\theta)$, where new variable $g \in \mathbb{R}^3$ is a continuous regression variable for describing the relative 2D offsets between the depth pixel x and a barycenter of the body x_0, and the geodesic distance $d_G(p_x, p_{x_0})$. By using the chain rule, we rewrite the joint distribution as $p_t(l, g | f_\theta) = p_t(l | f_\theta) p_t(g | f_\theta, l)$, where we assume that $p_t(g | f_\theta, l)$ is a multivariate normal distributions. That is, $p_t(g | f_\theta, l) \sim \mathcal{N}(\mu_{g|l}, \Sigma_{g|l} | g, f_\theta, l)$ is one distribution per class label l. We actually define the joint objective function $\mathcal{J}$ as follows:

$$\mathcal{J}(\mathcal{Q}, \phi) = \sum_{p \in \{L, R\}} \sum_{x \in \mathcal{Q}_p} \frac{|\mathcal{Q}_p|}{|\mathcal{Q}|} \psi_E(l, g; \mathcal{Q}_p), \tag{7}$$

$$\psi_E(l, g; \mathcal{Q}_p) = -\sum_{l \in \mathcal{C}} \int_{g \in \mathbb{R}^3} p_t(l, g | \boldsymbol{f}_\theta) \, \log(p_t(l, g | \boldsymbol{f}_\theta)) dg,$$

$$= \underbrace{-\sum_{l \in \mathcal{C}} p_t(l | \boldsymbol{f}_\theta) \log(p_t(l | \boldsymbol{f}_\theta))}_{\psi_E(l; \mathcal{Q}_p)}$$

$$+ \underbrace{\sum_{l \in \mathcal{C}} p_t(l | \boldsymbol{f}_\theta) \Big(\frac{1}{2} \log((2\pi e)^3 | \boldsymbol{\Sigma}_{g|l} |) \Big)}_{\psi_E(g; \mathcal{Q}_p | l)}, \tag{8}$$

where $|\boldsymbol{\Sigma}|$ denotes the determinant of a matrix.

Finally, the models of random forest are achieved by optimizing the joint objective function Eq. (8), including the conventional objective $\psi_E(l; \mathcal{Q}_p)$ for a discrete label l as well as the objective $\psi_E(g; \mathcal{Q}_p | l)$ for a continuous variable g given $\boldsymbol{f}_\theta$ and l variables. In here, the posterior that we are interested in is about the body parts classification. The overall prediction of the forest with T trees is estimated by averaging the individual predictions together and the output is predicted by inferring:

$$l^* = \arg\max_{l \in \mathcal{C}} p(l | \boldsymbol{f}_\theta) = \arg\max_{l \in \mathcal{C}} \frac{1}{T} \sum_{t=1}^{T} p_t(l | \boldsymbol{f}_\theta). \tag{9}$$

4   Body Joints and Skeleton Identification

Given a body-centric geodesic map, as with the way in [8,9,12], the extreme points are computed by incrementally maximizing geodesic distances on the body surface. Based on the classified body parts and the geodesic paths between the body's barycenter and its geodesic extrema (i.e., end-nodes of the human skeletal graph), we localize and identify the joint candidates lying on the paths. The joint candidates are selected with $\angle(\overrightarrow{\boldsymbol{x}_{k-1} \boldsymbol{x}_k}, \overrightarrow{\boldsymbol{x}_k \boldsymbol{x}_{k+1}}) > \epsilon)$. Here, $\angle(\overrightarrow{\boldsymbol{x}_{k-1} \boldsymbol{x}_k}, \overrightarrow{\boldsymbol{x}_k \boldsymbol{x}_{k+1}})$ is an angle between two vectors $\overrightarrow{\boldsymbol{x}_{k-1} \boldsymbol{x}_k}$ and $\overrightarrow{\boldsymbol{x}_k \boldsymbol{x}_{k+1}}$, where the three points $(\boldsymbol{x}_{k-1}, \boldsymbol{x}_k, \boldsymbol{x}_{k+1})$ being around the point $\boldsymbol{x}_k$, are on the same GPS $\boldsymbol{d}_g(\boldsymbol{x}_k)$. ϵ is a threshold depending on the body type, and it is empirically set to about 30 in our experiments. After obtaining the joint candidates set $\{\boldsymbol{x}_i\}$, the representative label l' is evaluated as Eq. (10) from the local window at each joint candidate, where the local patches are based on the already classified body parts. Fig. 3 describes the meta-examples generated at each step.

$$l'_{\boldsymbol{x}_i} = \arg\max_{l_u^*} \sum_{u \in \mathcal{W}_{\boldsymbol{x}_i}} \sum_{l \in \mathcal{C}} \delta(l_u^*, l), \tag{10}$$

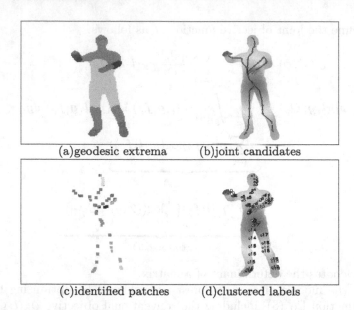

(a)geodesic extrema   (b)joint candidates

(c)identified patches   (d)clustered labels

Fig. 3. (a) Geodesic extrema (blue points in red regions); (b) joint candidates (red points) on five GPSs (black lines); (c) color-labeled patches on joint candidates; (d) labels set $\{l'\}$ classified into five sub-skeletons (i.e., each skeleton for four limbs and one trunk). (Color figure online)

where l_u^* is the label for the position $u \in \mathbb{R}^2$ being in the local window $\mathcal{W}_{x_i}$ centered at x_i. δ is a kronecker delta function. Our main idea is to match two graphs by comparing the labeled sets of ordered points on the paths between the body center and the geodesic extrema of the skeletal configuration under the assumption that there are meaningful joints for the human skeletal structure in the set of joint candidates. In here, the body center and the geodesic extrema labels are defined as 7 (center), 0 (head), 10 (right hand), 13 (left hand), 16 (right foot), and 19 (left foot), respectively. All joint candidates are identified and clustered as in Fig. 3(d) by matching with a given template graph as Fig. 4(a). In order to match the sub-skeletons (i.e., each skeleton for limbs and trunk) with the template graph, we consider a weighted bipartite graph such as illustrated in Fig. 4(b), which is with two vertex sets, a set of sub-skeleton labels and a set of joint labels, and the weight of each edge is defined as a DTW distance between two consecutive joint labels. Given the bipartite graph, the matching is performed by using the Hungarian method [14]. Finally, the skeletal graph with 15 labeled nodes is extracted, which correspond to the whole body skeleton (see Fig. 5(b)).

5   Numerical Experiments

We show the usefulness of our method, through the empirical comparison to different objective functions based on different types of features (i.e., our GPS

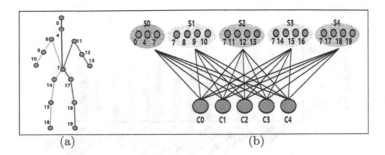

(a)   (b)

Fig. 4. (a) Template skeleton model consisting of four limb sub-skeletons and one trunk sub-skeleton; (b) bipartite graph with two vertex sets (s#: set of joint labels for each sub-skeleton in the template; c#: set of candidate joint labels for each geodesic path).

and depth comparison feature [1]). We applied our method to samples from our ground-truth data sets, consisting of five types of motions: forward/backward walking, hand waving1, hand waving2, sitting, and standing; each motion group has approximately 100 frames. As in a conventional leave-one-out training scheme, the sequences for each model is evaluated with the trained model from other models. For quantitative evaluation of estimated joint positions and skeleton accuracy, we present three different types of measurements: (a) we estimate the mean absolute error (MAE) Eq. (11) in order for the training error evaluation of the classified body parts; (b) the mean average precision (mAP) is evaluated by averaging the precision of the estimated 15 joints on each frame, which is to determine whether the position of the estimated joint is within a given threshold relative to the ground-truth (in here, the threshold is fixed to $max(\{|s_i|\}_{i=0}^{4})/10$); (c) the other is a new measurement of similarity between the estimated skeletons and the ground-truth skeleton by comparing their DTW score, which is referred to as mean average matching (mAM) and defined as Eq. (12).

$$MAE = \frac{1}{N}\sum_{i}\sum_{l \in \mathcal{C}}|l_i^* - l_i^G| \in [0,1], \tag{11}$$

where l_i^* and l_i^G represent the predicted label of data x_i and the corresponding ground-truth label, respectively.

$$AM(i) = \frac{1}{|\mathcal{F}|}\sum_{f \in \mathcal{F}}\frac{d_W(d_g(s_{fi}^*), d_g(s_{fi}^G))}{max(\{|s_i|\}_{i=0}^{4})} \in [0,1], \tag{12}$$

where $AM(i)$ is an average matching score between the estimated sub-skeleton s_i^* and the corresponding ground-truth sub-skeleton s_i^G, one body skeleton has five sub-skeletons, $\{s_i^*\}_{i=0}^{4}$, and the mean matching score is normalized by the maximum length of the five sub-skeletons. $\mathcal{F}$ is a set of target frames.

In [1,2], they apply a local mode-finding approach based on mean-shift with a weighted Gaussian kernel for each classified body part to infer the final positions of 3D skeletal joints. However, as shown in Fig. 5(a) and 6, the local modes

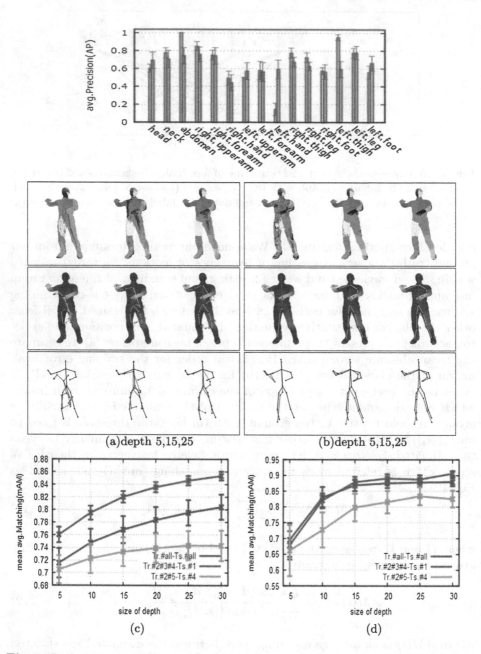

(a)depth 5,15,25   (b)depth 5,15,25

(c)   (d)

Fig. 5. Performance comparison: (top) APs for each joint at depth level 30, based on ours (blue) and [1] (red) with Training (T4)+Testing (T2, T5) sets; (mid) the predicted body parts for data in Fig. 1 and its skeletons with ground-truth (black lines); (bot) mAM values with different data sets, using [1] (a, c) and ours (b, d) methods, repectively. (Color figure online)

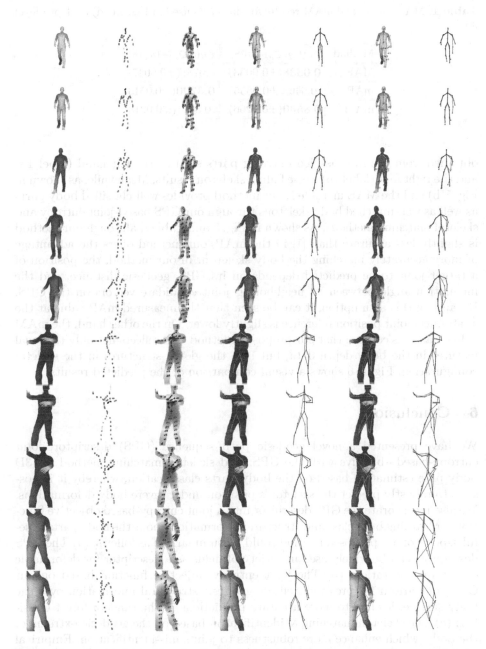

Fig. 6. Experimental results of estimated skeletons for forward-walking samples in Fig. 1: from the left, body depth, color-labeled patches on the joint candidates, identified & clustered joint candidates, body skeleton overlapping with the depth, skeletons from our method, and (the last two) skeletons from Shotton2011cvpr with ground-truth (black lines). (Color figure online)

Table 1. MAE, mAP, and mAM results for data sets used in Fig. 5(c, d) at depth level 30.

Method	$\psi_E(l, g; \mathcal{Q}_p)$+GPS	$\psi_E(l; \mathcal{Q}_p)$+Depth
MAE	**0.0338**(±0.0033)	0.0788(±0.0047)
mAP	0.6463(±0.0855)	**0.6700**(±0.0513)
mAM	**0.8560**(±0.0206)	0.8011(±0.0218)

obtained from the misclassified outlying parts such as the left hand (label 13) and the right foot (label 16) cause failing skeleton results. Meanwhile, as shown in Fig. 5(b) and the MAE in Table 1, our method provides well-classified body parts as well as well-matched body skeletons through our GPS based-joint entropy and skeletal matching methods. As shown in Fig. 5 and Table 1, although our method is slightly less accurate than [1] in the mAP, our method offers the advantage of more accurately matching the body skeleton. In our method, the position of a target joint to be predicted depends on its GPS, geodesic distance, and the inclination angle between its neighboring joint candidate vectors on the GPS. Because of this assumption, it can be seen that the measured mAP value at the anatomical joint position reference is slightly lower. On the other hand, the mAM value makes us confirm that our proposed method well reflects not only the local features in the body depth data, but also the global structures in the skeletal configuration. Figure 6 shows a visual comparison of the predicted results.

6   Conclusions

We have presented a novel geodesic path sequence (GPS) descriptor, joint entropy-based objective with the GPS, and skeleton matching method for 3D body pose estimation based on the body parts classification, whereby it is possible to robustly predict the skeleton's position under severe body deformations. We also incorporate the GPS descriptor into a joint entropy-based objective function for learning both class and structural information about the body parts. Useful aspects of our proposed method could be summarized as follows: (a) The GPS descriptors can be widely used in variety of fields as a descriptor for deformable object representation; (b) The joint entropy objective function based on our GPS comparison features well reflects geodesic structural information over the body surface, leading to more accurate predictions in the random forest classifier; (c) The skeleton matching & identificaton based on the geodesic extrema of the body, which enhance more robustness to joints mis-identification. Empirical comparison with the conventional solution, single entropy-based objective with depth comparison features, confirmed the high performance of our method.

Acknowledgments. This research is supported by Ministry of Culture, Sports and Tourism (MCST) and Korea Creative Content Agency (KOCCA) in the Culture Technology (CT) Research & Development Program R2016030043.

References

1. Shotton, J., et al.: Real-time human pose recognition in parts from a single depth image. In: Cipolla, R., Battiato, S., Farinella, G. (eds.) Proceedings of International Conference on Computer Vision and Pattern Recognition, pp. 1297–1304. Springer, Heidelberg (2011). https://doi.org/10.1007/978-3-642-28661-2_5
2. Shotton, J., et al.: Efficient human pose estimation from single depth images. IEEE Trans. Pattern Anal. Mach. Intell. **35**, 2821–2840 (2013)
3. Breiman, L.: Random forests. J. Mach. Learn. **45**, 5–32 (2001)
4. Gall, J., Lempitsky, V.: Class-specific hough forests for object detection. In: Criminisi, A., Shotton, J. (eds.) Proceedings of International Conference on Computer Vision and Pattern Recognition, pp. 1022–1029. Springer, London (2009). https://doi.org/10.1007/978-1-4471-4929-3_11
5. Tan, D.J., Ilic, S.: Multi-forest tracker: a chameleon in tracking. In: Proceedings of International Conference on Computer Vision and Pattern Recognition, pp. 1202–1209 (2014)
6. Dapogny, A., Bailly, K., Dubuisson, S.: Pairwise conditional random forests for facial expression recognition. In: Proceedings of International Conference on Computer Vision and Pattern Recognition, pp. 3783–3791 (2015)
7. Girshick, R., Shotton, J., Kohli, P., Criminisi, A., Fitzgibbon, A.: Efficient regression of general-activity human poses from depth images. In: Proceedings of International Conference on Computer Vision, pp. 415–422 (2011)
8. Schwarz, L., Mkhitaryan, A., Mateus, D., Navab, N.: Estimating human 3d pose from time-of-flight images based on geodesic distances and optical flow. In: Proceedings of International Conference on Automatic Face and Gesture Recognition, Santa Barbara, CA, pp. 700–706 (2011)
9. Baak, A., Müller, M., Bharaj, G., Seidel, H., Theobalt, C.: A data-driven approach for real-time full body pose reconstruction from a depth camera. In: Proceedings of International Conference on Computer Vision, pp. 1092–1099 (2011)
10. Kontschieder, P., Kohli, P., Shotton, J., Criminisi, A.: GeoF: geodesic forests for learning coupled predictors. In: Proceedings of International Conference on Computer Vision and Pattern Recognition, pp. 65–72 (2013)
11. Glocker, B., Pauly, O., Konukoglu, E., Criminisi, A.: Joint classification-regression forests for spatially structured multi-object segmentation. In: Proceedings of European Conference on Computer Vision, Florence, Italy, pp. 870–881 (2012)
12. Plagemann, C., Ganapathi, V., Koller, D., Thrun, S.: Real-time identification and localization of body parts from depth images. In: Proceedings of International Conference on Robotics and Automation, pp. 3108–3113 (2010)
13. Salvador, S., Chan, P.: FastDTW: toward accurate dynamic time warping in linear time and space. In: KDD Workshop on Mining Temporal and Sequential Data, pp. 70–80 (2004)
14. Kuhn, H.: The hungarian method for the assignment problem. Nav. Res. Logist. Q. **2**, 83–97 (1955)

Deep Learning

Analysis of Neural Codes
for Near-Duplicate Detection

Maurizio Pintus[(✉)]

CRS4, Science and Technology Park of Sardinia, Building 1, Pula, Italy
mpintus@crs4.it

Abstract. An important feature of digital asset management platforms and search engines is the possibility of retrieving near-duplicates of an image given by the user. Near-duplicates could be photos derived from an original photo after a certain transformation or different photos of the same scene. In this work we analyze the two cases, using convolutional neural networks for calculating the signatures of the images, introducing a new training set for model creation and some new datasets for performance evaluation. Results on these datasets and in standard datasets for image retrieval will be presented and discussed.

Keywords: Near-duplicate detection · Image retrieval
Deep learning · Convolutional neural networks

1 Introduction

Near-duplicate detection is a type of content-based image retrieval [1]. It includes two distinct applications: finding duplicates of images derived from the same digital source after applying some transformations, sometimes called identical near-duplicates (IND), and finding similar images, i.e. images of the same scene, the same object or the same landmark, but with possibly different viewpoints and illumination, sometimes called non-identical near-duplicates (NIND) [2]. The first application could be useful for retrieving several versions of the same original photo in a personal collection or in a photographer's collection, or for identifying copyright infringements. The second application could be used for retrieving photos taken in the same place, by the same user or by different users. Systems for near-duplicate detection should satisfy the following requirements [3]:

- high precision, a query image should be matched only with database images that are duplicates of it, or similar to it;
- high recall, all duplicate/similar images in the database should be found;
- efficiency, the time needed to query an image should be as low as possible.

Usually, a system for near-duplicate detection calculates a signature for the query and for each image in the database. Distance between signatures of duplicate or similar images should be smaller than distance between two unrelated images.

© Springer Nature Switzerland AG 2018
J. Blanc-Talon et al. (Eds.): ACIVS 2018, LNCS 11182, pp. 357–368, 2018.
https://doi.org/10.1007/978-3-030-01449-0_30

An example of an early algorithm for calculating the signature of an image is [4], which, adapting an earlier work for computing a signature for ID card photos [5], proposes to encode the relative brightness of different regions of the image, dividing the gray-scale version of the image into a 9×9 or 11×11 grid.

Up to a few years ago the most popular approach in image retrieval was that based on bag-of-words, or bag-of-features; [6] provides a survey of it. It is characterized by the following steps:

- detecting key points and describing local patches around them;
- quantizing local descriptors into visual words, with the generation of a vocabulary;
- indexing and search, possibly applying scalable indexing and fast search on the obtained vector space.

An example of a work that uses visual words is that described in [1]: here the SIFT features [7] are clustered by the k-means algorithm and the centroid of each cluster is used as a visual word in the given vocabulary. Afterwards, local features are extracted and then allocated onto the closest visual word. Next an image can be represented as a histogram of visual words, based on the number of visual words for each class.

Techniques developed for image classification can be harnessed also for near-duplicate image detection. In recent years, there has been a huge improvement in image classification thanks to deep convolutional neural networks. [8] suggests the use of the euclidean distance between the feature activation vectors produced in the upper layers of a CNN trained for image classification by two images as a measure of their similarity. In this case the image signatures are represented by this so called neural codes. The schema of such an approach is shown in Fig. 1.

The works in [9], using the features produced by a reimplementation of the system by [8], and in [10], using the features produced by the CNN OverFeat [11], demonstrate the competitiveness of neural codes in retrieval of similar images, particularly after retraining with a dataset similar to those used for testing. [12] shows that features from neural networks outperform SIFT features for retrieval of transformed images for all transformations but blur.

Other deep learning approaches use a siamese architecture [13], fine-grained image similarity models learned through *query-positive-negative* triplets [14] or attentive local features with geometric verification [15].

In this work the use of neural codes for near-duplicate detection is investigated, in a similar way to [9], but focusing more on duplicate detection and web images. To this end, we introduced new datasets for retrieval of both duplicate and similar images. This paper is organized as follows. Section 2 provides a description of datasets used for creation of the neural network models and for performance assessment. Section 3 provides a description of our experiments and of the metrics used for evaluation. Section 4 reports experimental results. Section 5 contains conclusions and future work.

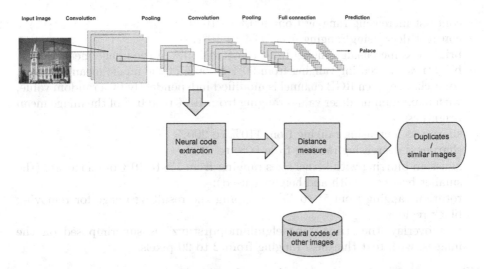

Fig. 1. Schema of a system for near-duplicate detection based on neural codes

2   Used Datasets

As in other works on near-duplicate detection [2,3,16], we created a dataset applying some transformations to a set of photos. We used, as original images, the **PASCAL VOC 2012** training and validation dataset [17], that contains 17 125 images. We applied to them a total of 215 different transformations of different types:

- image resizing with resize factors ranging from 2 to 20 (values equals for both x and y axes);
- JPEG compression with quality factors ranging from 75 to 3;
- cropping ranging from 80% to 10% of the image surface, 11 different parts of the image were used for each cropping level:
  - central part in both the horizontal and the vertical directions;
  - central part in the horizontal direction;
  - central part in the vertical direction;
  - left part;
  - righ part;
  - top part;
  - bottom part;
  - top-left part;
  - top-right part;
  - bottom-left part;
  - bottom-right part.

- contrast increasing ranging from 200% to 1000%;
- contrast decreasing ranging from 90% to 10%;
- brightness increasing ranging from 105% to 150% of the image mean value;
- brightness decreasing ranging from 90% to 50% of the image mean value;
- color change, each RGB channel is modified independently by a random value, with maximum modifier values ranging from 10 % to 100 % of the image mean brightness;
- sharpness increasing ranging from 110% to 200%;
- sharpness decreasing ranging from 90% to 0%;
- gaussian blurring with kernel sizes ranging from 1% to 10% of image size (the smaller between width and height is used);
- rotation ranging from 5° to 45°, cropping the resulting image for removing black regions;
- text overlay, the string "abcdefghilmnopqrstuvz" is superimposed on the images, with text thickness ranging from 2 to 20 pixels.

With respect to datasets used in other works, this dataset has some new transformations, in particular blurring, text overlay and several types of cropping, while usually only a cropping that preserves the center region of the image is used. The insertion of different types of cropping is important because we noticed that cropping is probably the most used transformation in web images.

We used 3 328 992 images created from 15 412 original images of **PASCAL VOC 2012** for training a widely used neural network, GoogLeNet [18], that obtained good results in the ILSVRC competition, for the task of image classification [19], using as classes the 15 412 original images. A class contains the original image with all the different transformations. 10% of these images have been used as validation set for training evaluation. Images are resized to a resolution of 224×224 pixels. We used the DeepDetect deep learning server[1], using Caffe [20] as framework. A SGD solver type, an initial learning rate of 0.01, that is decreased every epoch by using a drop factor of 0.9, a learning momentum of 0.9 and a weight decay of 0.0002 have been used. The model, obtained after 20 epochs, will be called **pascalvoc-nd-net**. Training the network took 12 days on one Nvidia Titan X GPU using a batch size of 64 images.

In order to evaluate the models and find the best extract layer, i.e. the inner layer to be used as code for every image, we considered different datasets. For testing retrieval of web images, we created two datasets by using the image retrieval function of the istella search engine[2], using italian translations of ILSVRC classes as keywords: **istella-duplicate-images** and **istella-similar-images**. The first one contains 266 original images (chosen among the groups of duplicate images) and 479 transformed images. Resolution of images ranges from 74×58 pixels to 5315×3543 pixels. The second one contains 271 images belonging to 78 groups. Resolution of images ranges in this case from 119×81

---

[1] www.deepdetect.com.
[2] www.istella.it.

pixels to 3929 × 4260 pixels. We also created a third dataset, **istella-1k**, containing 10 images for each of 530 ILSVRC classes, to be used as distractors. The classes are the ones among the 1000 ILSVRC classes for which at least 10 images were returned. Also in this dataset resolution of images varies greatly, ranging from 80 × 80 pixels to 5669 × 5170 pixels. Figures 2 and 3 show some examples from the two datasets **istella-duplicate-images** and **istella-similar-images**.

For testing retrieval of photos in a personal collection, we considered another dataset including 8 original photos and 8 enhanced versions, called **photo-retouch**. Figure 4 shows some examples.

We used also some standard datasets for image retrieval, the strong transformations of **INRIA Copydays** [16] for duplicates, **INRIA Holydays** [21] and **Oxford Buildings** [22] for similar images. We used the **Oxford5k** version of the last dataset, without distractors.

ORIGINAL   DUPLICATES

Fig. 2. Some images from **istella-duplicate-images** dataset

3   Test Procedure

In duplicate detection experiments, original images are used as query images and the distance with the other images in the dataset is calculated. The images obtained through the transformation of the original images are considered relevant images, all the other ones are considered irrelevant ones.

Fig. 3. Some images from **istella-similar-images** dataset

In similar image detection experiments, one image per each group is used as a query image and the distance with the other images in the dataset is calculated. In **istella-similar-images** and in **INRIA Holydays**, the other images belonging to the same group are considered relevant images, all the other ones are considered irrelevant ones. In Oxford Buildings, the three proposed degrees of relevance are used (*good*, *ok* and *junk*).

Image searching is carried out by using the Annoy library implementation of approximate nearest neighbors[3]. An angular metric, 100 trees and the default value for *search_k* are used as parameters.

We compared the results obtained by our model with those obtained by the GooLeNet network trained on the 1000 classes of the ILSVRC competition (**bvlc-googlenet**) [18] and those obtained by the Inception-v2 model trained on the 21 841 classes of the full ImageNet dataset [23] (**inception-21k**) [24]. We tested the last layer and the 4 outermost pooling layers as extract layers.

For each query, a precision/recall curve is obtained and the average precision, i.e. the area under the curve, is computed. The retrieval performance is evaluated by using the mean average precision (mAP), calculated as the mean of the average precisions for all the queries [22], and by using the mean of the true positive rates and false positive rates for all the queries.

---

[3] www.pypi.org/project/annoy/.

4   Results

Table 1 reports values of mAP obtained on the two datasets **istella-duplicate-images** and **INRIA Copydays** (strong transformations). With the first dataset, images from **istella-1k** were used as distractors. Table 2 reports values of mAP obtained on the three datasets **istella-similar-images, INRIA Holydays** and **Oxford5k**. With the first dataset, images from **istella-1k** were again used as distractors. Figure 5 reports ROC curves obtained in the two datasets **istella-duplicate-images** and **istella-similar-images**, using, for each model, the extract layer that showed the best mAP.

With **istella-duplicate-images, pascalvoc-nd-net, bvlc-googlenet** and **inception-21k** obtained very similar values for mAP. The ROC curve shows that our model is better for low values of false positive rate, while **inception-21k** is better for larger values. With **INRIA Copydays, pascalvoc-nd-net**, using **pool5/7x7_s1**, the second outermost layers, as extract layer, outdoes clearly the other models.

With **photo-retouch**, using as distractors the images from **PASCAL VOC 2012** together with their gray-level version, for a total of 34 266 images, a mAP of 1 was obtained with all models and layers.

With all datasets and according to both mAP and ROC curves, **inception-21k**, using **global_pool** as extract layer, is clearly superior. According to ROC curves, it is better than the other models for all values of false positive rate. The best extract layer proved to be again the second outermost among the considered layers. Retrieval performance is quite low in **Oxford5k**, with values of MAP up to about 50%: in this dataset buildings of interest often occupy only a small portion of the image and local approaches could behave better in this case.

Figure 6 shows an example of a search for duplicate images of a given query from **istella-duplicate-images** using **pascalvoc-nd-net** as model and **pool5/7x7_s1** as extract layer, while Fig. 7 shows an example of a search for similar images of a given query from **istella-similar-images** using **inception-21k** as model and **global_pool** as extract layer.

Signature calculation with the best model and extract layer, using one Nvidia Titan X GPU and a batch size of 32 images, varies between an average of about 0.05 seconds per image in **Oxford5k** and an average of about 1.6 seconds in

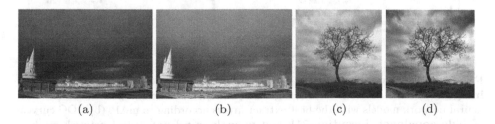

(a)   (b)   (c)   (d)

Fig. 4. Some images from **photo-retouch** dataset. (a) and (c) are original images, (b) and (d) are retouched versions.

istella-similar-images, while distance calculation varies between an average of about $1.6 \cdot 10^{-5}$ seconds per pair of images in **INRIA Copydays** and $5 \cdot 10^{-4}$ seconds per pair of images in **INRIA Holydays**.

Table 1. Results of experiments on retrieval of duplicate images

| MODEL | EXTRACT LAYER | MAP | |
		ISTELLA-DUP.	COPYDAYS
bvlc-googlenet	loss3/classifier	0.904	0.444
bvlc-googlenet	pool5/7x7_s	0.929	0.521
bvlc-googlenet	inception_5b/pool	0.950	0.561
bvlc-googlenet	inception_5a/pool	0.947	0.604
bvlc-googlenet	pool4/3x3_s2	0.929	0.566
inception_21k	fc1	0.915	0.566
inception_21k	global_pool	0.953	0.639
inception_21k	max_pool_5b_pool	0.946	0.648
inception_21k	ave_pool_5a_pool	0.945	0.627
inception_21k	max_pool_4e_pool	0.903	0.541
pascalvoc-nd-net	loss3/classifier	0.934	0.639
pascalvoc-nd-net	pool5/7x7_s	0.949	0.673
pascalvoc-nd-net	inception_5b/pool	0.938	0.645
pascalvoc-nd-net	inception_5a/pool	0.937	0.634
pascalvoc-nd-net	pool4/3x3_s2	0.925	0.617

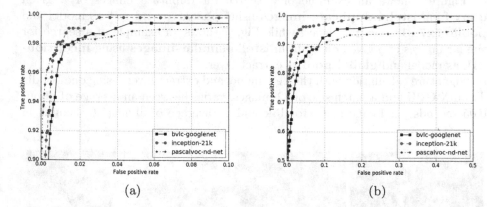

(a)   (b)

Fig. 5. ROC curves obtained on **istella-duplicate-images** and **istella-similar-images**. (a) ROC curves of **bvlc-googlenet**, **inception-21k** and **pascalvoc-nd-net** neural network models with the best extract layers according to mAP. (b) ROC curves of **bvlc-googlenet**, **inception-21k** and **pascalvoc-nd-net** neural network models with the best extract layers according to mAP.

Table 2. Results of experiments on retrieval of similar images

MODEL	EXTRACT LAYER	MAP		
		IST.-SIM.	HOLYD.	OXF5K
bvlc-googlenet	loss3/classifier	0.728	0.638	0.378
bvlc-googlenet	pool5/7x7_s	0.794	0.696	0.447
bvlc-googlenet	inception_5b/pool	0.800	0.691	0.433
bvlc-googlenet	inception_5a/pool	0.815	0.706	0.444
bvlc-googlenet	pool4/3x3_s2	0.750	0.660	0.408
inception_21k	fc1	0.867	0.761	0.458
inception_21k	global_pool	0.890	0.794	0.508
inception_21k	max_pool_5b_pool	0.813	0.752	0.440
inception_21k	ave_pool_5a_pool	0.809	0.758	0.434
inception_21k	max_pool_4e_pool	0.785	0.636	0.369
pascalvoc-nd-net	loss3/classifier	0.765	0.742	0.088
pascalvoc-nd-net	pool5/7x7_s	0.787	0.747	0.111
pascalvoc-nd-net	inception_5b/pool	0.774	0.718	0.137
pascalvoc-nd-net	inception_5a/pool	0.758	0.729	0.128
pascalvoc-nd-net	pool4/3x3_s2	0.733	0.689	0.157

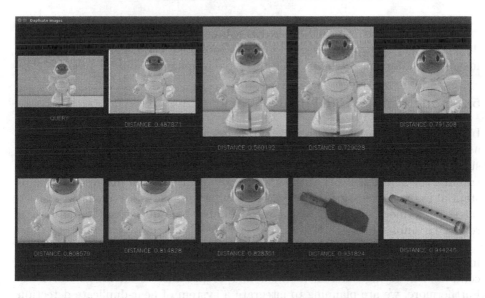

Fig. 6. Example of a search for duplicate images of a given query image from **istella-duplicate-images** using **pascalvoc-nd-net** with the **pool5/7x7_s** extract layer. The algorithm correctly retrieves duplicate images, followed by some images with a similar background.

Fig. 7. Example of a search for similar images of a given query image from **istella-similar-images** using **inception-21k** with the **global_pool** extract layer. The algorithm correctly retrieves photos of the same planetarium, followed by photos of other planetariums and buildings with a similar shape.

Both **pool5/7x7_s1** and **global_pool** layers provide signatures with a dimensionality of 1024, the second lowest size among the tried layers, after the **loss3/classifier** layer, whose signatures have a dimensionality of 1000.

5   Conclusions and Future Work

In this work the problem of retrieving images that are duplicates or similar to a given query image through neural codes is analyzed. Experimental results on standard datasets and two new datasets created using web images show that training the neural network with an artificially created set of images brings substantial improvements for the task of retrieving duplicates for low false positive rates, while models trained with natural photos remain better for the task of retrieving similar images.

Future work includes improving the retrieval of similar images by finding new suitable datasets and algorithms, investigating the effects of signature compression, trying different methods and/or parameters for image searching. Furthermore, we are planning to integrate a system of near-duplicate detection in a digital asset management platform that will be released in open source.

Acknowledgements. This work was supported by the ESSE3 project, funded by the P.O. FESR Sardegna 2007–2013 program. I would like to thank my collegues at the CONT group of CRS4 for their help in reviewing this paper and Maurizio Agelli for the photographies of the photo-retouch dataset.

References

1. Li, Z., Feng, X.: Near duplicate image detecting algorithm based on bag of visual word model. J. Multimed. **8**(5), 557–564 (2013)
2. Foo, J.J., Sinha, R., Zobel, J.: Discovery of image versions in large collections. In: Cham, T.-J., Cai, J., Dorai, C., Rajan, D., Chua, T.-S., Chia, L.-T. (eds.) MMM 2007. LNCS, vol. 4352, pp. 433–442. Springer, Heidelberg (2006). https://doi.org/10.1007/978-3-540-69429-8_44
3. Ke, Y., Sukthankar, R., Huston, L.: Efficient near-duplicate detection and sub-image retrieval. In: ACM Multimedia, pp. 869–876. ACM, New York (2004)
4. Wong, C.H., Bern, M., Goldberg, D.: An image signature for any kind of image. In: International Conference on Image Processing, pp. I-409–I-412. IEEE Press, New York (2002)
5. O'Gorman, L., Rabinovich, I.: Secure identification document via pattern recognition and public-key cryptography. IEEE Trans. Pattern Anal. Mach. Intell. **20**, 1097–1102 (1998)
6. Liu, J.: Image retrieval based on bag-of-words model (2013). arXiv:1304.5168
7. Lowe, D.: Distinctive image features from scale-invariant keypoints. Int. J. Comput. Vis. **60**(2), 91–110 (2004)
8. Krizhevsky, A., Sutskever, I., Hinton, G.: ImageNet classification with deep convolutional neural networks. Adv. Neural Inf. Process. Syst. **25**, 1106–1114 (2012)
9. Babenko, A., Slesarev, A., Chigoring, A., Lemptisky, V.: Neural codes for image retrieval. arXiv:1404.1777 (2014)
10. Razavian, A. S., Azizpour, H., Sullivan, J., Carlsson, S.: CNN features off-the-shelf: an astounding baseline for recognition. In: IEEE Conference on Computer Vision and Pattern Recognition Workshops, pp. 512–519. IEEE Press, New York (2014)
11. Sermanet, P., Eigen, D., Zhang, X., Mathieu, M., Fergus, R., Lecun, Y.: Overfeat: integrated recognition, localization and detection using convolutional networks (2013). arXiv:1312.6229
12. Dosovitskiy, A., Fischer, P., Springenberg, J. T., Riedmiller, M., Brox, T.: Discriminative unsupervised feature learning with exemplar convolutional neural networks (2015). arXiv:1406.6909
13. Chopra, S., Hadsell, R., LeCun, Y.: Learning a similarity metric discriminatively, with application to face verification. In: IEEE Conference on Computer Vision and Pattern Recognition, pp. 539–546. IEEE Press, New York (2005)
14. Wang, J., et al.: Learning fine-grained image similarity with deep ranking. In: IEEE Conference on Computer Vision and Pattern Recognition, pp. 1386–1393. IEEE Press, New York (2014)
15. Noh, H., Araujo, A., Sim, J., Weyand, T., Han, B.: Large-scale image retrieval with attentive deep local features (2016). arXiv:1612.06321
16. Douze, M., Jégou, H., Sandhawalia, H., Amsaleg, L., Cordelia, S.: Evaluation of GIST descriptors for web-scale image search. In: ACM International Conference on Image and Video Retrieval, Article no. 19. ACM, New York (2009)
17. Everingham, M., Ali Eslami, S.M., Van Gool, L., Williams, C.K.I., Winn, J., Zisserman, A.: The PASCAL visual object classes challenge: a retrospective. Int. J. Comput. Vis. **111**(1), 98–136 (2015)
18. Szegedy, C., et al.: Going deeper with convolutions. In: IEEE Conference on Computer Vision and Pattern Recognition, pp. 1–9. IEEE Press, New York (2015)
19. Russakovsky, O., et al.: Imagenet large scale visual recognition challenge (2014). arXiv:1409.0575

20. Jia, Y., et al.: Caffe: Convolutional Architecture for Fast Feature Embedding (2014). arXiv:1408.5093
21. Jegou, H., Douze, M., Schmid, C.: Hamming embedding and weak geometric consistency for large scale image search. In: Forsyth, D., Torr, P., Zisserman, A. (eds.) ECCV 2008. LNCS, vol. 5302, pp. 304–317. Springer, Heidelberg (2008). https://doi.org/10.1007/978-3-540-88682-2_24
22. Philbin, J., Chum, O., Isard, M. Sivic, J., Zisserman, A.: Object retrieval with large vocabularies and fast spatial matching. In: IEEE Conference on Computer Vision and Pattern Recognition, pp. 1–8. IEEE Press, New York (2007)
23. Deng, J., Dong, W., Socher, R., Li, L.-J., Li, K., Fei-Fei, L.: Imagenet: a large-scale hierarchical image database. In: IEEE Conference on Computer Vision and Pattern Recognition, pp. 248–255. IEEE Press, New York (2009)
24. Ioffe, S., Szegedy, C.: Batch normalization: accelerating deep network training by reducing internal covariate shift (2015). arXiv:1502.03167 (2015)

Optimum Network/Framework Selection from High-Level Specifications in Embedded Deep Learning Vision Applications

Delia Velasco-Montero[✉], Jorge Fernández-Berni, Ricardo Carmona-Galán, and Ángel Rodríguez-Vázquez

Instituto de Microelectrónica de Sevilla, Universidad de Sevilla-CSIC, Sevilla, Spain
`delia@imse-cnm.csic.es`

Abstract. This paper benchmarks 16 combinations of popular Deep Neural Networks for 1000-category image recognition and Deep Learning frameworks on an embedded platform. A Figure of Merit based on high-level specifications is introduced. By sweeping the relative weight of accuracy, throughput and power consumption on global performance, we demonstrate that only a reduced set of the analyzed combinations must actually be considered for real deployment. We also report the optimum network/framework selection for all possible application scenarios defined in those terms, i.e. weighted balance of the aforementioned parameters. Our approach can be extended to other networks, frameworks and performance parameters, thus supporting system-level design decisions in the ever-changing ecosystem of Deep Learning technology.

Keywords: Deep learning · Convolutional neural networks Embedded vision · Performance · High-level specifications

1 Introduction

The number of software frameworks, acceleration libraries, network architectures, hardware modules etc. related to the Deep Learning (DL) paradigm [7] increases virtually on a daily basis. Much better accuracy and a common approach for many different inference tasks are the primary reasons for the success of this paradigm.

In particular, concerning vision, Convolutional Neural Networks (CNNs) have become the core processing architecture underpinning most of the advances in the field [14]. As a result, the spectrum of DL technological components aiming at speeding up CNNs is vast. Each of these components comes along with claims about enhancements in certain aspects of performance, ease of use, compatibility etc. While this variety in principle covers a wide spectrum of specifications, there are no guidelines for system designers to perform an optimal selection meeting prescribed application-level requirements. This is especially an issue in embedded

© Springer Nature Switzerland AG 2018
J. Blanc-Talon et al. (Eds.): ACIVS 2018, LNCS 11182, pp. 369–379, 2018.
https://doi.org/10.1007/978-3-030-01449-0_31

platforms due to the heavy computational demands of CNNs [11]. In these cases, the adoption of non-optimal components could lead to the apparent impossibility of fulfilling some of those prescribed requirements.

In this paper, we analyze up to 16 combinations of popular network models and software frameworks from a practical point of view. We have implemented all of them in a Raspberry Pi 3 Model B, extracting three key performance parameters, namely accuracy, throughput and power consumption. But we do not assess them in a plain way. We propose a Figure of Merit (FoM) encoding the relative weight of each parameter according to specific high-level requirements. The results reveal that only 5 out of the 16 combinations perform the best in at least one point of the exploration domain. We also report the worst selection over all possible performance balances.

2   Baseline Performance Figures

Our starting point for this paper is the analysis reported in [13]. The embedded hardware platform picked for benchmarking is the Raspberry Pi (RPi) 3 Model B. It is inexpensive but still features enough computational power for real-time DL inference. Its Broadcom BCM2837 System On a Chip (SoC) comprises a Quad Core ARM Cortex-A53 1.2 GHz 64-bit CPU that can work at frequencies ranging from 700 MHz up to 1.2 GHz. It also features 1 GB RAM LPDDR2 900 MHz.

Concerning software tools, we chose 4 open-source frameworks widely adopted these days for DL, namely Caffe, TensorFlow, OpenCV and Caffe2. Each of them relies on particular encoding techniques and libraries for optimization of both training and inference. *Caffe* [6] is a DL software tool developed by the Berkeley Vision and Learning Center (BVLC) with the help of a large community of contributors on GitHub. It requires Basic Linear Algebra Subprograms (BLAS) as the back-end for matrix and vector computation. We compiled Caffe with OpenBLAS [15], an optimized implementation of BLAS supported by ARM processors. This library accelerates algebraic operations on CPU leveraging the 4 ARM cores of the RPi. *TensorFlow* [1] expresses computations as stateful dataflow graphs that manage tensors. This enables various optimizations, e.g. the elimination of redundant copies of the same operation in the graph or the implementation of careful operation scheduling that improves data transfer performance and memory usage. We installed pre-built TensorFlow 1.3.0 for RPi [2]. This version exploits ARM hardware optimizations for computational acceleration, namely ARM Neon and VFPv4 (Vector Floating Point Unit). ARM Neon supports Single Instruction Multiple Data (SIMD) instructions for exploiting data-level parallelism whereas VFPv4 provides high-efficiency floating-point operations. Concerning *OpenCV* [9] computer vision library, its DNN module enables the use of pre-trained models for inference from other frameworks such as Caffe, Torch or TensorFlow. OpenCV version 3.3.1 was compiled to exploit both ARM Neon and VFPv3 optimizations as well. Finally, *Caffe2* [3] also uses computation graphs for network definition. The building process of Caffe2 on the RPi admits ARM Neon optimizations too.

All of these software tools allow the deployment of state-of-the-art pre-trained models of Deep Neural Networks (DNNs) for inference. In particular, we benchmarked 4 well-known models trained for 1000-category image recognition, namely *Network in Network* [8], *GoogLeNet* [12], *SqueezeNet* [5] and *MobileNet* [4]. Although their architectures differ from each other, all of them contain layers performing point-wise convolutions − 1 × 1 filters operating across tensors − to reduce the number of model parameters. Other techniques aiming at lightening the computational load are the inclusion of Global Average Pooling layers before Fully-Connected (FC) layers − GoogLeNet −, the omission of FC layers − Network in Network and SqueezeNet − or the channel cutback and image resolution shrinking − MobileNet.

For each inference scenario − i.e. for each combination of DNN model on DL framework within the aforementioned set, running on the RPi − we have averaged the values of throughput and power consumption when processing a number of images extracted from the ImageNet dataset [10]. We have also measured the accuracy of the models over the 50k validation images of this dataset. The results are summarized in Table 1. Each triplet per entry corresponds to the benchmarked performance parameters, that is, [Top-5 Accuracy (%), Throughput (fps), Power Consumption (W)].

Table 1. Performance figures for the selected DL frameworks and DNN models running on a Raspberry Pi 3 Model B. Each triplet per entry corresponds to measured metrics: [Top-5 Accuracy (%), Throughput (fps), Power Consumption (W)]

	Network in Network	GoogLeNet	SqueezeNet v1.1	MobileNet v1 1.0-224
Caffe	[79.23, 2.08, 3.75]	[89.02, 0.80, 3.81]	[80.81, 3.11, 3.58]	[89.00, 1.14, 3.50]
TensorFlow	[79.23, 2.58, 3.49]	[89.53, 1.74, 3.58]	[80.81, 4.22, 2.95]	[89.67, 2.41, 3.26]
OpenCV	[79.23, 1.42, 3.08]	[89.02, 1.28, 3.62]	[80.81, 4.79, 3.20]	[89.00, 2.56, 3.23]
Caffe2	[79.23, 2.55, 3.44]	[89.02, 0.92, 3.00]	[79.92, 3.74, 3.11]	[89.00, 1.80, 2.88]

3 Weighted FoM

In [13], we defined the following FoM for the sake of merging the three measured performance parameters in a meaningful way:

$$\text{FoM} := \text{Accuracy} \cdot \frac{\text{Throughput}}{\text{Power}} \tag{1}$$

Note that this FoM can be expressed as 'number of correctly inferred images per joule', providing insight about the joint computational and energy efficiency of models and frameworks on the considered hardware platform. The results in [13] clearly show that SqueezeNet on OpenCV performs the best according to Eq. (1). The superiority of this combination arises from achieving a notable frame rate in a power-efficient way while trading off some accuracy in comparison with

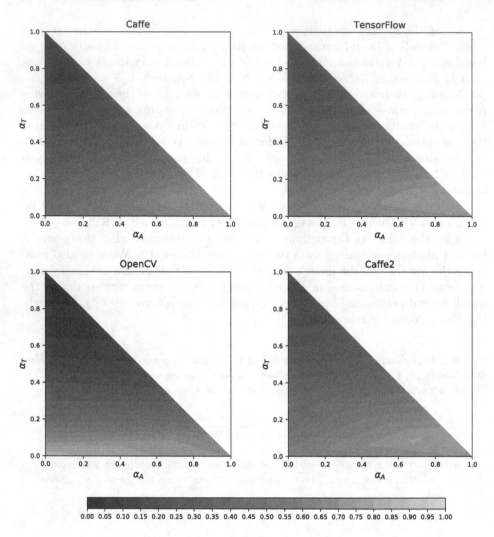

Fig. 1. Normalized weighted FoM for Network in Network model.

more precise networks like GoogLeNet or MobileNet. The question is: would SqueezeNet-OpenCV still be the best choice for a particular application where accuracy is more important, in relative terms, than throughput and/or power consumption?

In order to answer this question, we propose a re-definition of the FoM in such a way that the relative weight of each performance parameter on high-level specifications can be taken into account. Let $[\alpha_A, \alpha_T, \alpha_P]$ be the relative importance of the metrics [Accuracy, Throughput, Power] respectively, satisfying:

$$\sum_i \alpha_i = 1, \alpha_i \in [0, 1] \tag{2}$$

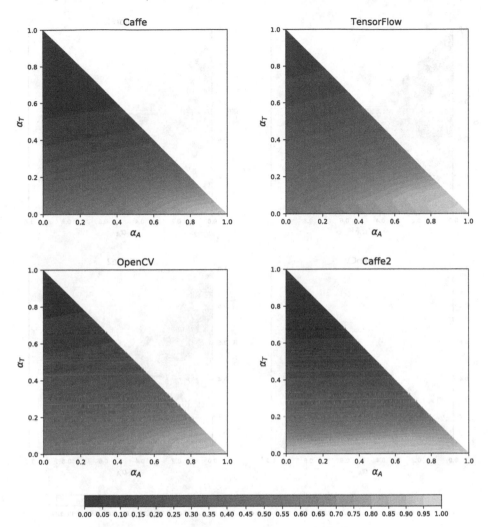

Fig. 2. Normalized weighted FoM for GoogLeNet model.

where i corresponds to A, T, P. It follows from Eq. (2) that, for instance, a vector $[\alpha_A, \alpha_T, \alpha_P] = [0.70, 0.15, 0.15]$ defines an application scenario where accuracy presents a relative preeminence of 70% on the targeted performance, being throughput and power consumption equally important with a weight of 15%.

A new weighted FoM can now be defined as:

$$\text{FoM}(\alpha_A, \alpha_T, \alpha_P) := \text{Accuracy}^{3\alpha_A} \cdot \frac{\text{Throughput}^{3\alpha_T}}{\text{Power}^{3\alpha_P}} \tag{3}$$

Hence, Eq. (1) is just a particular case of the generalized FoM in Eq. (3): when $\alpha_i = 1/3 \ \forall i$, we are assuming that the three performance parameters have the same impact on application-level requirements. Any other distribution of α_i

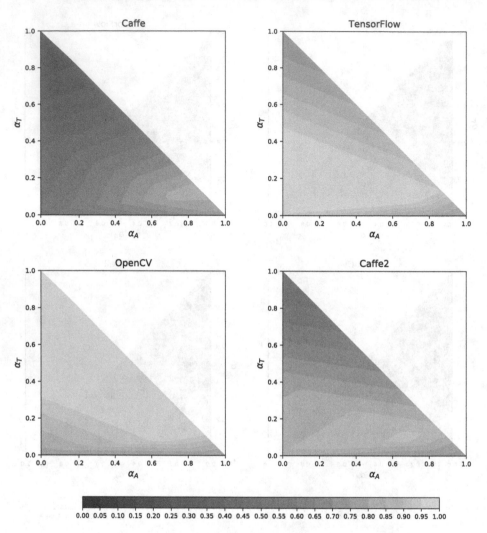

Fig. 3. Normalized weighted FoM for SqueezeNet model.

assigns a specific weight to their corresponding factor in Eq. (3) by setting its exponent to a value greater or less than unity. Extreme cases occur when any α_i is set to 1, boosting the influence of its associated performance parameter on the FoM at the cost of completely dismissing the other ones.

By introducing all the possible combinations of α_i in Eq. (3) together with the triplets in Table 1, we can find out the optimum network/framework choice per application scenario defined in terms of weighted balance of accuracy, throughput and power consumption. For each scenario – i.e. for each combination of α_i – we normalize the resulting FoM with respect to its maximum value obtained for the optimum network/framework pair. This permits to quickly identify the best

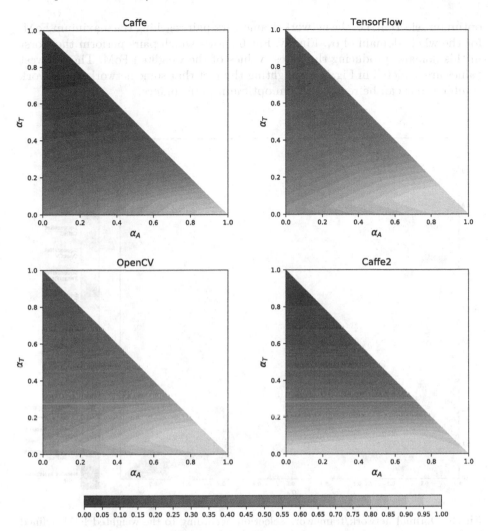

Fig. 4. Normalized weighted FoM for MobileNet model.

selection with a FoM equal to 1. Likewise, it allows to easily assess how far each pair is from the optimum point.

All in all, Figs. 1, 2, 3 and 4 depict the normalized values of the proposed weighted FoM for each DNN model per software framework. Note that we only consider α_A and α_T since, according to Eq. (2), once their values are established, α_P is automatically determined from $\alpha_P = 1 - \alpha_A - \alpha_T$. In these plots, the maximum is represented in yellow. Points getting far from this maximum turn green and eventually blue for the lowest values of the weighted FoM. A quick look permits to conclude that SqueezeNet-OpenCV keeps being the best choice for most application scenarios. This is confirmed by Fig. 5 where we represent the optimum selection, i.e. the network/framework pair reaching the maximum FoM, for the whole domain of α_i. Finally, Fig. 6 shows which pairs perform the worst in this domain, producing the lowest values of the weighted FoM. These lowest values are depicted in Fig. 7, highlighting the fact that some network/framework combinations can be really far from optimum performance.

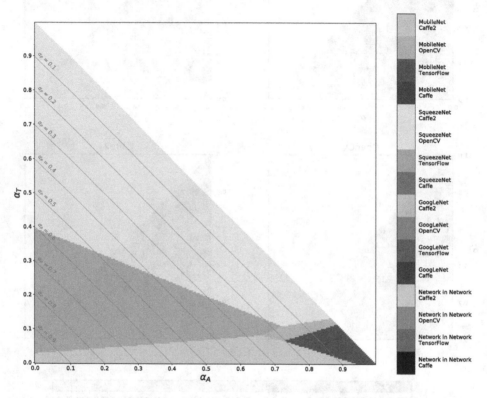

Fig. 5. Optimal network/framework selection according to the weighted FoM defined in Eq. (3). Only a reduced set of the analyzed combinations performs the best in at least one point of the exploration domain.

4 Discussion

The first significant outcome from these results is that SqueezeNet is the most suitable network for the majority of the explored performance weightings. For application scenarios where throughput is critical, SqueezeNet must be implemented on OpenCV. When energy efficiency gains relevance, TensorFlow makes the most of this network. For relaxed frame rate requirements, MobileNet performs the best on combining high accuracy and reduced power consumption. In fact, the lowest consumption – 2.88W – is attained when running this network

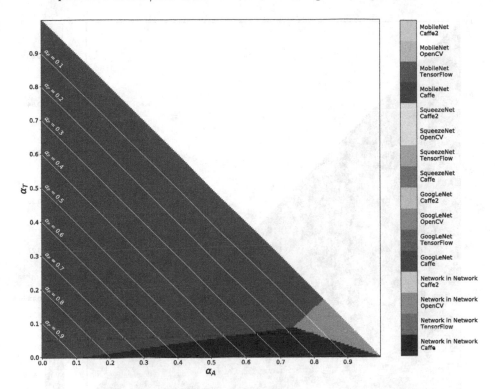

Fig. 6. Worst network/framework selection according to the weighted FoM. The normalized values of the FoM for each point in this plot are represented in Fig. 7.

on Caffe2 – both model and framework have been designed for mobile applications. When power loses importance, keeping relaxed constraints on throughput, MobileNet on TensorFlow is the best choice. In practical terms, we have narrowed down the original list of 16 network/framework pairs to 5 combinations to be selected.

Note that neither Network in Network nor GoogLeNet ever reach a maximum FoM, no matter the software framework considered. On the contrary, when running on OpenCV or Caffe, Network in Network performs the worst because of its lowest accuracy at moderate frame rate and high power consumption. Nevertheless, the combination covering most of the area in Fig. 6 is GoogLeNet on Caffe. The high accuracy from this network do not make up for their joint poorest performance in terms of throughput and power consumption. In particular, throughput is notably lower than the best one – 0.80 fps vs. 4.79 fps for SqueezeNet-OpenCV. This is why the normalized FoM in Fig. 7 degrades for increasing values of α_T.

We must emphasize that this comparative study is strictly quantitative in the sense that our discussion is exclusively focused on measurable performance. Thus, the FoM defined in Eq. (3) does not include any qualitative parameters

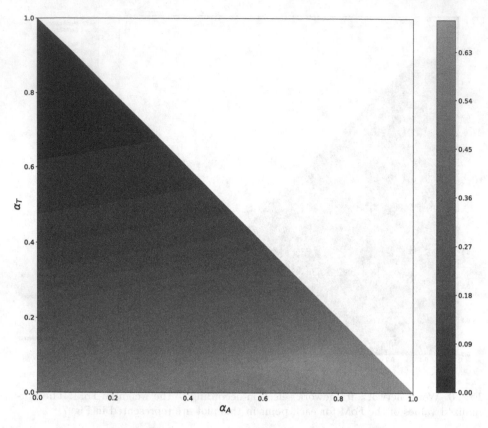

Fig. 7. Lowest values of the normalized FoM. They correspond to the network/framework pairs represented in Fig. 6.

encoding the suitability of a particular model, framework or hardware platform according to, for instance, a long-term technological strategy in a company or a product development roadmap. However, taking into account the current lack of standardization in the field of DL-based vision, and its rapid evolution pace, qualitative considerations should not have too much influence yet on the final decision to be made. The conclusions drawn in this manuscript should not therefore change significantly when such considerations were included in the analysis.

5   Conclusions

We have described a pragmatic approach to support system designers when making decisions in the vast ecosystem of DL components for embedded vision. The merit of a particular realization must be assessed according to the weighted relevance of key operation parameters on application-level specifications. The FoM proposed in this paper aims at facilitating this evaluation. When applied for 1000-category image recognition on a low-cost embedded platform, it has proved to be effective on filtering out most of the studied alternatives. Specifically, we have demonstrated that SqueezeNet-OpenCV presently constitutes the best choice to carry out this task in most application scenarios. In the future, we plan to extend our analysis to more hardware platforms, DNN models and software frameworks, also incorporating other essential performance parameters like memory.

Acknowledgments. This work was supported by Spanish Government MINECO (European Region Development Fund, ERDF/FEDER) through Project TEC2015-66878-C3-1-R, by Junta de Andalucía CEICE through Project TIC 2338-2013 and by EU H2020 MSCA ACHIEVE-ITN, Grant No. 765866.

References

1. Abadi, M. et al.: TensorFlow: a system for large-scale machine learning. In: 12th USENIX Symposium on Operating Systems Design and Implementation (OSDI), pp. 265–283 (2016)
2. DeftWork: a docker image for Tensorflow (2018). https://github.com/DeftWork/rpi-tensorflow. Accessed July 2018
3. Facebook Open Source: Caffe2 (2018). https://caffe2.ai/. Accessed July 2018
4. Howard, A. et al.: MobileNets: efficient convolutional neural networks for mobile vision applications. arXiv:1704.04861 (2017)
5. Iandola, F. et al.: SqueezeNet: AlexNet-level accuracy with 50x fewer parameters and <1 MB model size. arXiv:1602.07360 (2016)
6. Jia, Y., et al.: Caffe: convolutional architecture for fast feature embedding. arXiv:1408.5093 (2014)
7. LeCun, Y., Bengio, Y., Hinton, G.: Deep learning. Nature **521**(7553), 436–444 (2015)
8. Lin, M., Chen, Q., Yan, S.: Network in network. arXiv:1312.4400 (2013)
9. OpenCV Team: OpenCV (2018). https://opencv.org/. Accessed July 2018
10. Russakovsky, O., et al.: ImageNet large scale visual recognition challenge. Int. J. Comput. Vis. (IJCV) **115**(3), 211–252 (2015)
11. Sze, V., et al.: Efficient processing of deep neural networks: a tutorial and survey. Proc. IEEE **105**(12), 2295–2329 (2017)
12. Szegedy, C. et al.: Going deeper with convolutions. arXiv:1409.4842 (2014)
13. Velasco-Montero, D. et al.: Performance analysis of real-time DNN inference on Raspberry Pi. In: SPIE Commercial + Scientific Sensing and Imaging, April 2018
14. Verhelst, M., Moons, B.: Embedded deep neural network processing. IEEE Solid-State Circ. Mag. **9**(4), 55–65 (2017)
15. Xianyi, Z.: OpenBLAS, optimized BLAS library based on GotoBLAS2 1.13 BSD version (2018). https://github.com/xianyi/OpenBLAS. Accessed July 2018

Contour Propagation in CT Scans
with Convolutional Neural Networks

Jean Léger$^{(\boxtimes)}$, Eliott Brion, Umair Javaid, John Lee,
Christophe De Vleeschouwer, and Benoit Macq

Université catholique de Louvain, Louvain-la-Neuve, Belgium
{jean.leger,eliott.brion}@uclouvain.be

Abstract. Although deep convolutional neural networks (CNNs) have
outperformed state-of-the-art in many medical image segmentation
tasks, deep network architectures generally fail in exploiting common
sense prior to drive the segmentation. In particular, the availability of
a segmented (source) image observed in a CT slice that is adjacent to
the slice to be segmented (or target image) has not been considered
to improve the deep models segmentation accuracy. In this paper, we
investigate a CNN architecture that maps a joint input, composed of
the target image and the source segmentation, to a target segmentation.
We observe that our solution succeeds in taking advantage of the source
segmentation when it is sufficiently close to the target segmentation,
without being penalized when the source is far from the target.

Keywords: CNN · Segmentation · Computed tomography

1 Introduction

Computed tomography (CT) image segmentation is highly useful for a large
number of medical applications such as image analysis in radiology [1], treatment planning and patient follow-up in radiotherapy [2] or therapeutic response
prediction through radiomics [3]. Classical clinically used segmentation methods
include atlas-based methods [4] or active contours [5], but are generally failing in
cases where the regions of interest are subjected to large deformations or matter
income/outcome. Statistical shape models [6] allow to capture shape variations
but require landmarks definition. Pixel-wise classification has also been performed with random forest and SVM classifiers [7,8], potentially completed by
structure enhancing methods such as conditional random fields [9] or graph cuts.
However, those methods require a tedious and subjective definition of features.
Alternatively, numerous hybrid methods have proposed to combine several complementary segmentation approaches [10–12]. Nevertheless, none of them closes
the challenging task of CT image segmentation.

In parallel, the recent advances in computing capabilities and the availability
of representative datasets allowed deep learning approaches to reach impressive

---

J. Léger and E. Brion—Contributed equally.

© Springer Nature Switzerland AG 2018
J. Blanc-Talon et al. (Eds.): ACIVS 2018, LNCS 11182, pp. 380–391, 2018.
https://doi.org/10.1007/978-3-030-01449-0_32

segmentation performance, competing with state-of-the-art segmentation tools in many fields of medical imaging [13]. The main advantage of deep learning with respect to the aforementioned approaches is its high versatility and its capability to learn automatically a complex input-output mapping through the tuning of a large number of inner parameters. Fully convolutional neural networks [14] gained in popularity compared to patch-based approaches since they benefit from a higher computational efficiency and a better integration of contextual information. In particular, the u-net fully convolutional neural network architecture [15] resulted in good segmentation performances on 2D medical images, as well as its 3D extension [16] and 3D V-Net variation [17].

Deep learning segmentation algorithms have been successfully used to segment structures on CT images in the head and neck [18], the pelvic [19] and the abdominal [20,21] regions. Those deep models rely on the assumption that the learned filters are able to identify automatically, from the training data, the features that are relevant for the task at hand. However, in many applications, additional information, also named *prior information*, is available and should be exploited to constraint the segmentation process towards the desired output. Recently, shape prior constraints have been incorporated in deep neural networks through PCA [22], resulting in an improved segmentation robustness and accuracy. Anatomical constraints have also been accounted for through the combination of several deep neural networks [23].

Similarly, the availability of a coarse or approximate segmentation may provide a valuable prior information to guide a fine-grained segmentation process. Examples of practical applicative scenarios in which such approximate segmentation is available include dense 3D segmentation (for which only a subset of the slices have been segmented), but also the adaptation of radiotherapy treatments (for which the segmentation of an initial image should be propagated to subsequent images acquired along the treatment).

In this work, an approach has been developed to segment 2D CT image slices of the bladder using the manual segmentation (drawn by an experienced radiation oncologist) on an adjacent slice as prior knowledge. Our goal is to design a deep network architecture able to capture both the bladder appearance statistics and the deformation occurring between adjacent slices. In that sense, it allows to *propagate* the segmentation between adjacent slices. To demonstrate the relevance of our approach, we analyze quantitatively how the segmentation result improves with the proximity of the adjacent slice. Note that slices segmentation of the bladder is not a trivial task due to the high shape and size variation of the bladder, the presence or not of contrast material in it and the poor contrast between its wall and surrounding soft tissues.

Section 2 introduces the CT images that have been considered in our study and their preprocessing. Section 3 explains how to turn the segmentation problem into a learning problem, by defining pairs of input/output training samples based on the CT images and their manual annotations by medical doctors. The network architecture is also given in the same section. In Sect. 4, the results are provided and discussed. Conclusions are given in Sect. 5.

2 Materials and Preprocessing

Our data consist of 3D CT volumes (planning CTs, also named PCTs) from patients with prostate cancer. Each patient underwent External Beam Radiation Therapy (EBRT) at CHU-UCL-Namur (198 patients) or at CHU-Charleroi – Hôpital A. Vésale (141 patients). The images coming from both hospitals are shuffled in our dataset and images with and without contrast agent are present in it. For each patient, the bladder was manually delineated in the PCT by an experienced radiation oncologist. This has been considered as the ground truth in this work. The use of these retrospective, anonymized data for this study as been approved by each hospital ethics committee. The data were pre-processed according to the following steps: data re-sampling, slice selection and cropping and intensity range thresholding.

The 3D CT volumes (stacks of 2D CT slices) extracted in the hospitals under DICOM format have a different pixel spacing and slice thickness across different patients. In order to ensure data uniformity over the entire dataset, all the 3D CT volumes have been re-sampled through a linear interpolation according to a $1 \times 1 \times 1$ mm regular grid. The bladder contours drawn by the clinicians are stored as 3D binary masks with same sampling as the CT volumes. The DICOM format manipulation and re-sampling are performed using the opensource platform OpenReggui.[1]

For every patient, a single 2D axial slice where the bladder is present is randomly selected, together with several adjacent slices, from the 3D CT volume. This is further discussed in Sect. 3.1. Every selected slice is cropped to 192×192 pixels tile centered on the bladder. The reason for cropping is that it allows faster experimentation. Our investigations (not described in this paper) have shown that running the algorithm on the entire slice instead of an image tile centered on the region of interest leads to comparable segmentation accuracy.

All pixels intensities above 3000 HU (Hounsfield unit) and below -1000 HU are respectively set to 3000 HU and -1000 HU to avoid artifacts in the high Hounsfield unit range. The lower and upper bounds are respectively determined from the Hounsfield values of air and cortical bone.

3 Formulating the Contour Propagation as a Learning Problem

Our goal is to segment target 2D CT image slices of the bladder using the manual delineation on an adjacent slice as prior knowledge. In this section, the target and adjacent slices selection is firstly explained. Then, we present the CNN architecture used for the segmentation and how the prior information is included in it.

---

[1] https://openreggui.org/.

3.1   Prior Definition and Computation

Each 3D CT volume is made of a stack of 2D slices along the axial axis. For every patient, the bladder only appears on a set of slices indexed by $p \in [0, 1, ..., P-1]$ where the bladder top and bottom slices are respectively indexed by $p = 0$ and $p = P - 1$, with P the number of slices containing the bladder. For every patient, a single *target* slice indexed by p_t has been randomly selected such that $p_t \in [0, (P-1) - 10]$.

Then, for every target slice indexed by $p = p_t$, the corresponding prior knowledge is extracted from an *adjacent* slice indexed by $p_q = p_t + q$ with $q \in [0, 1, ..., 10]$. Hence, the indices p_q of the prior slices are always such that $p_q \in [0, (P-1)]$. The prior knowledge used in this work is the manual delineation (contour) drawn by the clinician on the qth adjacent slice. This contour is represented as a binary mask $\mathbf{Y}^{(k)[q]} \in \{0, 1\}^{M \times M}$ defined as

$$Y_{m,n}^{(k)[q]} = \begin{cases} 1 & \text{if } (m, n) \text{ is in the bladder} \\ 0 & \text{otherwise} \end{cases} \tag{1}$$

where (m, n) is the pixel position within the binary mask $\mathbf{Y}^{(k)[q]}$, k is the index of the binary mask in the dataset and M is the image tile width.

3.2   Network Architecture and Learning Strategy

The u-net fully convolutional neural network has been used for the bladder segmentation. The architecture of the network is shown in Fig. 1 with the hyperparameters used in this work. The network input is expanded from one to two channels. The additional channel is used to enter the prior knowledge into the network.

More precisely, let us assume that the network input for the kth data sample is denoted by $\mathbf{X}^{(k)} \in \mathbb{R}^{M \times M \times C}$, where C is the number of input channels. The corresponding labels are a binary mask $\mathbf{Y}^{(k)} \in \{0, 1\}^{M \times M}$ equal to one within the bladder and equal to zero everywhere else. The notation $\mathbf{X}_{:,:,c}^{(k)}$ is used in order to denote the cth network input channel. The input is the concatenation of the target image $\mathbf{I}^{(k)} \in \mathbb{R}^{M \times M}$ with the prior mask given on the qth adjacent slice above the target slice: $\mathbf{X}_{:,:,1}^{(k)} = \mathbf{I}^{(k)}$ and $\mathbf{X}_{:,:,2}^{(k)} = \mathbf{Y}^{(k)[q]}$ as shown in Fig. 2. Eleven different networks are trained, each for a different distance between the target slice and the adjacent slice. Formally, for $q \in [0, 1, ..., 10]$, eleven different models $\mathcal{M}_q$ are trained, each of them with a training set $\mathcal{S}_{train}^{[q]}$ given by $\mathcal{S}_{train}^{[q]} = \left\{ (\mathbf{X}^{(j)}, \mathbf{Y}^{(j)}) \mid j \in [0, 1, ..., J_{train} - 1], \mathbf{X}_{:,:,1}^{(j)} = \mathbf{I}^{(j)} \text{ and } \mathbf{X}_{:,:,2}^{(j)} = \mathbf{Y}^{(j)[q]} \right\}$, where J_{train} is the number of target tiles in the training set.

The network architecture, inspired from [15], is illustrated in Fig. 1. It takes as input either one channel (if the CNN is used without prior knowledge, see Sect. 4.1) or two channels (if the CNN is used with prior knowledge) and outputs a prediction for the target tile bladder segmentation. The input goes through a contracting path (left side) to capture context and an expanding path (right side) to enable precise localization.

In the contracting path, a collection of hierarchical features are learned thanks to successive convolutions and max-pooling operations. Successively, two series of 3×3 convolutions, followed by a ReLu activation and batch normalization are applied before applying a max-pooling 2×2. After each max-pooling step, the number of feature maps is doubled in order to allow the network to learn many high level features. From the features learned in the contracting path, the expanding path increases the resolution via successive 2×2 transposed convolutions (i.e. a 2×2 up-sampling followed by a 2×2 convolution), 3×3 convolutions and ReLu activations in order to recover the original tile size.

In the last layer, a sigmoid is applied and the network outputs the probability for each pixel to belong to the bladder. To obtain the final segmentation, a threshold of 0.5 is chosen. Hence, every pixels with a score above the threshold is predicted to belong to the bladder, while other pixels are predicted not to belong to the bladder.

The network is trained with the Dice loss. For the jth training example, the Dice loss between the predicted segmentation $\hat{\mathbf{Y}}^{(j)} \in [0, 1]^{M \times M}$ and the reference segmentation $\mathbf{Y}^{(j)} \in \{0, 1\}^{M \times M}$ is given by

$$
\mathcal{L}\left(\hat{\mathbf{Y}}^{(j)}, \mathbf{Y}^{(j)}\right) = -\frac{2 \sum_{m=0}^{M-1} \sum_{n=0}^{M-1} Y_{m,n}^{(j)} \hat{Y}_{m,n}^{(j)}}{\sum_{m=0}^{M-1} \sum_{n=0}^{M-1} \left(Y_{m,n}^{(j)} + \hat{Y}_{m,n}^{(j)}\right)}. \tag{2}
$$

The fact that each training tile has at least one pixel belonging to the bladder ensures that the denominator is non-zero. This loss drives the output segmentation to have a large overlap with the ground truth segmentation. This network is implemented in Keras, with TensorFlow back-end. The optimization algorithm used is Adam with a learning rate of $1e{-}4$, a batch size of 20 and the model is trained for 200 epochs. The other parameters (such as initialization strategy and learning rate decay) are left to Keras default. All the tiles in the dataset are shifted and scaled by the mean and the standard deviation of the tiles intensities over the training set. We use the validation set to earlystop the training. Training data is augmented using rotation, shift, shear, zoom and horizontal flip.

The target slices and their set of adjacent slices are randomly assigned to the training set (179 patients), the validation set (80 patients) and the test set (80 patients). Note that the patient distribution across the training, validation and test sets is the same for the eleven different models.

4   Results and Discussion

In this section, the validation metric and the comparison baselines are first presented. Then, our results are discussed.

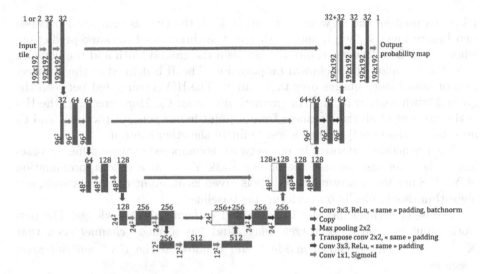

Fig. 1. The network architecture used is the u-net. Each blue rectangle represents the feature maps resulting from a convolution operation, while white rectangles represent copied featured maps. For the convolutions, the zero padding is chosen such that the image size is preserved ("same" padding). This figure is adapted from [15]. (Color figure online)

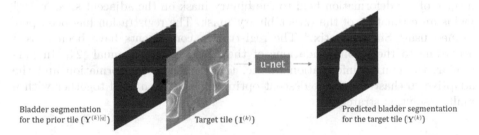

Fig. 2. The proposed u-net network takes as input the bladder segmentation for the prior tile and the target tile intensities. It outputs a prediction of the bladder segmentation for the target tile.

4.1 Validation Metrics and Comparison Baselines

In order to evaluate our results, we use three metrics: the Dice similarity coefficient (DSC) and the Jaccard index (JI) measure the overlap between two binary masks, while the Hausdorff distance (HD) assesses the distance between the contours (i.e. the sets of points located at the boundary of the binary masks) extracted from those binary masks. Most results are discussed based on the DSC since it is quite popular in the literature.

The DSC and the JI both reach one in case of perfect overlap and zero if there is no overlap between both binary masks. The DSC is slightly different from the

Dice loss used for the network training. Indeed, the DSC is computed between two binary images, the ground truth and the thresholded network prediction, whereas the Dice loss was computed between the ground truth and the network prediction to allow the backward propagation. The JI is defined as the intersection of both binary images over their union. The HD is computed between the ground truth contour $\mathcal{C}_1$ and the prediction contour $\mathcal{C}_2$. More precisely, the HD is the greatest of all the distances from a point in one contour (both $\mathcal{C}_1$ and $\mathcal{C}_2$ need to be considered) to the closest point in the other contour.

The prediction performed by our network is compared to three different baselines. The *copy* baseline uses the binary mask $\mathbf{Y}^{(k)[q]}$ as a valid approximation of $\mathbf{Y}^{(k)}$. Since the binary mask $\mathbf{Y}^{(k)[q]}$ is given as an input to the network, our algorithm should ideally outperform this baseline.

The u-net network has also been trained without prior knowledge. The network input is simply the target image and has a single channel such that $\mathbf{X}^{(k)}_{:,:,1} = \mathbf{I}^{(k)}$. This is shown in Fig. 1 and denoted as the *CNN without prior* baseline.

The results obtained with our CNN have been compared to the classical registration-based contour propagation. It has been denoted as the *registration* baseline. Such an approach computes a deformation field from the target image $\mathbf{I}^{(k)}$ (fixed image) to the image on the qth adjacent slice $\mathbf{I}^{(k)[q]}$ (the moving image). Applying the inverse of this deformation field to the moving image allows to align the moving image with the fixed image. In turn, applying the inverse of the deformation field to the binary mask on the adjacent slice $\mathbf{Y}^{(k)[q]}$ yields an estimation of the target binary mask. The registration has been performed using SimpleElastix.[2] The registration components have been chosen according to the recommendations of the Elastix's user manual [24]. In particular, the mutual information metric, the B-splines transformation and the adaptive stochastic gradient descent optimizer have been used together with a multi-resolution strategy.

4.2   Discussion

In Fig. 3, the DSC and HD between the output segmentation of the CNN with prior and the ground truth segmentation are computed and averaged over the test set for several distances (i.e. from 1 mm to 10 mm) between the target and adjacent slices. This distance is named the inter-slices distance below. The average DSC and HD obtained with the CNN with prior is also compared to the results obtained with the three aforementioned baselines. Table 1 provides the mean and standard deviation of the DSC and JI for all the considered methods and an inter-slices distance equal to 2, 4 and 8 mm. The following observations can be done based on Fig. 3 and Table 1.

- For a large inter-slices distance (i.e. $8 \leq q$), the CNN with prior does not clearly outperform the CNN without prior. This means that the CNN with

---

[2] http://simpleelastix.github.io.

prior struggles to improve the target segmentation when the prior is not sufficiently correlated with the target segmentation. However, the CNN with prior clearly outperforms the result of the registration-based approach in this region. This is illustrated in the first column of Fig. 4.

- For an intermediate inter-slices distance (i.e. $4 \leq q < 8$), the CNN with prior slightly outperforms the CNN without prior. However, the DSC of the copy is still low in this interval. This means that the CNN with prior is able to exploit jointly the information present in the target image and the prior mask. Furthermore, the prior exploitation increases on average as the prior mask becomes more relevant (i.e. for decreasing q). This is illustrated in the second column of Fig. 4.

- For a small inter-slices distance (i.e. $q < 4$), the CNN with prior performs like the registration-based approach, which is known to work efficiently when the moving (adjacent) and the fixed (target) images are close to each other. This means that the CNN with prior is able to capture information in the prior mask when it is relevant for the segmentation task. This is illustrated in the third column of Fig. 4.

From those observations, we can conclude that the CNN with prior reaches or exceeds the segmentation performance of alternative approaches for all the considered inter-slices distances. Furthermore, the prior knowledge is more intensively exploited as its correlation with the ground truth increases.

From Table 1, it appears that the variance of the considered metrics is relatively high compared to the difference in their mean computed with the different baselines. A finer analysis of the statistical distribution of our segmentation results reveals that, even if the average performance is only moderately better with prior than without prior, the exploitation of the segmentation mask in an adjacent slice (i.e. the prior) prevents a complete failure of the segmentation in worst cases. Indeed, the DSC is always larger than 0.78 in presence of prior, while being lower than 0.8 for about 15% of the images in absence of prior, with worst cases below 0.6 and reaching 0.2. Hence, we conclude that our proposed framework has a limited impact on the mean DSC performance but significantly increases the robustness of the approach, which is certainly important in practice (e.g. regarding organ segmentation for radiotherapy treatment).

Figure 3 and Table 1 also include the segmentation performance of a single CNN (in opposition to the eleven CNNs) trained over all the training samples, hence corresponding to different inter-slices distances. This CNN is named *CNN with prior single*. This unique CNN is then tested on the different inter-slices separately. It appears that this network improves the segmentation performance with respect to the CNN without prior for all the considered inter-slices distances, even if it is outperformed by the CNNs trained for a specific inter-slide distance for $q < 4$. This result is encouraging for applications where we cannot define a proper inter-slices distance (e.g. regarding the propagation of a contour over a temporal sequence of images).

388   J. Léger et al.

The training time was respectively 446 s and 488 s for the CNN without prior and for the CNN with prior. This training time is an average over the training times obtained for the three inter-slices distances considered in Table 1. The inference time was respectively ∼0 s, 3.58 s, 0.057 s and 0.184 s for the copy, the registration, the CNN without prior and the CNN with prior. This inference time is the time necessary to predict the segmentation of a single image, averaged over the three inter-slices distances and over all the test images. It appears that the CNN with and without prior have comparable training times. It can be also observed that the inference time of CNNs is one to two orders of magnitude below the inference time of the registration-based method. The CNNs have been trained and tested with nVidia cuDNN acceleration on a computer equipped with a GPU nVidia GeForce GTX 1080 Ti 11Gb.[3]

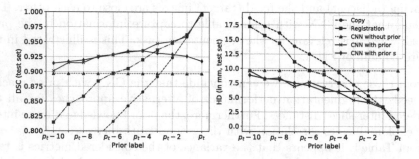

Fig. 3. Influence of the prior on the DSC and on the HD for test set examples. DSC: Dice similarity coefficient. HD: Hausdorff distance.

Table 1. Comparison of the different segmentation algorithms regarding to their DSC and JI on test set. See text for details. DSC: Dice similarity coefficient. JI: Jaccard index. Reg.: Registration, CNN wop: CNN without prior, CNN wp: CNN with prior (eleven different networks are trained, each one for a given inter-slice distance), CNN wps: CNN with prior single (one single network is trained on all different inter-slice distances).

Algorithm	$p_t - 8$		$p_t - 4$		$p_t - 2$	
	DSC	JI	DSC	JI	DSC	JI
Copy	.761 ± .176	.645 ± .214	.866 ± .122	.781 ± .166	.924 ± .077	.867 ± .120
Reg.	.858 ± .147	.775 ± .186	.920 ± .101	.864 ± .130	.948 ± .048	.905 ± .080
CNN wop	.897 ± .159	.840 ± .191	.897 ± .159	.840 ± .191	.897 ± .159	.840 ± .191
CNN wp	.915 ± .103	.856 ± .140	**.935 ± .071**	**.885 ± .104**	**.951 ± .036**	**.908 ± .062**
CNN wps	**.919 ± .088**	**.861 ± .128**	**.935 ± .055**	.883 ± .088	.929 ± .063	.873 ± .099

---

[3] https://developer.nvidia.com/cudnn.

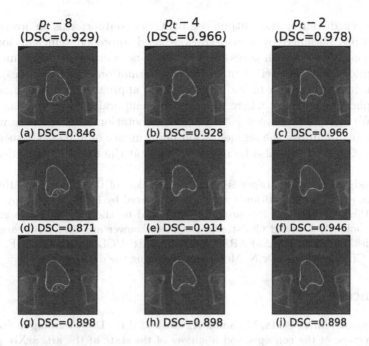

Fig. 4. Ground truth and predicted contours for a given patient at different inter-slices distances. This figure compares the CNN with prior (in green) with the ground truth (in yellow) and with the baselines (in red, each line showing a different baseline algorithm). In the first row, the baseline is the "copy" algorithm, with the DSC between the ground truth and this baseline reported in (a), (b) and (c). In the second line, the baseline is the registration algorithm. In the third row, the baseline is the CNN without prior. The DSC between brackets are computed between the ground truth and the CNN with prior. The first, second and third columns respectively correspond to an inter-slices distance of 8, 4 and 2 mm. DSC: Dice similarity coefficient. (Color figure online)

5   Conclusion

We propose a deep CNN network able to segment 2D CT image slices of the bladder, using the manual delineation on an adjacent slice as prior knowledge. This is done by training the u-net network with two input channels. The first channel is used to enter the target image and the second one is used to enter the manual delineation of the adjacent slice given as a binary mask.

The network is trained and tested for an increasing distance between the target and adjacent slices (i.e. from 1 to 10 mm). The Dice similarity coefficient between the thresholded network prediction and the ground truth exceeds 0.9 for all the considered distances and increases as the distance between the slices decreases.

For every distance, the output segmentation outperforms in average the segmentations obtained by a registration-based approach (efficient for small deformations between both slices) and by the u-net network trained and tested without prior knowledge (efficient for large deformations between slices).

In the future, we plan to use the proposed approach on other organs and other applicative contexts where a spatial or temporal contour propagation is needed. We also plan to investigate how the annotation of a few slices within a 3D volume can improve the segmentation performance of a 3D CNN such as 3D u-net [16]. Our code will also be made available at the time of the conference.

Acknowledgments. Jean Léger is a Research Fellow of the Fonds de la Recherche Scientifique - FNRS, Eliott Brion's work was supported by FEDER-RW project User-MEDIA, Umair Javaid is a Research Fellow funded by the FNRS Televie grant no. 7.4625.16, John A. Lee and Christophe De Vleeschouwer are Senior Research Associates with the Belgian F.R.S.-FNRS. We thank CHU-UCL-Namur (Dr J.-F. Daisne) as well as CHU-Charleroi (Dr N. Meert) for providing the data.

References

1. Mazurowski, M.A., Buda, M., Saha, A., Bashir, M.R.: Deep learning in radiology: an overview of the concepts and a survey of the state of the art. arXiv preprint arXiv:1802.08717 (2018)
2. Sharp, G., et al.: Vision 20/20: perspectives on automated image segmentation for radiotherapy. Med. Phys. **41**(5), 050902 (2014)
3. Cha, K.H., et al.: Bladder cancer treatment response assessment in CT using radiomics with deep-learning. Sci. Rep. **7**(1), 8738 (2017)
4. Iglesias, J.E., Sabuncu, M.R.: Multi-atlas segmentation of biomedical images: a survey. Med. Image Anal. **24**(1), 205–219 (2015)
5. Cremers, D., Rousson, M., Deriche, R.: A review of statistical approaches to level set segmentation: integrating color, texture, motion and shape. Int. J. Comput. Vis. **72**(2), 195–215 (2007)
6. Heimann, T., Meinzer, H.P.: Statistical shape models for 3D medical image segmentation: a review. Med. Image Anal. **13**(4), 543–563 (2009)
7. Polan, D.F., Brady, S.L., Kaufman, R.A.: Tissue segmentation of computed tomography images using a random forest algorithm: a feasibility study. Phys. Med. Biol. **61**(17), 6553 (2016)
8. Luo, S., Hu, Q., He, X., Li, J., Jin, J.S., Park, M.: Automatic liver parenchyma segmentation from abdominal CT images using support vector machines. In: ICME International Conference on Complex Medical Engineering, CME 2009, pp. 1–5. IEEE (2009)
9. Hu, Y.C.J., Grossberg, M.D., Mageras, G.S.: Semi-automatic medical image segmentation with adaptive local statistics in conditional random fields framework. In: 30th Annual International Conference of the IEEE on Engineering in Medicine and Biology Society, EMBS 2008, pp. 3099–3102. IEEE (2008)
10. Tong, T., et al.: Discriminative dictionary learning for abdominal multi-organ segmentation. Med. Image Anal. **23**(1), 92–104 (2015)
11. Gao, Y., Shao, Y., Lian, J., Wang, A.Z., Chen, R.C., Shen, D.: Accurate segmentation of CT male pelvic organs via regression-based deformable models and multi-task random forests. IEEE Trans. Med. Imaging **35**(6), 1532–1543 (2016)

12. Oda, M., et al.: Regression forest-based atlas localization and direction specific atlas generation for pancreas segmentation. In: Ourselin, S., Joskowicz, L., Sabuncu, M.R., Unal, G., Wells, W. (eds.) MICCAI 2016. LNCS, vol. 9901, pp. 556–563. Springer, Cham (2016). https://doi.org/10.1007/978-3-319-46723-8_64

13. Litjens, G., et al.: A survey on deep learning in medical image analysis. Med. Image Anal. **42**, 60–88 (2017)

14. Long, J., Shelhamer, E., Darrell, T.: Fully convolutional networks for semantic segmentation. In: Proceedings of the IEEE Conference on Computer Vision and Pattern Recognition, pp. 3431–3440 (2015)

15. Ronneberger, O., Fischer, P., Brox, T.: U-net: convolutional networks for biomedical image segmentation. In: Navab, N., Hornegger, J., Wells, W.M., Frangi, A.F. (eds.) MICCAI 2015. LNCS, vol. 9351, pp. 234–241. Springer, Cham (2015). https://doi.org/10.1007/978-3-319-24574-4_28

16. Çiçek, Ö., Abdulkadir, A., Lienkamp, S.S., Brox, T., Ronneberger, O.: 3D U-Net: learning dense volumetric segmentation from sparse annotation. In: Ourselin, S., Joskowicz, L., Sabuncu, M.R., Unal, G., Wells, W. (eds.) MICCAI 2016. LNCS, vol. 9901, pp. 424–432. Springer, Cham (2016). https://doi.org/10.1007/978-3-319-46723-8_49

17. Milletari, F., Navab, N., Ahmadi, S.A.: V-net: fully convolutional neural networks for volumetric medical image segmentation. In: 2016 Fourth International Conference on 3D Vision (3DV), pp. 565–571. IEEE (2016)

18. Ibragimov, B., Xing, L.: Segmentation of organs-at-risks in head and neck CT images using convolutional neural networks. Med. Phys. **44**(2), 547–557 (2017)

19. Kazemifar, S., et al.: Segmentation of the prostate and organs at risk in male pelvic CT images using deep learning. arXiv preprint arXiv:1802.09587 (2018)

20. Roth, H.R., et al.: Hierarchical 3D fully convolutional networks for multi-organ segmentation. arXiv preprint arXiv:1704.06382 (2017)

21. Larsson, M., Zhang, Y., Kahl, F.: Robust abdominal organ segmentation using regional convolutional neural networks. In: Sharma, P., Bianchi, F.M. (eds.) SCIA 2017. LNCS, vol. 10270, pp. 41–52. Springer, Cham (2017). https://doi.org/10.1007/978-3-319-59129-2_4

22. Milletari, F., Rothberg, A., Jia, J., Sofka, M.: Integrating statistical prior knowledge into convolutional neural networks. In: Descoteaux, M., Maier-Hein, L., Franz, A., Jannin, P., Collins, D.L., Duchesne, S. (eds.) MICCAI 2017. LNCS, vol. 10433, pp. 161–168. Springer, Cham (2017). https://doi.org/10.1007/978-3-319-66182-7_19

23. Trullo, R., Petitjean, C., Ruan, S., Dubray, B., Nie, D., Shen, D.: Segmentation of organs at risk in thoracic CT images using a sharpmask architecture and conditional random fields. In: 2017 IEEE 14th International Symposium on Biomedical Imaging (ISBI 2017), pp. 1003–1006. IEEE (2017)

24. Klein, S., Staring, M.: Elastix, the manual (2018). http://elastix.isi.uu.nl/download/elastix-4.9.0-manual.pdf

Person Re-identification Using Group Context

Yiqiang Chen[1]([⊠]), Stefan Duffner[1], Andrei Stoian[2], Jean-Yves Dufour[2], and Atilla Baskurt[1]

[1] Univ Lyon, INSA-Lyon, CNRS, LIRIS, 69621 Villeurbanne, France
yiqiang.chen@insa-lyou.fr
[2] Thales Services, ThereSIS, Palaiseau, France

Abstract. The person re-identification task consists in matching person images detected from surveillance cameras with non-overlapping fields of view. Most existing approaches are based on the person's visual appearance. However, one of the main challenges, especially for a large gallery set, is that many people wear very similar clothing. Our proposed approach addresses this issue by exploiting information on the *group* of persons around the given individual. In this way, possible ambiguities are reduced and the discriminative power for person re-identification is enhanced, since people often walk in groups and even tend to walk alongside strangers. In this paper, we propose to use a deep convolutional neural networks (CNN) to extract group feature representations that are invariant to the relative displacements of individuals within a group. Then we use this group feature representation to perform group association under non-overlapping cameras. Furthermore, we propose a neural network framework to combine the group cue with the single person feature representation to improve the person re-identification performance. We experimentally show that our deep group feature representation achieves a better group association performance than the state-of-the-art methods and that taking into account group context improves the accuracy of the individual re-identification.

Keywords: Person re-identification · Convolutional neural network
Group association

1 Introduction

Person re-identification consists in matching identities across images captured from disjoint camera views. Increasing attention has been dedicated to person re-identification algorithms in the past few years as they have important applications in video surveillance and are necessary for cross-camera tracking, multi-camera event detection, and pedestrian retrieval.

Despite the recent progress in the performance of person re-identification methods, several difficulties remain since a person's appearance often undergoes

© Springer Nature Switzerland AG 2018
J. Blanc-Talon et al. (Eds.): ACIVS 2018, LNCS 11182, pp. 392–401, 2018.
https://doi.org/10.1007/978-3-030-01449-0_33

large variations across different cameras due to varying view points and illumination conditions. Different human poses and partial occlusions further make the task challenging. Moreover, the problem becomes increasingly difficult when there is a large number of candidate persons since, in practice, many persons have similar appearance as they share the same visual attributes or wear very similar clothes and colours.

(a)

(b)

Fig. 1. (a) Single person images. (b) Corresponding group images of (a). Even for a human, it may be difficult to tell if the three top images belong to the same person or not. Using the context of the surrounding group, it is easier to see that the middle and right images belong to the same person and the left image belongs to another person.

Most existing approaches only use the visual appearance of a *single* person for its re-identification in different images. However, this can lead to strong ambiguities, for example when people wear similar clothes, as shown in Fig. 1. To address this problem, context information about the surrounding group of persons can be used. In realistic settings, people often walk in groups rather than alone. Thus, the appearance of these groups can serve as visual context and help to determine whether two images of persons with similar clothing belong to the same individual.

However, matching the surrounding people in a group in different views is also challenging. On the one hand, it undergoes the variations of single person appearances. On the other hand, the number of persons and their relative position within the group can vary over time and across cameras. Further, partial occlusions among individuals are very likely in groups.

In this paper, we propose to extract group feature representations using a deep convolutional neural network. First, we train the model with single-person re-identification data. Then, transfer it to the group association problem. In order to cope with the relative displacements of persons in a group, we applied a Global Max-Pooling (GAP) operation of CNN activations to achieve translation invariance in the resulting representation. Furthermore, we measure a group context distance with this representation and then combine it with the distance measure based on single-person appearance to enhance the re-identification accuracy.

The main contributions of this paper are the following:

- we learn a deep feature representation with displacement invariance and apply it to the group association problem. Our experiments show that this approach outperforms the state-of-the-art on group association.
- we propose a novel way to combine group context and single-person appearance and experimentally show that the group information can improve the person re-identification performance.

2   Related Work

Person Re-identification. Existing person re-identification approaches mainly concentrate on two aspects: building a robust feature representation for a person's visual appearance and learning a distance metric. Some specifically designed features are proposed in order to address the changes in view-point and pose, for example ELF [4], SADALF [3], LOMO features [10]. The main metric learning methods like KISSME [7], LFDA [12] and XQDA [10]. Recently, many methods based on deep convolutional neural network(CNN) have been proposed for person re-identification, like [1,9,16]. The advantage of these deep learning approaches is that they incorporate feature representations and the distance metric into an integrated model that is jointly learnt from the data.

Group Association. In the literature, there are several group association (or re-identification) approaches. Zheng et al. [19] extracted visual words which are the clusters of SIFT+RGB features in a group image. Then they built two descriptors that describe the ratio information of visual words between local regions to represent group information. Cai et al. [2] used covariance descriptor to encode group context information. And Lisanti et al. [11] proposed to learn a dictionary of sparse atoms using patches extracted from single person images. Then the learned dictionary is exploited to obtain a sparsity-driven residual group representation. These approaches can be severely affected by background clutter, and thus a preprocessing stage is necessary. For example, in [2,19] a background subtraction is performed before feature extraction. And in [11], three pedestrian detectors based on respectively deformable part models, aggregated channel features and RCNN were used to weight the contribution of each pixel in the histogram computation.

Some other approaches use trajectory features to describe group information. Wei et al. [17], for example, presented a group extraction approach by clustering the persons' trajectories observed in a camera view. They introduced person-group features composed of two parts: SADALF features [3], extracted after background subtraction and representing the visual appearance of the accompanying persons of a given individual, and a signature encoding the position of the subject within the group. Similarly, Ukita et al. [15] determined for each pair of pedestrians whether they form a group or not, using spatio-temporal features of their trajectories like relative position, speed and direction. Then, the group features composed of the trajectory features (position, speed, direction) of individuals in each group, the number of persons as well as the mean colour histograms of the individual person images. However, when people walk in group, the position and speed are not always uniform. Thus, the trajectory-based features may not be precise and change significantly over time.

Unlike these methods, the advantage of our approach is that there is no need for a pre-processing stage of person detection or background subtraction. Our model is pre-trained on single-person re-identification data to learn the discriminative features that distinguish identities in images. The applied global max-pooling operation captures maximum activations over feature maps, which correspond to salient discriminative patches in the input image. Thus the proposed model is, by design, invariant to displacements of individuals within a group. Moreover, the deep neural network that we employed can provide a richer feature representation to describe groups than the colour and texture features used by existing methods.

3   Proposed Method

In this section, we first describe our group association method and further introduce how we use the group information to improve person re-identification performance.

3.1   Group Association

In the first step, we train a neural network predicting the identities of the images, given an input image resized to 64×124. The model can be a CNN pre-trained on ImageNet, like Alexnet [8] or ResNet-50 [5]. The final fully-connected (FC) layer is replaced by another FC layer with an output dimension of N, with N being the number of identities in the training set.

Then, the CNN is fine-tuned in a supervised way, using images and identity labels from a separate person re-identification dataset. To this end, we minimise the following softmax cross-entropy loss on the given classification task:

$$E_{identification} = -\sum_{k=1}^{N} y_k log(P(y_k = 1|x)), \tag{1}$$

$$\text{with} \quad P(y_j = 1|x) = \frac{e^{W_j^T x + b_j}}{\sum_{k=1}^{N} e^{W_k^T x + b_k}}, \tag{2}$$

where y is the one-hot coded identity label, x is the input to the last fully-connected layer, W and b are weights and bias of the last fully-connected layer and $P(y_j = 1|x)$ is the predicted probability that the input x corresponds to identity j. The intuition of this supervised training is that the resulting feature representation can be used for learning similarities between arbitrary pedestrian images and thus be transferred to the task of group re-identification.

After training the model, we discard the FC layer and represent the activation map of the last convolutional layer as a set of K 2D feature channel responses

$\mathcal{X} = \mathcal{X}_i, i = 1...K$, where $\mathcal{X}_i$ is the 2D map representing the responses of the ith feature channel. A ReLU activation function is applied as a last step to guarantee that all elements are non-negative. A final location-invariant representation, called Maximum Activations of Convolutions (MAC) [14], is constructed by a spatial max-pooling over all locations concatenated in a K-dimensional vector:

$$f = [f_1...f_i...f_K]^\top, with \, f_i = \max_{x \in \mathcal{X}_i}(x) \,. \tag{3}$$

When applying the model for group association, a group image resized to 224×244 pixels is given as input (corresponding to a single-person size of roughly 64×128 pixels in the image). The distance between two images is measured with the cosine distance between the feature vectors produced as described above. This feature representation does not encode the location of the activations unlike the activations of fully connected layers, due to the max-pooling operated over the whole last convolution layer feature map. It encodes the maximum local response of each of the convolutional filters. Thus, it offers translation invariance to the resulting representation.

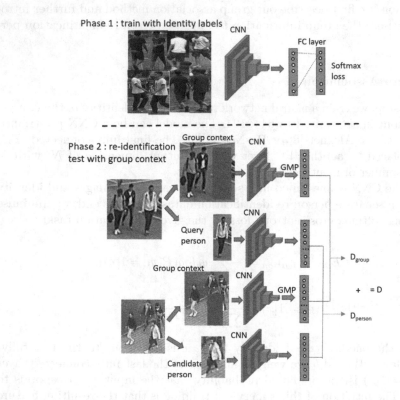

Fig. 2. Overview of our group association assisted re-identification method. The CNN is first trained with person images. Then, group context distance and single-person distance are computed and summed to obtain the final distance.

3.2   Group-Assisted Person Re-identification

In our setting, the input data is composed of group images with annotated individual identities and corresponding bounding boxes. As shown in Fig. 2, to explicitly capture both *person* and *group context* features, we divide the input image **I** into two input images to process them separately.

First, a query person image **P** is obtained from the raw group image by using the given annotated bounding box. Second, its group context image **G** is obtained from the raw group image by covering the query person image region with the pixels of the mean image colour. Then, these two images are given to the CNN model explained in Sect. 3.1. Two parallel branches of this network are employed to extract the feature embeddings for respectively the group context input image and the person image. The two branches are almost identical, except for the last layer, where the MAC feature representation is used for the (larger) group context image and the input vector of the discarded FC layer is used for the single-person image. As illustrated in Fig. 2, after processing the query image, the same procedure is applied to the gallery images in order to compute a distance measure on the two representation between query and candidate image (resulting in 4 feature vectors in total). The advantage of this method is that it can be easily combined with any CNN-based single person re-identification approach. For a given query and candidate image, the cosine distance is used to separately compute a group context distance D_{gr} between the two group images and a person distance D_{id} between the two single-person images. The final distance measure is simply the sum:

$$D(I_i, I_j) = D_{id}(P_i, P_j) + D_{gr}(G_i, G_j) \ . \tag{4}$$

This equation can also be formulated as a weighted sum. Since in our test, we don't have a validation set with group images to determine the weight which is a hyper-parameter, we use just a simple sum to combine these two distances.

4   Experiments

4.1   Datasets

The **Market-1501 Dataset** [18] is one of the largest publicly available datasets for human re-identification. It contains 32668 images of 1501 subjects captured on a campus. The dataset is split into 751 identities for training and 750 identities for test. In our experiments, we uses only the training set of Market-1501 to train the CNN model.

The **Ilidis-group Dataset** is extracted by Zheng et al. [19] from the i-LIDS MCTS dataset. It contains 274 images of 64 groups taken from airport surveillance cameras. Most of the groups have 4 images, either from different camera views or at different times. Some example images are shown in Fig. 3.

OGRE dataset   Ilidis-group dataset

Fig. 3. Some example images from people group datasets.

The **OGRE Dataset** [11] contains 1279 images of 39 groups acquired by three disjoint cameras pointing at a parking lot. This is a challenging dataset with many different viewpoints and self-occlusions. We manually annotated a subset of this dataset with 450 bounding boxes and 75 identities.

The Cumulative Match Curve (CMC) is employed as evaluation measure for both group association and person re-identification. The CMC curve shows the probability that a query identity appears in the top-k of the ordered candidate list with varying k.

For the group association test, we follow the test protocol in [11,19]. That is, for each group, one randomly selected image is included in the gallery, all the remaining images form the probe set. The test is repeated 10 times, then the average scores are computed. For the person re-identification test, the images with person bounding boxes are used. We take each person bounding box as query image in turn, and the rest of the images as gallery set. The final result is the average CMC score over all queries.

4.2   Experimental Setting

We used ResNet-50 as the CNN model and the weights pre-trained on the ImageNet dataset are used as initialization. For training, data augmentation is performed by randomly flipping the images and cropping central regions with random perturbation. Dropout is applied to the fully connected layers to reduce the risk of over-fitting. The optimization is performed by Stochastic Gradient Descent with a learning rate of 0.001, a momentum of 0.9 and a batch size of 50.

4.3   Group Association Results

The comparison with the state-of-the-art method on the Ilids-group and OGRE dataset is shown in the Table 1. We compared not only with the group association methods in [11,19], but also with two encoding techniques, namely IFV [13] and VLAD [6], applied by [11] in group association as well a CNN model with global average pooling (GAP). Our method outperforms the best state-of-the-art method PREF [11] in terms of the Rank 1 score by a margin of 5.6% and 6.1% points on Ilidis-group and OGRE datasets, respectively. This clearly shows the effectiveness of the deep feature representation. Compared to the GAP-based model, using GMP increased the Rank 1 score on the two datasets by 3.9% and 2.9% points, respectively. This demonstrates the benefit of the invariance property of the GMP for group association.

Table 1. Comparison with group association state-of-the-art methods on the Ilids-group and OGRE dataset. *: figures extracted from a curve.

Method	Ilids-Group			OGRE		
	Rank 1	Rank 10	Rank 25	Rank 1	Rank 10	Rank 25
CRRRO+BRO [19]	22.5*	57.0*	**76.0***	-	-	-
IFV [13]	26.1	60.2	75.8	14.6	43.3	76.8
VLAD [6]	26.0	57.0	75.0	13.0	41.1	74.3
PREF [11]	31.1	60.3	75.5	15.1	41.6	75.8
Ours with GAP	32.8	56.0	70.7	18.3	46.8	79.7
Ours with GMP	**36.7**	**60.5**	73.7	**21.2**	**50.4**	**82.2**

4.4 Group-Assisted Person Re-identification Results

The result of person re-identification is shown in Table 2. We compare the person re-identification results with some variants of our method. *Sum feature* and *Concatenate feature* represent variants that first sum or concatenate the single-person feature representation and the group feature representation and then compute the distance measure on these vectors. We compared also to a variant that retains the query or candidate person image in the group image without covering the corresponding region with the mean colour. The results show that the method proposed in this paper (i.e. covering the person in the group image and summing the person and group distance) achieved the best re-identification results. Covering the person image improves the Rank 1 score by 2.7% points. Since some persons from the same group share very similar context, covering the query or candidate person can better discriminate persons in the same or similar group context. Finally, the combined distance achieves better results than only using the single person distance and the group distance. Overall, our proposed method increases the result by 9.6% points with respect to only using single-person images. This clearly shows that group context has the ability to considerably reduce the appearance ambiguity.

Table 2. Person re-identification accuracy (in %) on the OGRE dataset.

Variant	Rank 1	Rank 5	Rank 10
Single person only	47.2	69.3	78.8
Group context only	26.2	57.2	66.3
Sum features	41.1	69.9	77.7
Concatenate features	51.9	**75.1**	81.1
Distance sum without mean image cover	54.1	73.7	80.8
Distance sum with mean image cover	**56.8**	73.7	**81.7**

5   Conclusion

In this paper, we presented an effective deep learning-based group association and group-assisted person re-identification approach. The deep group feature representation is extracted by a CNN and global max-pooling is applied to achieve location-invariance of individuals in group images. We also proposed a method improving single-person re-identification by incorporating the group context, defining a combined distance metric. This method can be combined with any CNN-based single person re-identification approach. We experimentally showed that our method outperforms the state-of-the-art in group association and that the deep group feature representation considerably enhances the person re-identification performance.

Acknowledgement. This work was supported by the Group Image Mining (GIM) which joins researchers of LIRIS Lab. and THALES Group in Computer Vision and Data Mining. We thank NVIDIA Corporation for their generous GPU donation to carry out this research.

References

1. Ahmed, E., Jones, M., Marks, T.K.: An improved deep learning architecture for person re-identification. In: Proceedings of CVPR, pp. 3908–3916 (2015)
2. Cai, Y., Takala, V., Pietikainen, M.: Matching groups of people by covariance descriptor. In: 2010 20th International Conference on Pattern Recognition (ICPR), pp. 2744–2747. IEEE (2010)
3. Farenzena, M., Bazzani, L., Perina, A., Murino, V., Cristani, M.: Person re-identification by symmetry-driven accumulation of local features. In: Proceedings of CVPR, pp. 2360–2367 (2010)
4. Gray, D., Tao, H.: Viewpoint invariant pedestrian recognition with an ensemble of localized features. In: Forsyth, D., Torr, P., Zisserman, A. (eds.) ECCV 2008. LNCS, vol. 5302, pp. 262–275. Springer, Heidelberg (2008). https://doi.org/10.1007/978-3-540-88682-2_21
5. He, K., Zhang, X., Ren, S., Sun, J.: Deep residual learning for image recognition. In: Proceedings of the IEEE Conference on Computer Vision and Pattern Recognition, pp. 770–778 (2016)

6. Jégou, H., Douze, M., Schmid, C., Pérez, P.: Aggregating local descriptors into a compact image representation. In: CVPR, pp. 3304–3311 (2010)
7. Koestinger, M., Hirzer, M., Wohlhart, P., Roth, P.M., Bischof, H.: Large scale metric learning from equivalence constraints. In: Proceedings of CVPR, pp. 2288–2295 (2012)
8. Krizhevsky, A., Sutskever, I., Hinton, G.E.: Imagenet classification with deep convolutional neural networks. In: Advances in Neural Information Processing Systems, pp. 1097–1105 (2012)
9. Li, W., Zhao, R., Xiao, T., Wang, X.: Deepreid: deep filter pairing neural network for person re-identification. In: Proceedings of CVPR, pp. 152–159 (2014)
10. Liao, S., Hu, Y., Zhu, X., Li, S.Z.: Person re-identification by local maximal occurrence representation and metric learning. In: Proceedings of CVPR (2015)
11. Lisanti, G., Martinel, N., Del Bimbo, A., Foresti, G.L.: Group re-identification via unsupervised transfer of sparse features encoding. In: Proceedings of ICCV (2017)
12. Pedagadi, S., Orwell, J., Velastin, S., Boghossian, B.: Local fisher discriminant analysis for pedestrian re-identification. In: Proceedings of CVPR, pp. 3318–3325 (2013)
13. Sánchez, J., Perronnin, F., Mensink, T., Verbeek, J.: Image classification with the fisher vector: theory and practice. Int. J. Comput. Vis. 105(3), 222–245 (2013)
14. Tolias, G., Sicre, R., Jégou, H.: Particular object retrieval with integral max-pooling of CNN activations. In: International Conference on Learning Representations (2016)
15. Ukita, N., Moriguchi, Y., Hagita, N.: People re-identification across non-overlapping cameras using group features. Comput. Vis. Image Underst. 144, 228–236 (2016)
16. Varior, R.R., Haloi, M., Wang, G.: Gated siamese convolutional neural network architecture for human re-identification. In: Proceedings of the ECCV, pp. 791–808 (2016)
17. Wei, L., Shah, S.K.: Subject centric group feature for person re-identification. In: CVPR Workshops, pp. 28–35 (2015)
18. Zheng, L., Shen, L., Tian, L., Wang, S., Wang, J., Tian, Q.: Scalable person re-identification: a benchmark. In: IEEE International Conference on Computer Vision (2015)
19. Zheng, W.S., Gong, S., Xiang, T.: Associating groups of people. In: BMVC (2009)

Fingerprint Classification Using Conic Radon Transform and Convolutional Neural Networks

Dhekra El Hamdi[1,2(✉)], Ines Elouedi[2], Abir Fathallah[3], Mai K. Nguyen[1], and Atef Hamouda[2]

[1] Laboratoire Equipes de Traitement de l'Information et Système (ETIS)- Université de Cergy-Pontoise/ENSEA/CNRS UMR 8051, 95000 Cergy-Pontoise, France
`dhikra.hamdi@ensea.fr`
[2] Laboratoire d'Informatique, Programmation, Algorithmique et Heuristiques (LIPAH), Faculté des sciences de Tunis, Université de Tunis EL Manar, Tunis, Tunisia
[3] ENISO, Université de Sousse, Sousse, Tunisia

Abstract. Fingerprint classification is a useful technique for reducing the number of comparisons in automated fingerprint identification systems. But it remains a challenging issue due to the large intra-class and the small inter-class variations.

In this work, we propose a novel approach to perform fingerprint classification based on combining the Radon transform and the convolutional neural networks (CNN). The proposed approach is based on the Conic Radon Transform (CRT). The CRT extends the Classical Radon Transform (RT) to integrate an image function f(x,y) over conic sections. The Radon technique enables the extraction of fingerprint's global characteristics which are invariant to geometrical transformations such as translation and rotation.

We define an expansion of convolutional neural networks input features based on CRT. Thus, we perform first RT over conic sections on source image, and then use the Radon result as an input for convolutional layers.

To evaluate the performance of this approach, we have driven tests on the NIST SD4 benchmark database. The obtained results show that this approach is competitive to other related methods in terms of accuracy rate and computational time.

Keywords: Fingerprint classification · Conic Radon Transform · Convolutional neural networks

1 Introduction

Fingerprint has been considered the most common and trusted biometric organs for personal identification.

© Springer Nature Switzerland AG 2018
J. Blanc-Talon et al. (Eds.): ACIVS 2018, LNCS 11182, pp. 402–413, 2018.
https://doi.org/10.1007/978-3-030-01449-0_34

In fact, the fingerprint recognition has been studied for many years and a great number of fingerprint identification algorithms have been proposed in the literature [9]. A fingerprint is a pattern of ridges and valleys. This pattern enables the extraction of several features which are useful for persons recognition. Most of the identification methods are constituted of two main modules: features extraction and matching step.

The analysis of fingerprint matching algorithms when dealing with large databases needs a high computational time. Fingerprint classification into several classes allows to achieve this goal easier. After classifying the query fingerprint, it is only necessary to compare that print with the other fingerprints which have the same class.

According to Henry [9], fingerprints could be classified into five main categories: Arch, Tented arch, Left loop, Right loop, and Whorl. (Fig. 1)

Source: Fingerprint Classification, Maor Sharf

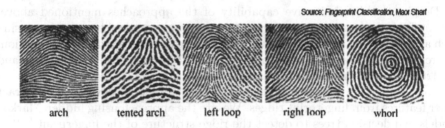

| arch | tented arch | left loop | right loop | whorl |

Fig. 1. Fingerprints classified into 5 fingerprints classes.

Generally, four classes are used, Arch and Tented arch are regrouped in the same class. In this paper, the four-class classification problem is taken into account.

A typical fingerprint classification algorithm usually extracts the global-level features of the fingerprint images that can lead to a high discrimination between the classes. After that, these features are given to different classifiers to determine the fingerprint class. These traditional classification algorithms are highly dependent on a preprocessing phase that can lead to a false classification.

Deep neural networks [4] techniques have reached a lot of success along the last few years due to their high capability to recognize complex patterns. In fact, various applications based on the Deep neural networks were implemented such as image classification [7] and digit recognition [13].

In this paper, we propose a novel system of fingerprint classification which extracts global fingerprints features by the use of the CRT. The CRT is applied in a first stage to detect the directional characteristics of fingerprints which are afterwads assigned as an input to the Convolutional Neural Networks (CNN).

The main contribution of this work is to associate the performance of Radon transform to extract fingerprints global features and the success of the deep learning techniques in the classification task. Our main motivation is to achieve both of the two following objectives: Evaluating our approach accuracy against that of state-of-the-art's fingerprint classifiers, and reducing CNN computational time compared to algorithms applying CNN directly on input images.

This paper is organized as follows: Sect. 2 presents a review of the literature related works. Afterwards, Sect. 3 describes the conic Radon transform and the Convolutional Neural Networks. we detail our approach and present the architecture of the adopted network. Then, we present the experimental evaluation of our method in Sect. 4. In final, Sect. 5 concludes the paper.

2   Related Works

Various feature extraction approaches have been proposed which are based on different features such as singular points [8] which are generally detected by the Poincaré method and the Orientation maps which are usually extracted by gradient-based methods [3].

Despite the discriminative capability of the approaches mentioned above, they suffer from some issues as they require a complex and ad hoc pre-processing such as binarisation and thinning which can cause false classification. In addition, the results are susceptible to geometrical transformations like the rotation and the translation of the fingerprint images.

To overcome the above lacks, several methods have been developed to extract other features from fingerprint ridges. We cite the approach using hidden Markov models and decision trees to detect the ridge structure of the fingerprint [14].

Further, the RT was used to extract straight lines from fingerprints. A fingerprint identification method [5] applied the RT and used correlation level between Radon profiles as a measure of the similarity between fingerprints.

In this paper, we propose to use CRT to extract parabolic and circular curves from fingerprint images. In fact, the structure of fingerprint ridges contains different parabolic or circular forms which can be directly detected by the CRT. Besides Radon, The CNN have received a great success in various classification applications. However, despite the efficiency of CNN approaches on many classification problems, they have been lightly applied on the fingerprint problem. Wang et al. [15] used a 3-hidden layer neural network for the classification. They achieved a 93.1% of accuracy over NIST-DB4, with a 1.8% rejection. In [12], the authors used a deep convolutional neural network to improve the quality of fingerprint images before minutiae extraction.

Peratla et al. [11] used the convolutional neural networks directly on the input images fingerprint without using an explicit feature extraction step. They proposed an architecture similar to CaffeNet a variant of the famous AlexNet which achieved the best accuracy rate on a challenging dataset ImageNet. They achieved an accuracy of 90.24% over NIST-DB4.

Michelsanti et al. [10] proposed also to use convolutional neural networks directly on the input images fingerprint by transfer learning with VGG-f architecture. They reported an accuracy of 94.4% over NIST-DB4. But theses two previous CNN approaches suffer from the high complexity level of the training stage.

Our motivation in this work is to investigate the association of Conic Radon transform for feature extraction and the convolutional neural networks applied on Radon space.

3   Proposed Approach

3.1   Conic Radon transform

Let us first recall the classical Radon transform (RT) [1]. The RT in Euclidean space integrates a function $f(x, y)$ over lines as defined in this equation:

$$Rf(\rho, \phi) = \int_{-\infty}^{+\infty} \int_{-\infty}^{+\infty} f(x, y)\delta(\rho - xcos(\phi) - ysin(\phi))dxdy, \qquad (1)$$

where $\delta(.)$ is the Dirac delta function, $\rho \in]-\infty, +\infty[$ is the distance between the origin of the coordinate system and the line and $\phi \in [0, \pi[$ is the orientation of the line.

The result of the RT is a (ρ, ϕ) space where a peak implies the presence of a line in the original image by extracting the corresponding line parameters ρ and ϕ. (Fig. 2).

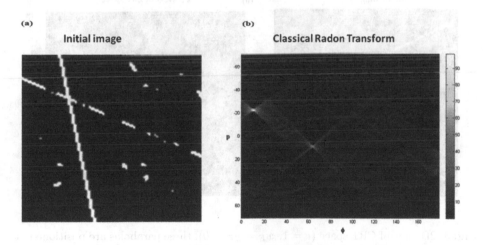

Fig. 2. The classical RT of the image represented in (a): the coordinates of each peak corresponds to polar parameters of lines.

Despite the success of RT for linear features detection, it can not detect arbitrary curves. For this purpose, The CRT was defined to detect the conic sections from the image independently of its locations, parameters and orientations [6].

A conic section in polar coordinates system of focus at the origin , is defined as: for $M(r, \theta)$:

$$r = \frac{\rho}{1 + e\ cos(\theta - \phi)}, \qquad (2)$$

where ρ is the conic parameter, ϕ is an angle corresponding to the orientation of the conic and e is the eccentricity. The curve is an ellipse, a parabola, or a hyperbola, according to whether its e value is less than, equal to, or greater than 1.

The generalized Radon transform represents the integration of a function $f(x, y)$ over conic sections in the plane. It is defined as:

$$R_c f(x_F, y_F, \rho, \phi, e) = \int_\gamma$$
$$f\left(\frac{\rho}{1 + e\ cos(\gamma)} cos(\gamma + \phi) + x_F, \frac{\rho}{1 + e\ cos(\gamma)} sin(\gamma + \phi) + y_F\right)$$
$$\rho \frac{\sqrt{1 + e^2 + 2\ e\ cos(\gamma)}}{(1 + e\ cos(\gamma))^2} d\gamma. \tag{3}$$

where $\gamma = \theta - \phi$. Therefore the CRT space is represented by five parameters which are the coordinates of the focus x_F and y_F, the conic parameter ρ, the orientation angle ϕ and the eccentricity e.

Figures 3 and 4 illustrate the Conic Radon transform on synthetic images which contains parabolas and circles.

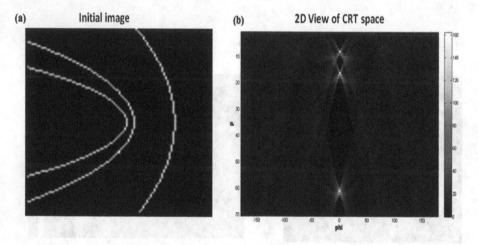

Fig. 3. 2D view of CRT space ($e = 1$, $x_F = y_F = 0$). three parabolas are positioned at the center row. The result of the CRT on (a) is the $(x_F, y_F, \rho, \phi, e)$ parameters space where (b) present 2D view of CRT space ($e = 1$, $x_F = y_F = 0$). and the coordinates of each of the peaks corresponds to the parameters (ϕ, ρ) of the curves. ρ is the distance of focus to directrix and ϕ is the orientation of parabola.

In this work, the basic idea of our proposed method is to use the CRT in order to extract parabolas and circles from fingerprint images which could be used further as an input for CNN.

According to the great presence of parabolic and circular forms in fingerprint image, we have chosen to integrate the image function along a class of parabolas

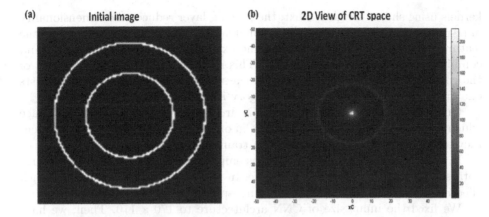

Fig. 4. (a) Initial image where two concentric circles are presented, (b) 2D view shows the maximum values of CRT at the points corresponding to the position of the center of circle corresponding to fixed radius ($p = 40$).

which corresponds to the special case of $e = 1$ and a fixed reference focus (x_F, y_F). We integrate also along a class of circles ($e = 0$) with fixed center of circles.

We apply CRT directly on original images with no added processing.

Looking at the overall structure of a fingerprint, we note that the parabolic and circular forms are associated to the core point of the fingerprint. Therefore, We have chosen to put the core point as the reference focus of parabolas as well as the center of circle. This core point is located in the top area of the innermost ridge line. We have adopted the algorithm presented in [16] to detect this core point. We integrate further along lines to detect the discontinuities of lines existing in fingerprint images.

The CRT over parabolas, circles offers excellent properties like the classical RT which are useful for fingerprint classification. The most important ones are the invariance of the Radon space to geometric transformations such as rotation and translation which could enhance the classification task.

The fingerprint is hence represented by its related linear, parabolic and circular Radon spaces where the goal of the concatenation step is to extract discriminant information that improves the fingerprint classification accuracy. The concatenation of these Radon spaces is then used as the input image for the CNN instead of the original image.

3.2   Fingerprint Classification with Deep Neural Networks

The Convolutional Neural Networks (CNN) architecture consists of a multi-stage processing of an input image to extract high-level feature representations.

Feature extraction stage is achieved by convolutional and pooling layers. These features are weighted and combined in the fully-connected layer to realize the classification. Finally there exists one output neuron for each object category in the output layer. The input are convolved with a predefined number of kernels using shared weights. Next, the pooling layer reduces the dimensionality of responses produced by the convolutional layers. These two layers compose the feature extraction stage. Afterwards, the extracted features are weighted and combined in the fully-connected layer. This represents the classification part of the convolutional network following by classification layer. Finally there exists one output neuron for each object category in the output layer.

Our purpose is to define a new CNN architecture applied on Radon space which is derived from fingerprint images in order to reach a better level of accuracy and reduce the computing time on training phase.

In general, deep learning applications employ complex architectures on huge data sets to get a great recognition rate. In this work, we have aimed to reach effective rates via small CNN architecture applied on only 4000 images.

We fixed the input size of CNN architecture to 170×170. Then, we have trained our architecture in order to generate our model. The input image of the network is sized of $170 \times 170 \times k$, where $k = 1$ for gray patches.

The proposed CNN architecture is composed of four convolutional layers with three max-pooling layers followed by fully-connected layer. The softmax output activation presents 4 fingerprint classes. The output is one of the 4 fingerprint classes. The proposed architecture of the convolutional neural network is shown in Fig. 5.

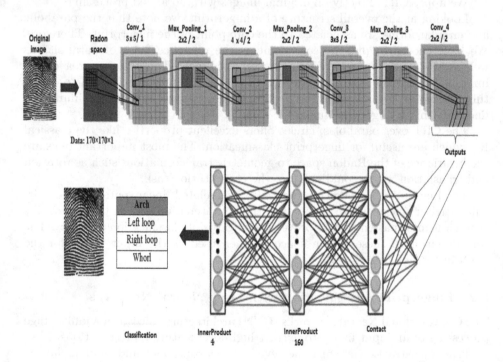

Fig. 5. Architecture of the convolutional neural network.

For the network configuration, the convolutional and max_Pooling layers are applied with different kernel sizes (5×5, 4×4, 3×3, 2×2) and different strides $(1, 2)$. The Dropout layers are used in order to avoid the issue of over fitting to the training data.

Table 1 shows the network configuration, the patch size of the input data is 170×170. Then, the convolutional and max_Pooling layers are applied with different kernel sizes (5×5, 4×4, 3×3, 2×2) and different strides $(1, 2)$.

4   Experimental Results

To evaluate the performance of our approach for fingerprint classification, we carried out experiments on the NIST Special Database 4 (NIST SD4) which is the most used benchmark for the fingerprint classification. The data base is composed of 4000 grayscale fingerprint images divided into five classes: Arch (A), Tented arch (T), Left loop (L), Right loop (L), and Whorl (W) (Table 2).

In order to perform a qualitative analysis, we have input to our system the same repartition of datasets as those used in previous methods. The data set is composed of 2000 fingerprints. Two impressions per fingerprint are provided in the data set. The fingerprints are then numbered from f0001 to f2000 and from s0001 to s2000.

Table 1. Network configuration.

Layer type	Size/Stride	Output
		Dropout probability
Data	170×170	-
Convolution_1	$5 \times 5/1$	20
Max_Pooling_1	$2 \times 2/2$	-
Convolution_2	$4 \times 4/2$	40
Max_Pooling_2	$2 \times 2/2$	-
Convolution_3	$3 \times 3/2$	60
Convolution_4	$2 \times 2/2$	80
Max_Pooling_3	$2 \times 2/2$	-
Fully connected	-	160 Dropout $= 0.5$
Fully connected	-	4

Table 2. Training parameters.

Algorithm	Parameters
SGD (Stochastic Gradient Descent)	Batch size $= 20$, Iterations $= 100000$, Learning rate $= 0.001$, Momentum $= 0.9$

The images from f0001 to f1000 and s0001 to s1000 are selected for the training set. The remaining instances constitute the classification set. However, there exist fingerprints which are considered ambiguous since they are assigned to two different labels. Therefore, we used only one label for those fingerprints.

In order to verify the performance of the proposed approach, either the accuracy and the error rate are measured. They are defined by:

$$Error\ rate = \frac{fp \times 100}{tp}\% \tag{4}$$

$$Accuracy = 100\% - Error\ rate \tag{5}$$

where fp is the number of misclassified fingerprints and tp is the total number of fingerprints.

For the experiments we worked on CAFFE framework deployed on ubuntu 14.04 LTS and a GeForce GT 525M GPU, which has 2 GB of memory. We trained the dataset in 100000 iterations, a training error of around 2% has been reached.

Figure 6 illustrates the confusion matrix on NIST SD4. We achieved 95% as recognition rate. We can observe that our proposed method achieves good performance on the four classes. But it can be seen that the network has the worst results in classifying right-loop fingerprints correctly (around 7% of them are misclassified).

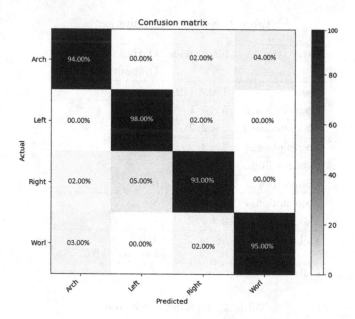

Fig. 6. Confusion matrix on NIST SD4 data set.

To evaluate our method we have carried out comparisons with similar works in the literature [3,10,15]. We give the results of error rate reported by other researchers along with the result of our proposed approach in Table 3. Our method is competitive with all the previous approaches. It has shown nearly similar accuracy results compared to other successful methods. It can be observed from the driven results that the features learned through the association of RT and CNN contain discriminative information for fingerprint classification.

Table 3. Error rate of fingerprint classification methods on NIST SD4

Approach	Error rate
OI+NN (PCASYS) [2]	11.4 %
HMM+DT+ PCASYS [14]	5.1 %
RF+OI+ CF+KNN+SVM [3]	2.8%
OI+DNN [15]	6.9 (rejection = 1.8 %)
CNN (CaffeNet model) [11]	9.7%
CNN (VGG-F model) [10]	5.6 %
Our proposed method	5 %

RF: Ridge Line Flow, OI = Orientation Image, SP: Singular Point, CF: Complex Filter, HMM: Hidden Marcov Model, KNN: K-Nearest Neighbor, SVM: Support Vector Machine, NN: Neural Network, DT: Decision Tree, DNN: Deep Neural Network.

It can be seen that the work reported by Cao et al. [3] achieves the best error rate on NIST SD4. It outperforms the other methods in terms of accuracy. It is composed of fives stages. Each stage distinguishes one class from others using different features extraction methods and finally SVM is adopted to make the final classification. However, the main issue with this method is that it takes a slow time computing: feature extraction time of 880 ms and an average classification time of 3.43 s on a 3.4 GHz Intel Pentium 4 processor.

The main contribution of our work is to achieve both a good accuracy and a low computational complexity. In fact, We have got a good accuracy with our small CNN architecture (only 8 layers) applied on only 4000 images and without augmentation data.

Those promising results driven from a small architecture are due to the Radon operation which concentrates patterns into points. It is then simpler for a classifier to learn about points than about full patterns. Our method is able to classify a fingerprint image in 60 ms within our hardware configuration. We take into account that our test hardware configuration is the less powerful among the hardware configurations of compared methods. We obtain an average Radon feature extraction time of 20 ms and 40 ms for CNN architecture. We have reduced the CNN layers number which counts 8 layers versus 17 in CNN architectures using transfer learning with VGG-F model presented in [10]. This method reported an average run-time for one fingerprint test of 71 ms on NVIDIA GTX 950M GPU which is eventually faster than our configuration.

Nevertheless, our method rates could be improved if some pre-processing on the Radon space are performed to enhance the extracted peaks representing the parabolic features.

5   Conclusion

In this work, we have proposed a new approach for fingerprint classification that combines the Conic Radon transform and the Convolutional neural networks. We showed that the performance obtained with our fingerprint method are close to the state-of-the-art. The main advantage of our approach is that it does not require a heavy computing time in both training and test phase as in the other related works, even though other methods are directly applied on original images.

References

1. Radon, J.: Über die Bestimmung von Funktionen durch ihre Integralwerte längs gewisser Mannigfaltigkeiten. Akad. Wiss. **69**, 262–277 (1917)
2. Candela, G.T., Grother, P.J., Watson, C.I., Wilkinson, R., Wilson, C.L.: Pcasys - a patternlevel classification automation system for fingerprints. NIST technical report NISTIR, 5647 (1995)
3. Cao, K., Pang, L., Liang, J., Tian, J.: Fingerprint classification by a hierarchical classifier. Pattern Recognit. **46**(12), 3186–3197 (2013). http://www.sciencedirect.com/science/article/pii/S0031320313002124
4. Goodfellow, I.J., Bengio, Y., Courville, A.C.: Deep Learning. Adaptive Computation and Machine Learning. MIT Press, Cambridge (2016). http://www.deeplearningbook.org/
5. Haddad, Z., Beghdadi, A., Serir, A., Mokraoui, A.: Fingerprint identification using radon transform. In: Image Processing Theory, Tools and Applications 2008, pp. 1–7 (2008)
6. Hamdi, D.E., Nguyen, M.K., Tabia, H., Hamouda, A.: Image analysis based on radon-type integral transforms over conic sections. In: Proceedings of the 13th International Joint Conference on Computer Vision, Imaging and Computer Graphics Theory and Applications (VISIGRAPP 2018), vol. 4, VISAPP, Funchal, Madeira, Portugal, 27–29 January 2018, pp. 356–362 (2018). https://doi.org/10.5220/0006613403560362
7. Krizhevsky, A., Sutskever, I., Hinton, G.E.: Imagenet classification with deep convolutional neural networks. Commun. ACM **60**(6), 84–90 (2017). https://doi.org/10.1145/3065386
8. Li, J., Yau, W., Wang, H.: Combining singular points and orientation image information for fingerprint classification. Pattern Recognit. **41**(1), 353–366 (2008). https://doi.org/10.1016/j.patcog.2007.03.015
9. Maltoni, D.: Fingerprint recognition, overview. In: Li, S.Z., Jain, A.K. (eds.) Encyclopedia of Biometrics, pp. 510–513. Springer, Boston (2009). https://doi.org/10.1007/978-0-387-73003-5
10. Michelsanti, D., Guichi, Y., Ene, A.D., Stef, R., Nasrollahi, K., Moeslund, T.: Fast fingerprint classification with deep neural network. In: VISAPP - International Conference on Computer Vision Theory and Applications (2017)

11. Peralta, D., Triguero, I., García, S., Saeys, Y., Benítez, J.M., Herrera, F.: On the use of convolutional neural networks for robust classification of multiple fingerprint captures. Int. J. Intell. Syst. **33**(1), 213–230 (2018). https://doi.org/10.1002/int.21948

12. Schuch, P., Schulz, S., Busch, C.: De-convolutional auto-encoder for enhancement of fingerprint samples. In: Sixth International Conference on Image Processing Theory, Tools and Applications, IPTA 2016, Oulu, Finland, 12–15 December 2016, pp. 1–7 (2016). https://doi.org/10.1109/IPTA.2016.7821036

13. Schwenk, H., Barrault, L., Conneau, A., LeCun, Y.: Very deep convolutional networks for text classification. In: Proceedings of the 15th Conference of the European Chapter of the Association for Computational Linguistics, EACL 2017, Valencia, Spain, 3–7 April 2017, vol. 1: Long Papers, pp. 1107–1116 (2017). https://aclanthology.info/papers/E17-1104/e17-1104

14. Senior, A.: A combination fingerprint classifier. IEEE Trans. Pattern Anal. Mach. Intell. **23**, 1165–1174 (2001). https://doi.org/10.1109/34.954606

15. Wang, R., Han, C., Wu, Y., Guo, T.: Fingerprint classification based on depth neural network. CoRR abs/1409.5188 (2014). arXiv:1409.5188

16. Zhu, E., Guo, X., Yin, J.: Walking to singular points of fingerprints. Pattern Recognit. **56**, 116–128 (2016). http://www.sciencedirect.com/science/article/pii/S0031320316000868

NoiseNet: Signal-Dependent Noise Variance Estimation with Convolutional Neural Network

Mykhail Uss[1], Benoit Vozel[2(✉)], Vladimir Lukin[1], and Kacem Chehdi[2]

[1] Department of Information and Communications Technology, National Aerospace University, Chkalova Street, Kharkov 61070, Ukraine
uss@xai.edu.ua,lukin@ai.kharkov.com
[2] University of Rennes 1, IETR UMR CNRS 6164, 22300 Lannion, France
{benoit.vozel,kacem.chehdi}@univ-rennes1.fr

Abstract. In this paper, the problem of blind estimation of uncorrelated signal-dependent noise parameters in images is formulated as a regression problem with uncertainty. It is shown that this regression task can be effectively solved by a properly trained deep convolution neural network (CNN), called NoiseNet, comprising regressor branch and uncertainty quantifier branch. The former predicts noise standard deviation (STD) for a 32×32 pixels image patch, while the latter predicts STD of regressor error. The NoiseNet architecture is proposed and peculiarities of its training are discussed. Signal-dependent noise parameters are estimated by robust iterative processing of many local estimates provided by the NoiseNet. The comparative analysis for real data from NED2012 database is carried out. Its results show that the NoiseNet provides accuracy better than the state-of-the-art existing methods.

Keywords: Blind noise parameter estimation
Signal-dependent noise · Convolutional neural network
Experimental data · Deep regression with uncertainty

1 Introduction

Images obtained with consumer digital cameras or specialized remote sensing sensors are subject to degradations deteriorating their quality. These degradations include but are not limited to image blur, compression artifacts, demosaicing artifacts, and sensor noise [3,10]. Knowing the sensor noise parameters is needed for successful image quality improvement via post-processing, e.g. image deblurring [2] or filtering [6]. However, such parameters may not be provided by a sensor manufacturer, may be dependent on unknown factors during image acquisition, or vary from camera to camera. In such a condition, noise parameters are to be estimated directly from degraded images in the so-called blind manner [3,4,7,12].

© Springer Nature Switzerland AG 2018
J. Blanc-Talon et al. (Eds.): ACIVS 2018, LNCS 11182, pp. 414–425, 2018.
https://doi.org/10.1007/978-3-030-01449-0_35

It should be stressed from the very beginning that there are different types of noise (additive, signal-dependent, multiplicative, impulsive, mixed, spatially uncorrelated or correlated) and, thus, different noise parameters have to be determined by a blind noise parameter estimation (BNPE) method. For example, in the case of synthetic aperture radar images, noise is pure multiplicative. Other examples are images formed by modern hyperspectral sensors or raw data from digital cameras for which it is commonly accepted nowadays that noise is signal-dependent [3,7,11,13]. For signal-dependent noise, its variance depends on image intensity as $\sigma_n^2(I) = \sigma_{SI}^2 + k_{SD}\,I$, where σ_{SI}^2 is signal-independent noise variance component, $k_{SD}\,I$ is signal-dependent noise variance component, k_{SD} is proportionality coefficient. The function $\sigma_n^2(I)$ is called Noise Level Function, NLF [4]. Below we concentrate on considering the practically important signal-dependent spatially uncorrelated noise NLF.

There exist several groups of BNPE methods. A first group performs in spatial domain and tries to get a set of normal local estimates in image homogeneous regions with further rejection of possible abnormal estimates and estimation of signal-dependent noise parameters by robust regression. Such methods are usually fast enough but they have problems with highly textural images for which estimates of the noise parameters can be sufficiently biased [1]. A second group of techniques performs in spectral domain and it is based on ability of orthogonal and other transforms to represent data sparsely [5]. These methods are able to separate noise and signal better and have less problems with textural images. Finally, a third group of model-based methods employs maximum likelihood estimation of noise and texture parameters in blocks [12,13]. It produces the most accurate estimation [14] but requires quite extensive computations.

The majority of BNPE methods operates on image patches (blocks, fragments): a processed image is split into overlapping or non-overlapping patches; in each patch, noise is modeled as pure additive; patches that are not suitable for noise STD estimation (according to some tests or rules) are detected and rejected; noise STD for the remaining patches is estimated; noise NLF parameters are estimated by a robust regression to diminish the influence of possible outliers (Fig. 1a). The key challenges with this approach is (1) how to find image patches that are suitable for noise STD estimation and (2) how to effectively estimate noise STD for the selected patches.

In our previous work [14], it was proposed to solve problems (1) and (2) using the model-based iterative approach (Fig. 1b). According to this approach, in each patch, image texture parameters are estimated using a suitable model (Fig. 1c). The estimated texture model is used to both estimate noise STD and uncertainty of this estimate (in the form of Cramer-Rao lower bound, CRLB). Noise STD uncertainty estimation allows identifying so called Noise Informative (NI) patches that are the most suitable for noise STD estimation. This approach is necessarily iterative as characterization of a patch uncertainty requires knowledge of both noise and image texture parameters. The NI+DCT BNPE proposed in [14] is an implementation of such an iterative scheme. It uses Discrete Cosine Transform (DCT) for patch noise STD estimation and Fractal Brownian motion (fBm) model for uncertainty calculation. The NI+DCT was shown to be the

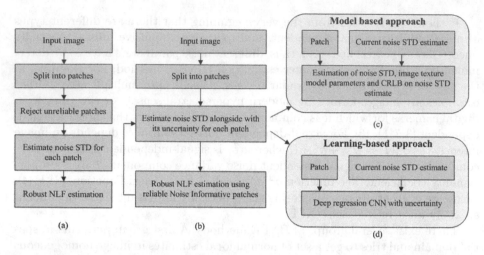

Fig. 1. The basic and proposed image noise NLF estimation pipeline: non-iterative approach (a); iterative approach (b); model-based patch processing (c); patch processing with the NoiseNet CNN (d)

most effective BNPE of signal-dependent NLF for NED2012 database of real images introduced in [14]. Its drawback is high computational complexity and reduced efficiency for images with textures different from assumed fBm model.

The motivation behind the present work is to benefit from high degree of adaptivity of CNN to image content [8] to improve signal-dependent NLF estimation accuracy. Our contribution is to replace model-based processing of a single patch in Fig. 1b to CNN-based processing (Fig. 1d). This task is challenging as it requires solving with CNN not only regression problem but also characterization of regression uncertainty. To solve it, we use recent advances in deep regression uncertainty calculation [9] using multitask CNN. The proposed patch processing CNN is called NoiseNet. It is integrated in the same processing pipeline as in Fig. 1b. The resulting signal-dependent noise parameter estimator will be later referred to as NI+NoiseNet (similar to NI+DCT).

The remaining part of the paper is organized as follows. Section 2 introduces the NoiseNet CNN structure, training dataset and validates the training results. Experimental Sect. 3 details the process of signal-dependent noise parameters estimation using NI+NoiseNet and application to NED2012 image database. It is shown that for NED2012 database, the NI+NoiseNet favorably compares to the NI+DCT methods with respect to both accuracy and time complexity. Finally, in Sect. 4, we summarize our findings and give few remarks on future work.

2   Noise STD and Its Uncertainty Estimation with the NoiseNet CNN

This section introduces structure of the proposed NoiseNet CNN that estimates additive noise STD and this estimate uncertainty from a noisy image patch.

The usage of the NoiseNet for more complex signal-dependent noise model is explained in the next Section.

2.1   The NoiseNet CNN Architecture

The proposed NoiseNet CNN comprises feature extraction part, regression branch and uncertainty quantifier branch (Fig. 2). Feature extraction is done with three convolutional blocks connected by 2×2 max pooling layers. Each convolutional block has two convolutional layers with ReLu (Rectified Linear Unit) activation function. With input patch of 32 by 32 pixels, the output feature vector is $128 \times 2 \times 2$. It is flattened to 512 feature vector before being fed to regressor and uncertainty quantifier branches. The regressor branch estimates noise STD $\hat{\sigma}_n$. It comprises two dense fully convolutional (FC) layers with ReLu activation (assuring positiveness of noise STD estimates). It was found that estimation of noise STD instead of variance significantly stabilizes NoiseNet training process. The uncertainty quantifier has the same structure as the regressor except of usage of an external estimate of noise STD, $\hat{\sigma}_{n.ext}$, which is the true noise STD $\sigma_{n.GT}$ during training. This estimate is fed to FC layer and added to the input of the last quantifier FC layer. A small constant ϵ (ϵ is set to 0.001 in our experiments) is added to the uncertainty quantifier output to prevent division by zero in loss function. By noise STD estimate uncertainty we understand relative STD of $\hat{\sigma}_n$: $\sigma_{uncert} = STD(\hat{\sigma}_n/\hat{\sigma}_{n.GT})$. Absolute STD of $\hat{\sigma}_n$ is $STD(\hat{\sigma}_n) = \sigma_{uncert} \hat{\sigma}_{n.GT}$.

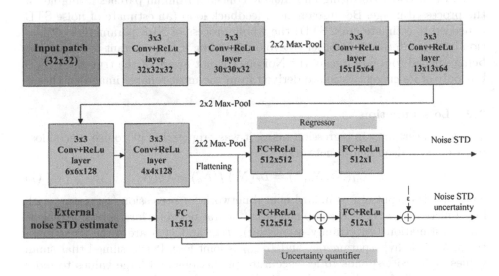

Fig. 2. The NoiseNet structure

Input image patch is preliminary normalized to zero mean and unit variance by $I_{norm} = \frac{I - \bar{I}}{\sigma_I}$, where $\bar{I}$ is mean patch intensity and σ_I is patch standard deviation. To reverse patch normalization, $\hat{\sigma}_n$ should be multiplied by σ_I. Uncertainty quantifier output is not affected by input patch normalization.

Selection of the input patch size is governed by two contradictory factors: image homogeneity and amount of information about image texture. A smaller patch size increases its homogeneity, but provides less information about image texture needed to accurately estimate noise STD and its uncertainty. Experimentally it was found that 32 by 32 pixels patch provides a satisfactory compromise.

The importance of supplying an external noise STD estimate into uncertainty branch is explained as follows. Uncertainty quantifier helps selecting those image patches that are Noise Informative, i.e. provide reliable noise STD estimates. These patches are mostly homogeneous with noise level higher or comparable to image texture level. Unreliable patches are heterogeneous with image texture dominating over noise level. Uncertainty of noise STD estimation depends on both noise STD and image texture parameters in a complex way [14]. If external noise STD is not provided, uncertainty quantifier branch has to estimate it by itself. For heterogeneous patches, noise STD cannot be accurately estimated making uncertainty calculation unreliable. As a result, uncertainty branch will fail for exactly those patches for which its output is the most valuable. Situation can be improved if we take into account that image NLF is estimated using a lot of patches. Among these patches there exist both homogeneous and heterogeneous ones; the former are used to estimate noise NLF. Feeding this estimate into the uncertainty quantifier branch makes it more reliable. It can be said that external noise STD estimate helps regularizing processing of a single image patch by the NoiseNet using information collected from all patches available for the processed image. Because of these feedback loop (an estimate of noise STD is needed to estimate noise STD) the whole process of NLF parameters estimation becomes iterative. In order to avoid confusion, it is important to stress that being applied to a real image, the NoiseNet is not aware of the true noise STD but only of its current estimate derived from the most reliable image patches.

2.2   Loss Function

To learn uncertainty in regression tasks, it was proposed in [9] to use special loss function with deep neural networks:

$$L_{joint}(N_r, N_q, \lambda) = L_r(N_r)\, f(N_q) + \lambda\, g(N_q), \tag{1}$$

where N_r is output of a standard neural network for regression (regressor) with loss function L_r, N_q is a complementary neural network for regression uncertainty estimation (uncertainty quantifier), $f(z)$ and $g(z)$ are some fixed functions, $\lambda > 0$ is hyperparameter, and L_{joint} is joint loss. It is assumed that small values of N_q corresponds to less accurate predictions and large values to more accurate predictions. As it is shown in [9], by training jointly the regressor and uncertainty quantifier with loss function (1), the uncertainty quantifier learns to predict the expected regressor's loss as:

$$expected\,regressor's\,loss = -\frac{\lambda\,g^{'}(N_q)}{f^{'}(N_q)}, \tag{2}$$

In the NoiseNet, we use a particular form of the loss function (1) with $f(z) = z$, $g(z) = -log(z)$, $\lambda = 1$. In this case, N_q corresponds to inversed variance of regressor output: $N_q = (\sigma_{uncert} \sigma_{n.GT})^{-2}$. The NoiseNet loss function takes the following form:

$$L = \left(\frac{\hat{\sigma}_n - \sigma_{n.GT}}{\sigma_{uncert} \sigma_{n.GT}} \right)^2 + 2 \, ln(\sigma_{uncert} \sigma_{n.GT}) \tag{3}$$

where $N_r = \hat{\sigma}_n$ is output of regressor, i.e. noise STD estimation, σ_{uncert} is predicted STD of $\hat{\sigma}_n$ relative to $\sigma_{n.GT}$.

2.3 Training Process

Training and validation dataset consist of 158 raw images taken by Nikon D80 DSLR camera. Ground truth signal-dependent NLF parameters for this particular camera were estimated using calibration images (defocused images of white sheet of paper taken at different ISO values and illumination levels). For each raw image, 3120 non-overlapping patches of 32 by 32 pixels size were selected from red, green and blue channels (demosaicking procedure was not applied). Ground truth noise STD for each patch was estimated as $\sigma_{n.GT} = \sqrt{\sigma_{SI.GT}^2 + k_{SD.GT} \, \bar{I}}$. To account for very sharp image textures, patches were also taken from raw images scaled with a factor $s = 2$ and 4 (scaling was performed by applying s by s averaging filter and subsampling of filtered image with step size s. Ground truth noise STD was correspondingly reduced by a factor of s). Patches affected by clipping effect were not included in the training data. In average, 3540 patches were collected from each raw image, and 552093 in total from the whole dataset. In addition, 138023 (25%) pure noise patches were added. This amount was randomly split into 552092 (80%) training and 138024 (20%) validation patches.

The NoiseNet training was performed on NVIDIA GeForce GTX 960M GPU. One epoch takes approximately 5 min to complete. We used Adam optimizer with learning rate $5 \cdot 10^{-6}$ and decay factor 10^{-6}. To stabilize convergence, 12 kernel regularizers were used for all convolutional and dense layers with weight regularization penalty of 0.025. It was found that training process stabilizes after 250 epochs. The total training time is, therefore, about 22 hours.

Ideally, the regressor should predict noise STD with error $\hat{\sigma}_n - \sigma_{n.GT}$ that follows normal distribution with zero mean and variance predicted by uncertainty quantifier as $\sigma_{uncert}^2 \sigma_{n.GT}^2$. Other way to express this is that relative regressor error $e_{rel} = (\hat{\sigma}_n - \sigma_{n.GT})/\sigma_{n.GT}$ follows $N(0, \sigma_{uncert}^2)$ distribution or normalized error $e_{norm} = (\hat{\sigma}_n - \sigma_{n.GT})/(\sigma_{n.GT} \sigma_{uncert})$ follows $N(0,1)$ distribution.

Probability density function (pdf) of relative error as a function of σ_{uncert} is shown in Fig. 3a (each image row corresponds to the relative error pdf for a particular σ_{uncert} value). It is seen that for $\sigma_{uncert} < 0.2$, the relative error closely follows normal distribution with zero mean and standard deviation proportional to σ_{uncert}. For $\sigma_{uncert} > 0.2$, the distribution becomes non-Gaussian with mean shifted from zero. In the experimental part, we treat all noise STD estimates with $\sigma_{uncert} > 0.2$ as unreliable and discard them from consideration.

Loss function (1) does not implicitly require the normalized error to follow N(0,1) but only its variance to be close to unity. Actual experimental pdf of the normalized error calculated for validation data with $\sigma_{uncert} < 0.2$ is given in Fig. 3b. The pdf is close to N(0.035, 0.96). This result validates that the NoiseNet uncertainty quantifier is able to correctly predict accuracy of noise STD estimates provided by the NoiseNet regressor branch.

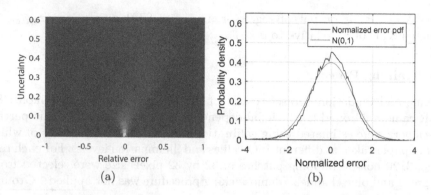

Fig. 3. Experimental distribution of the regressor error on validation data: the relative error pdf (a), the normalized error pdf (b)

Let us also check performance of the NoiseNet for pure homogeneous image patches with additive noise with unit STD and no image texture. The mean value of noise STD estimated by the NoiseNet averaged over 10000 samples is 0.998, i.e. the estimation bias is -0.2%. The corresponding mean value of σ_{uncert} is 0.02057. This is close to STD of sample STD estimate for 32 by 32 image patch equal to $1/(32\sqrt{2}) = 0.0221$.

3   Experimental Analysis

3.1   Test Data

Performance of the NoiseNet was tested on NED2012 image database. This database comprises 25 raw images from the same D80 camera that was used to collect training and validation data (however, training, validation and test sets do not intersect). The images of NED2012 cover a wide variety of scenes and textures: nature, urban scenes, aerial images, sky, land, water, trees, grass, sand, asphalt, stones. The database includes both homogeneous and highly textured images.

3.2   Patch-Based Estimation of Signal-Dependent Noise Parameters for a Single Image

As it was mentioned in the introduction, the NI+NoiseNet borrows general noise parameter estimation framework from the NI+DCT and replaces patch processing by the NoiseNet CNN. The particular implementation of this iterative estimation approach for 12bit D80 raw images includes the following stages. Image is split into overlapping 32 by 32 patches with step size 9 pixels w.r.t. rows and columns. Patches affected by clipping effect are discarded. All remaining patches are grouped into 50 bins uniformly covering image intensity variation range. In each bin, a particular number of the most uniform patches (with the lowest STD) are drawn. We used in total 2000 patches or 80 patches for each bin in average. Using more patches increases processing time but have no significant effect on NLF estimation accuracy.

Initial noise STD for each patch is set equal to 100 (upper bound of expected noise STD for D80 camera). The NoiseNet is applied to each patch to estimate noise STD and its uncertainty. Using estimates with $\sigma_{uncert} < 0.2$, the NLF parameters, $\hat{\sigma}_{SI}$ and $\hat{k}_{SD}$, are refined by robust linear weighted regression (regression problem is heteroscedastic as each STD estimate is characterized with its own uncertainty). Application of the NoiseNet and robust regression are iteratively repeated using previous NLF estimate as external noise STD source. We limited maximum number of iterations to 5. Source code of NI+NoiseNet for both training and evaluation is available at https://github.com/radiuss/NoiseNet

3.3   Analysis of the NI+NoiseNet Performance on NED2012 Database

In [14], it was demonstrated that the NI+DCT method is the best performing BNPE method for NED2012 database. Estimation results for NI+DCT and NI+NoiseNet are compared in Fig. 4 and Table 1. Quantitatively, accuracy of signal-dependent noise parameter estimation is characterized as standard deviation of relative errors, STD_{REL} and robust standard deviation of relative errors, STD_{REL}^{MAD}. The latter is measured as Median Absolute Deviation (MAD) of relative errors multiplied by normalization factor of 1.48. Here, relative error for signal-independent component is defined as $e_{SI.rel} = (\hat{\sigma}_{SI}^2 - \sigma_{SI.GT}^2)/\sigma_{SI.GT}^2$ and for signal-dependent component as $e_{SD.rel} = (\hat{k}_{SD}^2 - k_{SD.GT}^2)/k_{SD.GT}^2$. The difference between STD_{REL} and STD_{REL}^{MAD} is that the former one takes into account both normal and outlying errors, while the latter one measures standard deviation for only normal errors being robust to outliers.

According to Table 1, compared with the NI+DCT, the NI+NoiseNet improves SI component estimation accuracy by about 40% with respect to both STD_{REL} and STD_{REL}^{MAD}. Accuracy of SD component is improved by 55% with respect to STD_{REL}, and by 34% with respect to STD_{REL}^{MAD}.

These results reveal that NI+NoiseNet outperforms NI+DCT in two ways: it reduces number of outlying estimates and variance of normal estimates of noise components. For SI component, magnitude of the largest relative error seen for the NI+DCT is about 300%, while for the NI+NoiseNet the magnitude of the largest error does not exceed 120%. For SD component, the NI+NoiseNet relative error magnitude is less than 12%, while for the NI+DCT this threshold is exceeded for many images with the largest relative error reaching 40%. By comparing histograms of relative errors for SI noise component (Fig. 4a and c)

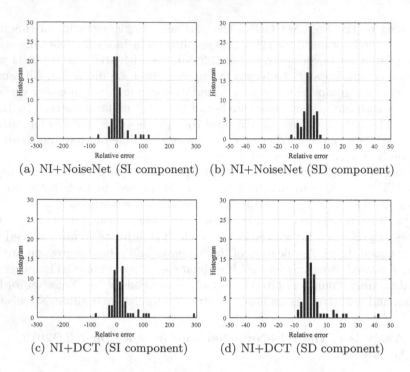

(a) NI+NoiseNet (SI component)   (b) NI+NoiseNet (SD component)

(c) NI+DCT (SI component)   (d) NI+DCT (SD component)

Fig. 4. Comparison of signal-dependent NLF estimation accuracy for NI+NoiseNet and NI+DCT methods for NED2012 database: signal-independent component (a), (c), signal-dependent component (b), (d)

Table 1. Relative error of signal-dependent noise parameters estimation by NI+NoiseNet and NI+DCT methods for NED2012 database

Method	SI component		SD component	
	STD_{REL}	STD_{REL}^{MAD}	STD_{REL}	STD_{REL}^{MAD}
NI+DCT	44.8538	20.0276	7.3907	3.6046
NI+NoiseNet	25.4431	12.4357	3.3497	2.3730

and SD component (Fig. 4b and d), it is seen that relative errors for the NI+NoiseNet are more concentrated around zero as compared to the NI+DCT.

Detailed comparison of NI+NoiseNet and NI+DCT for two challenging images from NED2012 is given in Fig. 5. For image ♯22 (Fig. 5e), the NI+DCT shows the largest SI component estimation error of about 300%. This is caused by the lack of reliable noise STD estimates for patches representing the darkest image parts - heterogeneous tree texture. Only few estimates are found by the NI+DCT for intensity less than 500 (individual estimates are shown as black dots in Fig. 5c). In contrast, the NI+NoiseNet identifies a larger amount of reliable patches that stabilizes the NLF estimate (Fig. 5a). For brighter areas of image ♯22 (e.g. sky), the NI+NoiseNet identifies a larger number of reliable patches and

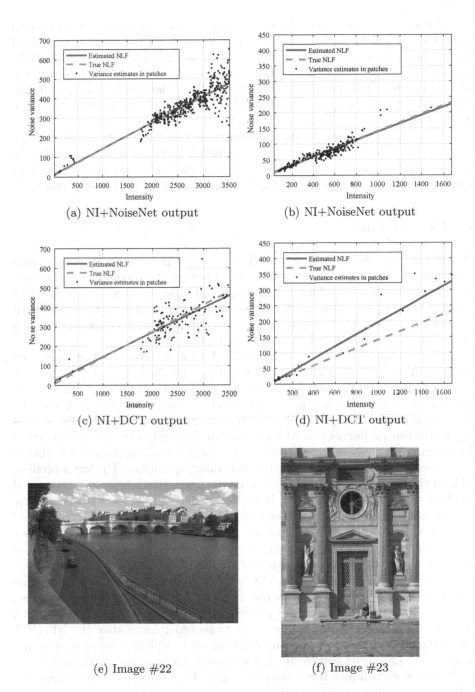

Fig. 5. Comparison of signal-dependent noise estimation accuracy by NI+NoiseNet and NI+DCT methods for challenging images from the NED2012 database: NI+NoiseNet output (a), (b); NI+DCT output (c), (d); original images, (e), (f)

provides more accurate noise STD estimate for them (more tightly concentrated around the true NLF) as compared to NI+DCT. Image ♯23 (Fig. 5f) is covered with stone texture and has no homogeneous areas. For this image, the NI+DCT shows the largest SD component estimation error of about 40%. In this case, the NI+DCT fails to find a sufficient number of reliable patches (Fig. 5d), while the NI+NoiseNet identifies plenty of them (Fig. 5b) in the darker parts of the image that correspond to shadows.

The provided quantitative data demonstrate that for a wide range of image textures, the NI+NoiseNet finds a sufficient number of reliable patches and accurately estimates noise STD for them. This allows it to improve accuracy of signal-dependent noise parameter estimates as compared to state-of-the-art.

4   Conclusions

This paper deals with the problem of blind noise parameter estimation using only noisy images. To estimate noise STD from a mixture of noise and image texture, the estimator should be robust to wide variety of image textures. While previous methods rely on parametric image texture modeling or simple assumptions about texture properties, we have proposed to benefit from high adaptivity of deep convolutional neural networks to image content.

From existing highly efficient NI+DCT BNPE method, we have borrowed the general estimation idea: image is spitted into patches; starting from an estimate of noise NLF, for each patch, the noise STD is estimated alongside with uncertainty of this estimate; noise NLF is refined using the most accurate estimates; the process is repeated iteratively improving the NLF estimate. In this paper, the central part of this NLF estimation scheme has been implemented as deep regression CNN called NoiseNet. The main feature of NoiseNet is that it comprises two branches: regressor and uncertainty quantifier. The latter predicts STD of the regressor output for a given image patch. Both branches are learned together using joint loss function. Iterative estimation method with patch processing using the NoiseNet has been called NI+NoiseNet.

The NoiseNet has been trained on calibrated Nikon D80 camera images. Using validation images, it has been shown that uncertainty quantifier branch of the NoiseNet is able to predict regressor error STD with an error less than 5%. Compared to state-of-the-art, the NI+NoiseNet reduces signal-dependent noise parameters estimation error by about two times on test NED2012 database. Its processing time is about 5.5 s for 3872 by 2592 pixels image on NVIDIA GeForce GTX 960M GPU. This is at least one order faster than the NI+DCT single-thread CPU implementation. While exact complexity comparison is not possible, NI+NoiseNet complexity is low enough for practical applications.

Future work will investigate the processing of not only single component but color images and the spatially correlated noise case.

References

1. Abramova, V., Abramov, S., Lukin, V., Vozel, B., Chehdi, K.: Scatter-plot-based method for noise characteristics evaluation in remote sensing images using adaptive image clustering procedure. In: Proceedings of the SPIE, vol. 10004, pp. 10004–10004-11 (2016)
2. Almeida, M.S.C., Figueiredo, M.A.T.: Parameter estimation for blind and non-blind deblurring using residual whiteness measures. IEEE Trans. Image Process. **22**(7), 2751–2763 (2013)
3. Alparone, L., Selva, M., Aiazzi, B., Baronti, S., Butera, F., Chiarantini, L.: Signal-dependent noise modelling and estimation of new-generation imaging spectrometers. In: First Workshop on Hyperspectral Image and Signal Processing: Evolution in Remote Sensing, WHISPERS 2009, pp. 1–4 (2009)
4. Ce, L., Szeliski, R., Kang, S.B., Zitnick, C.L., Freeman, W.T.: Automatic estimation and removal of noise from a single image. IEEE Trans. Pattern Anal. Mach. Intell. **30**(2), 299–314 (2008)
5. Danielyan, A., Foi, A.: Noise variance estimation in nonlocal transform domain. In: International Workshop on Local and Non-Local Approximation in Image Processing, pp. 41–45 (2009)
6. Fevralev, D., Ponomarenko, N., Lukin, V., Abramov, S., Egiazarian, K.O., Astola, J.T.: Efficiency analysis of DCT-based filters for color image database. In: Image Processing: Algorithms and Systems IX, vol. 7870, p. 78700R. International Society for Optics and Photonics (2011)
7. Foi, A., Trimeche, M., Katkovnik, V., Egiazarian, K.: Practical poissonian-gaussian noise modeling and fitting for single-image raw-data. IEEE Trans. Image Process. **17**(10), 1737–1754 (2008)
8. Goodfellow, I., Bengio, Y., Courville, A., Bengio, Y.: Deep Learning, vol. 1. MIT Press, Cambridge (2016)
9. Gurevich, P., Stuke, H.: Learning uncertainty in regression tasks by deep neural networks. arXiv preprint arXiv:1707.07287 (2017)
10. Keelan, B.: Handbook of Image Quality: Characterization and Prediction. CRC Press, Boca Raton (2002)
11. Tsin, Y., Ramesh, V., Kanade, T.: Statistical calibration of ccd imaging process. In: Proceedings of the Eighth IEEE International Conference on Computer Vision, ICCV 2001, vol. 1, pp. 480–487 (2001)
12. Uss, M., Vozel, B., Lukin, V., Abramov, S., Baryshev, I., Chehdi, K.: Image informative maps for estimating noise standard deviation and texture parameters. EURASIP J. Adv. Signal Process. **2011**(1), 806516 (2011)
13. Uss, M., Vozel, B., Lukin, V., Chehdi, K.: Maximum likelihood estimation of spatially correlated signal-dependent noise in hyperspectral images. Opt. Eng. **51**(11), 111712-1–111712-11 (2012)
14. Uss, M., Vozel, B., Lukin, V.V., Chehdi, K.: Image informative maps for component-wise estimating parameters of signal-dependent noise. J. Electron. Imaging **22**(1), 013019–013019 (2013)

Effective Training of Convolutional Neural Networks for Insect Image Recognition

Chloé Martineau[1]([⊠]), Romain Raveaux[1], Clément Chatelain[2],
Donatello Conte[1], and Gilles Venturini[1]

[1] LIFAT EA 6300, 37200 Tours, France
chloemartineau99@gmail.com
[2] LITIS EA 4108, 76800 Saint-Etienne du Rouvray, France

Abstract. Insects are living beings whose utility is critical in life sciences. They enable biologists obtaining knowledge on natural landscapes (for example on their health). Nevertheless, insect identification is time-consuming and requires experienced workforce. To ease this task, we propose to turn it into an image-based pattern recognition problem by recognizing the insect from a photo. In this paper state-of-art deep convolutional architectures are used to tackle this problem. However, a limitation to the use of deep CNNs is the lack of data and the discrepancies in classes cardinality. To deal with such limitations, transfer learning is used to apply knowledge learnt from ImageNet-1000 recognition task to insect image recognition task. A question arises from transfer-learning: is it relevant to retrain the entire network or is it better not to modify some layers weights? The hypothesis behind this question is that there must be part of the network which contains generic (problem-independent) knowledge and the other one contains problem-specific knowledge. Tests have been conducted on two different insect image datasets. VGG-16 models were adapted to be more easily learnt. VGG-16 models were trained (a) from scratch (b) from ImageNet-1000. An advanced study was led on one of the datasets in which the influences on performance of two parameters were investigated: (1) The amount of learning data (2) The number of layers to be finetuned. It was determined VGG-16 last block is enough to be relearnt. We have made the code of our experiment as well as the script for generating an annotated insect dataset from ImageNet publicly available.

1   Introduction

Insects are a class of invertebrates within the arthropods that have an exoskeleton, a three-part body (head, thorax and abdomen), three pairs of jointed legs, compound eyes and one pair of antennae. With 1,5 millions of species, they are more representative for wholesale organism biodiversity than any other group. Arthropods have been recognized as efficient indicators of ecosystem function and recommended for use in conservation planning [16]. Building accurate knowledge of the

© Springer Nature Switzerland AG 2018
J. Blanc-Talon et al. (Eds.): ACIVS 2018, LNCS 11182, pp. 426–437, 2018.
https://doi.org/10.1007/978-3-030-01449-0_36

identity, the geographic distribution and the evolution of insect species is essential for a sustainable development of humanity as well as for biodiversity conservation. Finding automatic methods for such identification is an important topic with many expectations. One of the most common data that can be used in this context are images. Images of arthropods can be acquired and further processed by an image classification system. The literature about insect image captures can be split into two broad categories: lab-based setting and field-based setting. In a lab-based setting there is a fixed protocol for image acquisition. This protocol governs the insect trapping, its placement and the material used for the acquisition (capture sensor, lighting system, etc.). Lab-based setting is mainly used by entomologists bringing the insects to the lab to inspect them and to identify them. Field-based settings means insect images taken directly in cultivated fields, without any particular constraints to the image capture system. An intermediate image type is multi-individuals which show many individuals at the same time, in a lab-based environment. Deep neural networks (DNN) have been extensively used in pattern recognition and have outperformed the conventional methods for specific tasks such as segmentation, classification and detection. However, to our knowledge, DNN have never been applied to insect image recognition [14].

Considering the hierarchical feature learning fashion in DNN, first layers are expected [20] to learn features for general simple visual building blocks, such as edges, corners and simple blob-like structures, while the last layers learn more complicated abstract task-dependent features. In general, the ability to learn domain-dependent high-level representations is an advantage enabling DNNs to achieve great recognition capabilities. However, DNN require large labeled dataset to efficiently learn their huge number of parameters, and insect image databases only contains a few hundred of labeled samples. To overcome this lack of data, transfer learning has been proposed which aims at transferring the knowledge from a more or less related task that has already been learned on a large dataset. Although transfer learning has been shown to be very efficient on many applications, its limits and its practical implementation issues has not been much studied. For example, it would be practically important to determine how much data on the target domain is required for domain adaptation with sufficient accuracy for a given task, or how many layers from a model fitted on the source domain can be effectively transferred to the target domain. Or more interestingly, given a number of available samples on the target domain, what layer types and how many of those can we afford to learn. Moreover, there is a common scenario in which a large set of annotated data is available (ImageNet-arthropod subset), often collected in a time-consuming and costly process. To what extent these data can contribute to a better analysis of new datasets is another question worth investigating.

In this study, we aim towards answering the questions discussed above. To tackle the insect image recognition problem, we use transfer learning methodology for domain adaptation of models trained on scene images. The contribution of this paper is an effective learning methodology for the insect recognition problem. The role of pretraining in the DNN accuracy is detailed and the question is raised about how many insect samples are required to achieve a sufficient accuracy from the biologist viewpoint.

2 Problem Definition: Image-Based Arthropod Classification

This section introduces the problem tackled in this article: Image-based arthropod classification.

Image-based arthropod classification could be seen as an application of image classification. Based on some photograph depicting the specimen, its biological identity is to be determined. The peculiarities of the problem are three-fold: rarity of images, image variations and large discrepancies among class cardinalities.

Rarity of images. Only experts such as taxonomists and trained technicians can identify accurately insect classes, because it requires special skills acquired through extensive experience. In lab-based setting, most of the acquisition systems are manually manipulated which increase the workforce amount (see examples on Fig. 1(a) and (b)). **Image variations.** Aside from classical object image variations (such as rotation, scale, parallax, background or lighting), insect images have more particular properties such as pose (because specimen appearance varies with the orientation they are been shown) and deformation (because the specimens are most of the time composed of articulated parts). These aforementioned variations can be referred to as *capture-specific* variations in the sense they only depend on capture factors. About the objects themselves (*object-specific* variations), age, sex and genetic mutations are the main factors of visual variations. The most instructive example is that of lepidoptera (commonly referred to as butterflies) which can have extremely different visual aspects along time, being successively caterpillars, chrysalises and butterflies. **Large discrepancies among class cardinalities.** Insect capture campaigns are season-dependent impacting the number and the type of the captured insects. This fact can be translated in the pattern recognition domain as an imbalanced classification problem.

3 State of the Art

Image-based insect recognition is not a newly addressed problem. A detailed study was lead [14]. This study focused successively on capture protocols, feature extraction methods and finally classification. The last two points are of first interest here while the first point is to be considered as an input constraint and depends fully on the biologists workflow. Moreover, the datasets and the classes that constitute them are very different and motivated by different biological scopes and problems.

(a)   (b)

Fig. 1. Lab-based samples.

Regarding feature extraction, the first pieces of work tackle the problem in very ad-hoc ways. [8] extract dimensions on the insects wings using venations as keypoints. Others [17] use geometrical attributes from the region of interest as features. Then, studies began to emerge using standard local and global features such as SIFT [18]. On top of these standard handcrafted features are learnt higher level features using MLPs [1], Bag of Words [12] or Sparse Coding [19]. The next step is to introduce hierarchically learnt descriptors, with many levels of abstraction. [18] uses a stack of Denoising Autoencoders to this extent. The observation that can be done is the features are no more particular to the problem itself and is about learning hierarchical representations to get satisfying feature spaces that suit the biologists goals through learning on their images. In such a frame, [1] applies MLPs on raw pixel even though the problem is a very simple one (two classes: harmful/harmless insects).

Deep neural networks are machine learning models that are now the state-of-the-art models for almost every pattern recognition tasks. Their main strength is their ability to learn suitable features for a given task, thus avoiding the need for handcrafted features design. Convolutional Neural Networks (CNN) are an instantiation of DNN dedicated to image processing. Through a mechanism of shared weights, their input layers are convolutional filters that automatically learn features from multidimensional signals [10]. CNN now outperform most of traditional approaches (e.g. based on handcrafted features) in various image analysis and recognition domains, such as natural scene recognition [10], Medical imaging [3], Image segmentation [11] or handwriting recognition [15]. However, although very efficient, deep CNN models require a huge amount of labeled data to avoid overfitting, which can be compared to "learning by heart" the training database. In order to circumvent these issues, the architectures are generally trained using many tricks such as regularization, dropout or data augmentation for improving their generalization ability. But all these hints do not replace a reasonable amount of labeled data for training the architectures.

Recently, the transfer learning idea has been proposed to train huge models with small labeled datasets. It is based on the exploitation of pre-trained models on a huge datasets from another domain. The model is then fine tuned on a specific, smaller database of the domain considered. Models are often trained on imagenet, a natural scene database of more than 14M images [10]. Even if the specific domain strongly differs from the imagenet database, the transfer learning have shown very impressive results on many tasks such as handwriting recognition [7], signature identification [9], and medical imaging [2].

4   Proposed Approach

This article presents a method to easily train a deep convolutional neural network architecture using transfer learning, with application to the insect image recognition problem. The deep architecture has been adapted to be easily trained on low volume datasets and the feeding, learning and hyper-parameter optimization procedures are detailed in the next subsections.

4.1 Transfer Learning Adapted VGG-16 Architecture

The fine-tuning CNN experiments involve VGG-16 instances pretrained on ImageNet 1000 [5]. Although more recent models exist (for example GoogLeNet or ResNet), preliminary experiments shown these architectures yielded similar results on the problem tackled here. Moreover, VGG-16 was chosen for its simplicity and relatively small number of layers.

VGG-16 has undergone a crucial modification in its convolutional end. The original model end (see Fig. 3 on the left) consists of a three-layered MLP (FC1-3) which takes as inputs every coefficients from the last 2D activation map (block5-pool): The $7 \times 7 \times 512$ volume is flattened to obtain 25088 input values for the MLP. This has been replaced by a global average pooling filter which keeps only the averages of each of the 512 7×7-slices of the $7 \times 7 \times 512$ volume [13]. This reduces the input size of the MLP from 25088 to 512. A visualization of such a transformation is pictured in Fig. 2. This modification acts as a regularization that limits overfitting as it reduces the number of neurons in the first part of the MLP, and takes more advantage to the structure of the last convolutional feature maps [13]. The remains of the MLP consist of a single fully-connected layer with 256 neurons and the prediction layer.

The overall transfer learning architecture is described in Fig. 3.

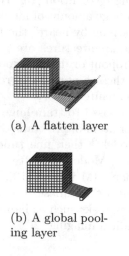

(a) A flatten layer

(b) A global pooling layer

Fig. 2. Layer translating volumes into 1-D vector to take as input of the fully-connected end of a CNN

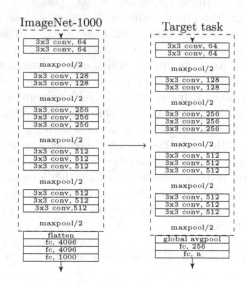

Fig. 3. Transfer learning architecture with modified VGG-16

4.2 Data Preprocessing

Preparing the data to feed the neural network is critical to the learning process. It consists of two steps. The first one is about adapting the input image (size $n \times m \times 3$, n an m being respectively the height and width of the input image) to the receptive field of the neural network (size $224 \times 224 \times 3$, the size of the ImageNet input on the three color channels). The most straight-forward approach to this size adaption is to resize the image. The pros of this method lies in the fact that all the information of the image is fed to the network. The main cons is that the ratio of the image is not preserved. In the case of one dataset in use in this article, the images ratios are significantly various and thus resizing the images would end up in bad performance. The method used here consists in cropping the images in their center so that the image is of size $k \times k \times 3$ with $k = \min(m, n)$. Cropping instead of resizing keeps the image ratio as is. Finally, a resize operation can be performed from size $k \times k \times 3$ to size $224 \times 224 \times 3$. Once the size normalization is done, the images must be standardized across their dataset featurewise by subtracting the mean image and dividing by the standard deviation image. This is done separately for each color channel. Using the image unstandardized results in inability to train the system (the loss stagnating at its initial level). This behavior can be explained in the following manner: if the image is not standardized, the differences in scales causes the gradient descend to be applied at these different scales thus differently depending on the location or pixel value. Finally, data augmentation was used. The training images, each time they were fed to the network, were modified using a set of randomly chosen transformations (sheer, flip and zoom). It virtually augments the size of the set as many times as the number of epochs.

4.3 Learning Rule

The data that is being dealt with is not only in small amount but also unequally distributed among classes. Neural networks are known to be very sensitive to cardinality discrepancies. There are two main solutions to address this problem. The first one consists of sampling on low-cardinality classes and the latter is to apply the gradient descent more or less depending on the cardinality of the examples classes. This latter method is the one being used here for the sake of simplicity. For each class $c \in Y$, the following weights have been applied: $w_c = \frac{\max_{c' \in Y} \text{Card}(c')}{\text{Card}(c)}$ where $\text{Card}(c)$ is the number of observations that fall into class c. w_c are used during the backpropagation step as weights to the error gradient descent term. The higher w_c, the stronger the error gradient is descended. The gradient descent method is vanilla Stochastic Gradient Descent with 0.9 Nesterov momentum [4]. On the contrary, the learning rate is divided by two if the loss hasn't been reduced in the span of 10 epochs. This intensification process enables the learning rate to be adapted depending on the scale of the gradient and therefore to avoid oscillations onto the search space. This behaviour is likely to lead to a local minimum. However in the case of multi-layer networks it is not a major issue. [6] not only states this but also that global minima are prone to lead to overfitting in practice.

4.4 Hyper-parameter Optimization

The main limitation to the use of deep models is the hyper optimization process it requires in order to get satisfying results. Additionally, using transfer learning during the training phase adds a new hyper-parameter: the number of layers to be trained on the target task. Deep CNN are composed of low level features that are independent from the problem. These low level features are located in the first layers. An efficient learning methodology should allow not spending time optimizing these layers. In our proposal, we show that only the last convolutional block (block5) and obviously the last dense layer which has not been pretrained (fc1) are of first interest to train at ease the model while achieving a good classification rate. Conceptually speaking, block5 represents the high level features that are insect-dependent, while fc1 is a decision phase to fit at best the classification task in term of number of classes and class distribution. Reducing the search space as much as possible is therefore critical when learning with small amount of data.

5 Experimental Work

This section details the lead experiments. Data in use is presented and both experiments and their respective results are described as well as their interpretations.

5.1 Datasets

The target domain is that of insect images and the task is to recognize the class they have been labeled in. Two sets have been used to that extent. The first set (IRBI) is a lab-based insect images set. Insects are captured into soapy liquid traps put at soil level. The captured specimens are then conserved in an alcohol solution before being identified. These same specimens are laid down on a plain background and under a constant and controlled lighting. Multiple shots are taken for a single individual: they are taken along 3 different orientation of the individual and with 7 different smartphones which makes up 21 shots for a single insect. The dataset split into train/valid/test sets has been performed at the insect level, in order to prevent the presence of an insect in two (or more) sets with different orientations, which would make the evaluation unfair. Besides, due to the cardinalities discrepancies, a stratified split method must be used at the class level. To have an analogous dataset with pictures taken in field-based settings, a subset of ImageNet was extracted: each leaf-synsets under the synset "arthropods" was used which make up 501 classes. In order to emulate the same constraints as in a real entomology set, the average class cardinality was lowered down to the average class cardinality of the IRBI set (see Table 1). The statistical validation method in use in this study is a 5-fold stratified cross-validation. To be able to apply such a split while keeping individuals from every class in the two sub-sets, classes with less than 5 individuals were removed.

Table 1. The image datasets

Dataset	Nb of classes	|Card(c)|	σ(Card(c))	min(Card(c))	max(Card(c))
IRBI	30	85	71	33	370
ImageNet-arthropods	443	96	78	6	392

5.2 Experiments

The first part of the experimental study is about comparing together (i) a traditional method based on handcrafted features, (ii) CNN-based models trained from scratch, and (iii) these same models which exploit the transfer learning trick. Table 2 shows results on both image datasets and for the following models: **(1) SIFTBoW:** a SIFT codebook representation combined with a SVM classifier with an RBF kernel. The SVM is optimized with a grid-search approach on the validation dataset for optimizing the penalty parameter ($C \in \{10^0, \ldots, 10^6\}$) and the kernel parameter ($\gamma \in \{10^{-6}, \ldots, 10^0\}$). **(2) Several VGG-16 based CNN classifiers,** as described in the previous section: **(a)** *frsc:* learnt from scratch with initial random weights (during 200 epochs); **(b)** *fitu:* pre-trained network on ImageNet-1000 (during 50 epochs), and fine tuned on IRBI/Arthropods; **(c)** *fitu/w:* same as **fitu**, with example weighting based on each class cardinality (see Sect. 4.3). **(d)** *fitu7/w:* same as **fitu/w** but with only 7 layers fine-tuned. All the code in use in this article can be retrieved here: https://github.com/prafiny/deep-insect-experiments/. A script to recreate the ImageNet-arthropods set is provided. In Table 2, the column "mean time" refers to the average time in seconds by epoch during the training. Additionally, the training curves (representing the recognition rate on validation set at each epoch) for VGG-16 both frsc and fitu/w are shown on Fig. 4.

Several observations can be made on these experiments results. First, the traditional approach based on handcrafted features (SIFTBow) performs moderately. Second, the VGG-16 architecture trained from scratch (VGG16-frsc) on the target dataset slightly outperforms the traditional BoW approach. On the other hand, the VGG-16 fine-tuned with transfer learning (VGG-16-fitu) gives significantly higher results compared to those of the traditional approach Additionally, weighting training examples based on class cardinalities (VGG16-fitu/w) made the model better on average regarding the top-1 figures.

In a second time, an advanced study was conducted on the behavior and efficiency of transfer learning and CNN. In this study two factors are investigated: the number of learning examples and the number of layer to be learnt. The results for one fold are shown in Fig. 5. The dataset used in this frame is IRBI. While the effect of changing the number of training example is not so surprising, it

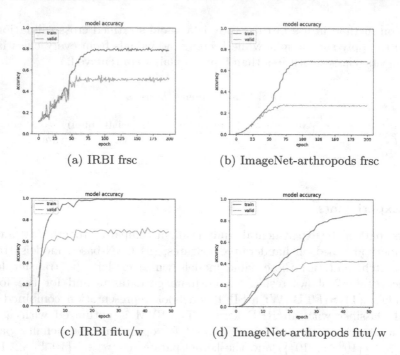

(a) IRBI frsc

(b) ImageNet-arthropods frsc

(c) IRBI fitu/w

(d) ImageNet-arthropods fitu/w

Fig. 4. Learning curves for VGG-16 (blue for train accuracy, orange for valid accuracy). (Color figure online)

is interesting to note that learning only on the last 7 layers seems enough to reach a 70% accuracy score. This observation can be interpreted as follows: the ImageNet-1000 features are generic enough (up to the $22 - 7 = 15$th layer) to learn on the lab-based insect identification problem. Learning only the 7 last layers is in fact learning the last VGG16 block (block5) and the two dense layers. Also, it is equivalent to learn about half of the network weights. This experiment corroborates the statement made in Section 1 that the first layers can be considered as a generic feature extractor while the last convolutional block and the dense layer represents the task-dependent features. Last line of

Table 2. Recognition rates on 5-fold experiments and mean epoch times

Model	IRBI			ImageNet-arthropods		
	Top-1	Top-5	Mean time (s)	Top-1	Top-5	Mean time (s)
SIFTBoW	52.3% ± 3.7	82.7% ± 3.3	–	11.7% ± 0.2	25.9% ± 0.4	–
VGG16-frsc	54.0% ± 5.0	84.9% ± 3.0	101	26.9% ± 0.7	50.1% ± 0.7	1470
VGG16-fitu	72.0% ± 3.2	92.1% ± 1.1	104.4	42.7% ± 0.9	69.4% ± 0.6	1473.6
VGG16-fitu/w	73.6% ± 1.8	92.4% ± 2.2	102.6	43.5% ± 1.1	71.3% ± 0.8	1473.2
VGG16-fitu7/w	72.4% ± 2.8	92.6% ± 2.1	52.4	43.3% ± 0.6	71.8% ± 0.4	721.6

Table 2 shows results learning only 7 layers for both IRBI and ImageNet. Last but not the least, the training phase is sped up by a factor two by learning only the 7 last layers instead of the whole network.

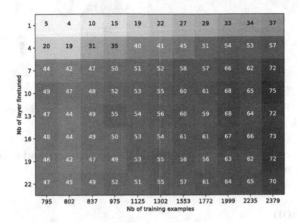

Fig. 5. Recognition rates on the testing set depending on the number of training examples and the number of layer finetuned

5.3 Misclassification Analysis

Finally, some observations were made at the query level to have an insight on what are the possible causes of misclassification. The problem is that it is very tedious to get an idea of how CNNs discriminate between classes. In [21] two main methods classes are listed: (1) to display the convolution filters, (2) to use the filter responses as a tool to generate images. The drawback of such a method is that the produced images are often abstract and therefore hard to interpret. A possible solution to the interpretation issue is to rely on the database images rather than trying to make up images. In this extent, we adopted the following method: for each misclassified image, we retrieve the output vector of the CNN and try to find the images whose output vectors are nearest in the sense of vector space distance. This enables to work in a Content-Based Image Retrieval-like framework and listing the k-nearest neighbors images for a given query. Examples of query results produced by using this method are shown on Fig. 6. Figure 6(a) shows a misclassification occurrence which is likely to be caused by the small apparent size of the insect. Each of the nearest images were from different classes yet quite resembling due to the low resolution of the capture. Figure 6(b) shows an example of a specimen mimicking another insect: syrphid flies colors enable them to be confused with bees or wasps suggesting they could sting a predator.

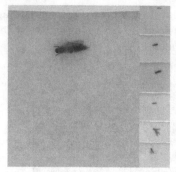

(a) Small specimen confused
with other small specimens

(b) Resembling classes

Fig. 6. Misclassified samples. Left big image is the query image and right small images are the 5-nearest neighbors.

6   Conclusion

This study uses fine-tuned deep convolutional neural networks on unbalanced and low-volume image datasets. Transfer learning definitely helps to compensate the low amount of data as performances clearly outperform traditional approach when using the transfer learning trick. Also, we have found that training only the last convolutional block of VGG16 is sufficient to find an almost optimal solution. This can lead to the conclusion that ImageNet-1000 features are generic enough up to a high level of abstraction. Finally, weighting the examples to tackle the unbalanced learning problem did influence positively the results but only marginally.

Acknowledgments. This work was supported by a research grant from the Région Centre-Val de Loire, France.

References

1. Al-Saqer, S.M., Hassan, G.M.: Artificial neural networks based red palm weevil (Rynchophorus Ferrugineous, Olivier) recognition system. Am. J. Agric. Biol. Sci **6**, 356–364 (2011)
2. Bar, Y., Diamant, I., Wolf, L., Greenspan, H.: Deep learning with non-medical training used for chest pathology identification. In: Proceedings of SPIE, Medical Imaging: Computer-Aided Diagnosis, vol. 9414, 94140V-7 (2015)
3. Belharbi, S., et al.: Spotting L3 slice in CT scans using deep convolutional network and transfer learning. Comput. Biol. Med. **87**, 95–103 (2017)
4. Bengio, Y., Boulanger-Lewandowski, N., Pascanu, R.: Advances in optimizing recurrent networks. CoRR, abs/1212.0901 (2012)
5. Chollet, F., et al.: Keras (2015). https://github.com/fchollet/keras

6. Choromanska, A., Henaff, M., Mathieu, M., Arous, G.B., LeCun, Y.: The loss surfaces of multilayer networks. In: Artificial Intelligence and Statistics, pp. 192–204 (2015)
7. Cireşan, D.C., Meier, U., Schmidhuber, J.: Transfer learning for Latin and Chinese characters with deep neural networks. In: The 2012 International Joint Conference on Neural Networks (IJCNN), pp. 1–6. IEEE (2012)
8. Dietrich, C.H., Pooley, C.D.: Automated identification of leafhoppers (Homoptera: Cicadellidae: Draeculacephala Ball). Ann. Entomol. Soc. Am. **87**(4), 412–423 (1994)
9. Hafemann, L.G., Sabourin, R., Oliveira, L.S.: Writer-independent feature learning for offline signature verification using deep convolutional neural networks. CoRR, abs/1604.00974 (2016)
10. Krizhevsky, A., Sutskever, I., Hinton, G.: Imagenet classification with deep convolutional neural networks. In: Pereira, F., Burges, C.J.C., Bottou, L., Weinberger, K.Q. (eds.), NIPS, vol. 25, pp. 1097–1105 (2012)
11. Lai, M.: Deep learning for medical image segmentation. CoRR, abs/1505.02000 (2015)
12. Larios, N., et al.: Automated insect identification through concatenated histograms of local appearance features: feature vector generation and region detection for deformable objects. Mach. Vis. Appl. **19**(2), 105–123 (2008)
13. Lin, M., Chen, Q., Yan, S.: Network in network. CoRR, abs/1312.4400 (2013)
14. Martineau, C., Conte, D., Raveaux, R., Arnault, I., Munier, D., Venturini, G.: A survey on image-based insect classification. Pattern Recognit. **65**, 273–284 (2017)
15. Poznanski, A., Wolf, L.: CNN-N-gram for handwriting word recognition. In: CVPR, pp. 2305–2314 (2016)
16. Van Straalen, N.M.: Evaluation of bioindicator systems derived from soil arthropod communities. Appl. Soil Ecol. **9**(1), 429–437 (1998)
17. Wang, J., Lin, C., Ji, L., Liang, A.: A new automatic identification system of insect images at the order level. Knowl. Based Syst. **33**, 102–110 (2012)
18. Wen, C., Wu, D., Hu, H., Pan, W.: Pose estimation-dependent identification method for field moth images using deep learning architecture. Biosyst. Eng. **136**, 117–128 (2015)
19. Xie, C., et al.: Automatic classification for field crop insects via multiple-task sparse representation and multiple-kernel learning. Comput. Electron. Agric. **119**, 123–132 (2015)
20. Yosinski, J., Clune, J., Bengio, Y., Lipson, H.: How transferable are features in deep neural networks? In: Advances in Neural Information Processing Systems, pp. 3320–3328 (2014)
21. Yosinski, J., Clune, J., Nguyen, A.M., Fuchs, T.J., Lipson, H.: Understanding neural networks through deep visualization. CoRR, abs/1506.06579 (2015)

A Deep Learning Approach to Hair Segmentation and Color Extraction from Facial Images

Diana Borza[1]([✉]), Tudor Ileni[2], and Adrian Darabant[2]

[1] Faculty of Automation and Computer Science,
Technical University of Cluj-Napoca, 28 Memorandumului Street,
400114 Cluj-Napoca, Romania
diana.borza@cs.utcluj.ro
[2] Faculty of Mathematics and Computer Science, Babes-Bolyai University,
1 Mihail Kogalniceanu Street, 400084 Cluj-Napoca, Romania

Abstract. In this paper we tackle the problem of hair analysis in unconstrained images. We propose a fully convolutional, multi-task neural network to segment the image pixels into hair, face and background classes. The network also decides if the person is bald or not. The detected hair pixels are analyzed by a color recognition module which uses color features extracted at super-pixel level and a Random Forest Classifier to determine the hair tone (black, blond, brown, red or white grey). To train and test the proposed solution, we manually segment more than 3500 images from a publicly available dataset. The proposed framework was evaluated on three public databases. The experiments we performed together with the hair color recognition rate of 92% demonstrate the efficiency of the proposed solution.

Keywords: Color classification · Fully convolutional networks
Hair analysis · Multi-task learning · Visagisme

1 Introduction

Over the past decades, automatic face analysis has been one of the most studied problems by computer vision scientists, as it has numerous applications in various pluridisciplinary domains: medicine, human computer interaction, security, just to name a few. However, the majority of the works addressed the problem of internal face feature analysis (eye tracking, expression analysis etc.), while little research has been conducted on the external face features (hair, external face contour etc.).

Some studies from the field of neuroscience proved that the hair is an important cue in human face recognition. The work [17] presented an illusion which proved that internal face features are ignored in favor of external ones and the

© Springer Nature Switzerland AG 2018
J. Blanc-Talon et al. (Eds.): ACIVS 2018, LNCS 11182, pp. 438–449, 2018.
https://doi.org/10.1007/978-3-030-01449-0_37

overall head structure. Other studies proved that the facial features are perceived holistically [16], and that the hairstyle and hair color are an important recognition cues in cases of low resolution or degraded images.

Nowadays, physical appearance is an essential aspect of social life. In this context, a new branch of study has emerged from the field of aesthetics: visagisme. This concept helps humans enhance their appearance by choosing the appropriate accessories which are in harmony with their face. Let's take for example the use case of (sun/eye)glasses selection. Visagisme defines a complex set of rules which take into account the hair color, the hair style, the face shape and its color, the eyes color etc., which are combined to help the user choose the best pair of glasses. Amongst these factors, the hair color is one of the decisive factors, as it covers a major part of the upper side of the head. Also, in the field of on-line eyeglasses sales, several virtual-try on applications[1] were developed, which allow customers to experiment difference frames from the comfort of their homes. The virtual glasses from these systems rely on databases of 3D models of real glasses and frames. Usually, theses datasets comprise tens of thousands of samples and visagisme rules are used to decide which glasses should be first display to the user.

Automatic hair color analysis poses two major challenges. First of all, unlike for other face features, it is harder to establish the areas where the hair is likely to be present. Numerous (extravagant) hairstyles exist: symmetrical, rasta, curly etc., or the hair might not be present at all (bald individuals). The second challenge is related to the hair color tone. The color distribution of (natural) hair colors is not uniform - highlights might be presents or there might be different colors for the roots and locks - making it more difficult to analyse.

This paper proposes a novel hair analysis framework targeted for visagisme applications in the context of online eyeglasses sales. The proposed method will highlight the following contributions: (1) The tuning of two fully convolutional neural networks for the problem of hair and facial skin segmentation from facial images. The networks are multi-task, in the sense that they also decide if the subject pictured in the input image has hair (is bald or not). (2) A novel method for natural hair color classification. The proposed algorithm operates on histograms of colors extracted at super-pixel level and uses simple classifiers to distinguish the hair between black, blond, brown, grey/white or red hair. (3) The manual semantic segmentation of more than 3500 images from a publicly available face dataset. The image pixels are classified into one of the following classes: facial skin, hair or background. The annotation masks are made public to the scientific community. In addition, each one of these images is annotated with the corresponding hair tone: black, blond, brown, grey/white or red.

The remainder of this paper is organized as follows: Sect. 2 reviews the recent methods proposed for hair color analysis (segmentation and hair color recognition). In Sect. 3 we describe in detail the proposed solution and in Sect. 4 we report the experimental results. Finally, Sect. 5 concludes this paper and discusses future improvements and research directions.

---

[1] http://www.trylive.com/demos.

2   Related Work

The first work to address the problem of automatic hair analysis was [18]; the method analyses frontal face images and segments the hair area using color features, anthropometric proportions and region growing. Based on the hair mask, multiple (geometrical) features of the hair are extracted: volume, length, symmetry, dominant color, just to name a few. In [13] the hair pixels are determined by the intersection of two image masks: the first one is computed through color analysis, while the latter is computed through frequency analysis. In [3] the hair region is segmented in two steps: first, a simple hair shape model is fitted to upper hair region using active shape models. Secondly, segmentation at pixel level is performed based on the appearance parameters (texture, color) of the pixels from the first region.

The problem of hair segmentation with applications in caricature synthesis is addressed in [15]. First, in the training phase, the priors for hair's position and color are estimated from a labeled dataset of images. In the test phase, the hair pixels are determined through graph-cuts and k-means clustering.

All the segmentation methods presented so far, perform well on near-frontal images with simple backgrounds. Recently, more complex methods were proposed, which are able to cope with images captured in more difficult, unconstrained scenarios. The work [11] proposes a novel two-layer Markov Random Field architecture designed for hair segmentation in unconstrained scenarios. The first later operates at pixel level and the second layer works at macro, object level to guide the algorithm towards the possible solutions. In [9] a two-stage hair segmentation algorithm is proposed. First, a rough hair probability map is constructed from local image patches using image features extracted by a convolutional neural network and a random forest classifier. Next, starting from the probability map computed by the first module, a support vector machine classifier uses local ternary patterns to perform hair classification at pixel level.

Some works addressed the problem of hair color recognition with applications in soft biometrics. In ([6,10]) the hair area is extracted from video sequences. First, the head is located in the current frame using background subtraction and face detection. As skin color mask is determined using the flood-fill algorithm. Finally, the hair pixels are computed as the difference between the pixels from the head and skin set respectively. These hair pixels are further analysed to determine the hair tone (black, blond, brown, red and white/grey). A more simple hair color taxonomy is reported in [14]: the hair tone is classified either as "black" or "non-black". A machine learning classifier - both k-Nearest Neighbours (kNN) and support vector machines (SVMs) were tested - operating on several combined color features (value, mean and variance of each channel in the RGB and HSV color-spaces) in order to determine the hair tone.

3   Proposed Solution

A general overview of the proposed hair analysis system is depicted in Fig. 1. The first step is to detect the faces in the input image using a general face detector

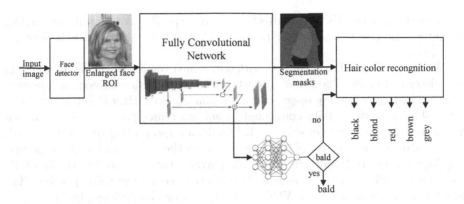

Fig. 1. Outline of the proposed hair analysis framework. In the segmentation mask, the black pixels represent the background, the red ones the face pixels, while the blue pixels represent the hair area. (Color figure online)

(*dlib* library) [5]. For each detected face, we establish a region of interest (ROI) around it and we feed this cropped image to the segmentation module. The ROI is established such that the face area is enlarged with a factor of 1.6 both on the horizontal and vertical dimensions.

The segmentation module assigns to each pixel one of the following classes: hair, facial skin and background. It also determines if the person pictured in the input image is bald or not.

Next, if the person is not bald, the input image and its corresponding segmentation mask are fed to the hair color recognition module, which determines the hair color of the subject: black, blond, brown, grey/white or red.

In the remainder of this section we detail each module of the proposed hair analysis framework.

3.1 Hair and Face Segmentation

The segmentation module is based on fully convolutional neural networks (FCN). We experimented with two FCN, inspired from the U-Net [12] and VGG architectures [8], respectively.

The layers of the U-Net network [12] are distributed in a symmetrically manner. Two paths are created: the contraction and the expansion path. The first layer has the dimension of $224 \times 224 \times 1$, corresponding to a 3 channel input image. Further, on the contraction part, at each step the number of output channels is doubled and its size is diminished by 3×3 pooling layers. On the expansion path, at each step the number of output filters is halved and the size is doubled. The upsampling is done by a 2D transpose convolution of stride 2. A cropped part of the corresponding contraction path layer is concatenated to each such convolutions. In this way the spacial information from the contraction path is not lost. In our implementation we replaced the classical convolutional layers by depthwise convolutions in order to reduce the computational cost. The loss function for the U-Net architecture is task specific: we aim at minimizing the intersection over union of the predicted class and the ground truth class, for each pixel.

We also performed experiments with a modified version of the VGG architecture described in [8]. The network starts with 5 convolution layers; each layer is followed by a max pooling operation and a non-linearity (ReLU activation function). The next three fully connected layers are reinterpreted as convolutions layers, that have outputs of size 1×1. Finally, an upsampling (deconvolutional) layer is attached to the end of the network. As in the U-Net architecture, several skip layers are introduced. The 4th pooling layer is concatenated to the last fully convolution (fc7) and this ensemble is concatenated to the third pooling. The loss function, we used for the VGG network, is a pixel-wise loss function.

Both architectures were trained to predict a three channel output, each of them corresponding to one of the classes: background, hair or face.

The VGG architecture is more complex than the proposed U-Net architecture (which uses depthwise convolutions); due to its complexity, it is likely that the VGG architecture attains (slightly) better results on the segmentation task. However, for real-word scenarios (for example, on mobile devices) its usage may not be feasible, due to the large memory requirements and computational costs. The segmentation results obtained by each network are reported and compared in Sect. 4.2.

3.2   Hair Detection

The hair analysis task also implies to determine if a person is bald or not. In order to solve this task, we extend the U-Net network with two fully connected layers of sizes 4096 and 2 (classification layers), respectively.

In the U-Net architecture, these layers were connected to the layer before the last upsampling operation. At this stage of the network, the features contain semantic information about the skin and hair pixels. Only the added fully connected layers are trained on problem of bald vs. no-bald detection, and all the other weights from the network remain unchanged.

3.3   Hair Color Recognition

An overview of the proposed hair color recognition module is depicted in Fig. 2. The hair color is classified as one of the five natural hair colors: black, blond, brown, red and white/grey. The proposed method operates on simple color features extracted at super-pixel level.

First, the pixels from the input images are grouped into super-pixels (using the Simple Linear Iterative Clustering or SLIC algorithm [1]), such that similar, neighbouring pixels are merged into the same entity. Only the super-pixels which overlap the hair segmentation mask are analysed by this module.

For each super-pixel s, we compute the average color of the comprising pixels: μ_s. In the offline learning stage, we perform kMeans clustering on all the

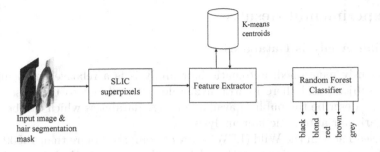

Fig. 2. Overview of the color recognition model.

average color values μ_s of the hair super-pixels from the training samples. We used kMeans++ [2] to choose the initial values of the cluster centers and NC clusters. Let $C = \{c_i | i \in \overline{1, NC}\}$ be the output of the kMeans algorithm, where c_i represents the coordinates of the ith cluster center.

We tested different values for the number of clusters NC. The cluster centers (i.e. colors) computed by the kMeans algorithm using $NC = 20$ clusters are depicted in Fig. 3.

The idea to use more clusters than the number of hair colors from the classification taxonomy is that the five natural hair colors comprise various tones. For example, red hair ranges from strawberry blond, to titian, copper, auburn or completely red. Basically, our algorithm assigns to each pixel one of the "color words" (the centroids computed by kMeans) to each super-pixel. This can be viewed as a feature space reduction step: for each superpixel, instead of working with its average color μ_{st} (which has $16.777.216 = 256 \cdot 256 \cdot 256$ possible values for a 8-bits per pixel RGB image), we compute its closest cluster center (which has only NC possible values).

In the test phase, for each hair super-pixel st we determine its average color μ_{st} and we compute its closest cluster: c_{st} from C; the value c_{st} is the color feature extracted from the super-pixel st, and we compute a normalized histogram on these features. This histogram is fed to a Random Forest Classifier in order to classify the hair color.

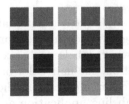

Fig. 3. Clusters computed by the kMeans algorithm. Each square represents a cluster and the square color encoded the centroid (or the mean color) of samples belonging to that cluster.

4   Experimental Results

4.1   Hair Analysis Databases

As previously mentioned, automatic hair analysis is a relatively new and less-studied subject, and there is little available data related to this task. In this section we describe the publicly available image databases which can be used as benchmarks for automatic hair analysis.

Labeled Faces in the Wild (LFW) dataset comprises more than 13000 "in the wild", face images randomly gathered from the Internet. Each image is annotated with the identity of the pictured person; the only constraint imposed on the images from LFW is that the face is detectable with the Viola-Jones face detector. Later, this database was extended with Part Labels Database [4], which contains annotations about the hair, skin and background pixels of 2927 image from LFW.

Another image database proposed for hair analysis is Figaro1k. It contains more than one thousand hair images, and each image is annotated with one of one the seven defined hairstyles (straight, wavy, curly, kinky, braids, dreadlocks, short). In addition, for each sample from the database, a manually segmented hair mask is provided. However, as this database was developed for general hair detection and hair style classification, often the face is not visible in the images; for example, in multiple samples the subject is pictured with the back at the camera, such that only the hair is visible. In this paper, we used only a subset of this dataset to test the proposed hair segmentation module. This subset was created by applying a general face detector [5] and by keeping only the samples in which faces were detected. From the total of 1050 images contained in the Figaro1k dataset, only 171 images were selected.

To the best of our knowledge, there is no image database intended just for automatic hair color recognition. CelebA [7] is a large scale face database, comprising more than 200k samples (from more than 10000 identities) captured in unconstrained environments. Each image is annotated with 40 binary, facial attributes, among which the hair color. However, only the following classes: black, blond, brown and grey hair (annotations: Black_hair, Blond_Hair, Brown_Hair and Grey_Hair, respectively) are defined for the hair tone, and the red hair is not considered. In addition, we noticed several miss-labellings of the hair color in the CelebA dataset.

One contribution of this work is the semantic labeling of a subset of CelebA dataset (more than 3500 images). Each image from this subset is linked with a manually annotated mask which labels each pixel into one of the *hair, skin* or *background* classes. In addition, the hair color of the person pictured in each sample is specified. The hair color information, the annotation masks and the sources of the ground-truth labeling application are made publicly available to the scientific community[2].

---

[2] Please send an email to diana.borza@cs.utcluj.ro to request the annotations.

4.2   Hair Segmentation

Intersection Over Union (IoU) is one of the most commonly used metrics to evaluate the performance of semantic segmentation algorithms. The metric is scale invariant and it computes the similarity of two sets by normalizing the cardinal number of the intersection with the cardinal number of union:

$$J(A, B) = |A \cap B|/(|A| + |B| - |A \cap B|).$$

Other commonly used variants of this metric are the Mean IoU ($mIoU$) and the frequency Weighted IoU ($fwIoU$), formally defined as:

$$mIoU = \frac{1}{n_{cl}} \sum_i \frac{n_{ii}}{t_i + \sum_j n_{ij} - n_{ii}}$$

$$fwIoU = (\sum_k t_k)^{-1} \sum_i \frac{t_i n_{ii}}{(t_i + \sum_j n_{ji} - n_{ii})}.$$

In the above equations, n_{cl} stands for the number segmentation classes, n_{ij} is the number of pixels of class i predicted to be in class j, and t_i the total number of pixels in ground truth segmentation of class i.

We also report the pixel accuracy or precision ($pixelAcc$) and the mean pixel accuracy ($meanPixelAcc$) metrics:

$$pixelAcc = \sum_i n_{ii}/\sum_i t_i; meanPixelAcc = \frac{1}{n_{cl}} \cdot \sum_i n_{ii}/\sum_i t_i.$$

We report our results for hair and face segmentation in Table 1.

The inference time for VGG is 0.064s and for U-Net 0.006s per image, so the U-Net network is 10 times faster. The small fluctuation of performance (in favor of VGG), comparing to the large difference of inference time (in favor of U-Net), make U-Net architecture more feasible to be run end devices like smartphones and tablets, that have a small memory and battery budget.

Some face and hair segmentation samples are depicted in Fig. 4.

Table 1. Hair segmentation performance of LFW, CelebA and Figaro1k datasets

	LFW		CelebA		Figaro1k	
	VGG	U-Net	VGG	U-Net	VGG	U-Net
mIoU	**87.09%**	83.46%	**92.03%**	88.56%	**77.79%**	77.75%
fwIoU	94.67%	92.75%	94.42%	91.79%	82.66%	83.01%
pixelAcc	97.01%	95.83%	97.06%	95.54%	90.15%	90.28%
meanPixelAcc	91.81%	88.84%	95.77%	93.61%	84.75%	84.72%

Fig. 4. Samples of hair and face segmentations.

4.3   Hair Detection

The task of hair (baldness detection) was trained on more than 6500 images from the CelebA dataset. To evaluate the method we selected 600 test images (300 for the class bald, 300 for the class not bald) from the same dataset. Table 2 shows the confusion matrix for this test experiment.

Table 2. Confusion matrix for hair vs no-hair classification

	Hair	No-Hair
Hair	269	9
No-Hair	31	291

The overall accuracy for the baldness detection task is **93.33%**. We report other metrics for this task as: precision: 0.933, recall 0.935 and f1-score 0.935. Some wrong non-bald/bald classification are presented in Fig. 5.

4.4   Hair Color Recognition

To train and test the proposed hair color recognition module, we used images from the CelebA dataset. CelebA dataset contains binary annotations about the hair color, but these annotations don't include the red hair class and often contain inconsistencies. Starting from these annotations, three independent human labelers were asked to annotate each image with the hair color (five classes). In cases of disagreements between the labellings, we used a simple voting procedure to determine the ground truth class. The annotations are made public to the scientific community.

Fig. 5. Bald vs. non-bald detection results.

Table 3. Hair color recognition performance

Colorspace	NC	Accuracy%
LAB	10	70.8
LAB	20	90.08
LAB	**30**	**92**
HSV	10	83.2
HSV	20	84
HSV	30	82.8

The training data consists of 1000 representative samples for the defined hair colors. The test dataset comprises 500 images (100 for each class). We chose balanced train and test datasets and the red hair class has the fewest samples.

The number of clusters NS in which to discretize the colors using kMeans algorithm is a hyper-parameter of the hair color recognition module. We also tested the performance of our algorithm using different color-spaces to represent the images.

In Table 3 we report the classification performance under the most commonly used color-spaces and with different values for the number of clusters NC.

The confusion matrix corresponding to the best classification performance (LAB colorspace, $NC = 30$) is reported in Table 4.

Table 4. Confusion matrix of the proposed hair color classification algorithm

	Black	Blond	Brown	Grey	Red
Black	99	0	1	0	0
Blond	0	94	2	4	0
Brown	4	3	89	3	1
Grey	1	1	0	98	0
Red	0	4	16	0	80

The majority of confusions occurred between the Red and Brown classes. We observed the same issues when merging the color annotations of the three independent human labellers.

5   Conclusions and Future Work

This work tackled the problem of automatic hair analysis in facial images. We proposed a framework which performs the following tasks: hair and face segmentation, hair (baldness) detection and, if the case, hair color recognition. The segmentation module relies on fully convolutional neural networks to determine the hair, face and background pixels. We performed experiments with two different architectures. The hair color recognition module operates on color features extracted at super-pixel level in order to distinguish the natural hair colors: black, blond, brown, grey/white or red hair.

The proposed hair analysis framework was tested on several publicly available databases, which contain images captured in unconstrained scenarios. In order to train and evaluate the proposed algorithms, we also manually segmented more than 3500 images from a public face database. The annotations are made public to the scientific community.

As future work, we plan to extend the proposed hair color taxonomy with other unnatural hair tones: blue, green, violet etc. Also, we plan to modify the segmentation FCN, such that is able to perform other classification tasks, such as hairstyle and face shape classification.

References

1. Achanta, R., Shaji, A., Smith, K., Lucchi, A., Fua, P., Süsstrunk, S.: SLIC superpixels compared to state-of-the-art superpixel methods. IEEE Trans. Patt. Anal. Mach. Intell. **34**(11), 2274–2282 (2012)
2. Arthur, D., Vassilvitskii, S.: k-means++: the advantages of careful seeding. In: Proceedings of the Eighteenth Annual ACM-SIAM Symposium on Discrete Algorithms, pp. 1027–1035. Society for Industrial and Applied Mathematics (2007)
3. Julian, P., Dehais, C., Lauze, F., Charvillat, V., Bartoli, A., Choukroun, A.: Automatic hair detection in the wild. In: 2010 20th International Conference on Pattern Recognition (ICPR), pp. 4617–4620. IEEE (2010)
4. Kae, A., Sohn, K., Lee, H., Learned-Miller, E.: Augmenting CRFs with Boltzmann machine shape priors for image labeling. In: Proceedings of the IEEE Conference on Computer Vision and Pattern Recognition, pp. 2019–2026 (2013)
5. King, D.E.: Dlib-ml: a machine learning toolkit. J. Mach. Learn. Res. **10**, 1755–1758 (2009)
6. Krupka, A., Prinosil, J., Riha, K., Minar, J., Dutta, M.: Hair segmentation for color estimation in surveillance systems. In: Proceedings of 6th International Conference on Advanced Multimedia, pp. 102–107 (2014)
7. Liu, Z., Luo, P., Wang, X., Tang, X.: Deep learning face attributes in the wild. In: Proceedings of International Conference on Computer Vision (ICCV), December 2015

8. Long, J., Shelhamer, E., Darrell, T.: Fully convolutional networks for semantic segmentation. In: Proceedings of the IEEE Conference on Computer Vision and Pattern Recognition, pp. 3431–3440 (2015)

9. Muhammad, U.R., Svanera, M., Leonardi, R., Benini, S.: Hair detection, segmentation, and hairstyle classification in the wild. Image Vis. Comput. **71**, 25–37 (2018)

10. Prinosil, J., Krupka, A., Riha, K., Dutta, M.K., Singh, A.: Automatic hair color de-identification. In: 2015 International Conference on Green Computing and Internet of Things (ICGCIoT), pp. 732–736. IEEE (2015)

11. Proença, H., Neves, J.C.: Soft biometrics: Globally coherent solutions for hair segmentation and style recognition based on hierarchical MRFs. IEEE Trans. Inf. Forensics Secur. **12**(7), 1637–1645 (2017)

12. Ronneberger, O., Fischer, P., Brox, T.: U-Net: convolutional networks for biomedical image segmentation. In: Navab, N., Hornegger, J., Wells, W.M., Frangi, A.F. (eds.) MICCAI 2015. LNCS, vol. 9351, pp. 234–241. Springer, Cham (2015). https://doi.org/10.1007/978-3-319-24574-4_28

13. Rousset, C., Coulon, P.Y.: Frequential and color analysis for hair mask segmentation. In: 2008 15th IEEE International Conference on Image Processing, ICIP 2008, pp. 2276–2279. IEEE (2008)

14. Sarraf, S.: Hair color classification in face recognition using machine learning algorithms. Am. Sci. Res. J. Eng. Technol. Sci. (ASRJETS) **26**(3), 317–334 (2016)

15. Shen, Y., Peng, Z., Zhang, Y.: Image based hair segmentation algorithm for the application of automatic facial caricature synthesis. Sci. World J. **2014**, 1–10 (2014)

16. Sinha, P., Balas, B., Ostrovsky, Y., Russell, R.: Face recognition by humans: nineteen results all computer vision researchers should know about. Proc. IEEE **94**(11), 1948–1962 (2006)

17. Sinha, P., Poggio, T.: 'united' we stand. Perception **31**(1), 133 (2002)

18. Yacoob, Y., Davis, L.: Detection, analysis and matching of hair. In: 2005 Tenth IEEE International Conference on Computer Vision, ICCV 2005, vol. 1, pp. 741–748. IEEE (2005)

Learning Morphological Operators
for Depth Completion

Martin Dimitrievski[✉], Peter Veelaert, and Wilfried Philips

IMEC-IPI-Ghent University, Ghent, Belgium
{martin.dimitrievski,peter.veelaert,wilfried.philips}@ugent.be

Abstract. Depth images generated by direct projection of LiDAR point clouds on the image plane suffer from a great level of sparsity which is difficult to interpret by classical computer vision algorithms. We propose a method for completing sparse depth images in a semantically accurate manner by training a novel morphological neural network. Our method approximates morphological operations by Contraharmonic Mean Filter layers which are easily trained in a contemporary deep learning framework. An early fusion U-Net architecture then combines dilated depth channels and RGB using multi-scale processing. Using a large scale RGB-D dataset we are able to learn the optimal morphological and convolutional filter shapes that produce an accurate and fully sampled depth image at the output. Independent experimental evaluation confirms that our method outperforms classical image restoration techniques as well as current state-of-the-art neural networks. The resulting depth images preserve object boundaries and can easily be used to augment various tasks in intelligent vehicles perception systems.

1 Introduction

Recent advances in active depth sensing technologies such as high resolution LiDAR and Time of Flight cameras have extended the applications where robustness and accuracy have been a limiting factor in the past. This is especially true in the field of robotics and computer vision where solving high level problems such as autonomous navigation relies on a rich, multi-modal information. Until recently, most of such applications reinforced visible light images with per-pixel depth obtained using stereo cameras. Even though, stereo reconstruction has been widely researched and many high *performance* solutions do exist, e.g. close to 100 submissions on the KITTI Stereo 2015 benchmark[1], it still suffers from the effect of measurement correlated noise. As the 3D point is further away from the sensor, the perceived image disparity between the stereo cameras drops exponentially and so does the accuracy of it's estimate. Furthermore, the quality of reconstructed depth is directly coupled to the camera baseline which creates another set of challenges. A small autonomous robot cannot carry a wide baseline stereo rig, limiting the efficiency of any stereo reconstruction algorithm applied

---

[1] http://www.cvlibs.net/datasets/kitti/eval_scene_flow.php?benchmark=stereo.

© Springer Nature Switzerland AG 2018
J. Blanc-Talon et al. (Eds.): ACIVS 2018, LNCS 11182, pp. 450–461, 2018.
https://doi.org/10.1007/978-3-030-01449-0_38

on the data. On the other hand, wide baseline stereo setups suffer greatly from disparity artifacts around object boundaries that are close by. Solving for these issues is a challenging task which more often than not requires additional computational load on the perception system. Active depth sensors contrast the principle of depth reconstruction by sending out well controlled infra-red pulses into the surroundings. This infra-red light reflects back from the environment and can be correlated to distances in a systematic way. At the time of writing, commercial real-time LiDAR sensors can reliably measure depth in the range of 50 cm to 80 m while at the same time sustain low noise levels that are uncorrelated with the measurements. Data is being produced in the form of 3D point clouds which have excellent depth resolution, but rather low spatial resolution. In order to achieve real time operation, the LiDAR electronics sample depth at predefined sparse azimuth and elevation latices. Usually, the spatial resolution is lower in the elevation axis, which compared to the contemporary camera sensors varies in the range of 1 depth scan-line for each 5–10 lines of RGB data. See the left image in Fig. 1 for illustration of this effect.

Consequently, a direct projection of a LiDAR point cloud on the camera image produces a very sparse depth image. Classical computer vision algorithms such as visual odometry, scene understanding, segmentation, object detection, etc. have difficulties extracting useful information from this sparse input. Many of the processing steps need to be specifically tuned to incorporate sparse depth pixels and ignore missing depth values which reduces the efficiency and in turn usefulness of state-of-the-art algorithms. The problem of achieving equal sampling resolution of the RGB camera and depth sensor is called depth completion and is the main topic of this paper. A non-sparse data cube consisting of reconstructed depth pixels, middle image on Fig. 1, can be easily interpreted by classical computer vision algorithms. We have shown that pedestrian detection, in particular, can achieve much higher performance when operating on a RGB-D data reconstructed from a camera-LiDAR pair [1]. Even though many of the proposed depth completion methods produce dense and visually pleasing depth

RGB-D input   Proposed output   Ground truth

Fig. 1. Illustration of the depth completion problem. Left: input data is given in the form of registered RGB image and projected LiDAR points. Middle: Fully reconstructed depth image, output of the proposed method. Right: dense ground truth data used for training.

images, the depth completion problem is not entirely solved. A seemingly under explored track is the exploitation of contextual information in the camera image in order to produce more accurate depth in a semantically meaningful manner. In this paper we propose a novel neural network architecture which is capable to complete missing depth pixels by employing a mixture of morphological and convolutional layers. Learnable morphological operators provide robustness to the input sparsity, while multi-resolution convolutional layers extract contextual information about object shapes and boundaries. Our network reconstructs a depth image which is complete, accurate and preserves object edges.

In the following section we will make a brief overview of how the state-of-the-art handles context in the depth completion problem, then in Sect. 3 we will define a novel neural network architecture suited for fusion of RGB information with sparse depth input, furthermore in Sect. 4 we perform large scale evaluation of the proposed method and report the accuracy and performance, and we discuss the effectiveness and possible downsides of using our method in Sect. 5.

2   Overview

One of the pioneering depth completion methods, [2] considered to estimating each pixel location in the sparse depth image by means of local interpolation within a square window. The authors analyzed various classical reconstruction techniques which rely on depth information alone such as inverse distance weighting, Shepard's Method, ordinary Kriging, Delaunay triangulation and bi-lateral filtering. Furthermore they introduced a modified bilateral filter which also considers depth dispersion within the interpolation window. This method can crudely model the appearance of an object edge or boundary into two categories: foreground and background. A local segmentation is performed on the depth pixels which produces two clusters from which only the points that belong to the dominant cluster contribute to the bi-lateral filter. These authors found out that even simple techniques such as the minimum and median filter can complete missing depth with comparable accuracy to the more complex bi-lateral filter. A major drawback in this work is the simple model of the environment which doesn't take into account the geometrical and contextual structure of objects.

Following the success of the bi-lateral filter, we have proposed a semantically aware multi-lateral filter, [1] that is guided by a segmentation image. The segmentation image is computed by segmenting the LiDAR point cloud in a pre-processing step and is independent on the filtering window size. In our modified multi-lateral filter we use both IR-reflectance and depth pixels originating from the dominant object within the reconstruction window. Although we report state-of-the-art performance on a small scale database, we encountered difficulties obtaining an accurate segmentation image due to the point cloud sparsity. This effect is especially pronounced in distant objects that are sampled by only a few points. Contextual information originating from the RGB camera image remained untapped.

Ku et al. [3] propose a surprisingly simple yet efficient depth completion method using a sequence of morphological operations on the sparse depth image. In their experiments they show that a small set of fine tuned dilations and erosions is enough to reconstruct a high quality depth image. By experimenting with various kernel sizes and shapes they come to the conclusion that 5×5 diamond shaped morphological operators are able to outperform even some neural network based methods. However, higher level information about object types and shapes is completely ignored, which can potentially lead to even better reconstructions.

Recently, [4] proposed a method for semantically guided depth completion by means of local plane fitting. They use the assumption that the environment is locally smooth and can be piece-wise modeled by 3D planes. With the intention of preserving depth discontinuities and tiny structures, they introduce an novel edge and semantics aware geodesic distance metric. Additionally, they propose an outlier rejection scheme by utilizing labels from the state-of-the-art semantical segmentation algorithm, FCN [5]. Their reported qualitative results are promising, however, the method is not monolithic as it relies on different technologies and has been outperformed by special purpose neural networks.

Uhrig et al. [6] propose a depth completion method by processing the raw RGB-D data cube using a novel neural network. They propose a sparsity invariant convolutional layer which is built using an additional sampling mask. The mask holds binary information about which pixel is scanned by the LiDAR and is used to normalize the convolutional operations. Therefore, the network can easily handle varying degrees of input data sparsity without any adjustments or tweaking of the parameters. One downside of this method is that the network is based on the Fully Convolutional architecture which has a high computational load. Each inference produces a single depth pixel value and thus can not be employed in real time applications.

3   Method

3.1   General Architecture

Our depth completion system is a transfer function which takes sparse depth images and corresponding camera frames as inputs and outputs fully sampled depth images. The system consists of a fixed and a trainable part. In the fixed part, we make full use of previous state-of-the-art interpolation knowledge, as implemented in [3], and for the trainable part we propose a novel morphological CNN based on the principles of U-Net [7], SegNet [8] and ResNet [9]. We argue that local depth information alone is not enough to fill-in the missing depth values in a semantically meaningful way. Therefore, our CNN processes RGB-D information at multiple scales, learning an optimal depth reconstruction function that in part is guided by color information. The fixed part of the pipeline can be seen as a pre-processing step used to better initialize the later CNN. Even though all of the network weights are actually initialized using the Xavier method [10], we noticed that convergence is reached much faster when the preprocessed data is introduced.

Standard 2D convolution operations have difficulties in learning sparse data input problems [6, 11]. This is especially true when it is necessary to distinguish between actual measurement values and invalid pixels. Therefore, the entry point of our CNN is a series of trainable morphological filters operating on the sparse depth images alone. The purpose of this morphological sub-network is to separate the low level RGB and depth processing pipelines in order to better learn an initial depth image estimate. We approximate morphological dilation and erosion operations by utilizing the limit behavior of the Contraharmonic Mean Filter (CHM). These filters can be easily implemented in most contemporary deep learning frameworks by means of standard convolutional layer building blocks. In the later layers, morphologically processed depth and RGB information is fused using standard convolutional layers. Instead of operating directly on the depth values, our morphological network operates on inverse depth (disparity) values which are then converted back to depth when computing the loss function. The network is designed to output a depth patch with equal spatial size of the input patch and can thus process high resolution data in reasonable time.

3.2   Morphological Filtering with a Contraharmonic Mean Filter

Morphological operators are the foundation of many image segmentation algorithms. Using so called "structuring elements" they represent non-linear operations which compute the minimum, maximum or the combination of both within the element. Morphological operations are also invariant to translation and are strongly related to Minkowski addition. In the context of depth completion, it is of interest for the system to learn the shape and the operation type that fits best the data. However, due to the non-differentiable nature of minimum and maximum filtering, only few approaches have been found to succeed in the literature. To this end, we find that the approximation of morphological operators by the contraharmonic mean (CHM) filter in [12] is the best founded technique which can easily be integrated in a deep learning framework. In this paper, we also use the CHM to approximate our basic learn-able morphological block. Following the analysis in [12–14], we model the contraharmonic mean filter function $\psi_k(x)$ as the power-weighted 2D convolution of the image $f(x)$ and a filter w representing the structuring element:

$$\psi(x; W, k) = \psi_k(x) = \begin{cases} \dfrac{(f^{k+1} * w)(x)}{(f^k * w)(x)} & if\, k \in \mathbb{R} \\ \min(f_W(x)) & if\, k = -\infty \\ \max(f_W(x)) & if\, k = +\infty \end{cases} \tag{1}$$

where the order k of the filter defines the desirable properties such as morphological erosion if $k \ll 0$ or dilation if $k \gg 0$, W defines the set of values in the support of w and f^k is simply the input image raised to the power k. When k is large, the filter tends to select the pixels x_i with the largest values within

Algorithm 1. MATLAB snippet for initializing the morphological operator using AutoNN and MatConvNet framework

```
function  out = morphFilter(x,weights,biases,k)
          x1 = x.^(k+1);
          x1 = vl_nnconv(x1,weights,biases);
          x2 = x.^k;
          x2 = vl_nnconv(x2,weights,biases);
          out = x1./x2;
end
```

the support region W which in the limit case, $k \rightarrow \infty$ equates to the supremum, i.e. morphological dilation:

$$\lim_{k \to \infty} \psi_k (x) = \max_{x_i \in W} (f (x - x_i)) \equiv \psi_\infty (f). \tag{2}$$

Otherwise, when k is sufficiently small, the CHM filter will tend to select the smaller valued pixels which in the limit case $k \rightarrow -\infty$ equates to the infimum, i.e. the morphological erosion:

$$\lim_{k \to -\infty} \psi_k (x) = \min_{x_i \in W} (f (x - x_i)) \equiv \psi_{-\infty} (f). \tag{3}$$

In practice, the choice of k, and thus computing the derivative, will be limited by the computer number precision, but we found that a value of $k = 5$ produces the desired morphological filtering effect using single-precision floating point filter and pixel values. For a more detailed analysis of the filter properties and their proofs we advise the reader to the works of van Vliet [13] and Angulo [14]. This filter formulation is differentiable with respect to the input data x and both the filter mask w and the filter order k, as given in [12]. However, in our problem where we only encounter empty regions which we need to complete, we fix the order k to a positive value and are mainly focused on learning the structuring element w, hence the partial derivative of the filter with respect to the input is:

$$\frac{\partial \psi_k (x)}{\partial f} = \frac{f (x) - f (x) \left(f^{k+1} * w \right) (x)}{\left(f^k * w \right) (x)}, \tag{4}$$

and is used while applying the chain rule in back propagation. In practice we use the MatConvNet [15] framework with the AutoNN implementation of automatic differentiation API which successfully handles the inference and backpropagation operations. Formally, the CHM filter is implemented using two convolution layers representing the denominator and numerator in (1). The convolution layers share the same filters and biases and have the same learning rates. An extract of our initialization procedure for a single morphological operator is given in Algorithm 1. The learned morphological operators can thus be visualized by taking the logarithm $m_i (x) = \log (w)$. We note that the input data for our morphological operators is the inverse depth (disparity) image rather than the raw depth values. This is done because of the nature of the morphological dilation operation in gray level images, where pixels with larger values are extended by the shape of the structuring element. In cases where an area to be dilated is completely filled with measurements, the resulting dilation will accentuate objects that are nearer to the camera (lower depth, greater disparity), rather than the background. This result is more desirable since it is safer to assume that no object with size less than half of the structuring element will be completely lost by applying the dilation.

3.3   Proposed Network Architecture

Depth completion using a regression neural network can be performed in three different ways. Depth can be reconstructed per pixel, per patch or per entire frame. While processing each pixel individually enables us to use a very deep CNN, in reality it's deployment is intractable due to long computing times. On the other hand, reconstructing the entire image in a single pass is most desirable, but it is difficult to achieve acceptable results since processing is limited by image resolution and GPU memory. Single image inference networks must be kept relatively simple which limits their performance in terms of error rates. In this paper we propose a medium sized CNN architecture which can process sparse patches of the input RGB-D images. A complete dense depth image is thus formed by processing overlapping patches over the input image. The general structure of the proposed network is loosely based on the auto-encoder with information leak proposed in [7,8] for high resolution segmentation tasks.

We exploit the knowledge from [3] and start with a fixed pre-processing morphological block consisting of a sequence of dilation, closing and two hole filling operators. In a separate step we use two learnable CHM filter layers to infer the optimal morphological operations from the sparse input, each with a support of 11×11 pixels. In the end, the RGB channels are concatenated with the output of [3] and the output of our CHM filters and are passed through a three stage U-Net, Fig. 2.

The novelty of this paper comes from the introduction of morphological layers before the contracting part which helps to eliminate the sparsity in the input. By applying a succession of convolutional filters we then double the number of channels and reduce the spatial resolution by half. Having learned the optimal dilation structuring elements in the CHM filter, the U-Net channels have no sparsity and can easily adapt to produce the desired output. At the end of each contracting block we specifically chose to use average pooling layers because they capture all of the information necessary for later depth reconstruction. In the expanding part, the network employs standard series of "deconvolution" layers that upsample their input by a factor of 2× and concatenate outputs from the respective contracting layers. The purpose of the expanding blocks is to reduce the number of contextual information stored within the channels while at the same time increase the spatial resolution. This way, information over varying resolutions and abstraction levels in both RGB and input depth is used to form high resolution depth output which adheres to object boundaries.

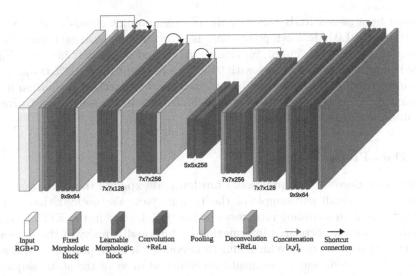

Fig. 2. General structure of the proposed network.

In order to learn the shape of the morphological and convolutional filters we use dense ground truth depth images to which we compare the output of our proposed network. Deviations from the ground truth can be quantified by a multitude of different metrics, such as absolute error, squared error, inverse absolute error, inverse squared error, absolute and squared relative error, percentage of outliers, etc. but in this paper we stick to the classical mean squared error as our loss function. We are motivated to do so because we are interested in accurate depth reconstructions which can later be used for the tasks of autonomous driving. Therefore, object distances need to be accurate regardless of their absolute distance to the sensor, i.e. we equally penalize error whether it is for a distant or a close object. For training the entire network we use the backpropagation chain rule and employ the stochastic gradient descent by adaptive moment estimation (ADAM) technique, [16]. This method computes individual adaptive learning rates for different parameters from estimates of first and second moments of the gradients. We set an initial learning rate $(\alpha = 10^{-5})$ and two hyper-parameters: decay for the first moment vector $(\beta_1 = 0.9)$ and decay for the second moment vector $(\beta_2 = 0.999)$.

4 Experiments

4.1 General

Learning a robust set of depth completion filters requires a large and variable set of input and ground truth data. The very recently published KITTI depth completion[2] data [6,17,18] is an excellent example of such dataset in the context of depth sensing for autonomous navigation. It consists of video sequences

---

[2] www.cvlibs.net/datasets/kitti/eval_depth.php?benchmark=depth_completion.

captured by a stereo RGB camera pair as well as point clouds from the Velo-
dyne HDL-64E LiDAR. Each point cloud is projected on a virtual camera image
creating a sparse depth image, previously visualized on the left image in Fig. 1.
There is a total of 151 sequences with 93505 frames split into ~92% training and
the remaining ones for model validation. Independent method evaluation is also
provided by means of an on-line server which tests uploaded results to frames
with unknown ground truth data.

4.2   Data Preparation

In order to reduce the computational burden and expedite the training process
we only use a small sub-sample of the training set. We noticed that, due to
the relatively high sampling rate, most of the 93K frames in the KITTI dataset
contain temporally correlated information. Additionally, many of the sequences
are recorded from a static vehicle and thus contain a large portion of the same
content. Thus, in all our experiments we removed most of the static sequences
and only sample every 6th frame from the remaining data. We ended up with a
training set of ~4.3K samples. Input images are padded to a fixed resolution of
1280×384 pixels from which we randomly sample rectangular patches of size
96×96. Since our network uses 3 stage contraction, the lowest resolution of the
input image inside the network is 12×12 with a channel depth of 256.

4.3   Training Procedure

Learning of the optimal network parameters is done by presenting the network
with sub-batches of the labeled training set. After each inference, batch-average
MSE is calculated from ground truth and the value is used as a loss to adjust
the convolution filter parameters. Each successive layer is updated by backpropagation using the chain rule. We employ the ADAM optimization method and,
since we train using small patches of images, we train until convergence for ~200
epochs. During training we keep the hyper-parameters α, β_1 and β_2 fixed, but
adaptively change the batch size, starting from 4 increasing to 64. After each
epoch, we also perform validation using a small *sub-set* of the validation dataset.

4.4   Analysis

We deployed our trained neural network on the 1000 test samples from the
KITTI depth completion benchmark. The accuracy of our method in terms of
iRMSE, iMAE, RMSE and MAE is independently measured by the KITTI online server and compared to various techniques. This benchmark contains various
anonymous submissions, summarized on Table 1, to which we can't fully compare
since we don't know their exact details. To that end, we match our results only
to results from publicly available techniques. In terms of raw RMSE error we
outperform classical methods such as [3,4], as well as the only published CNN
method [6]. Qualitatively, our method also better preserves object boundaries

Fig. 3. Typical use case scenario in an urban environment. From top to bottom: (a) RGB camera frame, (b) completed depth image from [3], (c) completed depth image from [6] and (d) completed depth image from the proposed method.

which is visible from the results shown on Fig. 3. Using the RGB information in the contracting and expanding network architecture, we are able to effectively fill in missing object parts with the relevant depth information. This is especially noticeable in transparent objects such as house and car windows and glass displays. The inclusion of morphological layers makes the network flexible enough so that sparse data is handled in the initial layers, while the rest of the network is dedicated to better extracting contextual information.

Table 1. Depth completion results on the KITTI benchmark.

Method	iRMSE	iMAE	RMSE	MAE	Runtime	Source
HMS-Net_v2	3.90	1.90	911.49	310.14	0.02 s / GPU	n/a
Sparse-to-Dense-2	3.21	1.35	954.36	288.64	0.04 s / GPU	n/a
HMS-Net	3.25	1.27	976.22	283.76	0.02 s / GPU	n/a
Morph-Net	**3.84**	**1.57**	**1045.45**	**310.49**	**0.17 s / GPU**	**Proposed**
IP-Basic	3.78	1.29	1288.46	302.60	0.011 s / 1 core	Ku [3]
ADNN	59.39	3.19	1325.37	439.48	0.04 s / GPU	n/a
NN+CNN	3.25	1.29	1419.75	416.14	0.02 s /	Uhrig [6]
SparseConvs	4.94	1.78	1601.33	481.27	0.01 s /	Uhrig [6]
NadarayaW	6.34	1.84	1852.60	416.77	0.05 s / 1 core	Uhrig [6]
SGDU	7.38	2.05	2312.57	605.47	0.2 s / 4 cores	Schneider [4]
NiN CNN	4.60	2.15	2378.79	685.53	0.01 s /	n/a
NiN+Mask CNN	4.63	2.40	2534.26	848.25	0.01 s / GPU	n/a

5   Conclusion

Depth completion from sparse inputs has traditionally been solved by local image processing that handles sparsity using fine tuned filters. However, in instances where the level of sparsity varies spatially or parts of objects are completely missing, local processing is unable to accurately reconstruct depth information. Contextual information from the entire scene or parts of objects must be considered to better fill-in missing depth. We explored the idea of adding learnable morphological filters to a state-of-the-art multi-scale neural network in order to reduce the input sparsity. At the same time, these filters are adjusted for optimal reconstruction quality. We implemented morphological operators using the CHM filter which can be constructed out of standard deep learning building blocks. In terms of reconstruction accuracy, our method outperforms all published classical and neural network based approaches. It operates on image stripes which are concatenated to form the final depth image. The run-time for completing a single image (inference) of size 1280×384 pixels is on average 1.75 s. A future direction of our research will be to investigate which additional steps need to be taken in order to reduce the complexity of the network while not sacrificing reconstruction accuracy.

Acknowledgements. "The Titan Xp used for this research was donated by the NVIDIA Corporation through the Academic Grant Program."

References

1. Dimitrievski, M., Veelaert, P., Philips, W.: Semantically aware multilateral filter for depth upsampling in automotive LiDAR point clouds. In: 2017 IEEE Intelligent Vehicles Symposium (IV), pp. 1058–1063, June 2017
2. Premebida, C., Garrote, L., Asvadi, A., Ribeiro, A.P., Nunes, U.: High-resolution LiDAR-based depth mapping using bilateral filter. CoRR, abs/1606.05614 (2016)
3. Ku, J., Harakeh, A., Waslander, S.L.: In defense of classical image processing: fast depth completion on the CPU. CoRR, abs/1802.00036 (2018)
4. Schneider, N., Schneider, L., Pinggera, P., Franke, U., Pollefeys, M., Stiller, C.: Semantically guided depth upsampling. In: Rosenhahn, B., Andres, B. (eds.) GCPR 2016. LNCS, vol. 9796, pp. 37–48. Springer, Cham (2016). https://doi.org/10.1007/978-3-319-45886-1_4
5. Long, J., Shelhamer, E., Darrell, T.: Fully convolutional networks for semantic segmentation. CoRR, abs/1411.4038 (2014)
6. Uhrig, J., Schneider, N., Schneider, L., Franke, U., Brox, T., Geiger, A.: Sparsity invariant CNNs. CoRR, abs/1708.06500 (2017)
7. Ronneberger, O., Fischer, P., Brox, T.: U-net: Convolutional networks for biomedical image segmentation. CoRR, abs/1505.04597 (2015)
8. Badrinarayanan, V., Kendall, A., Cipolla, R.: SegNet: a deep convolutional encoder-decoder architecture for image segmentation. CoRR, abs/1511.00561 (2015)
9. He, K., Zhang, X., Ren, S., Sun, J.: Deep residual learning for image recognition. CoRR, abs/1512.03385 (2015)
10. Glorot, X., Bengio, Y.: Understanding the difficulty of training deep feedforward neural networks. In: Teh, Y.W., Titterington, M. (eds.) Proceedings of the Thirteenth International Conference on Artificial Intelligence and Statistics. Proceedings of Machine Learning Research, PMLR, vol. 9, Chia Laguna Resort, Sardinia, Italy, 13–15 May 2010, pp. 249–256 (2010)
11. Chen, X., Ma, H., Wan, J., Li, B., Xia, T.: Multi-view 3D object detection network for autonomous driving. CoRR, abs/1611.07759 (2016)
12. Masci, J., Angulo, J., Schmidhuber, J.: A learning framework for morphological operators using counter-harmonic mean. CoRR, abs/1212.2546 (2012)
13. Vliet, L.J.V.: Robust local max-min filters by normalized power-weighted filtering. In: 2004 Proceedings of the 17th International Conference on Pattern Recognition, ICPR 2004, vol. 1, pp. 696–699, August 2004
14. Angulo, J.: Pseudo-morphological image diffusion using the counter-harmonic paradigm. In: Blanc-Talon, J., Bone, D., Philips, W., Popescu, D., Scheunders, P. (eds.) ACIVS 2010. LNCS, vol. 6474, pp. 426–437. Springer, Heidelberg (2010). https://doi.org/10.1007/978-3-642-17688-3_40
15. Vedaldi, A., Lenc, K.: MatConvNet - convolutional neural networks for MATLAB. In: Proceedings of the ACM International Conference on Multimedia (2015)
16. Kingma, D.P., Ba, J.: Adam: a method for stochastic optimization. CoRR, abs/1412.6980 (2014)
17. Geiger, A., Lenz, P., Urtasun, R.: Are we ready for autonomous driving? the KITTI vision benchmark suite. In: 2012 IEEE Conference on Computer Vision and Pattern Recognition, pp. 3354–3361, June 2012
18. Geiger, A., Lenz, P., Stiller, C., Urtasun, R.: Vision meets robotics: the KITTI dataset. Int. J. Robot. Res. (IJRR) 32, 1231–1237 (2013)

Dealing with Topological Information Within a Fully Convolutional Neural Network

Etienne Decencière[1(✉)], Santiago Velasco-Forero[1], Fu Min[2], Juanjuan Chen[2], Hélène Burdin[3], Gervais Gauthier[3], Bruno Laÿ[3], Thomas Bornschloegl[4], and Thérèse Baldeweck[4]

[1] MINES ParisTech, PSL Research University, Centre for Mathematical Morphology, Fontainebleau, France
etienne.decenciere@mines-paristech.fr
[2] L'Oréal Research and Innovation, 550 Jinyu Road, Pudong, Shanghai, China
[3] ADCIS SA, 3 rue Martin Luther King, 14280 Saint-Contest, France
[4] L'Oréal Research and Innovation, 1 avenue Eugène Schueller, 93601 Aulnay-sous-Bois, France

Abstract. A fully convolutional neural network has a receptive field of limited size and therefore cannot exploit global information, such as topological information. A solution is proposed in this paper to solve this problem, based on pre-processing with a geodesic operator. It is applied to the segmentation of histological images of pigmented reconstructed epidermis acquired via Whole Slide Imaging.

Keywords: Histological image segmentation
Convolutional neural network · Geodesic operators
Mathematical morphology

1 Introduction

Image processing and analysis has been revolutionized by the rise of deep learning. For semantic segmentation, deep learning approaches use mainly convolutional neural networks (CNN) [3,4]. In the biomedical field, U-Net [5] has become the state-of-the-art method for this task, but other solutions exist, such as SegNet [1]. These networks are fully convolutional. Their receptive fields are of limited size. Therefore, they cannot intrinsically process global information, such as topological information [6]. We recall that, unlike in networks containing fully connected layers, where the value of each unit depends on the entire input to the network, a unit in a convolutional networks only depends on a region of the input. This region in the input is the receptive field for that unit.

In the following, we present a practical real-world situation where the segmentation result depends on topological information. We show that a classical fully convolutional CNN does not give satisfactory results and propose a solution to this problem.

© Springer Nature Switzerland AG 2018
J. Blanc-Talon et al. (Eds.): ACIVS 2018, LNCS 11182, pp. 462–471, 2018.
https://doi.org/10.1007/978-3-030-01449-0_39

During the past years, the way of working in the histological field has changed due to the emergence of Whole Slide Imaging solutions that are now available and useful for pathologists but also for dermatologists and cosmetologists. These automated scanners improve digital histology by allowing several hundred slides a day to being acquired through imaging process, and stored, used for second opinions or for automated analysis. This well-established technology provides large sets of image data that need to be automatically processed whatever the amount of generated data. Image analysis methods based on Deep Learning have proved to be extremely useful in this field [2, 8].

To circumvent this, we propose a method that allows neural networks based on convolutional layers to take into account non-local information. It is based on a geodesic operator, a popular tool from mathematical morphology. We apply it to the segmentation of whole slide images of pigmented reconstructed epidermis, as described hereafter.

2 Material

The images that were collected include pigmented reconstructed epidermis samples used to evaluate and identify the de-pigmenting or pro-pigmenting efficiency of cosmetic ingredients. They have been colored using a Fontana Masson staining, a silver stain that is used to highlight melanin and to also reveal skin layer morphology. Their sizes are diverse and can reach up to 20 million pixels. The ground truth (GT) has been obtained by an automatic method developed by the ADCIS company, whose results were manually edited and modified by L'Oréal experts, when needed. On those images, the goal of the segmentation is to identify two regions corresponding to two specific skin layers: the *stratum corneum* (SC) and the living epidermis.

The boundary between the SC and the background is usually rather difficult to determine, mainly because it can be composed of different layers separated by gaps due to the desquamation process that happens in the SC. Such layers are only considered as part of the SC if they constitute an unbroken boundary between the background and the sample. This feature is highly non local, and as such a convolutional neural network cannot enforce it.

The resulting database contains 120 color images, coming from 46 different slides. The original images are of variable size, and can contain more than 20 million pixels. The test database has been built with approximately 20% of the images. The remaining images were used for learning and validation. We took care to transfer all the images from a given slide into the same database.

Given the layered structure of the images, segmenting them into four regions is equivalent to finding three frontiers (see Fig. 1). The top one, between background and SC, has a variable appearance. In some images it corresponds to a contrasted contour; in other to a soft irregular contour. More importantly, its exact position depends on non-local information. Indeed, when a layer of the SC is separated from the rest of the tissue, it will belong to the SC region only if it is connected to the rest of the SC, or if it constitutes an unbroken frontier,

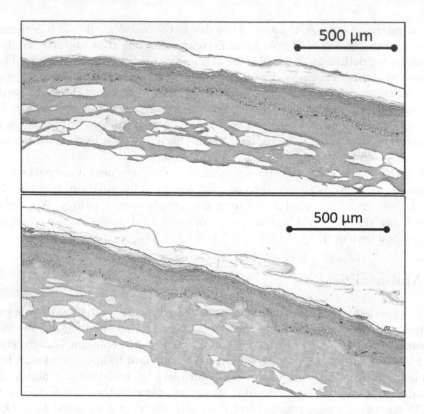

Fig. 1. Examples of original images with overlayed contours of the reference segmentation (best viewed in colour). Red/top contour: frontier between background and SC. Note that its position can be completely shifted if the detached layer is unbroken (top) or broken (bottom). Green/middle contour: frontier between SC and living epidermis. Cyan/bottom contour: frontier between living epidermis and collagen scaffold (here considered as background). (Color figure online)

going from the left side of the image to the right. Therefore it is not possible to make that decision based uniquely on local information. This kind of situation is illustrated by Fig. 1.

The second frontier separates SC from living epidermis. On our images, the distinction between those regions raises from different textures. The third frontier corresponds to the limit between living epidermis and collagen scaffold. Note that the fourth region to be segmented is made up of two compartments: the collagen scaffold that supports the reconstructed skin, and background. The collagen scaffold contains some large "holes" that can locally look as the "holes" within the SC.

The ground truth was generated using an automatic method developed by the ADCIS company, whose results were manually edited and modified by L'Oréal experts, when needed. Given that the top and bottom regions of the ground truth contained both a large white region, which could not be locally differentiated, we decided to only consider three labels: label 1 corresponds to the background (both at the top and the bottom of the images) and collagen scaffold; label 2 to the SC; label 3 to the living epidermis.

3   Methods

It was decided to use convolutional neural networks to tackle this problem. During the learning phase, we worked with crops of a given size. After running some tests, the final size of the crops was 512×512.

The ground truth segmentations, as they contained three labels, were classically represented as an image with three binary channels. Given that background white regions covered the majority of the images, for training we only used the crops that contained at least label 2 or label 3. This procedure is illustrated in Fig. 2. The resulting set of crops contains 1458 elements. 80% are randomly picked for learning; the other 20% are used for validation.

Concerning the loss functions, based on our experience with segmentation using convolutional neural networks, we chose the following loss function between two same length vectors X and Y, containing values included between 0 and 1:

$$J_2 = 1 - \frac{XY + \epsilon}{X^2 + Y^2 - XY + \epsilon}$$

where ϵ is a small constant, used for numerical reasons. This loss is based on the Jaccard index, also called "intersection over union" similarity measure, often used to evaluate the quality of a segmentation.

3.1   Neural Network Architecture and First Results

After testing several network architectures, the one that gave the best results was U-Net [5] (using a sigmoid in the last activation layer, and using zero-padding in the convolutional layers). Full details are given in Sect. 3.3. The validation loss of the resulting model was 0.085.

The networks output contains 3 channels. Each one can be interpreted as the probability of a given pixel of belonging to each region. In order to obtain a segmentation, we naturally gave to each pixel the label corresponding to the channel with the highest probability.

A qualitative analysis of the first results showed that the resulting frontiers between SC and living epidermis, on the one hand, and between living epidermis and collagen scaffold, on the other hand, were very satisfactory. However, the frontier between background and SC, in some cases, contained errors. In Fig. 2 (middle) we see that this frontier is incorrectly detected and that the gap between the detached SC layer, on the right, is incorrectly considered as belonging to the SC. These errors span from the fact, as we previously said, that the definition of this frontier is based on non local information.

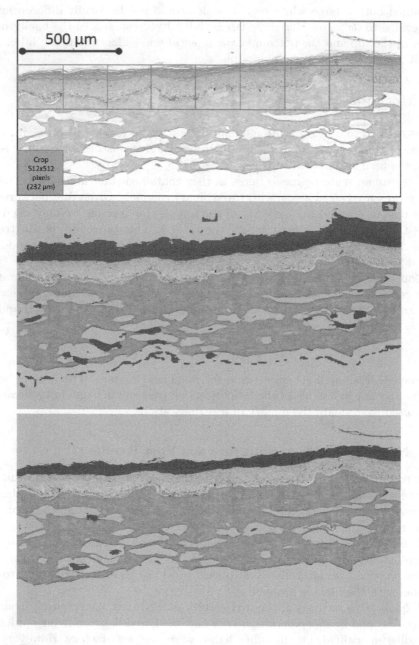

Fig. 2. Top: original image, showing the selected crops. Middle: results without using global information. Bottom: result using global information, thanks to the presented method. These segmentation results have not been postprocessed. They are overlayed on the original data using the following colour code: SC (magenta), living epidermis (orange) and other regions (cyan). (Color figure online)

3.2 Taking into Account Non Local Information in a Convolutional Neural Network

A new method is proposed here to cope with non local information within convolutional neural networks. It is based on a geodesic reconstruction of the input image from the top and bottom of the image, channel-wise.

We recall that the geodesic reconstruction [7] of a grey level image I from an image J_0 (often called marker) is obtained by iterating the following operation until idempotence:

$$J_{n+1} = \delta(J_n) \bigwedge I$$

where δ is a morphological dilation, here with the cross structuring element, corresponding to the 4-connectivity. Our marker image J_0 is equal to I on the first and last rows of the image domain, and to zero elsewhere. The process will "fill" the holes within the tissue, and preserve the intensity of the pixels that belong to the connected components of the background that touch the top and bottom of the image. This geodesic operation thus allows to bring topological information, which is essentially global, to a local scope.

In order to recover some of the bright details of the tissue sample, the result of this reconstruction is combined with the initial image by computing their mean. If we call J the result of the above reconstruction, the output image is simply:

$$F = (J + I)/2.$$

This operator is illustrated in Fig. 3. All images follow the same preprocessing (before computing the crops). Learning is done as before, with the same parameters.

The new CNN suppresses the segmentation errors due to the lack of global information on the background/SC boundary. Figure 2 clearly shows this improvement.

The validation loss of the model is 0.028, to be compared with the previous value of 0.085.

3.3 Hyper-parameters Optimization and Data Augmentation

We tuned the hyper-parameters of our system through manual grid search using the validation dataset. The parameters of the final model are: Optimizer: adadelta [9], with default parameters (learning rate: 1; rho: 0.95; epsilon: 10-8; decay: 0); epochs: 200; patience: 20, batch size: 4. The initial number of convolutional filters of the U-Net network is 16 (instead of 64 in the original paper), resulting in a network with 1,962,659 parameters.

We also tested several standard data augmentation methods, but they did not bring any improvement. We believe that this result means that our database constitutes a representative sampling of our image domain.

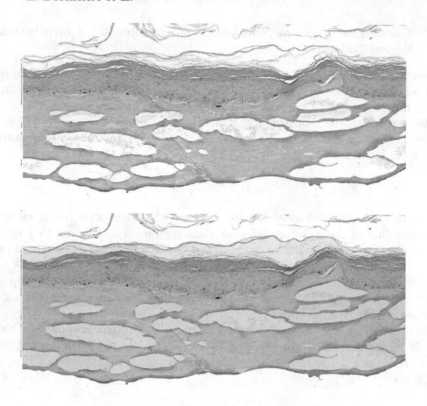

Fig. 3. Top: original image. Bottom: image after pre-processing based on the geodesic reconstruction. Differences are mainly visible on the holes within the tissue sample.

3.4 Post-processing

The current results are already satisfactory. There are however a few defects in the resulting segmentation (as can be seen in Fig. 2), most of which can be corrected with the following post-processing method:

1. For the SC and the living epidermis, keep only the largest connected component; for the background, keep the connected components that touch the top and the bottom of the image.
2. Pixels without label are given the label of the closest labelled pixel.

4   Results

It is interesting to note that once a convolutional neural network has been trained (with crops of constant size, as previously stated) it can be applied to images of almost arbitrary sizes. There are only two limitations: the system memory has to

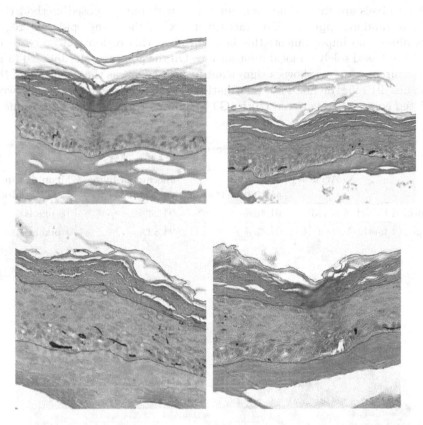

Fig. 4. Zoom-in on some test images to illustrate the results, as well as its robustness to acquisition artifacts. The contours of the segmentation computed with the final model are overlayed on the original images.

be large enough and neural network architectures that use downsampling layers impose that the dimensions be multiple of some $2n$ (where n is the number of such layers, supposing that the sampling steps are equal to 2). This approach is interesting not only for practical reasons (no need to compute any more crops and stitch them back together at prediction time) but also significantly alleviates border effects.

There are 23 images in the test database. Globally, the results were considered as very good by the final users. They are illustrated in Fig. 4. Only two errors were visible at first sight among the 23 images. They are shown in Fig. 5. Other errors are less visible. They correspond most of the time to a slight displacement of the obtained contour[1].

Table 1 gives some quantitative results on the test database. Accuracy values (the proportion of pixels that are correctly classified) show that incorrectly

---

[1] One image in Fig. 4 contains such an error; let the reader try to find it!.

classified pixels are three times less numerous with our proposed method than with the standard approach. The Jaccard index[2] of the living epidermis region shows almost no improvement: this is natural, as this region can be correctly segmented based solely on local information. On the contrary, the Jaccard index of the *stratum corneum* shows a significant improvement, as the definition of this region heavily relies on non-local information. Finally the mean distance between predicted contours and ground truth (GT) contours confirms this improvement.

Table 1. Quantitative results on test database.

	Accuracy	Jaccard of *stratum corneum*	Jaccard of living epidermis	Mean distance to GT contour
Standard U-Net	98.33%	91.46%	94.24%	18 pixels
Proposed method	99.49%	97.43%	94.82%	4 pixels

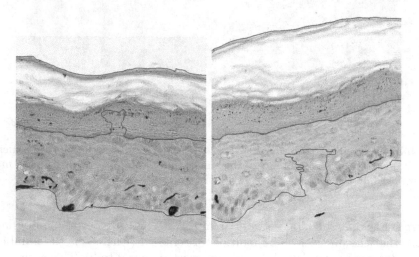

Fig. 5. Zoom-in onto the two more significant errors found on the 23 images of the test database.

Processing times are as follows. The standard U-Net takes 171 seconds to process the full 23 test images on a conventional laptop with a NVidia GeForce GTX 980M graphics card. The improved method, including the geodesic reconstruction, takes 407 s. We think that the pre-processing could be optimized, but the current version is already fast enough for the application at hand.

---

[2] The Jaccard index of two sets is the ratio between the size of their intersection and the size of their union.

5   Conclusion

A novel method to utilize global information within a convolutional neural network has been introduced. Based on the morphological reconstruction by dilation, it allows the network to take advantage of geodesic information.

This method has been successfully applied to the segmentation of histological images of reconstructed skin using a U-Net architecture. We believe that a similar improvement should be obtained with other fully convolutional neural networks, such as SegNet.

The method is being integrated in a complete software in order to use it in routine practice.

As a perspective, it would be interesting to explore other ways to use global information within convolutional neural networks, and compare them.

References

1. Badrinarayanan, V., Kendall, A., Cipolla, R.: SegNet: a deep convolutional encoder-decoder architecture for image segmentation. arXiv:1511.00561 [cs], November 2015. http://arxiv.org/abs/1511.00561
2. Bejnordi, B.E.: The CAMELYON16 consortium: diagnostic assessment of deep learning algorithms for detection of lymph node metastases in women with breast cancer. JAMA **318**(22), 2199–2210 (2017). https://jamanetwork.com/journals/jama/fullarticle/2665774
3. Fukushima, K.: Neural network model for a mechanism of pattern recognition unaffected by shift in position- neocognitron. Electron. Commun. Japan **62**(10), 11–18 (1979)
4. LeCun, Y., et al.: Backpropagation applied to handwritten zip code recognition. Neural Comput. **1**(4), 541–551 (1989). http://www.mitpressjournals.org/doi/10.1162/neco.1989.1.4.541
5. Ronneberger, O., Fischer, P., Brox, T.: U-Net: convolutional networks for biomedical image segmentation. In: Navab, N., Hornegger, J., Wells, W.M., Frangi, A.F. (eds.) MICCAI 2015. LNCS, vol. 9351, pp. 234–241. Springer, Cham (2015). https://doi.org/10.1007/978-3-319-24574-4_28
6. Rosenfeld, A.: Digital topology. Amer. Math. Monthly **86**, 621–630 (1979)
7. Vincent, L.: Morphological gray scale reconstruction in image analysis: applications and efficient algorithms. IEEE Trans. Image Process. **2**(2), 176–201 (1993)
8. Wang, D., Khosla, A., Gargeya, R., Irshad, H., Beck, A.H.: Deep learning for identifying metastatic breast cancer. arXiv:1606.05718 [cs, q-bio], June 2016. http://arxiv.org/abs/1606.05718
9. Zeiler, M.D.: ADADELTA: an adaptive learning rate method. CoRR abs/1212.5701 (2012). http://arxiv.org/abs/1212.5701

Coding and Compression

L-Infinite Predictive Coding of Depth

Wenqi Chang, Ionut Schiopu$^{(\boxtimes)}$, and Adrian Munteanu

Department of Electronics and Informatics, Vrije Universiteit Brussel,
Pleinlaan 2, 1050 Brussels, Belgium
ischiopu@etrovub.be

Abstract. The paper introduces a novel L_∞-constrained compression
method for depth maps. The proposed method performs depth segmentation and depth prediction in each segment, encoding the resulting information as a base layer. The depth residuals are modeled using a Two-
Sided Geometric Distribution, and distortion and entropy models for the
quantized residuals are derived based on such distributions. A set of optimal quantizers is determined to ensure a fix rate budget at a minimum
L_∞ distortion. A fixed-rate L_∞ codec design performing context-based
entropy coding of the quantized residuals is proposed, which is able to
efficiently meet user constraints on rate or distortion. Additionally, a
scalable L_∞ codec extension is proposed, which enables encoding the
quantized residuals in a number of enhancement layers. The experimental
results show that the proposed L_∞ coding approach substantially outperforms the L_∞ coding extension of the state-of-the-art CALIC method.

Keywords: L-Infinite norm · Optimized fixed-rate quantization
Depth map compression · Context modeling

1 Introduction

The technological advances in the field of 3D technologies are now materialized
in a large diversity of sensors for a wide range of applications. Since depth
images are used to obtain the shape of objects, the development of algorithms
specifically devised for depth map compression becomes a critical problem. Many
applications from different domains, such as computer vision, gaming, medical,
etc., do require accurate depth map information or can accept only a bounded
maximum distortion incurred by lossy compression of the input depth maps.

The state-of-the-art methods for depth compression are driven using the
classical L_2 distortion, which offers only an approximation of the *global* coding
error over the entire image. The drawback of using the L_2 norm is that the local
error is not bounded and does not offer any control over the pixel error. In this
paper, we propose an alternative L_∞ constrained codec for depth. The proposed
method controls the local error on each pixel by employing optimal quantizers
designed under the L_∞-constraint. In literature, the local reconstructed error

The work in this paper has been supported by Innoviris (3DLicorneA) and FWO.

J. Blanc-Talon et al. (Eds.): ACIVS 2018, LNCS 11182, pp. 475–486, 2018.
https://doi.org/10.1007/978-3-030-01449-0_40

is denoted as the maximum absolute difference (MAXAD), which is computed between the pixel values in the original and decoded images.

In the near-lossless image compression domain, several L_∞ coding approaches have been proposed over the years. In [1], the authors proposed a differential pulse-code modulation (DPCM) coding method based on the use of uniform quantizers to ensure that the L_∞ distortion is not greater than 1. In [2], the authors modified the state-of-the-art method in lossless image compression, namely the Context-based, Adaptive, Lossless Image Codec (CALIC) of [3], by extending it towards near-lossless compression. To achieve L_∞-scalability, a wavelet-based L_∞-constrained embedded coding scheme is proposed in [4], generalizing the method from [5] to any dimension n of any (non-integer) lifting-based wavelet transform.

In the depth map compression domain, several lossy compression methods have been proposed. Early approaches include [6], where a quadtree-based coding scheme for depth maps is proposed. In [7], the depth maps are encoded based on the segmentation of the corresponding color images. In [9], the depth maps are losslessly encoded using a chain-code representation strategy. In [10], a segmentation algorithm is proposed by generating a number of lossy versions obtained by introducing a target distortion in the initial depth image.

In this paper, a L_∞-constrained predictive coding scheme operating in the spatial domain is proposed. For any given MAXAD value, an optimal quantizer is devised such that the reconstruction error is locally bounded at a minimum bitrate. Vice-versa, the corresponding local error can also be minimized for any given bitrate budget. To this end, an analysis of the prediction error modeling for depth images is carried out and optimal scalar quantizer designs for the resulting model are proposed. A context-based entropy codec is proposed for fixed rates, together with a scalable L_∞ codec extension which allows for truncating the bitstream at a desired bitrate or target distortion.

The paper is organized as follows. Section 2 describes the proposed method. Section 3 analyzes the performance of the proposed codec compared to a state-of-the-art codec equipped with L_∞ coding functionality. Section 4 draws the conclusions of this work.

2   Proposed Method

This section is organized as follows. Section 2.1 presents the prediction error model for the proposed codec and establishes the mathematical formulations for the MAXAD and the entropy of the quantized prediction error for this model. Section 2.2 determines the optimal quantizers for a target bit-rate or target distortion. Sections 2.3 and 2.4 introduce the fixed and embedded codec designs respectively.

2.1   Prediction Error Modeling for Depth

This section models the prediction error between the actual and the predicted depth values produced by the base layer codec. The modeling methodology is generic, being applicable to any depth prediction method. The distribution of the prediction error is found by making use of the Kullback-Leibler divergence [11], denoted by $D_{KL}(P||Q)$, defined as follows:

$$D_{KL}(P||Q) = \sum_i P(i) \cdot \log \frac{P(i)}{Q(i)}, \tag{1}$$

where the discrete probability distributions P and Q are the modeled and reference probability distributions respectively. The prediction error model is chosen from a known set of parametrized distributions as the distribution Q with the smallest D_{KL} computed between the discrete probability distribution P of the prediction residuals and a known parametrized distribution Q.

Extensive experiments using various models (e.g., Laplace Distribution) have shown that the smallest KL divergence is obtained for the Two-Sided Geometric Distribution (TSGD). The probability distribution function (PDF) of TSGD is defined as:

$$f(x) = -\frac{ln(\theta)}{2}\theta^{|x|}, \tag{2}$$

where x is the prediction residual, and $\theta \in (0, 1)$ is the parameter which controls the two-sided exponential decay rate.

The prediction error x is quantized by applying deadzone scalar quantizers Q_ξ, where ξ is the parameter which controls the width of the deadzone. The prediction error x is quantized into q_ξ using the quantizer Q_ξ as follows:

$$q_\xi = Q_\xi(x) = \begin{cases} sign(x) \cdot \left\lfloor \frac{|x|}{\Delta} + \xi \right\rfloor, & \text{if } \frac{|x|}{\Delta} + \xi > 0 \\ 0, & \text{otherwise} \end{cases}, \tag{3}$$

where $\Delta > 0$. In this paper, we will set $\xi = \frac{1}{2}$ and we will define the uniform scalar quantizers $Q_{\frac{1}{2}}$.

By taking into account that for depth images the prediction error x can be modeled as a TSGD, the entropy of the quantized residuals can be now modeled. The entropy of a variable X with possible values $\{x_1, x_2, \ldots, x_M\}$ is defined as follows:

$$H(X) = -\sum_{i=1}^{M} p_i \cdot \log_2(p_i). \tag{4}$$

In our case, X stands for the quantized values of prediction errors x, $p_i = \int_{b_{i-1}}^{b_i} f(x)dx$ is the probability that x falls into the cell $[b_{i-1}, b_i)$ of the quantizer, and M is the number of bins of the quantizer. One notes that the overload region is neglected due to its extremely low probability. Equation (4) can be rewritten for the TSGD and the scalar quantizers Q_ξ as follows:

$$H(Q_\xi(x)) = -\sum_{i=1}^{M} \left(\int_{b_{i-1}}^{b_i} -\frac{ln(\theta)}{2}\theta^{|x|}dx \right) \log_2 \left(\int_{b_{i-1}}^{b_i} -\frac{ln(\theta)}{2}\theta^{|x|}dx \right). \tag{5}$$

By carrying out the derivations, Eq. (5) simplifies to:

$$H(Q_\xi(x)) = -2B(\Delta) \sum_{i=0}^{\frac{M-3}{2}} \theta^{i\Delta} \log_2 \left(B(\Delta)\theta^{i\Delta}\right) + 2A(\Delta)log_2\left(A(\Delta)\right), \quad (6)$$

where $B(\Delta) = \frac{(1-\theta^\Delta)\theta^{\frac{\Delta}{2}}}{2}$ and $A(\Delta) = \frac{\theta^{\frac{\Delta}{2}}-1}{ln(\theta)}$.

Since the quantization is the only part of the proposed method which introduces distortions, the distortion in the reconstructed depth maps dependents only on the reconstruction error introduced by the quantizer. The uniform scalar inverse-quantizer Q_ξ^{-1} is using the quantized value, q_ξ, to obtain the reconstructed residual, $\hat{x}_\xi$, as follows:

$$\hat{x}_\xi = Q_\xi^{-1}(q_\xi) = \begin{cases} sign(q_\xi) \cdot (|q_\xi| - \xi + \delta)\,\Delta, & q_\xi \neq 0 \\ 0, & q_\xi = 0 \end{cases}, \quad (7)$$

where $0 \leq \delta < 1$ controls the placement of the reconstruction points in their respective cells. Since in case of the L_∞ norm we have:

$$\|x - \hat{x}\|_\infty = \sup_n |x_n - \hat{x_n}|,$$

we choose mid-point reconstruction, which corresponds to the optimal reconstruction point in each quantization cell. The MAXAD of the reconstruction error is then given by:

$$D = \frac{\Delta}{2}, \quad (8)$$

where Δ is the bin size of quantizer Q_ξ.

In conclusion, the entropy of the quantized residual for depth is modeled by Eq. (6), whereas the L_∞ distortion incurred by the uniform quantizer Q_ξ is expressed by Eq. (8).

2.2   Optimized Fixed-Rate Quantizer Design

For depth images, pixel values correspond to points in 3D that are located at specific distances relative to the camera. The question one needs to answer is how should the depth values be quantized such that the overall distortion is minimal for a given bit budget or viceversa.

Denote the distribution of depth values in an image as $f_d(x)$; for depth values in the range $[x_n, x_{n+1})$, we apply a quantizer $Q_{\xi,n}$ with corresponding bin-size denoted by Δ_n. The total distortion introduced in the depth map D_{tot} can be formulated as:

$$D_{tot} = \frac{1}{2} \sum_{n=1}^{N} p_n \Delta_n, \quad (9)$$

where $p_n = \int_{x_n}^{x_{n+1}} f_d(x)dx$ describes the number of values that are quantized with $Q_{\xi,n}$. Similarly, the total rate, R_{tot}, is computed as follows:

$$R_{tot} = \sum_{n=1}^{N} p_n H(\Delta_n) + R_{base}, \tag{10}$$

where R_{base} is the bitrate used to encode the base layer. As detailed in Sect. 2.3, the base layer encodes a prediction of the depth map which is computed based on a segmentation of the input depth image.

The design of the optimized fixed-rate quantizers is formulated as follows: *Find the optimal set of quantizers $\{Q_{\xi,n}\}, 1 \le n < N$, subject to the constraint of either a target L_∞ distortion, D_t, or encoding the image while not exceeding a maximum budget bitrate, R_t.* Let us introduce the following cost functions:

(i) J_1, is the cost function for the optimization problem resulting from imposing a total bitrate constraint;

(ii) J_2, is the cost function for the optimization problem resulting from imposing an L_∞ distortion constraint.

The two cost functions are defined as follows:

$$J_1(p_n, \Delta_n) - \frac{1}{2} \sum_{n=1}^{N} p_n \Delta_n + \lambda_1 \left(\sum_{n=1}^{N} p_n H(\Delta_n) - R_t \right) + \gamma_1 \left(\sum_{n=1}^{N} p_n - 1 \right), \tag{11}$$

$$J_2(p_n, \Delta_n) = \sum_{n=1}^{N} p_n H(\Delta_n) + \lambda_2 \left(\frac{1}{2} \sum_{n=1}^{N} p_n \Delta_n - D_t \right) + \gamma_2 \left(\sum_{n=1}^{N} p_n - 1 \right). \tag{12}$$

where λ_i and $\gamma_i, i = 1, 2$ are Lagrange multipliers. Minimizing the two functionals is obtained by solving:

$$\begin{cases} \frac{\partial J}{\partial p_n} = 0 \\ \frac{\partial J}{\partial \Delta_n} = 0 \end{cases}.$$

The following solution is obtained for the cost function J_1:

$$\begin{cases} \gamma_1 = -\dfrac{\Delta_n}{2} - \lambda_1 H(\Delta_n), & (13a) \\ \dfrac{\partial H(\Delta_n)}{\partial \Delta_n} = -\dfrac{1}{2\lambda_1}, & (13b) \end{cases}$$

while for the cost function J_2 we obtain:

$$\begin{cases} \gamma_2 = -\lambda_2 \dfrac{\Delta_n}{2} - H(\Delta_n), & (14a) \\ \dfrac{\partial H(\Delta_n)}{\partial \Delta_n} = -\dfrac{\lambda_2}{2}. & (14b) \end{cases}$$

From Eqs. (13b) and (14b), one can notice that for any given quantization interval n, Δ_n must be constant. This implies that using an unique quantizer throughout the entire depth map is the optimal solution, i.e., $N = 1$ and $p_n = p = 1$. Hence, when given a target bitrate R_t or a fixed distortion constraint D_t, the optimal quantizers Q_ξ with corresponding Δ can be easily determined.

2.3   Fixed-Rate Codec Design

In this paper, the initial depth map image, I, is encoded using a base layer and an enhancement layer, as depicted in Fig. 1. The base layer is encoded using a lossless codec and contains the segmentation-based prediction, $\hat{I}$. Firstly, the initial image is segmented using the Greedy Rate-Distortion Slope Optimization with merging (GSOM) algorithm of [10] operating with a user-defined target distortion. Secondly, the pixels are predicted as the average depth values in their corresponding segment, yielding the predicted depth image. The predicted depth image is subsequently encoded in the base layer using the Anchor Points Coding (APC) algorithm of [9].

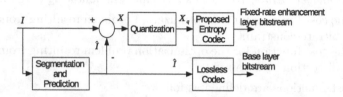

Fig. 1. Schematic description of the fixed-rate codec.

The enhancement layer is encoded using the proposed L_∞-constrained fixed-rate codec. The proposed residual codec first determines the uniform quantizer for a specific target rate or distortion. The uniform quantizer is applied on the residual image $X = I - \hat{I}$ to obtain the quantized residual image, denoted by X_q. In this paper, a context-based adaptive arithmetic codec is used to encode X_q by first obtaining the binary representation of the quantized residuals and by encoding each bitplane at a time. For each bitplane k, the image is encoded in raster order using the context generated from the corresponding bit of the northern pixel, denoted N, and the western pixel, denoted W. Figure 2 details the adopted context conditioning scheme which is very simple: if the corresponding bit of N and W are the same, then select either $C1$ (both bits are 1) or $C2$ (both bits are 0), else select $C3$.

	C1	C2	C3	
N	1	0	1	0
W	1	0	0	1

Fig. 2. Context conditioning using the corresponding bit of the northern N and western W pixels respectively.

2.4 Embedded Extension

An embedded extension for the codec is also proposed, enabling what is called scalability in L_∞-sense [8]. The proposed scalable extension offers the possibility to truncate the bitstream at any desired bitrate or to offer a number of potential decodable points for which the L_∞ reconstruction error is accurately known.

To enable L_∞ scalability, one employs embedded uniform quantizers, denoted by $\{Q^e_{\xi,n}\}$, defined as follows:

$$q^e_{\xi,n} = Q^e_{\xi,n}(x) = \begin{cases} sign(x) \cdot \left\lfloor \frac{|x|}{3^n \Delta} + \frac{\xi}{3^n} \right\rfloor, & \text{if } \frac{|x|}{3^n \Delta} + \frac{\xi}{3^n} > 0 \\ 0, & \text{otherwise} \end{cases}. \tag{15}$$

Figure 3 shows the relation between two consecutive quantization levels, each corresponding to an enhancement layer. Let us denote C^n the quantization intervals at level n. The figure shows how the partition cells of Q^e_n are embedded in one of the partition cells of Q^e_{n+1}. The relation between the bin size Δ_{n+1} of quantizer Q^e_{n+1} and the bin size Δ_n of Q^e_n is defined as follows:

$$\Delta_{n+1} = 3\Delta_n. \tag{16}$$

Since Q^e_{n+1} is available before Q^e_n, Q^e_n is reconstructed only by encoding the transitions from each C^{n+1} to the corresponding cells of Q^e_n, selected among three possible cases. Hence, the quantized value q_n can be encoded by refining the previously encoded quantizer index q_{n+1} using a symbol s_n, where $s_n \in \{0, 1, 2\}$ is the symbol pointing to the corresponding partition cell of Q^e_n.

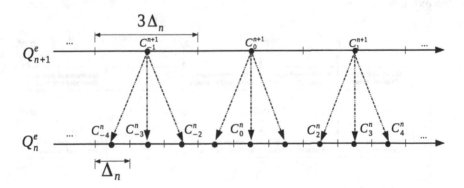

Fig. 3. The relation between quantization levels n and $n+1$ in the embedded uniform quantizer.

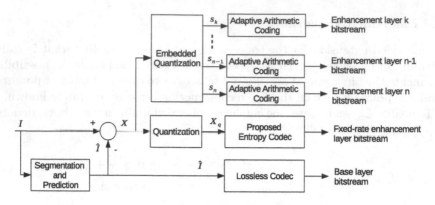

Fig. 4. The structure of the embedded extension of the proposed codec.

Figure 4 depicts the architecture of the proposed embedded L_∞ codec design. After losslessly encoding the base layer, the fixed-rate codec described in Sect. 2.3 is used to encode X_q at the largest MAXAD value D_{n+1} and to obtain the fixed-rate enhancement layer. The process continues with the progressive encoding of each enhancement layer i, where $i = n, n-1, \ldots, k$. The current enhancement layer i yields the MAXAD value D_i, while the previous enhancement layer $i-1$ corresponds to the MAXAD value $D_{i-1} = 3D_i$. Hence, with each enhancement layer the distortion is decreased up to the target distortion, which corresponds to the minimal MAXAD value that the k-enhancement-layers structure can achieve. Figure 5 depicts the structure of the bitstream generated by the embedded L_∞ codec extension.

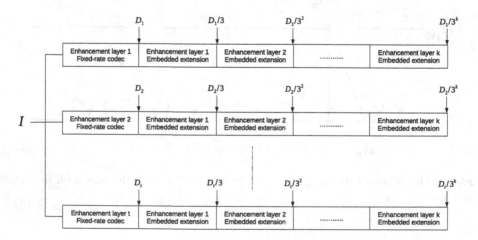

Fig. 5. The bitstream structure of the embedded codec extension.

3   Experimental Results

The experimental evaluation was carried out on a set of depth images from the Middlebury dataset [12], available online [13].

Figure 6 shows the effect incurred by the depth errors on the object reconstruction. The figure shows the point clouds obtained for the set of quantizers $\{Q_{\xi,n}\}$ with MAXAD values $MAXAD \in \{10, 50, 100\}$. One can notice that if the distortion becomes too large, the point cloud will lose its original shape. This illustrates the importance of controlling local errors in depth maps.

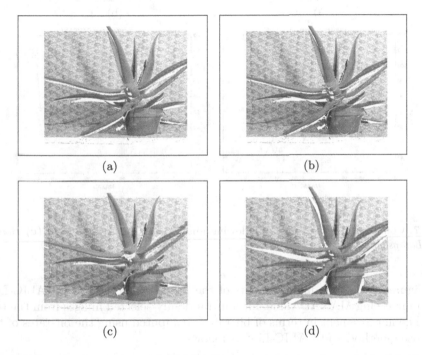

Fig. 6. Point clouds of *Aloe* with MAXAD set to: **(a)** 0 (no distortion); **(b)** 10; **(c)** 50; **(d)** 100.

The results of the proposed method are compared with a reference codec built based on the CALIC codec and equipped with the L_∞ coding functionality, denoted here as CALIC-Linf. The codec was modified to apply the proposed quantization to the CALIC prediction errors. Note that in CALIC-Linf, the quantized values are replacing the prediction errors and are further modeled in the same way as CALIC's prediction errors are modeled in the original CALIC codec.

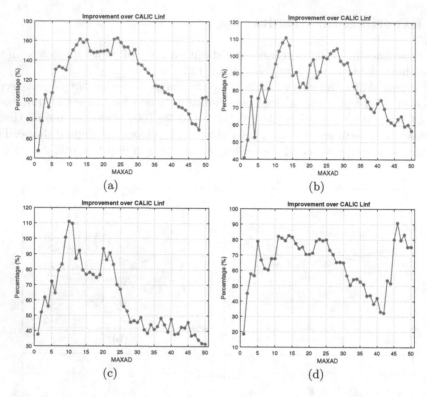

Fig. 7. Comparison results for the following images: **(a)** *Aloe*; **(b)** *Baby3*; **(c)** *Books*; **(d)** *Lampshade*.

Figure 7 shows the improvements of the proposed method over CALIC-Linf at an increasing MAXAD value, for four randomly-selected images from the test set. The improvement in terms of bit rate is computed using the bit-rates of the proposed method and CALIC-Linf as follows:

$$\Delta_{BR} = \frac{BR_{CALIC-Linf}}{BR_{Proposed}} - 1.$$

One can see that the proposed method yields much better results compared to CALIC-Linf; e.g., for *Aloe* the proposed method obtains an average rate improvement of around 120%.

Figure 8 shows the comparative results of the fixed-rate codec vs. its embedded extension. In our experiments, we use three enhancement layers for the embedded extension. This design is based on the value range of quantized residuals. One notes that if we design the embedded extension to use too many enhancement layers by setting a quantizer with a large Δ, then the maximum absolute value of the residuals might turn out to be smaller than the imposed Δ. In other words, the number of enhancement layers is limited by the maximum value range of the quantized residuals.

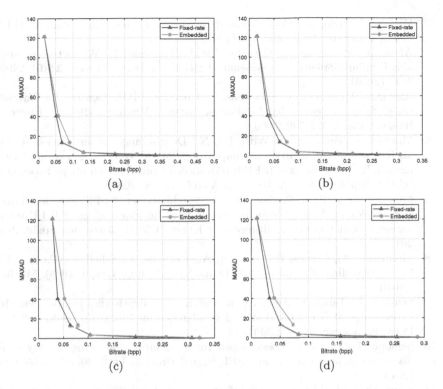

Fig. 8. Comparison results for the proposed fixed-rate codec and the embedded version for the following images: **(a)** *Aloe*; **(b)** *Baby3*; **(c)** *Books*; **(d)** *Lampshade*.

4   Conclusions

Novel fixed-rate and embedded L_∞ predictive codec designs for depth maps are proposed in this paper. The codecs combine segmented-based depth prediction with optimized scalar quantization of the residuals. In this context, the distributions of the prediction errors for depth maps are modeled based on which optimized fix-rate quantizers are designed. The quantizers are integrated in a fixed-rate codec design which performs context-based adaptive entropy coding of the quantized residuals. Moreover, an embedded extension of the codec is proposed which employs embedded uniform quantizers and enables scalability in L_∞ sense. The results demonstrate an outstanding performance of the proposed method compared to an L_∞ extension of the state-of-the-art CALIC codec.

References

1. Ke, L., Marcellin, M.W.: Near-lossless image compression: minimum-entropy, constrained-error DPCM. IEEE Trans. Image Process. **7**(2), 225–228 (1998)
2. Wu, X., Choi, W.K., Bao, P.: L_∞-constrained high-fidelity image compression via adaptive context modeling. In: Proceedings of Data Compression Conference (DCC), Snowbird, UT, USA, pp. 91–100 (1997)

3. Wu, X., Memon, N.: Context-based, adaptive, lossless image coding. IEEE Trans. Commun. **45**(4), 437–444 (1997)
4. Alecu, A., Munteanu, A., Cornelis, J.P.H., Schelkens, P.: Wavelet-based scalable L-infinity-oriented compression. IEEE Trans. Image Process. **15**(9), 2499–2512 (2006)
5. Alecu, A., Munteanu, A., Schelkens, P., Cornelis, J.P.H., Dewitte, S.: Wavelet-based fixed and embedded L-infinite-constrained image coding. J. Electron. Imaging **12**(3), 522–539 (2003)
6. Morvan, Y., Farin, D., de With, P.H.N.: Depth-image compression based on an R-D optimized quadtree decomposition for the transmission of multiview images. In: Proceedings of IEEE International Conference on Image Processing (ICIP), San Antonio, TX, USA, pp. V-105–V-108 (2007)
7. Milani, S., Zanuttigh, P., Zamarin, M., Forchhammer, S.: Efficient depth map compression exploiting segmented color data. In: Proceedings of IEEE International Conference on Multimedia & Expo (ICME), Barcelona, Spain, July 2011
8. Munteanu, A., Cernea, D.C., Alecu, A., Cornelis, J., Schelkens, P.: Scalable L-infinite coding of meshes. IEEE Trans. Vis. Comput. Graph. **16**(3), 513–528 (2010)
9. Schiopu, I., Tabus, I.: Anchor points coding for depth map compression. In: Proceedings of IEEE International Conference on Image Processing (ICIP), Paris, France, pp. 5626–5630 (2014)
10. Schiopu, I., Tabus, I.: Lossy depth image compression using Greedy rate-distortion slope optimization. IEEE Signal Process. Lett. **20**(11), 1066–1069 (2013)
11. Kullback, S., Leibler, R.A.: On information and sufficiency. Ann. Math. Stat. **22**(1), 79–86 (1951)
12. Hirschmuller, H., Scharstein, D.: Evaluation of cost functions for stereo matching. In: Proceedings of IEEE Conference on Computer Vision and Pattern Recognition (CVPR), Minneapolis, MN, USA, June 2007, pp. 1–8
13. Middlebury Dataset. http://vision.middlebury.edu

An Application of Data Compression Models to Handwritten Digit Classification

Armando J. Pinho[✉] and Diogo Pratas

IEETA/DETI, University of Aveiro, Aveiro, Portugal
{ap,pratas}@ua.pt

Abstract. In this paper, we address handwritten digit classification as a special problem of data compression modeling. The creation of the models—usually known as training—is just a process of counting. Moreover, the model associated to each class can be trained independently of all the other class models. Also, they can be updated later with new examples, even if the old ones are not available anymore. Under this framework, we show that it is possible to attain a classification accuracy consistently above 99.3% on the MNIST dataset, using classifiers trained in less than one hour on a common laptop.

Keywords: Compression-based classification · Handwritten digits
Kolmogorov complexity · Finite-context models
Probabilistic algorithms · Algorithmic information theory

1 Introduction

Since the reborn interest on the use of artificial neural networks, many pattern classification problems have seen a significant boost in performance. Among other application areas, deep and convolutional architectures have been successfully proposed for tackling computer vision problems, such as image classification. Handwritten digit recognition was probably the seed for these advances, planted by the seminal work of Yan LeCun and colleagues [16]. Since then, the now well-known MNIST dataset of handwritten digits has been used by many researchers as one of the datasets of choice for benchmarking image-related pattern classification algorithms.

Despite the undeniable success of deep learning and convolutional neural networks, there are some aspects in these approaches that deserve attention and, possibly, a search for alternative and/or complementary methods. The main purpose of this paper is to give evidence of the suitability of a different approach to the problem of pattern classification. To illustrate the proposed method, we use the problem of handwritten classification instantiated on the MNIST dataset. At least, two websites have been maintained with performance information regarding the MNIST dataset [1,17], showing, on one hand, the diversity of methods

© Springer Nature Switzerland AG 2018
J. Blanc-Talon et al. (Eds.): ACIVS 2018, LNCS 11182, pp. 487–495, 2018.
https://doi.org/10.1007/978-3-030-01449-0_41

used, and, on the other hand, the impressive evolution of the recognition accuracy of the methods.

The idea behind the approach described here is quite simple:

1. For each class of interest, build a compression model, using the "training" data;
2. Determine the amount of information that each model needs for representing the object to be classified;
3. Assign the class corresponding to the model that requires the least amount of information.

Some early work on this core idea was carried on by several researchers [31–33, 39, 42, 43]. The foundations that support this approach lay on the notion of Kolmogorov complexity [4, 14, 35, 36] and associated information distances. Because the Kolmogorov complexity is uncomputable, approximations based on compression models are usually sought. Therefore, the models to which we refer to are typically those that are found in compression algorithms.

The main characteristics of the approach that we present in this paper are the following:

1. The whole process is able to run on a common laptop computer—no GPUs needed;
2. Training times of all the models are in the order of a few tens of minutes;
3. For each class, the model is trained independently of all the other class models;
4. Training is incremental, in the sense that we can add more examples to a model previously trained, without having to re-train the complete model again. In other words, each training example needs to be seen by the model only once;
5. Classification of the 10,000 digits of the test set is done in less than 40 s.

2   The Proposed Approach

The MNIST dataset (which we downloaded from [17]) comprises two sets: one, with 60,000 samples, is used for training; another one, with 10,000 samples, known as the test set, is used for benchmarking. According to [16], these digits were obtained from the original bilevel images, by normalization into a 20×20 square (preserving the original aspect ratio) and embedding them in a 28×28 image (centered according to the center of mass of the digit). One of the effects of the normalization process, used in the construction of the MNIST dataset, was the conversion of the original bilevel images into graylevel ones.

We performed two preprocessing steps in these images:

1. They were deskewed, essentially using the algorithm described in [25];
2. They were segmented (i.e., transformed back to bilevel images), using Otsu's adaptive thresholding procedure [28], as provided by OpenCV (https://opencv.org/).

In [29], we proposed an information measure, the Normalized Relative Compression (NRC), defined as

$$\text{NRC}(x\|y) = \frac{C(x\|y)}{|x|}, \tag{1}$$

where $C(x\|y)$ represents the amount of information needed by an hypothetical compressor that, for compressing a target object x, can only use information from a reference, y. In other words, this compressor is not able to find (and use) redundancies inside x. We denote by $|x|$ the length of x.

The NRC is not a distance (fails some of the properties of a metric). Nevertheless, we claim that it is appropriate for answering questions of the type "How well is x representable by y?", which seems to be a reasonable question to ask when we are dealing with classification problems. In fact, we argue that finding the y_d that minimizes $\text{NRC}(x\|y_d)$ corresponds to finding the class of x. Another kind of question, "How similar is x to y?", can be answered by other information measures, such as those resulting from approximations to the Normalized Information Distance (NID) [3,20]. However, recent results suggest the superiority of the NRC over the NID for pattern classification [29].

The classification approach that we have followed is sketched in Fig. 1. The image of the digit to be classified is first deskewed, then segmented using Otsu's algorithm and finally made available to ten different models, denoted as $M_0, M_1, \ldots, M_9$, each one providing as output a measure, $m_0, m_1, \ldots, m_9$, of the amount of information they need to represent the input image. This measure can correspond, for example, to number of bits. We use M_d as a shorthand notation for the $C(\cdot\|\cdot)$ that appears in (1), which represents *relative* compression, such that $M_d(x) := C(x\|y_d)$, where "d" indicates the digit class and y_d refers to the training data associated to digit "d". The model that requires the smallest amount of information, i.e.,

$$\tilde{d} = \underset{d=0,1,\ldots,9}{\arg\min} \{m_d\},$$

determines the class, $\tilde{d}$, of the digit being classified.

The models, M_d, rely on finite-context modeling. For each (bilevel) pixel of the image, x_k, the finite-context model assigns probabilities to both possible values, $x_k \in \Sigma = \{B, W\}$, of the pixel at position k, according to the current context, c_k. This probability is calculated using a Krichevsky-Trofimov estimator [15], according to

$$P_d(\sigma|c_k) = \frac{n_d(\sigma|c_k) + 1/2}{n_d(B|c_k) + n_d(W|c_k) + 1}, \tag{2}$$

where $n_d(\sigma|c_k)$ denotes the number of times that a pixel with value σ was observed in the reference data of digit d having context c_k. In practice, c_k is a one-to-one identifier, obtained by mapping vectors of pixel values, drawn according to a predefined context template (or configuration), to the non-negative integers.

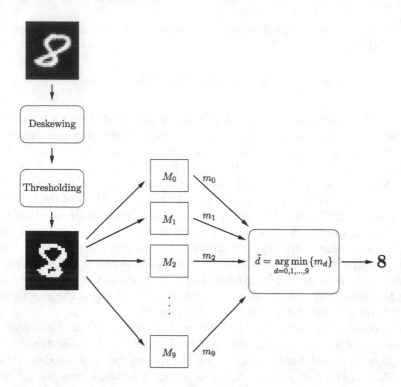

Fig. 1. The classification approach: the M_d are models that, given the input image of the digit being classified, give a measure, m_d, of the amount of information they need to represent the image. The class assigned, $\tilde{d}$, corresponds to the identity of the model requiring the smallest amount of information.

Fig. 2. The context template used to obtain the experimental results presented in this paper. It is composed of 93 points, sampling a region of 21×19 pixels. This template is not necessarily the best one—during the development of the work, several were tested, including some of them obtained randomly. Although some variations can be found in the classification accuracy, usually they are not much significant. Nevertheless, the determination of good context templates is a subject that deserves further attention.

The differences among the models M_d arise due to their dependence on the values conveyed by the n_d counters, which are collected from the corresponding training data. The template (or templates) used define the contexts, and in practice are vectors of coordinates specifying pixel positions relative to the pixel under analysis. Figure 2 shows the context template used to obtain the

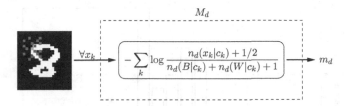

Fig. 3. Details regarding the operation of one of the class models. Using the predefined context template (represented by red pixels), all pixels of the digit are addressed (the green pixel indicates that processing was addressing that pixel), and the logarithms of the values given by (2) are accumulated. The output value, m_d, represents the amount of information required by model M_d to describe the digit. The model requiring the smallest amount of information is the winning one and, hence, the one that designates the class of the digit under classification. (Color figure online)

results described in this paper. The measures of information, m_d, are obtained by accumulating the negative logarithms of the estimated probabilities, i.e.,

$$m_d = -\sum_k \log P_d(x_k|c_k), \qquad (3)$$

with k going through all pixels of the image, and where x_k and c_k are, respectively, the pixel value and context value at position k (see Fig. 3 for an illustration). We recall that an event with probability p can be described optimally, on average, using an amount of information proportional to $-\log p$.

3   Experimental Results

A constraint that we have set to this work, was to be capable of running all required computations on a common laptop. Therefore, everything presented in this paper was done using a laptop with a Intel(R) Core(TM) i7-6500U CPU @ 2.50 GHz, 16 GB of RAM, running Ubuntu 16.04.3 LTS. This CPU allows running a maximum of four threads in parallel. In order to fulfill this self-imposed constraint, we had to resort to efficient probabilistic data-structures and algorithms, namely, count-min sketches [10] and approximate counting [11,27].

3.1   Training Without Data Augmentation

As baseline, we first give classification results using only the 60,000 training examples included in the MNIST dataset, for creating the models M_d. With this setup, we attained a classification accuracy of 98.76% in the 10,000 digits of the test set. In this case, training of the whole classification system took about half a minute of real-time.

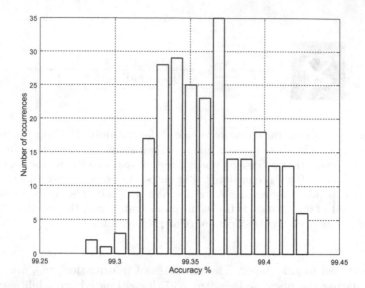

Fig. 4. Histogram of classification accuracy, obtained from 250 instances of the training set augmented using random affine transformations.

3.2   Training with Data Augmentation

It is known that data augmentation is able to improve the performance of the classifiers [34]. Hence, we defined another setup, where each digit of the training set was replicated using random affine transformations.

The distribution of results is shown in Fig. 4, for 250 instances of the augmented training set. For each instance, 200 distorted versions of each of the previously deskewed training digits were obtained by applying random affine transformations to the images. In this case, the complete classification system was trained, using the mentioned laptop computer, in roughly 50 min of real-time.

4   Discussion and Conclusion

Handwritten digit classification using compression-based approaches was previously addressed in [6], where an approximation to the Normalized Information Distance, known as the Normalized Compression Distance (NCD), was used. In this case, the classification relied on matrices of pairwise distances among the digits of the NIST Special Data Base 19, attaining an accuracy of 87%. More recently, Cohen et al. [9] used the first 1,000 digits of the MNIST training set to evaluate the classification performance of an extension of the NCD to multisets. Using leave-one-out cross validation, it was attained an accuracy of 85%. Moreover, combining the NCD extension to multisets with the normal NCD, and also using leave-one-out cross validation, 99.1% was attained. However, these results are difficult to compare with others obtained in the MNIST dataset, due to the different approach used in obtaining them (leave-one-out cross validation, using only a part of the dataset). Also, running this experiment required about five days on a cluster with 94 cores.

As far as we know, the best classification performance attained on the MNIST dataset was 99.79% [40], using the DropConnect scheme, a technique for regularizing large and deep neural network models. To attain this classification accuracy, convolutional neural networks were used for feature extraction, followed by a fully connected network to which DropConnect was applied. Also, it was used data augmentation based on cropping, scaling and rotation, as well as voting by a committee of five independent networks. Other methods, with error rates below 0.4%, have been reported, most of them relying on neural networks [5,7,8,12,18,19,21–24,26,30,34,41].

The approach that we present in this paper relies on a single classifier, i.e., it does not use a voting committee, and runs on a common laptop computer. Besides, each class can be trained independently of the other classes—new classes can be added to a classifier, without having to re-train the already existing classes. Also, because of the nature of the training process—which is based on counting—more training examples can be added to a class model, without the need of completely re-training an existing class model.

Dimensionality reduction in known to be beneficial to classification [13]. Recently, it was put forward a possible explanation regarding how deep neural networks and deep learning might work [38]. Is relies on the information bottleneck [37], a principle related to concepts of information theory and data compression. Also, autoencoders and stacked autoencoders have shown the importance that compression-related concepts have in pattern classification [2]. In our opinion, the results presented in this paper clearly reinforce this idea.

Acknowledgments. This work was partially funded by National Funds through the FCT-Foundation for Science and Technology, in the context of the projects UID/CEC/00127/2013 and PTDC/EEI-SII/6608/2014, and also by the Integrated Programme of SR&TD "SOCA" (Ref. CENTRO-01-0145-FEDER-000010), co-funded by Centro 2020 program, Portugal 2020, European Union, through the European Regional Development Fund.

References

1. Benenson, R.: Are We There Yet? (2013). http://rodrigob.github.io/are_we_there_yet/build/. Accessed 15 Sept 2017
2. Bengio, Y.: Learning deep architectures for AI. Found. Trends Mach. Learn. **2**(1), 1–127 (2009)
3. Bennett, C.H., Gács, P., Vitányi, M.L.P.M.B., Zurek, W.H.: Information distance. IEEE Trans. Inf. Theory **44**(4), 1407–1423 (1998)
4. Chaitin, G.J.: On the length of programs for computing finite binary sequences. J. ACM **13**, 547–569 (1966)
5. Chang, J.R., Chen, Y.S.: Batch-normalized Maxout network in network. Technical report arXiv: 1511.02583v1, November 2015

6. Cilibrasi, R., Vitányi, P.M.B.: Clustering by compression. IEEE Trans. Inf. Theory **51**(4), 1523–1545 (2005)
7. Cireşan, D., Meier, U., Schmidhuber, J.: Multi-column deep neural networks for image classification. In: IEEE Conference on Computer Vision and Pattern Recognition (CVPR), Providence, RI, USA, June 2012. Supp material in Ciresan-2012aa.pdf
8. Cireşan, D.C., Meier, U., Gambardella, L.M., Schmidhuber, J.: Deep, big, simple neural nets excel on handwritten digit recognition. Neural Comput. **22**, 3207–3220 (2010)
9. Cohen, A.R., Vitányi, P.M.B.: Normalized compression distance of multisets with applications. IEEE Trans. Pattern Anal. Mach. Intell. **37**(8), 1602–1614 (2015)
10. Cormode, G., Muthukrishnan, S.: An improved data stream summary: the count-min sketch and its applications. J. Algorithms **55**, 29–38 (2005)
11. Flajolet, P.: Approximate counting: a detailed analysis. BIT Numer. Math. **25**(1), 113–134 (1985)
12. Graham, B.: Fractional max-pooling. Technical report arXiv: 1412.6071v4, May 2015
13. Hinton, G., Salakhutdinov, R.R.: Reducing the dimensionality of data with neural networks. Science **313**, 504–507 (2006)
14. Kolmogorov, A.N.: Three approaches to the quantitative definition of information. Probl. Inf. Transm. **1**(1), 1–7 (1965)
15. Krichevsky, R.E., Trofimov, V.K.: The performance of universal encoding. IEEE Trans. Inf. Theory **27**(2), 199–207 (1981)
16. LeCun, Y., Bottou, L., Bengio, Y., Haffner, P.: Gradient-based learning applied to document recognition. Proc. IEEE **86**(11), 2278–2324 (1998)
17. LeCun, Y., Cortes, C., Burges, C.J.C.: The MNIST Database of Handwritten Digits (1998). http://yann.lecun.com/exdb/mnist/. Accessed 15 Sept 2017
18. Lee, C.Y., Gallagher, P.W., Tu, Z.: Generalizing pooling functions in convolutional neural networks: mixed, gated, and tree. In: Proceedings of the 19th International Conference on Artificial Intelligence and Statistics (AISTATS), Cadiz, Spain (2016)
19. Lee, C.Y., Xie, S., Gallagher, P.W., Zhang, Z., Tu, Z.: Deeply-supervised nets. In: Proceedings of the 18th International Conference on Artificial Intelligence and Statistics (AISTATS), San Diego, CA, USA (2015)
20. Li, M., Chen, X., Li, X., Ma, B., Vitányi, P.M.B.: The similarity metric. IEEE Trans. Inf. Theory **50**(12), 3250–3264 (2004)
21. Liang, M., Hu, X.: Recurrent convolutional neural network for object recognition. In: IEEE Conference on Computer Vision and Pattern Recognition (CVPR), Boston, MA, USA, June 2015
22. Liao, Z., Carneiro, G.: Competitive multi-scale convolution. Technical report arXiv: 1511.05635v1, November 2015
23. Liao, Z., Carneiro, G.: On the importance of normalisation layers in deep learning with piecewise linear activation units. In: Proceedings of the IEEE Winter Conference on Applications of Computer Vision (WACV), Lake Placid, NY, USA, March 2016
24. Mairal, J., Koniusz, P., Harchaoui, Z., Schmid, C.: Convolutional kernel networks. In: Proceedings of Neural Information Processing Systems (NIPS), Montreal, Canada, December 2014
25. Mallick, S.: Handwritten Digits Classification: An OpenCV (C++/Python) Tutorial (2017). https://www.learnopencv.com/handwritten-digits-classification-an-opencv-c-python-tutorial/. Accessed 15 Sept 2017

26. McDonnell, M.D., Vladusich, T.: Enhanced image classification with a fast-learning shallow convolutional neural network. In: Proceedings of the International Joint Conference on Neural Networks (IJCNN), Killarney, Ireland, July 2015

27. Morris, R.: Counting large numbers of events in small registers. Commun. ACM **21**, 840–842 (1978)

28. Otsu, N.: A threshold selection method from gray-level histograms. IEEE Trans. Syst. Man Cybern. **9**(1), 62–66 (1979)

29. Pinho, A.J., Pratas, D., Ferreira, P.J.S.G.: Authorship attribution using relative compression. In: Proceedings of the Data Compression Conference, DCC-2016, Snowbird, Utah, March 2016

30. Ranzato, M.A., Poultney, C., Chopra, S., LeCun, Y.: Efficient learning of sparse representations with an energy-based model. In: Proceedings of Neural Information Processing Systems (NIPS), vol. 19 (2006)

31. Rissanen, J.: Modeling by shortest data description. Automatica **14**, 465–471 (1978)

32. Rissanen, J.: A universal prior for integers and estimation by minimum description length. Ann. Stat. **11**(2), 416–431 (1983)

33. Rissanen, J.: Universal coding, information, prediction, and estimation. IEEE Trans. Inf. Theory **30**(4), 629–636 (1984)

34. Sato, I., Nishimura, H., Yokoi, K.: APAC: augmented PAttern Classification with neural networks. Technical report arXiv: 1505.03229v1, May 2015

35. Solomonoff, R.J.: A formal theory of inductive inference. Part I. Inf. Control **7**(1), 1–22 (1964)

36. Solomonoff, R.J.: A formal theory of inductive inference. Part II. Inf. Control **7**(2), 224–254 (1964)

37. Tishby, N., Pereira, F., Bialek, W.: The information bottleneck principle. In: Proceedings of the 37th Annual Allerton Conference on Communication, Control, and Computing, Illinois, USA, pp. 368–377 (1999)

38. Tishby, N., Zaslavsky, N.: Deep learning and the information bottleneck principle. In: IEEE Information Theory Workshop (ITW), Jerusalem, Israel, April 2015

39. Wallace, C.S., Boulton, D.M.: An information measure for classification. Comput. J. **11**(2), 185–194 (1968)

40. Wan, L., Zeiler, M., Zhang, S., LeCun, Y., Fergus, R.: Regularization of neural network using DropConnect. In: Proceedings of the 30th International Conference on Machine Learning, Atlanta, GA, USA, pp. 1058–1066 (2013)

41. Wang, D., Tan, X.: Unsupervised feature learning with C-SVDDNet. Pattern Recognit. **60**, 473–485 (2016)

42. Ziv, J.: On classification with empirically observed statistics and universal data compression. IEEE Trans. Inf. Theory **34**(2), 278–286 (1988)

43. Ziv, J., Merhav, N.: A measure of relative entropy between individual sequences with application to universal classification. IEEE Trans. Inf. Theory **39**(4), 1270–1279 (1993)

A Global Decoding Strategy
with a Reduced-Reference Metric Designed
for the Wireless Transmission of JPWL

Xinwen Xie[1,2,3]($\boxtimes$), Philippe Carré[1], Clency Perrine[1],
Yannis Pousset[1], Jianhua Wu[2], and Nanrun Zhou[2]

[1] Université de Poitiers, XLIM Institute,
86962 Futuroscope-Chasseneuil Cedex, France
xinwen.xie@univ-poitiers.fr
[2] Department of Electronic Information Engineering,
Nanchang University, Nanchang 330031, China
[3] Department of Electronic Engineering, Jiujiang University,
Jiujiang 332005, China

Abstract. A new global decoding strategy with Reduced-Reference (RR) metric is proposed to improve the Quality of Experience (QoE) in a wireless transmission context. The RR metric (FMRP) utilizes the magnitude and the relative phase information in the complex wavelet domain as the evaluation features. It determines the number of decoder layers to achieve the goal of evaluating the image in a consistent way with the Human Visual system (HVS). To evaluate the performance of the decoding strategy, we collected some distorted images in realistic channel attacks and recruited volunteers to do a large psychovisual test. The distorted images and the classification data of voluntary assessors are integrated into a database which is in a realistic wireless channel context quite different from the classic database. Experimental studies confirm that the decoding strategy is effective and improves the QoE while ensuring the Quality of Service (QoS).

Keywords: Reduced-reference image quality metric · Quality of experience
Decoding strategy · Support vector machine · K-nearest neighbors algorithm

1 Introduction

During the last few years, a number of researchers have contributed significant research in the design of Image Quality Assessment (IQA) algorithms and proposed many excellent algorithms such as SSIM [1], MS_SSIM [2], FSIM [3] and Wang Z. et al. method [4]. Meanwhile, some representative image databases like LIVE, TID2013, CSIQ and Toyama_LCD were built to test IQA methods. Most of the proposed IQA methods have good performance in these image databases. However, all these databases were built in the simulated environment, namely, the degradative images were distorted by the artificial noise, not by the realistic channel noise. To the best of our knowledge, very few image databases are based on the realistic channel environment.

© Springer Nature Switzerland AG 2018
J. Blanc-Talon et al. (Eds.): ACIVS 2018, LNCS 11182, pp. 496–505, 2018.
https://doi.org/10.1007/978-3-030-01449-0_42

Thus, the design of database based on real-time communication environments is of great significance.

In this paper, we focus on using Reduced-Reference (RR) IQA to optimize the decoding strategy based on Multiple Input Multiple Output (MIMO) system over the realistic spatiotemporal wireless channel. The global scheme considering a hierarchical content to be transmitted is shown in Fig. 1 (as recognized by the Long-Term Evolution standard).

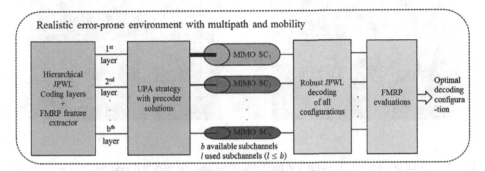

Fig. 1. Global scheme of realistic MIMO wireless channel

For the realistic error-prone environment, we utilize JPWL (JPEG2000 Wire-less) which is an extension of the JPEG 2000 baseline specification to enable the efficient transmission of JPEG 2000 code-stream over an error-prone network [5]. To overcome the instability of channel, the JPWL codec generates b code-streams sorted in a decreasing magnitude order, and the precoder solutions decouple a MIMO channel into hierarchical, parallel and independent sub-channels of different magnitudes and sorted in a decreasing SNR order. Therefore, we can transmit the image by hierarchical layers (1st layer by 1st sub-channel, 2nd layer by 2nd sub-channel, etc.) and decode a version of the transmitted image (quality scalability) even in the case of channel perturbation. The UPA strategy is applied to guarantee partial or total reception of the image depending on the channel status [6]. During the JPWL coding process, a small amount of data, extracted from the original image and called RR feature is embedded in the code-stream by the RR-IQA system FMRP, described in the next sections. At the decoding side, the system is jointly used with a robust JPWL decoder to provide the best decoding configuration to the user by exploiting the embedded information. To ensure the embedded information is not distorted in the transmission, the Error Protect Block (EPB) is used in the JPWL decoder.

As the sub-channels with higher index are very sensitive to the perturbations, it is not necessary to use them all (b sub-channels in total) when the channel conditions are unfavorable. In the previous work [6, 7], we have designed a decoding strategy to determine the number l of used SISO (Single Input Single Output) sub-channels/layers among the b available SISO sub-channels/layers. However, the decoding strategy (called total decoding) utilizes the bitrate control of JPEG 2000 to minimize the image distortion in the transmission. It is a purely Quality of Service (QoS) based strategy, not considering the Quality of Experience (QoE) of the users. Thus, in some case, the image decoded with the l-1 layers (partial decoding) has better quality than that decoded with l layers, as illustrated in Fig. 2. The reason is that the lth layer is attacked by channel noises and during the reconstruction false coefficients are associated with parasitic oscillations. It usually occurs when the bit error rate in the lth sub-channel has large value but is below the error-protection threshold.

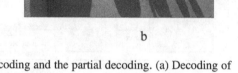

a   b

Fig. 2. Visual comparison between the total decoding and the partial decoding. (a) Decoding of l-1 (1st + 2nd + 3rd) layers of Caps, (b) Decoding of all l (1st + 2nd + 3rd + 4th) layers of Caps.

In this study, we introduced the FMRP metric to determine the number of decoder layers and improve the QoE of users. The FMRP evaluation should select better quality images in a consistent way with the users and define the proper decoding layers, for example, it would select Fig. 2 (a) as the better quality, and l-1 as the decoding layers.

For the machine algorithm FMRP, it needs a database for learning and testing. However, it is hard to find an appropriate database since most of the databases were built in the simulated environment. Therefore, we proposed to build the database based on a realistic wireless channel. The goal is to acquire the classification behavior of Human Visual System (HVS) to the total and partial decoding images. To do this, we collected many images of total and partial decoding and recruited volunteers to do the psychovisual test. The two types of images were displayed on the interactive interface with random order and the voluntary assessors were asked to select what they considered as better-quality images. The selections of all assessors were recorded and then processed (detailed in Sect. 3). The final results were used as learning and testing data for FMRP.

So, the contribution of this paper lies on three aspects. In first, we have implemented an RR-IQA system based on the magnitude and relative phase of DT-CWT coefficients. Secondly, we have built a new database to test the image quality system over a realistic MIMO channel. For most of the representative database like LIVE, CSIQ and TID2013, the distorted images are gotten from the experimental method. Therefore, building the database containing the distorted images which are acquired from the practical and realistic environment is significantly important. Thirdly, with the

RR metric we optimized the decoding configuration to achieve the improvement of QoE while ensuring the QoS.

The remainder of this paper is organized as follows: In Sect. 2, we introduce the RR metric and the model of magnitude and relative phase. Section 3 describes the details of the presented database, including the post-processing of raw data. Section 4 is devoted to evaluating the performance of the proposed decoding strategy with the new database. Finally, in Sect. 5, summaries and prospects are reached.

2 The Introduction of the FMRP

2.1 Outline of the FMRP

The strategy of FMRP lies in using the Dual-tree Complex Wavelet Transform (DT-CWT) as a decomposition tool, employing the relative phase and the magnitude information as the feature extraction source, utilizing Information Criterion (IC) to obtain an optimal histogram of the Probability Density Function (PDF) model and taking the parameters of the model as the image features. We name it FMRP since it is based on the Feature of Magnitude and Relative Phase. The outline is shown in Fig. 3.

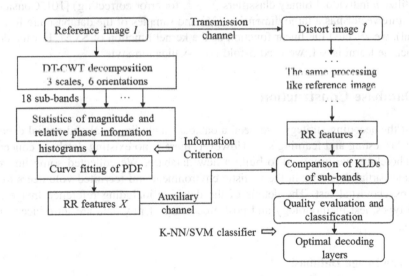

Fig. 3. Outline of the RR-IQA system FMRP.

We first utilize the 2-D DT-CWT [8] to decompose the reference image I into three scales where each scale has six orientations. The main advantages of employing the DT-CWT as a decomposed tool is that it has shift invariant property and better directional selectivity over the conventional Discrete Wavelet Transform (DWT) [8]. Secondly, we model the magnitude and the relative phase defined as the phase difference [9]. In this step, the IC is employed to reduce the error between the model and the distribution, and the Kullback-Leibler Divergences (KLDs) of all sub-bands are used as the RR features. Next, the image I and the RR features X will be transmitted to the receiver side through a transmission channel and an auxiliary channel (with EPB) respectively. At last, features X and Y are compared at the quality evaluation and classification unit, and the result is used to choose the optimal decoding layers.

2.2   Classification by K-NN/SVM Classifiers

After getting the features from the magnitude and the relative phase, it is necessary to design a classifier to conduct classification in a way that is consistent with the HVS. Two classifiers: K-Nearest Neighbors (K-NN) and multiclass Support Vector Machine (SVM) with the Error-Correcting Output Codes (ECOC) [10], were used in this step.

The K-NN is a simple yet popular approach for classification. For a given new example, it classifies the example as the class of the nearest training example to the observation. Here, we set the K to 5 and choose the Euclidean distance as the basis of classification.

The SVM is a binary classifier which can be extended to a multiclass classifier by using multiple classifiers. The standard approach is to learn k individual binary classifiers $c_1 \ldots c_k$, one for each class (k = classes). To improve the robustness, we use another advanced approach which employ an ECOC as the representation of k classes and utilize n individual binary classifiers ($n > k$, for error correcting) [10]. Considering the feature vector has a large dimension (36) and samples of the database are few (210 in total), we selected the linear function as the kernel function of SVM. For the design of machine learning set, we used 5-fold cross-validation style.

3   Database Construction

To test the decoding strategy, we need a database including total and partial decoding images as testing and learning sets. However, there is no existing database concerning this. Therefore, we proposed to build a new database. We collected some images of total and partial decoding in the realistic environment and recruited volunteers to do a large psychovisual test. The details of the database including reference images, distorted types, test methodology and post-processing of raw data are introduced in this section.

3.1   The Image Database

1. Reference and distorted image:

The reference images named caps, house, monarch and bikes are selected from the LIVE database. These images with representative features have different scenes, details, colors and contrasts.

All the distorted images are derived from the reference images. The process is as follows: (1) transmitting the reference image to the receiver through the realistic wireless channel, (2) decoding it by the JPWL coder (l-1 layers or l layers) with UPA strategy and then obtaining the distorted images (Referring Fig. 2). The realistic suburban environment is one part of campus of Poitiers in France. In the simulation scene, buildings are represented in red (see Fig. 4-a). The MIMO transmitter is fixed and the MIMO receiver moves with a speed of 5 m/s over a distance of 180 meters. We note LOS and NLOS for Line of Sight and Non-Line of Sight configurations. We alternate successively bad (NLOS in the area 1), moderately good (NLOS in areas 2 and 4) or good (LOS in the area 3) areas. The evolution of gain of MIMO channel is presented in the Fig. 4-b. The impulse responses at every position of the MIMO receiver are obtained thanks to a simulator of 3D rays [11]. Thus, different level distortions will occur in the images with different received locations. These distorted images will be submitted to subjective test with the aim of identifying the quality of both images (total and partial decoding).

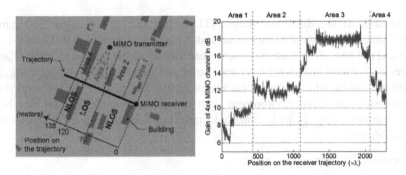

Fig. 4. (a) Topology of the transmission scene and (b) gain evolution of the MIMO channel

2. Image distortion types:

The distortions for the two types of images are JPEG 2000 compression and JPEG 2000 transmission errors. For the partial decoding images, the distortion is caused by the compression procedure in the transmitter. There are very few transmission errors because of using the EPB of JPWL decoder. For the total decoding images, the distortions may be compression and transmission errors since the lth sub-channel is instability and the lth layer is easily attacked by channel noise (the higher the channel index, the more sensitive to the perturbation). Thus, if the sub-channel has a good status without noise attack, the image with total decoding have better quality than that with partial decoding. Otherwise, the image with partial decoding may have better quality than the image with total decoding.

3.2 Test Methodology

1. Display configuration and equipment:

The experiment interface is a GUI interface designed by Matlab as illustrated in Fig. 5. Two types of images are shown in random order, and the assessors classify the images by discrete values −2, 0 and 2 representing respectively that the left one has better quality, they have the same quality, and the right one has better quality.

Fig. 5. The interface used in the experiment, −2, 0, 2 represent 3 classifications.

According to the Recommendation ITU-R BT.500-13, we chose 210 pairwise impaired images covering the range of all distortions for subjective test. To avoid visual fatigue and test time more than 30 min, we partitioned all the images to 7 sets. Each set contains 30 pairwise images with random order and each assessor complete 1or 2 sets test, and the number of assessors of each set is ensured more than 16. The workstations were placed in an office environment with normal indoor illumination levels. The display monitor was 14-inch notebook monitors displaying at a resolution of 1366 × 768 pixels.

2. Assessors, Training, and Testing:

Around 100 assessors recruited from three Universities: University of Poitiers (Ph. D. and Engineer students), Nanchang University (graduate students) and Jiujiang University (undergraduate students and University teachers) participated in the study, over a course of half year. Most of them have no experience with image quality assessment and image impairments. The demographic data of subjective assessors is shown in Table 1. The assessors were tested for vision problems, and people with poor eyesight (corrected eyesight) were forbidden to take part in the experiment. Each assessor was individually briefed about the goal of the test and given a demonstration of the experimental procedure.

Table 1. The demographic data of subjective assessors

Items	Nationality			Age			Sex	
	French	Chinese	Indian	19–29	30–45	46–55	Male	Female
Percentage (%)	15.38	81.54	3.07	46	50	4	66.67	33.33

3.3   Post-processing of the Raw Data

To reduce the accidental errors, we utilized a simple algorithm to conduct outlier detection and subject rejection, according to the suggestion given by Video Quality Experts Group (VQEG) [12]. The points satisfied the Eq. 1 is defined as the outlier points.

$$|c - c_{mean}| > 2\sigma \tag{1}$$

where c is the classified value of the image given by an assessor, c_{mean} and σ are respectively the average classified value and the standard deviation given by all assessors for one image. For assessor rejection, we utilized the similar solution used in LIVE database [13]. For any set, all quality evaluations of an assessor were rejected if more than 16.67% of his evaluations in that are outliers. Overall, a total of five assessors were rejected, and about 4.67% of the difference value were rejected as being outliers (we took all data points of rejected assessors as outliers).

After the outlier detection and assessor rejection, we used a majority rule to select the classified value occurring most frequently as the final value. The classified statistics of the database is shown in Table 2. Since the test images were selected randomly and most partial decoding images have same or better quality than the total decoding ones, the samples with the label of 2 are very few.

Table 2. Classified statistics of the database

Items	Total samples	Label −2	Label 0	Label 2
Number	210	160	46	4

4   Experimental Result

In the precedent study, we have compared the proposed metric FMRP with other representative metrics on the classic databases and the experimental results show it has higher accuracy (higher Pearson linear correlation coefficient) and better consistency (lower outlier ratio) than other metrics [submitted to "Signal Processing: Image Communication", in revision process]. In this paper, the test is about the realistic channel attacks. The results are the comparison of the proposed metric and other representative metrics, based on the new presented database.

We used 7 other image quality metrics, including 4 FR metrics: PSNR, SSIM [1], MS_SSIM [2] and FSIM [3], and three RR metrics: Wang et al. RR method [14] (we call it DWT), ADI [15] and Lin Z. et al. RR method [16] (we call it RP), to compare with the proposed metric. The implementation codes of the SSIM, MS_SSIM, Wang Z. et al. RR and FSIM methods are provided by the authors [17, 18]. For the other metrics, we implemented them according to the algorithms [15, 16, 19].

For the training and testing subset design, we utilize 5-fold cross-validation method introduced in Sect. 2.2. All the performance of the metrics is summarized in Table 3. The ratio of right classified samples (compared with the HVS) is utilized as the criterion. To increase the readability of the results, we mark the first and the second-best metrics by bold font. One can see that the proposed metric FMRP performs quite well for the 2 classifiers. It provides better prediction accuracy than any other metrics except MS_SSIM. Compared to the state-of-the-art FR metric MS_SSIM, it also performs quite well, with higher accuracy in the SVM classifier and the comparable accuracy in the K-NN classifier. Recall that the proposed metric FMRP is an RR metric, it uses very few information in image evaluation and more appropriate applying in the real-time communication system than the FR metrics.

Table 3. Performance of the objective quality metrics based on the proposed database

	Metric	PSNR	SSIM	MS-SSIM	FSIM	DWT	ICP	RP	FMRP
KNN	acc	78.10	80.48	**82.71**	80.95	76.67	79.52	75.71	**81.43**
	std	4.58	6.16	4.76	2.38	5.16	7.82	2.61	6.39
SVM	acc	75.71	77.62	**83.10**	83.05	76.19	80.95	74.76	**83.71**
	std	0.82	1.51	0.58	0.26	0.00	1.15	1.74	1.03

From the results, it is obvious that the classifier can give the right classification for most of the samples. For each subset (42 samples) of learning-testing procedure, there are about 6 or 7 misclassified samples. We found that these samples are consist of 2 categories. Category 1: the distortion of 2 images are very slight and it is a bit hard for the HVS to found the difference. Category 2: local distortion occurs in parts of the image. Most of the misclassified samples (5 or 6) are of category 1 and it furtherly manifests the classifier of FMRP is effective. Category 2 rarely appeared is a challenge to all of the metrics since the local distortion in different parts will lead to different levels perception of the HVS.

5 Conclusion

To optimize the QoE of the users in a real-time wireless communication system, we proposed a global decoding strategy with FMRP metric. The classifier with FMRP metric determines the number of decoder layers to achieve the goal of evaluating the image in a consistent way with the HVS.

Meanwhile, to provide testing and learning sets, we built a database in a realistic error-prone environment with multipath and mobility. The comparison of the FMRP metric and other representative metrics is based on the new database. The experimental results show that the FMRP metric is quality-aware and more correlated with the HVS, and the decoding strategy is effective.

References

1. Wang, Z., Bovik, A.C., Sheikh, H.R., Simoncelli, E.P.: Image quality assessment: from error visibility to structural similarity. IEEE Trans. Image Process. **13**, 600–612 (2004)
2. Wang, Z., Simoncelli, E.P., Bovik, A.C.: Multiscale structural similarity for image quality assessment. In: Conference Record of the Thirty-Seventh Asilomar Conference on Signals, Systems and Computers, vol. 1392, pp. 1398–1402 (2004)
3. Zhang, L., Zhang, L., Mou, X., Zhang, D.: FSIM: a feature similarity index for image quality assessment. IEEE Trans. Image Process. **20**, 2378–2386 (2011)

4. Wang, Z., Simoncelli, E.P.: Reduced-reference image quality assessment using a wavelet-domain natural image statistic model. In: Human Vision and Electronic Imaging X, pp. 149–160. International Society for Optics and Photonics (2005)
5. Dufaux, F., Nicholson, D.: JPWL: JPEG 2000 for wireless applications. In: Applications of Digital Image Processing XXVII, pp. 309–319. International Society for Optics and Photonics (2004)
6. Abot, J., Olivier, C., Perrine, C., Pousset, Y.: A link adaptation scheme optimized for wireless JPEG 2000 transmission over realistic MIMO systems. Sig. Process. Image Commun. **27**, 1066–1078 (2012)
7. Abot, J., et al.: A robust content-based JPWL transmission over a realistic MIMO channel under perceptual constraints. In: 2011 18th IEEE International Conference on Image Processing (ICIP), pp. 3241–3244. IEEE (2011)
8. Kingsbury, N.: Complex wavelets for shift invariant analysis and filtering of signals. Appl. Comput. Harmon. Anal. **10**, 234–253 (2001)
9. An, V., Oraintara, S.: A study of relative phase in complex wavelet domain: property, statistics and applications in texture image retrieval and segmentation. Sig. Process. Image Commun. **25**, 28–46 (2010)
10. Dietterich, T.G., Bakiri, G.: Solving multiclass learning problems via error-correcting output codes. J. Artif. Intell. Res. **2**, 263–286 (1994)
11. Chartois, Y., Pousset, Y., Vauzelle, R.: A SISO and MIMO radio channel characterization with a 3D ray tracing propagation model in urban environment. In: European Conference on Propagation and Systems, p. 8 (2005)
12. Tutorial, I.: Objective perceptual assessment of video quality: full reference television. In: ITU-T Telecommunication Standardization Bureau (2004)
13. Sheikh, H.R., Sabir, M.F., Bovik, A.C.: A statistical evaluation of recent full reference image quality assessment algorithms. IEEE Trans. Image Process. **15**, 3440–3451 (2006)
14. Wang, Z., Wu, G., Sheikh, H.R., Simoncelli, E.P., Yang, E.H., Bovik, A.C.: Quality-aware images. IEEE Trans. Image Process. **15**, 1680–1689 (2006)
15. Lin, Z., Tao, J., Zheng, Z.: Reduced-reference image quality assessment based on average directional information. In: IEEE International Conference on Signal Processing, pp. 787–791 (2015)
16. Lin, Z., Zheng, Z., Guo, R., Pei, L.: Reduced-reference image quality assessment based on phase information in complex wavelet domain, pp. 966–971 (2015)
17. http://ivc.uwaterloo.ca/ivc-code.php
18. http://sse.tongji.edu.cn/linzhang/IQA/FSIM/FSIM.htm
19. Traoré, A., Carré, P., Olivier, C.: Quaternionic wavelet coefficients modeling for a reduced-reference metric. Sig. Process. Image Commun. **36**, 127–139 (2015)

Reconfigurable FPGA Implementation of the AVC Quantiser and De-quantiser Blocks

Vijaykumar Guddad(✉), Amit Kulkarni, and Dirk Stroobandt

ELIS Department, Computer Systems Lab, Ghent University,
iGent, Technologiepark Zwijnaarde 15, 9052 Ghent, Belgium
{Vijaykumar.Guddad,Amit.Kulkarni,Dirk.Stroobandt}@UGent.be

Abstract. As image and video resolution continues to increase, compression plays a vital role in the successful transmission of video and image data over a limited bandwidth channel. Computation complexity, as well as the utilization of resources and power, keep increasing when we move from the H264 codec to the H265 codec. Optimizations in each particular block of the Advanced Video Coding (AVC) standard significantly improve the operating frequency of a hardware implementation. In this paper, we designed parametrized reconfigurable quantiser and de-quantiser blocks of AVC through dynamic circuit specialization, which is different from traditional reconfiguration of FPGA. We implemented the design on a Zynq-SoC board, which resulted in optimizations in resource consumption of 14.1% and 20.6% for the quantiser and de-quantiser blocks respectively, compared to non-reconfigurable versions.

Keywords: AVC · FPGA · Quantiser · De-quantiser
Parameterized reconfiguraton

1 Introduction

The Advanced Video Coding (AVC) standard is a video compression standard. AVC is a joint video project of the ITU-T Video Coding Experts Group (VCEG) and the ISO/IEC Moving Picture Experts Group (MPEG) standardisation organisation, working together in a partnership known as Joint Collaborative Team on Video Coding (JCT-VC). Video coding standards mainly evolved from ISO/IEC and ITU-T standards, ITU-T produced H.261 [1] and H.263, ISO/IEC produced MPEG-1 and MPEG-4, both organisations jointly produced the H.262, H.264, and H.265 coding standards. These joint standards have found their way into a wide variety of products that are paramount in our daily lives. Alongside this evolution, sustained efforts have been made to maximise the robustness, compression capability, and to improve data loss [2].

The dominant video coding standard directly pioneering the AVC project was MPEG-4 AVC/H.264, which was initially developed in the years between 1999 and 2003, and then was protracted in several vital ways from 2003–2009.

© Springer Nature Switzerland AG 2018
J. Blanc-Talon et al. (Eds.): ACIVS 2018, LNCS 11182, pp. 506–517, 2018.
https://doi.org/10.1007/978-3-030-01449-0_43

Now H.264 has been an enabling technology for digital video in almost every area that was not previously covered by H.262 video and has extensively displaced the older standards within existing application domains [1,3]. H.264 is widely used for many applications, including terrestrial transmission systems and broadcast of high definition (HD) TV signals over satellite, Blue-ray Disc, camcorders, internet and mobile network video, security applications, real-time conversational applications such as video conferencing, video chat, and telepresence systems.

However, an expanding diversity of services, the growing popularity of high definition video, and the evolution of beyond high definition formats (e.g., 8kx4k or 4kx2k resolution) are creating stronger needs for coding efficiency. To make high definition videos widely used in the world, as H.264, a great effort is currently being made in various organisations and research labs for the development of feasible high definition codecs. Because of the high computation cost for high definition videos, specially designed hardware components such as ASIC, DSP, FPGA, and GPU are crucial.

Nowadays the entire video industry is moving towards quad high definition (4K/QHD), which contains four times more video data than HD video. It is expected that by observing this current trend, high definition video codec will become the codec of choice in the future. This growing trend of HD and 4K video and the increased added resource utilisation, power consumption, and increased computation complexity from disadvantages of the high definition codec. Therefore hardware implementation of high definition encoders are encouraged especially for low power mobile applications. Hardware high definition encoders can come in two forms: an Application Specific Integrated Circuit (ASIC) or FPGA. In an ASIC implementation of the high definition encoder, the silicon chip is custom made, which results in high performance, but the nonrecurrent engineering cost associated with silicon chip development is exceptionally high. The only thing making them economical is producing a considerable volume [2,3].

On the other hand, FPGA based AVC implementations use existing configurable hardware platforms to implement specific hardware architectures. Where performance is lower than that of an ASIC, the cost is also much lower. Therefore it is more cost effective for a small volume, such as specialised broadcast equipment, and the reconfigurability of FPGA's is an added advantage. Furthermore, the ability to implement a design easily and port it across existing FPGA hardware makes it an interesting choice for industry and academic researchers who research video processing applications.

Many researchers and chip vendors come up with real-time QHD capable hardware architectures. However, all of them are mostly optimised for ASIC's as the same hardware architecture runs much slower on FPGA platforms. Kuon and Rose found that the critical path delay differs between FPGA and ASIC implementations and ranges from three to four times the smallest obtained delay [4]. So it requires careful analysis of resources and capabilities of an FPGA while designing real-time high definition encoder solutions on FPGA.

To reduce the critical path of the AVC design when implementing it on FPGA, we provide in this paper reconfigurable quantise and de-quantise hardware design blocks. These blocks offer lossy compression by shrinking the range of values to a single quantum value when the number of distinct symbols in a given stream is reduced; the stream then becomes more compressible. In our design, we turned the Quantisation parameter (QP) into a constant input (with temporarily changing constant value) rather than implementing it as a regular quantisation parameter input. This allows us to optimize the design for speed and lower resorce usage. When there is a change in the QP value, the required bit stream is evaluated using an on-chip processor or Micro-Blaze or Power PC, which internally micro-reconfigures the design using the Dynamic Circuit Specialization technique [5]. This reconfigurable design comes at the cost of reconfiguration time. To reduce this reconfiguration time, the MiCAP-pro reconfiguration controller is used, which reduces reconfiguration time to 64.1 μs per lookup table [6].

2   Quantiser of AVC

Video sequences are represented by a series of matrices; these matrices are often huge, requiring a significant amount of space or bandwidth. When resources are limited (bandwidth or storage space), it is essential to reduce the amount of data needed to represent the digital imagery. Data can be compressed either with some distortion/degradation of the data (lossy compression) or with no loss in data (lossless compression). In lossless compression, redundancy in data is removed, resulting in a smaller representation, but the ratio of compression is minimal. On the other hand, lossy compression techniques trade off the compression ratio against the tolerated distortion, since human visual perception can tolerate a particular amount of distortion in the presented visual data.

Lossy compression is achieved by the quantiser block in AVC, quantisation is performed using a variable scalar quantiser, where individual values are quantised. The quantisation process inserts the signal degradation and distortion by removing signal information from the coded representation. By careful control of the quantisation parameter (QP), a visible distortion can be minimized.

The AVC quantisation core is shown in Fig. 1, core quantiser composed of five main blocks. The first block is a nine deep and 32-bit wide array of registers. The first eight registers are used to store the sixteen residual values; each is of 16 bits wide. There are 8 control registers, which contains the following values, "go", "done", "reset", "count", "bypass DCT", "intra", "bypass Quant", and "QP" (not all signals shown in the figure). The 9'th register stores a count value. The second block is the DCT, which performs an integer DCT on all sixteen values stored in the first eight registers. The transform is done in one clock cycle, and the results are stored in its internal registers. The third block called "const" is composed of two sequential lookup tables which contain the multiplication factor. The lookup table return the value based upon the contents of register 8. The first lookup table is 64 entries deep and 30 bits wide. The second lookup table is eight entries deep by 42 bits wide. The fourth block performs the quantisation. The fifth and final block is the counter, this counter is a simple sixteen-bit adder, and it is used purely for performance measurements [7,8].

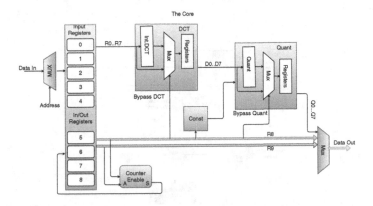

Fig. 1. Block diagram of AVC Quantiser

Dynamic Circuit Specialization (DCS). Members of our research group developed the Dynamic Circuit Specialization (DCS) technique which is a partial reconfiguration technique for parametrized applications [9]. The DCS tool flow provides a way for specializing a design for specific inputs using Run Time Reconfiguration (RTR). A parameterized application consists of a set of inputs which change infrequently compared to the other inputs in the design, occasionally changing inputs are called parameters. DCS uses the run-time-reconfiguration technique to specialize the parametrized design depending on the parameter values. Thus for every change in the input (parameter) value, a new bitstream is generated and the FPGA is reconfigured with newly generated specialized bit streams. Because the specialized bitstream is generated at runtime for reconfiguration, instead of compile time, this allows DCS for broader applicability with different possible implementation variants. Here generating a specialized configuration bitstream does not undergo time-consuming steps of the flow to solve computationally hard problems like placement and routing as in the conventional tool flow. These problems are already solved when the parametrized configuration is generated [6,10].

2.1 Two Staged Tool Flow for Parametrized Configuration

The conventional FPGA tool flow cannot be used to generate a parametrized configuration. Therefore, a new two staged tool flow is used, as shown in Fig. 2, consisting of a generic stage and a specialization stage. In the generic stage, an HDL design with parametrized inputs annotated by "–PARAM" in VHDL code is processed to yield a partial parametrized configuration. A partial parametrized configuration (PPC) is a part of the FPGA configuration that is expressed in terms of boolean functions. These boolean functions are evaluated on an embedded processor such as the Power PC or the ARM cortex-A9 present within the FPGA core.

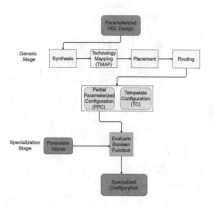

Fig. 2. Dynamic circuit specialization tool flow

The synthesis step in the DCS tool flow is very similar to the conventional tool flow, but the technology mapping step is substantially different. A conventional tool flow generates a network of logic blocks, each consisting of flip-flops and look-up tables. The truth table entries of look tables and the bit that controls the selection of the outputs are constant zeros and ones. We will further refer to logic blocks as look-up tables or LUTs. The mapper used in the DCS tool flow maps the parametrized design on a tunable circuit, which is a network of tunable logic blocks, in short, tunable LUTs or TLUTs, with truth table bits expressed as a boolean expression of the parameters. In the placement step, the mapped resources are placed to specific blocks of the target FPGA architecture. Extensive optimization is considered, so that interconnect delay and wire length is minimized. The router configures the switch blocks to achieve the required interconnect according to the circuit. As the parameters are already included in the TLUT functionality through boolean functions, they are no longer present in the physical implementation of the netlist, so the entire netlist can be placed and routed as if there are no parameters present in the design [11,12].

System Overview. In this experiment, we used a Xilinx SoC board, with Xilinx Platform Studio (XPS) as a design suite, and we designed a parameterized Quantiser/Dequantiser of AVC and implemented it using DCS. The parametrized design is executed on a self-reconfigurable platform.

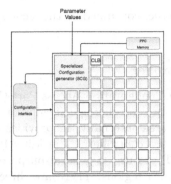

Fig. 3. Self reconfiguration controller For DCS implementation

3 Implementation of DCS on Xilinx SoC Board

For implementing DCS on the Xilinx SoC board, we integrated the TLUTMAP algorithm with the Xilinx tool flow as shown in Fig. 3. TLUTMAP is responsible for parametrization of TLUTs; a unique name is assigned to each TLUT so that each TLUT can be identified and reconfigured with correct specialized truth table entries. After the place and route step the bit stream containing Boolean functions is generated. Upon a change in input parameter values, specialized truth table entries are created, and the truth tables of each TLUT are modified. The configuration of a TLUT is spread over multiple frames, and therefore it is required to alter multiple frames during the reconfiguration process. With the help of the HWICAP configuration interface controller, the FPGA is reconfigured in the following steps:

- Read frame: the truth table entries of each TLUT, located within multiple frames of FPGA configuration memory, are read using the frame address through the configuration interface (HWICAP).
- Modify frames: the current truth table entries of each TLUT are replaced with specialized bits; therefore the frames contain specialized bitstreams.
- Write-back frames: the frames containing specialized bit streams are written back to the same configuration memory using the frame address, thus achieving micro-reconfiguration. Micro-reconfiguration is a fine grain form of reconfiguration used for DCS.

3.1 Cost of Micro-reconfiguration

The micro-reconfiguration comes with four primary costs:

- The PPC memory that is required to store all Boolean functions of the parametrized design.
- The time required to update all truth table entries of every TLUT
- The time required to assess all Boolean functions during the specialization stage.

- The dynamic and idle power consumed by the reconfiguration interface during micro-reconfiguration.

3.2   MiCAP-Pro Controller

The Micro-reconfiguration Configuration Access Port (MiCAP-Pro). It is a custom designed reconfiguration controller specifically for DCS [13]. As this design uses less implementation area internally, it provides a faster reconfiguration speed compared to the other controllers. The MiCAP-Pro is implemented on the zynq SoC. The MiCAP-pro uses a general purpose master port to transfer the data between PL and PS regions. There are three memory-mapped registers for data transfer. The input register is used to send ICAP commands to configuration memory through ICAPE2. The command register is used to give the write/read command to the MiCAP Pro and the output register is used to receive read frames from the configuration memory. An AXI-Lite bus establishes a connection between memory mapped registers and The MiCAP Pro [6,14]. The data transfer between the DRAM memory of the PS region and the MiCAP Pro preferably runs through a high communication bandwidth HPO port. This port provides a high reconfiguration speed, and establishes high-speed data transfer between the PS and PL regions during reconfiguration.

AXI DMA Engine: The AXI Direct memory access (DMA) provides high bandwidth data transfer between AXI-Stream type peripherals implemented on the programmable logic and the DRAM controller present in the PS. The initialization, status registers and management of the DMA engine can be accessed using an AXI-Lite interface. The DMA comes with an optional capability to offload the data transaction completely from the processor in a processor-based system. The high-speed data movement between PL and PS is achieved through a burst-capable AXI4 bus. The DMA supports high-speed data transfer between Stream to memory mapped (S2MM) type peripherals, and Memory mapped to Stream (MM2S) [6,9]. The high-speed data transfer is achieved using full duplex communication and allowing MM2S and the S2MM transfer in parallel. Some of the typical applications with high-speed data transaction between Ethernet and system, can be achieved using AXI DMA for efficient data transfer.

4   Experiment Setup

To design the reconfigurable quantiser/dequantiser blocks of AVC, we first set up a DCS system on a self-reconfigurable platform. In this section, we describe the experimental set up of the parametrized design of quantise and dequantise hardware blocks using the MiCAP-pro reconfiguration controller. We implemented the parametrized design of quantise and dequantise blocks and measured the reconfiguration time and resource utilization on a Xilinx SoC FPGA board.

We implemented the designed parametrized quantiser keeping the QP (Quantisation parameter) value constant, as shown in Fig. 4. For every change in the QP value, a specialized bitstream is generated and reconfigured accordingly.

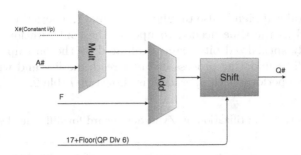

Fig. 4. Reconfigurable quantiser block of AVC

We have used the self-reconfigurable platform as shown in Fig. 5 for implementing the reconfigurable quantiser using DCS. Here we used the MiCAP-Pro reconfiguration controller, on a Zynq-SoC (XC7Z020-CLG484-1, ZedBoard) FPGA and Xilinx v14.7 for the project system builder. The PPC Boolean functions are stored in the DRAM memory of the processor system and all the actions of reconfiguration are controlled by the ARM Cortex-A9 processor. Therefore, the user can use a simple program to control the reconfiguration activity. The entire system is connected using the AXI bus for the data transfer.

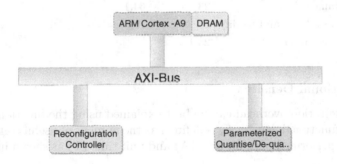

Fig. 5. Self reconfiguration controller for reconfigurable design implementation

5   Results and Discussion

In this section, we present the results of the reconfigurable quantiser and de-quantiser design on a Zynq SoC board. Table 1 shows the number of LUT's required to implement the regular design compared to the reconfigurable design. It can be observed that the number of LUT's improves by 14.0% and 20.60% for the reconfigurable quantiser and de-quantiser designs, respectively. Table also shows the number of tunable lookup tables (TLUT), i.e., the number of lookup tables that need to be micro reconfigured during runtime with the help of the MiCAP-pro reconfiguration controller. The reconfiguration time required for each reconfigurable design is also tabulated in Table 1, where the reconfiguration time is defined as the time needed to update the configuration of all TLUTs using the newly specialized bitstreams evaluated by the on-chip ARM-Cortex-A9 processor. The updating process involves read, modify and write steps, the time required to perform each operation is shown in Table 2.

Table 1. LUT utilization on Zynq SoC board for different designs

	LUT's	Reduction	TLUT 's	Reconfiguration time in ms
Quantiser	610	14.09%	0	0
Scalable Quantiser	524		91	5.83
De-quantiser	237	20.60%	0	0
Scalable De-quantiser	188		56	3.58

Table 2. Reconfiguration time for each operation in the Micap-Pro controller

Micro-reconfiguration task	Time (μs)	TLUT reconfiguration time (μs)
Read frames	23	64.1
Boolean evaluate and modify	18	
Write-back frames	23.1	

5.1 Functional Density

The reconfiguration overhead can be best explained using the functional density curve. The functional density F_d is defined as the number of useful computations that can be performed per unit area (A) and unit time (T) as shown in following equation.

$$F_d = \frac{N}{A \times T}$$

This equation can be calculated for both regular design implementation and reconfigurable design implementation.

5.2 Functional Density for a Regular Design

For a regular design the functional density simply results in:

$$F_d^{regular} = \frac{N}{A_{circ} \times T_{exec}}$$

where T_{exec} is the time it takes to perform N calculations using the total resource cost of the circuit A_{circ}.

5.3   Functional Density for Reconfigurable Design

For a reconfigurable design implementation, the functional density results in:

$$F_d^{reconf} = \frac{N}{(A_{circ} + A_{reconf}) \times (T_{exec} + n_{reconf} \times T_{reconf})}$$

where T_{reconf} is the overhead time of one reconfiguration procedure and A_{reconf} the cost of the hardware overhead (this mainly includes FPGA resources utilized to implement SCG and reconfiguration controller). The n_{reconf} represents number of times the design needs to be optimized for new conditions while running the parametrized application. Also T'_{exec} can be smaller than T_{exec} if the optimized design can be clocked at a higher clock frequency. A'_{circ} is smaller than A_{circ} since the optimized circuit needs less resources. The plot of functional

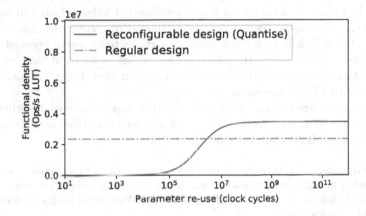

Fig. 6. Functional desnsity curve for reconfigurable quantise block

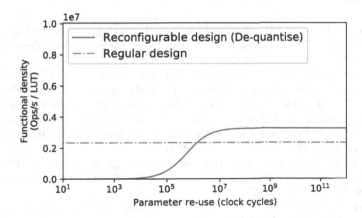

Fig. 7. Functional desnsity curve for reconfigurable de-quantise Block

density curve for a reconfigurable quantiser and de-quantiser is shown in Figs. 6 and 7. The functional density is plotted against the rate of change of parameter values. The functional density shows the efficiency of the implementation as a function of how frequent the parameter values change. Clearly, the magnitude of the functional density of reconfigurable quantiser and dequatiser design is higher that of the regular design when average time between parameter changes is higher than 10^6 clock cycles.

6 Conclusion and Future Work

In this paper, we successfully designed parameterized reconfigurable quantiser and de-quantiser for AVC codec, which is different from regular micro and modular reconfiguration. Here we smartly identified infrequently changing parameters in the design and considering them as constant while implementing it on FPGA, when there is a change in input parameters value which will be reconfigured in real time using custom designed Micap-Pro reconfiguration controller. From Figs. 6 and 7, it is very much clear that input parameters selected in these designs, change after every 10^6 clock cycles which is of ideal time, so reconfiguration time does not effect on video output in this design instead it gives optimization in FPGA resources.

In our future experiments, we would like to extend this work to other blocks of the AVC encoder, such as the DC transform and Coretransform. This would improve the overall AVC encoder design.

Acknowledgments. This work was supported by the Help Video! imec.icon research Project funded by imec and Agentschap innoveren and ondernemen. The authors would like to thank Vasileios Avramelos and Dr. Glenn Van Wallendael for their timely advice and valuable suggestions.

References

1. Sullivan, G.J., Ohm, J.R., Han, W.J., Wiegand, T.: Overview of the high efficiency video coding (HEVC) standard. IEEE Trans. Circuits Syst. Video Technol. **22**(12), 1649–1668 (2012)
2. Ohm, J.R., Sullivan, G.J., Schwarz, H., Tan, T.K., Wiegand, T.: Comparison of the coding efficiency of video coding standards—including high efficiency video coding (HEVC). IEEE Trans. Circuits Syst. Video Technol. **22**(12), 1669–1684 (2012)
3. Bossen, F., Bross, B., Suhring, K., Flynn, D.: HEVC complexity and implementation analysis. IEEE Trans. Circuits Syst. Video Technol. **22**(12), 1685–1696 (2012). https://doi.org/10.1109/TCSVT.2012.2221255
4. Kuon, I., Rose, J.: Measuring the gap between fpgas and asics. IEEE Trans. Comput.-Aided Des. Integr. Circuits Syst. **26**(2), 203–215 (2007). https://doi.org/10.1109/TCAD.2006.884574
5. Bruneel, K., Heirman, W., Stroobandt, D.: Dynamic data folding with parameterizable configurations. ACM Trans. Des. Autom. Electron. Syst. **16**(4) (2011)

6. Kulkarni, A., Kizheppatt, V., Stroobandt, D.: MiCAP: a custom reconfiguration controller for dynamic circuit specialization. In: 2015 International Conference on ReConFigurable Computing and FPGAs (ReConFig), pp. 1–6, December 2015
7. Abeydeera, M., Karunaratne, M., Karunaratne, G., Silva, K.D., Pasqual, A.: 4K real-time HEVC decoder on an FPGA. IEEE Trans. Circuits Syst. Video Technol. **26**(1), 236–249 (2016)
8. Miyazawa, K., et al.: Real-time hardware implementation of HEVC video encoder for 1080p HD video. In: 2013 Picture Coding Symposium (PCS), pp. 225–228, December 2013
9. Abouelella, F., Davidson, T., Meeus, W., Karel Bruneel, W., Stroobandt, D.: How to efficiently implement dynamic circuit specialization systems. ACM Trans. Des. Autom. Electron. Syst. **18**(3) (2013)
10. Amit, K., Dirk, S.: How to efficiently reconfigure tunable lookup tables for dynamic circuit specialization. Int. J. Reconfig. Comput. **2016**, 3 (2016). https://doi.org/ 10.1155/2016/5340318
11. Davidson, T., Bruneel, K., Stroobandt, D.: Identifying opportunities for dynamic circuit specialization. In: Becker, T., (ed.) Workshop on Self-Awareness in Reconfigurable Computing Systems, Proceedings, pp. 18–21 (2012). http://srcs12.doc.ic. ac.uk/docs/srcs_proceedings.pdf
12. Farisi, B.A., Bruneel, K., Cardoso, J.M.P., Stroobandt, D.: An automatic tool flow for the combined implementation of multi-mode circuits. In: 2013 Design, Automation Test in Europe Conference Exhibition (DATE), pp. 821–826, March 2013
13. Kulkarni, A., Stroobandt, D.: MiCAP-Pro: a high speed custom reconfiguration controller for dynamic circuit specialization. In: Design Automation for Embedded Systems, vol. 20 (4), pp. 341–359 (2016) https://doi.org/10.1007/s10617-016-9180-6
14. xilinx product guide, "Logicore ip axi," vol. v6.03a (2012)

Image Restoration and Reconstruction

Large Parallax Image Stitching Using an Edge-Preserving Diffeomorphic Warping Process

Geethu Miriam Jacob[(✉)] and Sukhendu Das

V.P. Lab, Department of Computer Science and Engineering,
Indian Institute of Technology Madras, Chennai, India
geethumiriam@gmail.com, sdas@iitm.ac.in

Abstract. Image Stitching is a hard task to solve in the presence of large parallax in video frames. In many cases, video frames shot using hand-held cameras have low resolution, blur and large parallax errors. Most recent works fail to align such a sequence of images accurately. The proposed method aims to accurately align image frames, by employing a novel demon-based, edge-preserving diffeomorphic registration for image stitching, termed as "DiffeoWarps". The first stage aligns the images globally using a mesh-based perspective (homography) transformation. At the second stage, an alternating method of minimization of correspondence energy and TV-regularization improves the alignment. The "diffeowarped" images are then blended to obtain good quality stitched results. We experimented on two standard datasets as well as on a dataset comprising of 10 sets of images/frames collected from unconstrained videos. Both qualitative and quantitative performance analysis show the superiority of our proposed method.

Keywords: Diffeomorphic demons · Total variation regularization
Panorama stitching

1 Introduction

Image Stitching is the process for obtaining a large field-of-view image from a set of smaller images with overlapping spatial neighborhood. Several existing image stitching algorithms such as Microsoft Photosynth, Adobe Photoshop and Autostitch [1], with built-in image stitching tools, fail to perform well when non-ideal data is provided as input. Parallax error, occlusions, motion blur and presence of moving objects, are the main challenges of any stitching algorithm. Particularly, when unconstrained videos shot from hand-held cameras are the source of images used for stitching, the occurrences of these issues are more prominent.

A common approach followed by many of the existing image stitching algorithms [4,10,13,14,21,27], has the following pipeline: estimate the transformations between the overlapping part of the frames, align the images using a warping model and combine the images to render them to a common canvas. Our

© Springer Nature Switzerland AG 2018
J. Blanc-Talon et al. (Eds.): ACIVS 2018, LNCS 11182, pp. 521–533, 2018.
https://doi.org/10.1007/978-3-030-01449-0_44

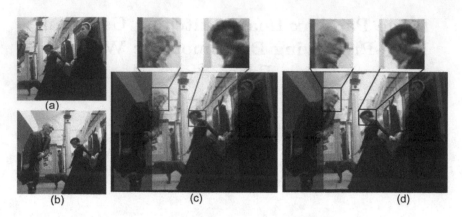

Fig. 1. DiffeoWarps: (a) Fixed and (b) Unaligned Images; Alignment by: (c) spatially varying warps [4], (d) proposed DiffeoWarp. The ghosting effects in (c) are subverted in (d).

proposed method also follows the same pipeline. When the images to be aligned have large parallax error, many of the existing methods fail to provide satisfactory results. Some recent parallax-tolerant methods [5,13,27] aim at producing an accurate seams between images instead of properly aligning (registering) them. Our method provides a solution to these problems by proposing a new warping model based on edge-preserving diffeomorphic registration, termed "DiffeoWarps", to obtain a perfectly aligned image even in the presence of large parallax.

Many existing methods compute spatially varying warps [4,15] to perfectly align images. For large parallax as shown in Fig. 1, non-linear deformation is required to align the images. "Diffeowarps" is a novel edge-preserving diffeomorphic warping model to align images, composed of a combination of mesh-based homography warp and edge-preserving diffeomorphic transformations. Diffeomorphic registrations have been used in medical image registration [19,23] extensively, but not for image stitching purposes. The reason is that diffeomorphism does not consider the displacement field discontinuities of the image into consideration, while generating the transformations. We propose the use of a TV based regularization along with the diffeomorphic process such that the warping process becomes adaptable for image stitching. Figure 1 shows the alignment using a mesh based warping model [4] and DiffeoWarps. The ghosting effects generated by the mesh based methods are alleviated in DiffeoWarps.

The main contributions of the proposed work are: (i) A 2-stage process of alignment involving computation of spatially varying mesh-based homography and diffeomorphic registration, followed by frame rendering, (ii) A combination of the frame alignment using a spatially-varying homography and a novel demon based diffeomorphic registration with TV regularization for image stitching within the warping model, and (iii) Formation of a new stitching dataset, containing 10 sets of 2–3 frames in each, created from a stabilization dataset of unconstrained videos with large parallax and moving objects. Section 2 provides a brief summary of the previous works, Sect. 3 describes the details of the proposed method, and in Sect. 4 experimental results are discussed, while Sect. 5 provides the conclusions.

2 Brief Literature Review

Warping Models. Several warping models have been proposed in the past few decades. Richard Szeliski has given a comprehensive study on the various image stitching methods and the different warping models used for warping in [20]. The traditional methods [1,21] use a single homography to warp images. Recently, variants of homography models [7,10] have also been proposed. Spatially varying warps were introduced to handle parallax errors. Many methods, such as Smoothly Varying Affine warps (SVA) [14], Content Preserving Warps (CPW) [15], Adaptive As Natural As Possible (AANAP) [12] and As Projective As Possible (APAP) [26] have been proposed in the past years. Shape Preserving Half Preserving (SPHP) [4] combines the projective and similarity models, whereas NIS [4] add a global similarity prior to the local warping models. All the above methods aim at aligning the images and blending them.

Recently, some parallax tolerant image stitching methods have been developed. Zhang and Liu [27] provide a solution for large parallax by combining homography and CPW, followed by seamcuts and blending. Lin *et al.* in [13] proposed multiple alignment hypotheses to locate a good seam for stitching. These methods aim at obtaining a good stitching results by computing optimal seams, rather than aligning the images accurately. Our method gives importance to both alignment as well as obtaining good stitching results.

Demons-Based Registration. Thirion in his work [22] proposed a diffusion model to perform image-to-image matching on medical images based on Maxwell's demons. The main idea was to consider one image (moving image) to be a deformable grid model which deforms to match the other image (fixed image). Diffeomorphic demons [23] was an extension to the Thirion's demons where the demons were adapted to the space of diffeomorphic transformations. It is shown that diffeomorphic demons algorithm is able to register the images faster and does it with a smooth invertible transformation. Inertial demons, a momentum based approach, proposed by Rebeiro *et al.* [19] reported improvement in the convergence and accuracy of the diffeomorphic demons registration. We propose a Nesterov Accelerated Gradient (NAG) based approach to further accelerate the estimation of the diffeomorphic transformations. We adapt the diffeomorphic transformation to image stitching with Total Variation (TV) regularization in the computation of demons energy [6,24] to preserve displacement field discontinuities at motion boundaries.

3   Proposed DIFFEOWARPS Framework

We propose a method to perfectly align images by performing a edge-preserving diffeomorphic registration. The aim of Diffeowarps process is to register adjacent images (frames) to a fixed image with minimal alignment error. The stages of our framework are the following:

1. Feature Matching and Mesh-based Frame Alignment: A mesh-based and spatially varying homography matrix is used to perform the initial alignment. This is the first stage of the warping model proposed.
2. Edge-Preserving Diffeomorphic Registration: A diffeomorphic registration process is adopted as the second stage of the warping model. A TV regularization based formulation imparts edge-preserving properties to the solution, thus making the method adaptable to the image stitching process.
3. Stitched Image Rendering: The stitched image is rendered using a blending process.

Each module, formulated using suitable optimization functions, is described in detail in the following sub-sections:

3.1   Feature Matching and Global Frame Alignment

Consider two images to be stitched and let one of them be the fixed (reference) image R and other be the unaligned image U. SIFT features [17] are extracted from the pair of images and a spatially varying homography matrix, $\{H_{ij}\}, i = 1, 2..m, j = 1, 2..n$ ($m \times n$ being the mesh dimension) is estimated using the method in [15], which minimizes an optimization function obtaining the deformed mesh vertices. The optimization function in [15] consists of two terms, one aligning the point correspondences and another ensuring the similarity transformation of the mesh grids. U is aligned with R by transforming it with $\{H_{ij}\}$, thereby performing a mesh-based alignment of the frames. However, this alignment fails to align all regions of the overlap accurately. This phenomenon is illustrated in Fig. 1(c), where the red boxes indicate the region which fails to align accurately. Thus, in addition to the mesh-based perspective (homography) transformation, we adopt a method for diffeomorphic registration/warping [23]. The diffeomorphic registration reduces the alignment errors caused by large parallax, as in Fig. 1(d).

3.2   Edge-Preserving Diffeomorphic Registration

Let F and M be the overlapping regions in R and U frames, after global alignment. The aim of this module is to perform an effective non-linear registration, such that the frames are perfectly aligned. Non-linear registration (spatially variant) generates spatial transformations at the pixel level that best aligns the two images. Inertial demons [19], an improved version of Thirion's demons [22] in terms of accuracy and convergence, is used for non-linear diffeomorphic registration. Any demons algorithm alternates between the computations of demons forces and regularization of the demons forces. We perform a novel improved version of Inertial demons and Total Variation (TV) function for regularization, which possesses edge-preserving properties.

Demons as a Minimization Function [23]: The diffeomorphic deformation (using non-parametric spatial transformation, s, as per-pixel displacement field) that best aligns M with F is estimated using the optimization function:

$$E(s) = Sim(F, M \circ s) + Reg(s) \tag{1}$$

The demons algorithm [23] is solved by the alternate minimization of the similarity function (Sim) and the regularization function (Reg). For the ease of solving Eq. 1, a latent correspondence variable (c) is introduced and the function to be solved is modified as:

$$E(c, s) = Sim(F, M \circ c) + \sigma||c - s||^2 + Reg(s) \tag{2}$$

This allows for alternate (separable) optimization of the similarity term, also known as the correspondence energy, $E_{corr} = Sim(F, M \circ c) + \sigma||c - s||^2$ and the regularization term, $E_{reg} = \sigma||c - s||^2 + Reg(s)$, where

$$Sim(F, M \circ c) = ||F - M \circ c||^2,$$
$$Reg(s) = \sum_{i=1}^{N} ||\nabla s_i||_1 \tag{3}$$

Here, $Reg(s)$ is the TV-based regularization function, where s_i is the deformation (transformation) of the ith pixel. $\circ$ is the warping function and $N = |F| = |M|$, where $|.|$ is the cardinality function. Unlike Gaussian smoothing used in [3, 23], we use Total Variation (TV) regularization. TV regularization has edge-preserving properties while smoothing the deformation, required to avoid the random deformation and wobble effects. The solution to the correspondence energy and the regularization energy follows.

Demons Forces, a Solution to the Correspondence Energy: Eq. 2 can be rewritten as a correspondence energy function (E_{corr}) using $u = c - s$, as

$$E_{corr}(u) = ||F - M \circ (s + u)||^2 + \sigma||u||^2 \tag{4}$$

The demon forces, $u(p)$, for each pixel p, is obtained using 2nd order minimization techniques [19]:

$$u(p) = \frac{F(p) - M \circ s(p)}{||J^p||^2 + \sigma^2(F(p) - M \circ s(p))^2} J^p \tag{5}$$

where $J^p = \nabla F(p) + \nabla M \circ s(p)$. The updated diffeomorphic demons forces are given as $exp(u)$ [23]. The demons forces are added to exhibit the diffeomorphic properties.

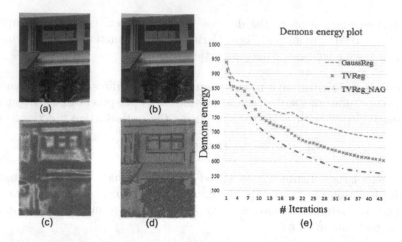

Fig. 2. (a) Fixed frame (F), (b) Globally aligned frame (M), (c) updates of image pixels using Gaussian regularization, (d) updates of image pixels using our TV regularization with edge-preserving property, (e) plots of convergence of demons energies (Eq. 1) with Gaussian regularization, TV regularization with and without Nesterov momentum (NAG) [18].

TV Regularization of Demons Forces: The regularization term (E_{reg}) to be minimized is given as:

$$E_{reg}(s) = Reg(s) + \sigma ||u||^2 \tag{6}$$

This expression is minimized using FASTA implementation [9] of forward-backward splitting algorithm [8] to solve the TV optimization function. Figure 2 qualitatively illustrates the enhancement in performance of the demon forces over the traditional 2-norm based regularization (see effect of Gaussian smoothing in Fig. 2(c) and better results of TV regularization in Fig. 2(d)). The diffeomorphic deformation (s) in TV regularization preserves the edges of the image while computing the demons forces, whereas the deformation for 2-norm regularization smooths the demon forces irrespective of the structural properties of the image. Thus, unwanted wobbling artifacts produced by the diffeomorphic warping are avoided by using TV regularization.

Nesterov Momentum for Accelerated Convergence: Demons algorithm is an iterative process, where at each iteration a momentum based update similar to that in [19] is used. We propose the usage of the Nesterov momentum [18] for faster convergence. Nesterov momentum ensures that the demons forces move in the same velocity direction as in the previous steps. It has been proved to converge faster [18] than other momentum techniques. Considering t to be the current iteration, the Nesterov momentum based update is given by the following expression:

$$v(p)^t = mom * v(p)^{t-1} - scale * \nabla u(p)^{t-1}$$
$$u(p)^t = u(p)^{t-1} + v(p)^t + mom * (v(p)^t - v(p)^{t-1})$$

(7)

Here, *mom* and *scale* are two parameters, set as 0.5 each. Figure 2(e) shows
the plots of demons energy with Gaussian regularization, inertial demons with
TV regularization, and proposed demons energy with TV regularization accelerated by Nesterov Accelerated Gradient (NAG) for aligning the image patches
in Fig. 2(a) and (b). Our proposed optimization function provides fastest convergence rate.

The diffeomorphic registration is done in a coarse-to-fine manner. The minimization of the variational problem in Eq. 2 is performed in a hierarchical (or
multiscale) approach, starting from the coarsest level and later proceeding to the
finest level, helping to improve the registration accuracy of the images. We have
used 3 levels and bicubic interpolation method for downsampling. The performance further improves if a patchwise registration, followed by a combined one
is performed. The estimated diffeomorphic deformations help to inpaint over the
non-overlapping regions of the frame U, which are then warped. The maximum
number of iterations at each level is set as 50.

3.3   Stitched Image Rendering

With moving objects and occlusions, the DiffeoWarps will not produce perfectly
aligned frames. Hence, we adopt a blending process with more weightage given to
image R. Our process is similar to the multi-band blending method of [2], where
only regions on either side of an edge from both R and U are blended. Thus, with
moving objects or occlusions, the stitched image is devoid of undesired artifacts.

The overall framework of Diffeowarps is a 3-stage process: the 1st stage comprises of a global alignment stage (using homography), the 2nd stage performs
diffeomorphic warping using the optimization function (Eq. 1) with TV-based
regularization (Eq. 6) using Nesterov-momentum based minimization (Eq. 7),
whereas the 3rd stage performs blending of the aligned images.

4   Experiments and Results

Experiments were performed on two standard image stitching datasets [13, 27]
and on a new dataset created from a stabilization dataset [16]. The parameters
were empirically estimated and were used uniformly for all experiments. The
comparing methods were executed using default parameters given in the code
published by the authors. For the qualitative results, the results were obtained
from the author's website. The standard image stitching dataset contains 30
sets of images [27]. These images have large parallax error among themselves.
The new dataset, PARALLAX-BLUR-STITCH has 10 sets of image sequences
with high parallax (2–3 frames in each set), taken from stabilization dataset

[16] (shot using handheld cameras). The images in each imageset have: (i) low resolution and blur, (ii) presence of moving objects and (iii) large parallax error. The performance of the proposed method is compared with 5 state-of-the-art techniques [4, 13, 15, 26, 27].

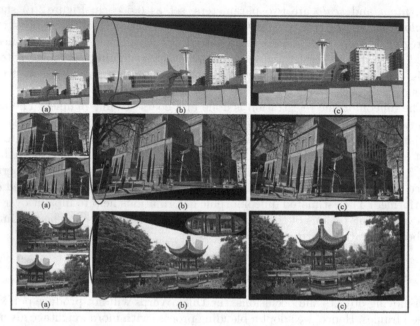

Fig. 3. (a) The pair of images to be aligned (samples of dataset of [27]), Output of (b) a parallax-tolerant method [27], and (c) proposed method. The method generates undesired distortions in the stitched images (marked by red oval shapes). (Color figure online)

4.1   Qualitative Results on Parallax-Tolerant Methods

Figures 3 and 4 show results on the examples used in the parallax-tolerant methods proposed in the works [13, 27], which gives importance to seam stitching rather than alignment of images. Most parallax tolerant methods [5, 13, 27] allow alignment error to occur in the images and aim at generating good quality seams. This is hazardous in case of stitching multiple images, where a global chain of transformations is difficult to implement for accurately stitching the images, resulting in a degrading performance in panorama creation. Our method performs well for both the aligning as well as the blending tasks. Results of stitching using [27] and the proposed method on few examples of the dataset [27] are shown in Fig. 3(b) and (c). The method [27] produces unwanted distortions and elongations towards the ends of the images in the stitched images, shown by red oval shapes. Our output shown in the third column shows results superior to that

Fig. 4. (a) The pair of images to be aligned (samples of the SEAGULL dataset [13]), Output of (b) SEAGULL [13], and (c) proposed method. The competing method elongates the stitched images with undesired distortions (marked by red oval shapes). (Color figure online)

of [27] in the second column. Similarly, Fig. 4(b) and (c) compares the results of image stitching using the method in [13] and the proposed method. The method [13] abnormally elongates the stitched image to a great extent (second column) towards the ends without proper alignment, whereas ours provide a good quality stitching result (third column).

Qualitative results for our dataset are given in Fig. 5. The red boxes indicate the zoomed in artifacts introduced in the images by the stitching algorithms [4,26], whereas the blue boxes indicate the corresponding high quality stitching provided by our proposed Diffeowarp process. Red oval shapes show the ghosting effects caused by the moving object in the input frames which is not present in our output. It can be seen that, very less artifacts and defects are introduced in our method compared to the other methods in all the images.

4.2 Quantitative Results

Quantitative evaluation of performance is done using the correlation (E_{corr}) error and the mean geometric error (E_{mg}) on points (similar to that in [25]). E_{corr} is defined as one minus the average of the Normalized Cross Correlation (NCC) over a neighborhood in the overlapped region, while the mean geometric error, E_{mg}, gives the average distance between the points of the warped (aligned) and fixed images. The parallax-tolerant methods [11,13,27] does not aim at

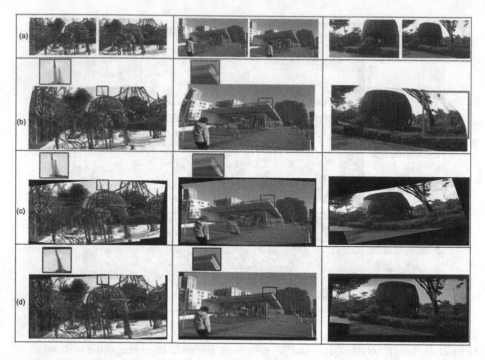

Fig. 5. Qualitative Comparison with the state-of-the-art aligning methods: (a) Two image pairs to be aligned, Output of (b) SPHP [4], (c) APAP [26] and (d) our proposed method. The areas exhibiting undesired artifacts are highlighted with red boxes and oval shapes in (b) and (c). (Color figure online)

aligning the images perfectly, but concentrate of obtaining the best seams and is not fair to compare their registration accuracy with our method. Thus, we have compared quantitatively with other recent methods which registers the images accurately.

$$E_{corr}(F, f(M)) = \sqrt{\frac{1}{N} \sum_{i=1}^{N} (1 - NCC(p_i, p_i'))^2},$$

$$E_{mg} = \frac{1}{K} \sum_{i=1}^{K} ||f(c_i) - c_i'|| \tag{8}$$

where p_i and p_i' are pixel grey-values in the fixed and aligned images respectively and c_i, c_i' are the ith pair of point correspondences in the original pair of images. f indicates the estimated warping, K denotes the total number of correspondences. K is estimated using SIFT and will be the same for all the methods. Smaller values of E_{corr} and E_{mg} indicate better quality of stitching results and alignment. Both the error measures reflect how well the alignment is done by the stitching algorithm, which in turn quantitatively evaluates the stitching quality of the algorithm.

Table 1. Comparison of the performances for aligning/warping using dataset [27]. *-Average over 30 imagesets

E_{mg}					E_{corr}				
ImSet	[26]	[4]	[15]	Ours	**ImSet**	[26]	[4]	[15]	Ours
PT05	2.36	2.25	1.94	**1.13**	PT05	0.34	0.35	0.33	**0.26**
PT07	2.53	2.41	2.01	**1.04**	PT07	0.32	0.28	0.28	**0.26**
PT10	3.48	5.75	4.45	**1.89**	PT10	0.45	0.44	0.43	**0.39**
PT12	2.82	9.35	3.24	**1.76**	PT12	0.35	0.35	0.36	**0.32**
PT13	6.98	46.98	40.98	**6.90**	PT13	0.38	0.36	0.37	**0.35**
PT25	2.90	6.73	3.25	**2.20**	PT25	0.46	0.43	0.43	**0.41**
PT26	3.20	2.91	2.31	**2.04**	PT26	0.26	0.26	0.25	**0.21**
PT27	2.50	17.04	3.97	**1.62**	PT27	0.34	0.40	0.38	**0.26**
PT28	5.03	2.70	7.28	**1.75**	PT28	0.33	0.31	0.32	**0.26**
PT30	3.66	3.07	5.02	**2.04**	PT30	0.41	0.43	0.40	**0.36**
Avg*	4.28	6.44	7.37	**2.31**	**Avg***	0.36	0.34	0.35	**0.32**

Table 1 shows the quantitative results for alignment in the standard image stitching dataset [27] for parallax tolerance. On an average, our method performs the best for the 30 sets of images. Table 2 gives the quantitative results for the dataset, PARALLAX-BLUR-STITCH. Our method performs superior (best results in bold) than the comparing methods for all sets of image sequences, by

Table 2. Comparison of the performances of alignment/warping for the dataset, PARALLAX-BLUR-STITCH.

E_{mg}					E_{corr}				
ImSet	[26]	[4]	[15]	Ours	**ImSet**	[26]	[4]	[15]	Ours
Ours01	119.25	85.82	72.25	**69.79**	Ours01	0.26	0.25	0.26	**0.21**
Ours02	2.20	2.54	2.11	**1.34**	Ours02	0.26	0.28	0.24	**0.17**
Ours03	5.87	6.88	7.33	**4.77**	Ours03	0.49	0.46	0.47	**0.43**
Ours04	4.09	3.19	6.06	**2.65**	Ours04	0.44	0.43	0.40	**0.33**
Ours05	3.73	2.89	2.98	**1.8**	Ours05	0.32	0.31	0.32	**0.29**
Ours06	4.71	6.94	10.45	**2.64**	Ours06	0.37	0.33	0.38	**0.31**
Ours07	2.46	3.43	2.99	**1.84**	Ours07	0.26	0.26	0.28	**0.25**
Ours08	5.54	2.24	4.20	**2.12**	Ours08	0.38	0.30	0.30	**0.29**
Ours09	1.86	1.23	1.98	**1.11**	Ours09	0.21	0.18	0.19	**0.16**
Ours10	15.25	3.91	4.37	**2.28**	Ours10	0.35	0.34	0.35	**0.30**
Avg	25.25	11.91	11.47	**9.03**	**Avg**	0.33	0.31	0.32	**0.27**

a considerable margin. For more results, visit our website www.cse.iitm.ac.in/ ~vplab/ACVIS_18/DiffeoWarps_supplementary.pdf.

5   Conclusion

An effective method of image warping for image stitching, known as "DiffeoWarps" has been proposed. A novel diffeomorphic demon-based registration with TV-based regularization is proposed to perfectly align the images. Process of TV regularization is modeled to maintain the structure of the edges while warping. We have experimented on an existing dataset and image sequences from a stabilization dataset, consisting of frames with high parallax. Quantitatively as well as qualitatively, our method aligns frames better than the state-of-the-art image alignment methods for both the datasets, providing enriched performance.

References

1. Brown, M., Lowe, D.G.: Automatic panoramic image stitching using invariant features. Int. J. Comput. Vis. **74**(1), 59–73 (2007)
2. Burt, P.J., Adelson, E.H.: A multiresolution spline with application to image mosaics. ACM Trans. Graph. (TOG) **2**(4), 217–236 (1983)
3. Cachier, P., Bardinet, E., Dormont, D., Pennec, X., Ayache, N.: Iconic feature based nonrigid registration: the PASHA algorithm. Comput. Vis. Image Underst. **89**(2), 272–298 (2003)
4. Chang, C.H., Sato, Y., Chuang, Y.Y.: Shape-preserving half-projective warps for image stitching. In: CVPR, pp. 3254–3261 (2014)
5. Chen, Y.-S., Chuang, Y.-Y.: Natural image stitching with the global similarity prior. In: Leibe, B., Matas, J., Sebe, N., Welling, M. (eds.) ECCV 2016. LNCS, vol. 9909, pp. 186–201. Springer, Cham (2016). https://doi.org/10.1007/978-3-319-46454-1_12
6. Demirović, D., Šerifović-Trbalić, A., Prljača, N., Cattin, P.C.: Total variation filtered demons for improved registration of sliding organs. ISRN Biomathematics 2013 (2013)
7. Gao, J., Kim, S.J., Brown, M.S.: Constructing image panoramas using dual-homography warping. In: CVPR, pp. 49–56. IEEE (2011)
8. Goldstein, T., Studer, C., Baraniuk, R.: A field guide to forward-backward splitting with a FASTA implementation. arXiv eprint arXiv:abs/1411.3406 (2014)
9. Goldstein, T., Studer, C., Baraniuk, R.: FASTA: a generalized implementation of forward-backward splitting, January 2015. http://arxiv.org/abs/1501.04979
10. Li, N., Xu, Y., Wang, C.: Quasi-homography warps in image stitching. arXiv preprint arXiv:1701.08006 (2017)
11. Li, Y., Monga, V.: SIASM: sparsity-based image alignment and stitching method for robust image mosaicking. In: ICIP, pp. 1828–1832. IEEE (2016)
12. Lin, C.C., Pankanti, S.U., Natesan Ramamurthy, K., Aravkin, A.Y.: Adaptive as-natural-as-possible image stitching. In: CVPR, pp. 1155–1163 (2015)
13. Lin, K., Jiang, N., Cheong, L.-F., Do, M., Lu, J.: SEAGULL: seam-guided local alignment for parallax-tolerant image stitching. In: Leibe, B., Matas, J., Sebe, N., Welling, M. (eds.) ECCV 2016. LNCS, vol. 9907, pp. 370–385. Springer, Cham (2016). https://doi.org/10.1007/978-3-319-46487-9_23

14. Lin, W.Y., Liu, S., Matsushita, Y., Ng, T.T., Cheong, L.F.: Smoothly varying affine stitching. In: CVPR, pp. 345–352. IEEE (2011)
15. Liu, F., Gleicher, M., Jin, H., Agarwala, A.: Content-preserving warps for 3D video stabilization. ACM Trans. Graph. (TOG) 28(3), 44 (2009)
16. Liu, S., Yuan, L., Tan, P., Sun, J.: Bundled camera paths for video stabilization. ACM Trans. Graph. (TOG) 32(4), 78 (2013)
17. Lowe, D.G.: Distinctive image features from scale-invariant keypoints. Int. J. Comput. Vis. 60(2), 91–110 (2004)
18. Nesterov, Y.: A method for unconstrained convex minimization problem with the rate of convergence o $(1/k^2)$. Dokl. AN USSR 269, 543–547 (1983)
19. Santos-Ribeiro, A., Nutt, D.J., McGonigle, J.: Inertial demons: a momentum-based diffeomorphic registration framework. In: Ourselin, S., Joskowicz, L., Sabuncu, M.R., Unal, G., Wells, W. (eds.) MICCAI 2016. LNCS, vol. 9902, pp. 37–45. Springer, Cham (2016). https://doi.org/10.1007/978-3-319-46726-9_5
20. Szeliski, R.: Image alignment and stitching: a tutorial. Found. Trends® Comput. Graph. Vis. 2(1), 1–104 (2006)
21. Szeliski, R., Shum, H.Y.: Creating full view panoramic image mosaics and environment maps. In: SIGGRAPH, pp. 251–258. ACM Press/Addison-Wesley Publishing Co. (1997)
22. Thirion, J.P.: Image matching as a diffusion process: an analogy with Maxwell's demons. Med. Image Anal. 2(3), 243–260 (1998)
23. Vercauteren, T., Pennec, X., Perchant, A., Ayache, N.: Diffeomorphic demons: efficient non-parametric image registration. NeuroImage 45(1), S61–S72 (2009)
24. Vishnevskiy, V., Gass, T., Szekely, G., Tanner, C., Goksel, O.: Isotropic total variation regularization of displacements in parametric image registration. IEEE Trans. Med. Imaging 36(2), 385–395 (2017)
25. Xiang, T., Xia, G.S., Bai, X., Zhang, L.: Image stitching by line-guided local warping with global similarity constraint. arXiv preprint arXiv:1702.07935 (2017)
26. Zaragoza, J., Chin, T.J., Brown, M.S., Suter, D.: As-projective-as-possible image stitching with moving DLT. In: CVPR, pp. 2339–2346 (2013)
27. Zhang, F., Liu, F.: Parallax-tolerant image stitching. In: CVPR, pp. 3262–3269 (2014)

A Wavelet Based Image Fusion Method Using Local Multiscale Image Regularity

Vittoria Bruni[1,2]($\boxtimes$), Alessandra Salvi[2], and Domenico Vitulano[2]

[1] Dip. Scienze di Base e Applicate per l'Ingegneria (SBAI), Sapienza - Universitá di Roma, Rome, Italy
vittoria.bruni@sbai.uniroma1.it
[2] Istituto per le Applicazioni del Calcolo "M. Picone" (IAC) — C.N.R., Rome, Italy
d.vitulano@iac.cnr.it

Abstract. This paper presents an image fusion method which uses an image dependent multiscale decomposition and a fusion rule which is based on the local image multiscale activity. The latter is used for determining a proper partition of the frequency plane where the max-based fusion strategy is applied; the same image activity is used for guiding the max-based fusion rule. Multiscale local image activity is computed using an estimation of the local Lipschitz regularity at different resolutions. Preliminary experimental results show that the proposed method is able to provide satisfying fusion results in terms of details preservation of the visible image, extraction of the important regions in the infrared image, reduction of artifacts and noise amplification. Presented results, even if in their preliminary form, are able to reach and outperform some of the most recent and performing fusion methods.

Keywords: Rational dilation wavelet transform · Image fusion
Interscale dependencies

1 Introduction

The increasing use of multisensors based analysis requires the development of novel and more precise methods oriented to merge the different acquired information. In particular, there is a growing interest in fusion methods for several purposes in the image analysis context, as for example surveillance, military and defense applications, robotics, remote sensing, medical imaging, biometric, just to mention some of them [5]. The aim of image fusion is to combine the different information that is captured by distinct sensors into a single image in order to increase the visual informative content. Hence, having N source images coming from different sensors or acquisition modes, the goal is to provide an image which is able to give a more accurate and detailed description of the scene under study. The final image has to be informative from the human vision point of view as it has to support successive or immediate analyses and decisions. Fusion methods can be broadly classified into two main categories [5]: pixel-level and features

© Springer Nature Switzerland AG 2018
J. Blanc-Talon et al. (Eds.): ACIVS 2018, LNCS 11182, pp. 534–546, 2018.
https://doi.org/10.1007/978-3-030-01449-0_45

level methods. The first ones mainly rely on pointwise operations which guarantee a certain loss of artifacts in the final image; on the contrary, the second class of methods combine image features such as edges, textures, regions so that the selection of the best fusion rule results more critical. It is worth observing that the selection of the best fusion method often strongly depends on the fusion purpose and kind of acquisition modes/sensors. In this paper we will focus on infrared (IR) and visible (VI) image fusion methods. The fusion of IR and VI images plays a fundamental role, for example, in surveillance applications whenever there are critical visual conditions (low contrast, low light, night acquisition, etc.). IR sensor captures the thermal radiation of an object or an area; hence, it is useful for the detection of hidden objects or for the identification of camouflages or obstacles. VI sensor represents the visible spectral reflection so that it is able to capture the visible characteristics of an object/area. Hence, the aim of IR and VI image fusion is to make richer the visible image by adding details and new components that are present in IR image. IR and VI image fusion methods can be classified as pixel-based or region based methods [5,9]. The first ones properly fuse the two source images in a proper projection space by using suitable expansion bases. The latter are in general originated by multiscale transforms having different spatial or spectral properties, as for example, laplacian pyramids [17], discrete cosine transform [6,15], wavelet transform (decimated, stationary, complex) [8,14,18] and the more recent contourlets [13] and shearlets [7]—see [5,9] for details. The region based fusion methods extract salient regions from the source images and then properly combine them. Independently of the class, the two main problems in image fusion are: the choice of image representation and the choice of the fusion rule. Multiscale approaches combine the different frequency content of source images according to the amplitude of the expansion coefficients: the higher the coefficients amplitude the more significant their contribution. That is why the selection of the transform represents a crucial step in image fusion. In IR/VI context, the transform is required to represent in a compact and fine way details in VI image as well as to capture the main details and frequency content of IR image. With regard to this point it is worth mentioning the paper in [11] where a total variation functional is defined for fusion purposes. The fidelity term involves IR image while the total variation one is related to VI image—as a result the aim is to reconstruct an image which is similar to IR but has VI details. In the multiscale context, fusion rules which involve local and global image features, such as local spectral features, interscale relationships and local contrast, are more promising [5,9]. That is why in this paper we are interested in a first study whose aim is to define a transform and a fusion rule able to combine all these features in a single measure. Based on the relationship between local image Lipschitz regularity and multiscale transforms [12] and by considering the great flexibility of the rational dilation wavelet transform [1], in this paper we present a fusion method which: (i) defines the multiscale image activity on the basis of interscale dependencies of wavelet coefficients; (ii) properly partitions the frequency axis on the basis of the estimated local multiscale activity of the source images; (iii) applies a max based fusion strategy in the detected bands using the activity value as a weight. To this aim the Lipschitz regularity and its relation to the decay of wavelet coefficients are used for determining the activity index while the rational dilation wavelet transform is used for defining a multiscale transform which provides a frequency axis partition having a non constant Q factor (frequency bandwidths do not obey a logarithmic law). Preliminary experimental results show that the proposed method is able to reach better results, in terms of some standard objective fusion metrics, than some recent literature. The computing time is very moderate thanks to RDWT implementation through iterated perfect reconstruction filter banks.

The outline of the paper is the following. Next section presents a brief description of the main theoretical foundations of the proposed method and gives some implementation details. Section 3 presents some preliminary experimental results and comparative studies, while the final section draws the conclusions.

2   The Proposed Fusion Method

Let I and V respectively be an infrared and a visible image. The aim of fusion is to define a new image F by suitable combining I and V. As mentioned in Introduction, the main issue in image fusion is to select the proper image transform where defining a suitable fusion rule in order to recover significant components (edges, textures, details) of the source images while preventing artifacts and noise amplification. In this paper the Rational-Dilation Wavelet Transform (RDWT) [1] using an image dependent dilation factor is employed.

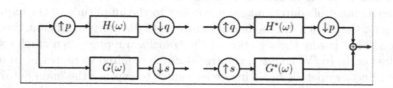

Fig. 1. Filter bank configuration for RDWT. H and G respectively are the low and high pass filter response functions, while (p, q, s) define the dilation factors.

2.1   Multiscale Transform

RDWT is an overcomplete discrete wavelet transform in which the dilation factor can be adjusted by the user. It can be set between 1 and 2 allowing a more gradual scaling between consecutive subbands as compared to the dyadic wavelet transform (DWT) [12]. However, it inherits some nice properties of DWT as the implementation using an iterated perfect reconstruction digital filter bank, the Parseval energy preservation property and a direct extension to the 2D case by means of filters having circularly symmetric frequency responses. The main difference with DWT is that RDWT filter bank has a different configuration of up- and down-samplers, as shown in Fig. 1; hence, a frequency domain implementation is more convenient. The integer sampling factors (p, q, and s) control the dilation factor of the RDWT and the Q-factor of the wavelet. In particular, whenever $s = 1$, the corresponding wavelet is not oscillatory. In addition, thanks to the perfect reconstruction property, RDWT allows us to vary the sampling factors p and q from scale to scale.

The use of a gradual scaling allows us to better separate image components having distinct features and then to apply to them a more adapted fusion rule. In fact, the contribution of those little objects in IR images which are not present or are slightly visible in the VI image is accounted for in more than one scale level, allowing for their better discrimination and representation.

The selection of the dilation parameter is based on the procedure proposed in [3] and it consists of finding the largest frequency resolution step (inverse of dilation factor) which provides the minimum perceivable loss of contrast between two consecutive frequency bands. As a result, it guarantees that two consecutive detail bands in the corresponding wavelet decomposition retain visible and distinguishable details. More precisely, it is derived from the optimal point of a visual rate-distortion curve, given by the Peli contrast C_n, which quantifies the image details loss as the resolution decreases. The desired resolution is such that

$$q_{\bar{n}} = argmin_{q_n \in \mathbf{N}} C_n + \alpha q_n, \tag{1}$$

where $\alpha \in \mathbf{R}$ balances the two competing terms in the functional in the rightmost term of previous equation. C_n is computed between two successive images of the sequence which is composed of blurred versions of the original image I using a low pass filter h_{q_n} with increasing support $q_n \times q_n$, with $q_n \geq 2$, $n \in \mathbf{N}$. It turns out that the optimal dilation factor to adopt in the RDWT is given by

$$p = q_{\bar{n}} - 1, \quad q = q_{\bar{n}}, \tag{2}$$

which represents the dilation factor to apply to the filter at resolution $\frac{1}{q_{\bar{n}-1}}$, as defined in Eq. (1), in order to get the successive resolution.

2.2 Multiscale Activity

One of the most interesting properties of the wavelet transform is its ability in measuring local signal singularities through the decay of wavelet coefficients as the scale increases. More precisely the following result holds [12]

Theorem 1. Let ψ be a wavelet with n vanishing moments and n derivatives having a fast decay. Let $f(x) \in L^2(\mathbf{R})$ be uniformly Lipschitz $\gamma \leq n$ over $[a, b] \subset \mathbf{R}$, then there exists $A > 0$ such that

$$|Wf(u, s)| \leq As^{\gamma+1/2}, \quad \forall \, (u, s) \in [a, b] \times \mathbf{R},$$

where $Wf(u, s)$ is the wavelet transform at scale s and time u of the function f. Conversely, if $Wf(u.s)$ satisfies previous inequality and if $\gamma < n$ is not an integer then f is uniformly Lipschitz γ on $[a - \varepsilon, b + \varepsilon]$, for any $\varepsilon > 0$.

A similar result holds for pointwise Lipschitz regularity and 2D functions [12]. This multiscale property has been further investigated in [2], where it has been proved that coefficients corresponding to a singularity point, namely atom, trace precise curves in the time-scale plane according to a proper evolution law. More precisely, atoms corresponding to isolated singularities dilate and grow along scales with the amplitude law given in the previous theorem; on the contrary, atoms corresponding to close singularity points interact in the time-scale plane and the amount of interaction increases with the scale till corresponding atoms completely interfere; from this point on they behave as if they were just one atom—see Fig. 2.

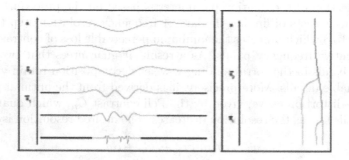

Fig. 2. (*Left*) Wavelet transform of a signal having close singularity points generating three atoms in the time-scale plane. (*Right*) Time-scale trajectories of atoms' centers of mass.

Based on these results, at each level of the transform, we define the activity index at scale s and spatial location (x, y), i.e.

$$S(x, y, s) = \left| \frac{W_{s+\Delta s}(x, y)}{W_s(x, y)} \right|, \quad \forall (x, y) \in \Omega,$$

where $W_s(x, y)$ denotes the 2D wavelet transform of the image computed at scale s and spatial location $(x, y) \in \Omega$ and Ω is the image domain[1]. The global activity index at scale s is then simply defined as the average value of the local activity indices, i.e.

$$S_s = \frac{1}{|\Omega|} \sum_{(x, y) \in \Omega} S(x, y, s). \tag{3}$$

It is worth observing that from [2], the result in Theorem 1 is an equality along modulus maxima chains, which, in turn, do not change their location in case of isolated singularities; hence, for those points we have

$$S(x, y, s) \approx \frac{A(s + \Delta s)^{\gamma + 1/2}}{A s^{\gamma + 1/2}} = \left(\frac{s + \Delta s}{s} \right)^{\gamma + 1/2}. \tag{4}$$

---

[1] A stabilizing constant is used in order to make $S(x, y, s)$ well defined.

More in general, the value of S_s and its variation with respect to the scale s provide information concerning the image variability, and then image activity, in a wavelet subband. In fact, for a simple regular signal with piecewise isolated singularities, we expect an almost flat S_s curve along scales; in addition, in stationary conditions, the higher the order of the local regularity the larger S_s. On the contrary, S_s changes its values along scales whenever two interfering atoms interact till their complete interference. As a result, in correspondence to more complicated regions, like textures, we expect a varying S_s which converges to a sort of flat profile at very large scales (all singularities atoms are in interference), as depicted in Fig. 3. Finally, it is worth observing that since noisy flat regions contribute with a negative γ value, their contribution to the global activity index is minor.

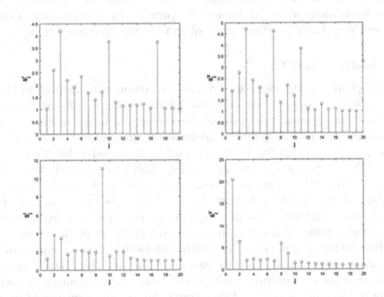

Fig. 3. Activity index as in Eq. (3) versus scale level j ($s = (q/p)^j$) for IR (*left*) and VI (*right*) images of Natocamp and Duine frames in Fig. 5.

2.3 Fusion Rule

Let us denote with $\{W_{I,p,q,j}\}_{j=1,...,J+1}$ and $\{W_{V,p,q,j}\}_{j=1,...,J+1}$ the RDWT decomposition, respectively of the infrared and visible source image, computed up to $J - th$ level with dilation factor equal to $\frac{q}{p}$, where p and q are defined as in Eq. (2). Let $\{W_{F,p,q,j}\}_{j=1,...,J+1}$ be the corresponding RDWT decomposition of the fused image we are going to define. The activity index in Eq. (3) can guide the max-based fusion rule. In particular, it can be used as a weight for the amplitude of wavelet coefficients of the source images with the aim of including some interscale information in the comparison procedure at a fixed decomposition level. More precisely, the weights are defined as follows

$$\lambda_{I,j} = \frac{S_{I,j}}{S_{I,j} + S_{V,j}} \qquad \lambda_{V,j} = 1 - \lambda_{I,j} = \frac{S_{V,j}}{S_{I,j} + S_{V,j}}. \tag{5}$$

$S_{*,j}$ is the activity index in Eq. (3) computed for the j-th level of 2D-RDWT of the source image $*$. The max rule is then applied to corresponding wavelet coefficients multiplied by the normalized activity weight of the whole subband, i.e.

$$W_{F,p,q,j}(x,y) = \begin{cases} W_{I,p,q,j}(x,y) & \text{if } \lambda_I|W_{I,p,q,j}(x,y)| > \lambda_V|W_{V,p,q,j}(x,y)| \\ W_{V,p,q,j}(x,y) & otherwise \end{cases}, \tag{6}$$

where $j = 1, ..., J + 1$, $(x,y) \in \Omega_j$ and Ω_j is the subband spatial domain. As a result the coefficients of the fused image are those having high amplitude and/or enough activity along scales. In this way, we promote the preservation of textured regions and singularity points of small objects in the final fused image.

2.4   Adaptive RDWT

To further prove the significance of S_s, its behaviour with respect to the scale levels has been considered and only the bands having a considerable activity index have been retained. For example, for the second image in Fig. 3, the scale levels carrying the more significant information with respect to the remaining ones are just 2, 3 and 9 for IR image and 1, 2, 8 and 9 for VI image. Let $J_I = \{j \in [1, J] : S_{I,j} > T_I\}$ and $J_V = \{j \in [1, J] : S_{V,j} > T_V\}$, where T_I and T_V are two thresholds —they have been fixed equal to the average values with respect to j of $S_{I,j}$ and $S_{V,j}$. Let set $J_{I,V} = J_I \bigcup J_V = \{j_1, j_2, ..., j_K\}$, $K \leq J$. If $K < J$, then the indexes composing $J_{I,V}$ do not provide a logarithmic law. However, they provide a nearly non redundant partition of the frequency axis for the two source images. In other words bandwidths having comparable and small activity index can be merged into a single one. The RDWT framework allows us to define a transform where the dilation factor changes at each level according to the indexes in $J_{I,V}$. Using some algebra, the sequence of successive dilation factors $(p_1, q_1), (p_2, q_2), ..., (p_K, q_K)$ can be defined as follows

$$\begin{cases} \frac{p_1}{q_1} = \left(\frac{p}{q}\right)^{j_1} \\ \frac{p_2}{q_2} = \left(\frac{p}{q}\right)^{j_2 - j_1} \\ \vdots \\ \frac{p_K}{q_K} = \left(\frac{p}{q}\right)^{j_K - j_{K-1}}. \end{cases}$$

Some constraints on p_k and q_k can also be added for computational purposes and for preventing numerical instability. As it will be shown in the section of experimental results, fusion results do not significantly change if the adaptive 2D-RDWT is used instead of the one having the fixed dilation factor defined in Eq. (2). This confirms the ability of the proposed activity index in determining a compact representation even along scale axis.

2.5 Fusion Algorithm

Summing up, the main steps of the proposed image fusion method are described in the following. Let I and V be the two source images.

1. Compute the optimal dilation factor for I as in Eq. (2).
2. Compute the 2D-RDWT of I and V up to level J using the dilation factor output of previous step. Let $\{W_{I,p,q,j}\}_{j=1,\ldots,J+1}$ and $\{W_{V,p,q,j}\}_{j=1,\ldots,J+1}$ be the corresponding subbands.
3. For each level $j = 1, \ldots, J+1$ and for each source image, compute the activity index as defined in Eq. (3) and the corresponding weight as in Eq. (5).
4. For each level $j = 1, \ldots, J+1$, compute the coefficients of the 2D-RDWT of the fused image F as defined in Eq. (6).
5. Let $\{W_{F,p,q,j}\}_{j=1,\ldots,J+1}$ be the coefficients output of the previous step, invert the 2D-RDWT and let F be the final fused image.

If the adaptive RDWT is used, as described in the previous section, step 3 is replaced by the following steps:

3 For each level $j = 1, \ldots, J$ and for each source image, compute the activity index as defined in Eq. (3) and estimate the set of dilation factors as described in Sect. 2.4.

3.1 Compute the adaptive 2D-RDWT of I and V up to level K using the estimated dilation factors. Let $\{W_{I,p_j,q_j}\}_{j=1,\ldots,K+1}$ and $\{W_{V,p_j,q_j}\}_{j=1,\ldots,K+1}$ be the corresponding subbands.

Next steps remain the same but they are referred to the adaptive RDWT.

3 Experimental Results

The proposed fusion method has been tested on several test images. In this paper results on some images extracted from the commonly used TNO image fusion dataset [19] are presented. TNO contains multispectral night-time registered images concerning different scenes. More precisely, the infrared and visible images of Tank, Sandpath, frame no. 1805 of Nato-camp sequence and frame no. 7417 of Duine sequence have been considered for fusion. The results have been evaluated in terms of Q^{IVF} index [16] and entropy E, which are commonly used for evaluating fusion results. E is the Shannon entropy of the fused image [4], while Q^{IVF} quantifies edge preservation and it is defined as follows

$$Q_{IVF} = \frac{\sum_x \sum_y (Q^{IF}(x,y)w_I(x,y) + Q^{VF}(x,y)w_V(x,y))}{\sum_x \sum_y (w_I(x,y) + w_V(x,y))},$$

where Q^{*F} measures the edge strength and orientation preservation values while w_* are suitable weights—see [16] for details. $Q^{IVF} \in [0,1]$; the larger Q^{IVF} the more edge information is kept in the final fused image. For comparative studies, the following methods have been considered: a standard wavelet based fusion (WT) method using the max fusion rule (a Daubechies with 4 vanishing moments and 5 decomposition levels have been used); the wavelet-based fusion method using the complex wavelet transform as in (DCWT) [8]; the laplacian pyramid based method with sparsity constraints (LP-SR) in [10] and the total variation based method (TV) in [11]. Results are shown in Table 1. The proposed method has been denoted with RDWT; the same table contains results achieved using the adaptive RDWT (Adapt RDWT) described in Sect. 2.4 and results achieved by applying the simple max fusion rule in the RDWT domain (m-RDWT). The maximum decomposition level J has been set equal to 20. As it can be observed, the proposed method provides high values of the adopted evaluation metrics for most of test images. This means that the proposed method is able to preserve fine details of sources images, i.e. edges and textures. It is also worth observing that the proposed weighting strategy allows us to better tune the max-based fusion rule providing better results than the simple m-RDWT. On the contrary, the adaptive RDWT (Adapt-RDWT) provides similar and sometimes slightly better results than the one provided by RDWT, thanks to a more adapted partitioning of the frequency plane. Table 2 contains the dilation parameters that have been automatically estimated from the source images and that have been used for building RDWT decomposition. As it can be observed, the dilation parameter is higher for less textured images. In addition, the number and the distribution of decomposition levels, which have been selected for the adaptive RDWT, depend on the image under study, as the adaptive dilation parameters show. Finally, Fig. 4 depicts the estimated weight values for each image pairs. Even in this case the weight, which represents the normalized activity for the infrared image, strongly depends on the image content at each subband of the adopted RDWT.

Table 1. Fusion results on Tank, Sand, Natocamp and Duine test images from TNO image fusion database. Results have been measured in terms of edge-based preservation measure Q^{IVF} (first row) and Shannon entropy (second row). Best results are in bold.

Image	WT	DCWT	LP-SR	TV	RDWT	Adapt RDWT	m-RDWT
Tank	0.42	0.49	0.51	0.46	**0.53**	**0.53**	0.48
	7.88	7.40	**7.92**	6.73	7.85	7.85	7.86
Sand	0.38	0.45	0.50	**0.60**	0.56	0.57	0.44
	6.94	6.41	7.11	7.05	**7.15**	**7.15**	6.90
Nato	0.39	0.41	0.47	0.55	0.54	**0.55**	0.45
	7.00	6.51	6.91	6.93	7.09	**7.12**	6.98
Duine	0.45	0.42	0.48	0.51	0.53	**0.54**	0.49
	6.17	6.47	**6.73**	6.29	6.39	6.38	6.13

Table 2. Tank, Sand, Natocamp and Duine test images. Estimated dilation factor for RDWT, as in Sect. 2.1; number of levels of the adaptive RDWT, as in Sect. 2.4, and the corresponding dilation parameters.

Image	p	q	K	Dilation p_k/q_k
Tank	7	8	14	7/8, 7/8, 7/8, 7/8, 7/8, 7/8, 7/8, 7/8, 7/8, 7/8, 7/8, 13/17, 2/3, 17/29
Sand	7	8	15	7/8, 7/8, 7/8, 7/8, 7/8, 7/8, 7/8, 7/8, 7/8, 7/8, 7/8, 7/8, 7/8, 7/8, 7/8
Nato	6	7	12	6/7, 6/7, 6/7, 6/7, 6/7, 6/7, 6/7, 11/15, 6/7, 6/7, 2/5, 17/27
Duine	6	7	12	6/7, 6/7, 6/7, 6/7, 6/7, 6/7, 6/7, 6/7, 6/7, 6/7, 6/7, 6/7

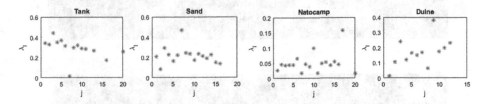

Fig. 4. Weights $\lambda_{I,j}$ as in Eq. (5) for Tank, Sand, Natocamp and Duine test IR images.

In order to allow the visual inspection of fused images, the two source images and the fused one by the proposed RDWT method are shown in Fig. 5. The fused image is very similar to the visible one except for regions where IR image has significant visible details—i.e. where the activity index is considerably higher than its counterpart in VI image. In addition, the proposed method allows to preserve VI image details and to introduce new ones from IR image by limiting halo and ringing effects near the new edge components as well as preventing noise amplification. Finally, the computational effort of the proposed RDWT-based fusion method is very moderate thanks to RDWT implementation with perfect reconstruction filters bank.

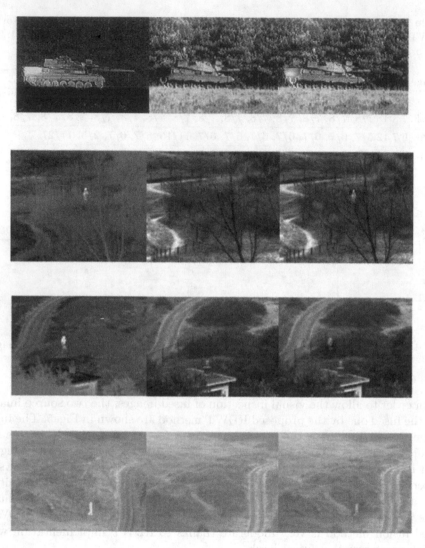

Fig. 5. From top to bottom: Tank, Sand, Natocamp and Duine test images. From left to right: infrared image, visible image and fused image using the proposed RDWT-based fusion method.

4   Conclusions

In this paper an image fusion method based on the multiscale image activity has been proposed. The latter takes into account the local multiscale behaviour of the coefficients of a suitable wavelet transform whose dilation factor depends on the frequency image content. The relationships between Lipschitz regularity and wavelet coefficients and the rational dilation wavelet transform are the main ingredients of the proposed method. The former are used to estimate the local complexity of the image through its multiscale behaviour; the latter allows for the use of a dilation parameter different from 2 while preserving a fast filter bank implementation. As a result, important visual information extracted from the two source images is retained in the final fused image, preventing artifacts and noise amplification. Preliminary experimental results demonstrate that the proposed method is able to reach and outperform some existing literature. Future research will be devoted to refine the proposed method. In particular, we will investigate the use of the pointwise activity in the fusion rule. Part of the research work will also be devoted to the study of the proposed multiscale image activity as an index for image sparsity along scale axis.

References

1. Bayram, I., Selesnick, I.W.: Frequency-domain design of overcomplete rational-dilation wavelet transforms. IEEE Trans. Sig. Process. **57**(8), 2957 (2009)
2. Bruni, V., Piccoli, B., Vitulano, D.: A fast computation method for time scale signal denoising. Sig. Image Video Process. **3**(1), 63–83 (2009)
3. Basile, M.C., Bruni, V., Vitulano, D.: A CSF-based preprocessing method for image deblurring. In: Blanc-Talon, J., Penne, R., Philips, W., Popescu, D., Scheunders, P. (eds.) ACIVS 2017. LNCS, vol. 10617, pp. 602–614. Springer, Cham (2017). https://doi.org/10.1007/978-3-319-70353-4_51
4. Cover, M., Thomas, J.A.: Elements of Information Theory. Wiley, New York (1991)
5. Jin, X., et al.: A survey of infrared and visual image fusion methods. Infrared Phys. Technol. **85**, 478–501 (2017)
6. Jin, X., et al.: Infrared and visual image fusion method based on discrete cosine transform and local spatial frequency in discrete stationary wavelet transform domain. Infrared Phys. Technol. **88**, 1–12 (2018)
7. Kong, W., Lei, Y., Zhao, H.: Adaptive fusion method of visible light and infrared images based on non-subsampled shearlet transform and fast non-negative matrix factorization. Infrared Phys. Technol. **67**, 161–172 (2014)
8. Lewis, J.J., O'Callaghan, R.J., Nikolov, S.G., Bull, D.R., Canagarajah, N.: Pixel- and region-based image fusion with complex wavelets. Inf. Fusion **8**(2), 119–130 (2007)
9. Li, S., K, X., Fang, L., Hu, J., Yin, H.: Pixel-level image fusion: a survey of the state of the art. Inf. Fusion **33**, 100–112 (2017)
10. Liu, Y., Liu, S., Wang, Z.: A general framework for image fusion based on multi-scale transform and sparse representation. Inf. Fusion **24**, 147–164 (2015)
11. Ma, Y., Chen, J., Chen, C., Fan, F., Ma, J.: Infrared and visible image fusion using total variation model. Neurocomputing **202**, 12–19 (2016)
12. Mallat, S.: A Wavelet Tour of Signal Processing. Academic Press, Orlando (1998)
13. Nencini, F., Garzelli, A., Baronti, S., Alparone, L.: Remote sensing image fusion using the curvelet transform. Inf. Fusion **8**(2), 143–156 (2007)
14. Pajares, G., De La Cruz, J.M.: A wavelet-based image fusion tutorial. Pattern Recogn. **37**(9), 1855–1872 (2004)
15. Paramanandham, N., Rajendiran, K.: Infrared and visible image fusion using discrete cosine transform and swarm intelligence for surveillance applications. Infrared Phys. Technol. **88**, 13–22 (2018)

16. Xydeas, C., Petrovic, V.: Objective image fusion performance measure. Electron. Lett. **36**(4), 308–309 (2000)
17. Toet, A.: Image fusion by a ratio of low-pass pyramid. Pattern Recogn. Lett. **9**(4), 245–253 (1989)
18. Zeeuw, P.: Wavelets and Image Fusion. CWI, Amsterdam (1998)
19. http://figshare.com/articles/TNO_Image_Fusion_Dataset/1008029

Optimising Data for Exemplar-Based Inpainting

Lena Karos, Pinak Bheed$^{(\boxtimes)}$, Pascal Peter, and Joachim Weickert

Mathematical Image Analysis Group, Faculty of Mathematics and Computer Science, Saarland University, Campus E1.7, 66041 Saarbrücken, Germany
{karos,bheed,peter,weickert}@mia.uni-saarland.de

Abstract. Optimisation of inpainting data plays an important role in inpainting-based codecs. For diffusion-based inpainting, it is well-known that a careful data selection has a substantial impact on the reconstruction quality. However, for exemplar-based inpainting, which is advantageous for highly textured images, no data optimisation strategies have been explored yet. In our paper, we propose the first data optimisation approach for exemplar-based inpainting. It densifies the known data iteratively: New data points are added by dithering the current error map. Afterwards, the data mask is further improved by nonlocal pixel exchanges. Experiments demonstrate that our method yields significant improvements for exemplar-based inpainting with sparse data.

Keywords: Exemplar-based inpainting · Texture
Spatial data optimisation · Error maps · Dithering · Densification

1 Introduction

Inpainting fills in missing parts of an image from available data [11,17]. Originating from the field of texture synthesis [7,12,15], the class of exemplar-based inpainting techniques developed during the last few decades. These methods are of a non-local nature in the sense that unknown image regions are reconstructed based on known data from different locations in the image domain [2,6,8]. During the inpainting process, information is exchanged between known and unknown image patches, which are usually disk- or rectangle-shaped regions. Moving away from the classical concept of completely known patches, Facciolo et al. [8] introduced a framework that acts upon patches that can contain unknown pixels. This allows the method to inpaint images with sparse known data, which makes it interesting for applications such as compression [19].

Most inpainting methods that are well-known to work with sparse image data rely on partial differential equations (PDEs) [4,17,22]. Numerous publications have shown that a careful selection of the known data is crucial for the reconstruction quality [3,5,10,13,16,21]. Despite the success of these data optimisation techniques for PDE-based inpainting, this idea has not yet been studied for

© Springer Nature Switzerland AG 2018
J. Blanc-Talon et al. (Eds.): ACIVS 2018, LNCS 11182, pp. 547–558, 2018.
https://doi.org/10.1007/978-3-030-01449-0_46

exemplar-based inpainting. The latter is far more challenging, because exemplar-based techniques are nonlocal and thus computationally much more expensive than PDE-based approaches. However, since they offer advantages for strongly textured images, it would be desirable to have suitable data optimisation strategies.

Our Contribution. In this paper, we introduce data optimisation to exemplar-based inpainting. To this end, we use the sparse inpainting method of Facciolo et al. [8]. Building upon the idea of iterative data selection [14,16], we come up with a densification approach that picks the most important pixels in a bottom-up fashion. In order to adapt our technique to the specific setting of exemplar-based inpainting in an efficient manner, we propose the concept of error maps. They measure the spatial distribution of the reconstruction error. We apply a modified form of the Floyd-Steinberg dithering technique [9] to convert error maps into binary images that specify the location of important known data. An evaluation of our experimental results shows that our new approach is well-suited for exemplar-based inpainting.

Related Work. Working in a patch-based setting, exemplar-based methods pursue different strategies for the search of patches and consolidation of information for the purpose of inpainting. Criminisi et al. [6] proposed a direct extension of the idea by Efros and Leung [7], where the priority order for patches to be inpainted depends on the local structural information. The encouraging results of both methods led to a popularisation of exemplar-based inpainting. Contrary to these methods, the approach of Facciolo et al. [8] relies on the idea of partial patches: Regardless of whether a patch contains unknown pixels, information is exchanged between a pair of known and unknown pixels in the compared patches. The resulting ability to inpaint from sparse known data makes the approach of Facciolo et al. particularly useful for our purpose.

Compression applications based on image inpainting are a classic motivation for data optimisation, since they need to select a sparse set of representative image points. Galić et al. [10], Schmaltz et al. [21], and Peter et al. [18] validated the potential of diffusion-based codecs with subdivision strategies for sparsifying data. These techniques restrict known locations to an adaptive regular grid that is easy to store. Their results are competitive to the quasi-standards JPEG and JPEG 2000. In particular, a proof-of-concept by Peter and Weickert [19] combined diffusion-based inpainting with exemplar-based methods to address textured images. However, they did not optimise the known data w.r.t. the exemplar-based inpainting operator due to runtime issues. The impact of this omission remains unknown.

The success of early PDE-based compression codecs [10] also sparked research on optimising inpainting data without constrained locations. For homogeneous diffusion inpainting, Belhachmi et al. [3] showed that optimal known data should be chosen according to the Laplacian magnitude. For practical purposes, they propose a dithering of this magnitude. An iterative strategy by Mainberger et al. [16] called probabilistic sparsification gradually removes pixels that are least significant for the image reconstruction. Thus, it selects the known data in a top-down manner. In contrast, the probabilistic densification method by Hoffmann et al. [14] adds significant pixels iteratively. As a postprocessing step to the sparsification approach, Mainberger et al. [16] introduced another iterative method called nonlocal pixel exchange in which known and unknown pixels pairs are swapped if the error of reconstruction from such an exchange improves. While the previous approaches rely strictly on local per-pixel errors, the densification method by Adam et al. [1] computes the global error in every iteration.

Since the probabilistic data optimisation methods [1,13,16] are least restricted to a specific inpainting operator such as homogeneous diffusion inpainting, our densification approach builds upon some of their core ideas. However, our adaptation to the computationally expensive setting of exemplar-based inpainting leads to key differences in judging the importance of known data: We capture the nonlocal nature of exemplar-based inpainting through global interactions by dithering of error maps.

Concerning dithering, a wide variety of image halftoning techniques are available. The method Floyd and Steinberg [9] is popular due to its simplicity: Based on the concept of error diffusion, it achieves dithering by distributing pixel errors in their respective neighbourhood. For alternative and more sophisticated dithering approaches, we refer to the monograph of Ulichney [23] and more recent papers such as that by Schmaltz et al. [20].

Structure of the Paper. First the essentials of the sparse exemplar-based inpainting method of Facciolo et al. [8] are presented in Sect. 2. Section 3 reviews spatial data optimisation for inpainting-based image compression. We introduce our novel approach in Sect. 4. Experiments and their results are presented in Sect. 5. Finally, we conclude the paper with a summary and give an outlook on future work in Sect. 6.

2   Exemplar-Based Inpainting of Sparse Data

Let us consider a greyscale image as a function $f : \Omega \to \mathbb{R}$, where Ω denotes some rectangular image domain in $\mathbb{R}^2$. Given the set of sparse known data $K \subset \Omega$, the so-called *inpainting mask*, the intent of inpainting is to find an image u that is a reconstruction of the image f in the inpainting domain $\Omega \setminus K$.

Exemplar-based inpainting methods exchange information between patches according to their similarity. In the case of sparse data, the idea of using complete patches fails. To this end, Facciolo et al. [8] introduced a similarity measure V for partial patches, that only relies on the known pixels within a patch. This similarity measure between two patches that are centred at x and x' is given by the weighted squared difference function

$$V\left(x, x'\right) = \int_D g\left(x, x', y\right) \left(u\left(x + y\right) - u\left(x' + y\right)\right)^2 \, dy. \tag{1}$$

Here D denotes the domain of a patch centred at the origin, usually taken as a disk or a rectangle. The vector y indicates the displacement w.r.t. to this centre.

The weighting function g is defined as

$$g\left(\boldsymbol{x}, \boldsymbol{x}', \boldsymbol{y}\right) = \frac{g_\sigma\left(\boldsymbol{y}\right)}{\rho\left(\boldsymbol{x}, \boldsymbol{x}'\right)} \left(\alpha \mathcal{X}_K\left(\boldsymbol{x} + \boldsymbol{y}\right) + \beta \mathcal{X}_K\left(\boldsymbol{x}' + \boldsymbol{y}\right)\right). \tag{2}$$

Its normalisation parameter $\rho\left(\boldsymbol{x}, \boldsymbol{x}'\right)$ allows for a similarity measure that does not depend on the amount of known data contained in the patches. The Gaussian function $g_\sigma\left(\boldsymbol{y}\right)$ with a standard deviation σ provides a spatial weighting which depends on the distance from the patch centre. The various combinations of the constants α and β for the characteristic function $\mathcal{X}_K$ result in different schemes, called A, B and AB as described in the original paper [8].

Following this definition of similarity, inpainting corresponds to the minimisation of the *data coherence term*

$$F\left(u, w\right) = \int_\Omega \int_K w\left(\boldsymbol{x}, \boldsymbol{x}'\right) V\left(\boldsymbol{x}, \boldsymbol{x}'\right) d\boldsymbol{x}' d\boldsymbol{x}, \tag{3}$$

where $w\left(\boldsymbol{x}, \boldsymbol{x}'\right)$ is called the *non-local weight function*. Such a weighting allows for a pixel $\boldsymbol{x}$ to be reconstructed from several different patches in the image. As w is unknown, the framework introduces a second term, namely the *weight entropy term*

$$H\left(\boldsymbol{x}, w\right) = -\int_K w\left(\boldsymbol{x}, \boldsymbol{x}'\right) \log w\left(\boldsymbol{x}, \boldsymbol{x}'\right) d\boldsymbol{x}'. \tag{4}$$

This term is maximised following the *principle of maximum entropy* from information theory which states that the maximum entropy distribution is the least biased estimate under the constraint of known information. Together these two terms form the energy functional

$$E\left(u, w\right) = \frac{1}{h} F\left(u, w\right) - \int_\Omega H\left(\boldsymbol{x}, w\right) d\boldsymbol{x}, \tag{5}$$

where h is a constant used to balance the influence of the two terms in the energy functional. The following constrained optimisation problem is then solved to find the solution u^* to the inpainting problem:

$$\left(u^*, w^*\right) = \arg\min_{u, w} E\left(u, w\right), \tag{6}$$

$$s.t.: \quad \int_K w\left(\boldsymbol{x}, \boldsymbol{x}'\right) d\boldsymbol{x}' = 1 \qquad\qquad \forall \boldsymbol{x} \in \Omega, \tag{7}$$

$$u\left(\boldsymbol{x}\right) = f\left(\boldsymbol{x}\right) \qquad\qquad \forall \boldsymbol{x} \in K. \tag{8}$$

Implementation. The minimisation of this energy functional is done by an alternate coordinate descent of two steps: the weight update and a step that updates the image u. Moving slightly away from the variational framework, Facciolo et al. [8] proposed the scheme O: Here they compute the patch distance V differently for different steps. The weight update step uses the scheme A, whereas the image update step uses the scheme AB.

In our experiments, we use the reference implementation and the default parameter settings provided by Facciolo et al. [8] for the scheme O.

3   Spatial Data Optimisation

In the following, we discuss the core ideas of iterative data selection that are useful to understand our method: a bottom-up [16] and a top-down [14] approach for probabilistic optimisation, and an important postprocessing strategy [16].

In the remainder of this paper we need discrete concepts. We use the symbol Ω for the discretised image domain, and K for its set of mask pixels. The notion $f = (f_{i,j})$ describes a discrete version of the image f, where the index (i, j) denotes the pixel. We represent the set K also by its binary mask function $c = (c_{i,j})$ with $c_{i,j} = 1$ for $(i, j) \in K$ and $c_{i,j} = 0$ elsewhere.

Probabilistic Sparsification by Mainberger et al. [16] is an iterative greedy algorithm that follows the natural idea of removing unimportant pixels. It starts with a full inpainting mask and removes a set of pixels in every iteration. After randomly removing a fraction p of all pixels, the so-called *candidate set*, an inpainting yields per-pixel errors for each candidate. Then a fraction q of the candidate pixels with the least error is permanently removed. This corresponds to removing $p \cdot q \cdot |K|$ pixels that are easily reconstructed in every iteration, where $|K|$ represents the total number of known pixels in the inpainting mask. The algorithm terminates when the desired density of pixels is achieved.

Probabilistic Densification. In contrast to probabilistic sparsification, the probabilistic densification method of Hoffmann et al. [14] starts with an empty mask. In every iteration, the algorithm first computes the inpainted image and then randomly selects a candidate set containing a fraction p of all non-mask pixels. A fraction q of pixels from the candidate set, which have the largest local error, are permanently added to the mask. This is repeated until the required density is reached. In our approach, we modify probabilistic densification.

Nonlocal Pixel Exchange. One key shortcoming of sparsification is that when a pixel is removed it never gets added back into the mask. As the algorithm relies on selecting pixels randomly for the candidate set, there exists a possibility of removing significant pixels. Therefore, Mainberger et al. [16] introduced the nonlocal pixel exchange to compensate for this disadvantage and to avoid getting trapped in a suboptimal local minimum. This iterative method exchanges a set of randomly selected known pixels with a set of unknown pixels that exhibit a large local error. The exchange is accepted if the global mean squared error (MSE) obtained by inpainting the new mask is reduced. Otherwise, the exchange is discarded and a new one is made. The inherent nature of the algorithm guarantees that any change in the mask yields a better mask. We use nonlocal pixel exchange as a postprocessing step to our method.

4   Our Approach: Densification by Dithering

While the probabilistic methods from Sect. 3 can be combined with any inpainting operator, they have a specific weakness with regard to exemplar-based inpainting: The error computation in classic sparsification and densification methods is highly localised due to the use of per-pixel errors, while patch-based inpainting is nonlocal. For PDE-based methods, Adam et al. [1] provide evidence that this local error computation can have a negative impact on the results of probabilistic optimisation. Instead, they propose to compute the global error associated with each candidate pixel. However, such a strategy is not feasible for exemplar-based inpainting because the non-locality of these methods leads to significantly larger runtimes than for PDE-based methods. In order to find a remedy for this issue, we consider the concepts of error maps and dithering.

Error Maps. As the two previously mentioned densification methods are iterative, they require to add mask pixels given a set of already known data in every step. In such a situation, a natural way to identify regions of importance for data selection is through error maps. Since we are interested in minimising the mean squared error of the inpainted image, we compute our error map $e = (e_{i,j})$ pixelwise as the squared difference between the original image $f = (f_{i,j})$ and the inpainted image $u = (u_{i,j})$.

The error map depends directly on the inpainting mask and its density, because it is computed from the inpainting result. In particular, it gives a global overview of the spatial error distribution within the entire image domain. Figure 1 shows an example. As we shall see next, the error map becomes particularly useful when we combine it with dithering concepts.

Fig. 1. Error map of Barbara image. *Left:* Original image. *Middle:* Random mask with a density of 3%. *Right:* Error map computed after exemplar-based inpainting. *Note:* For better visualisation, the error map is logarithmically rescaled such that the maximum error is denoted by black and the minimum error by white.

Modified Floyd-Steinberg Dithering. The established probabilistic sparsification and densification methods suffer from two drawbacks: 1. They select only a small subset of the error map as candidate pixels. 2. They only consider local per-pixel errors, which might result from noise or small-scale features, such that choosing them would lead to a bad global approximation.

Interestingly, binary dithering can solve both problems: 1. Dithering replaces a random sampling by a deterministic, global strategy that takes into account all error values. 2. Classical binary dithering methods such as Floyd-Steinberg dithering place a data point at a certain location if the accumulated values from a neighbourhood are sufficiently large. Thus, they overcome the drawback of purely local per-pixel errors. This motivates us to perform binary dithering on error maps. To this end, we regard the error map as a greyscale image, such that its dithering provides us with the inpainting mask.

Of the many dithering methods, the classical Floyd-Steinberg dithering [9] is popular due to its simplicity. In order to adapt it to our setting, we have to take into account that the locations of mask pixels from previous iterations are already set and should remain unaltered. The dithering should only create additional mask points at different locations.

The original Floyd-Steinberg algorithm sequentially scans the image from left to right and top to bottom. Each pixel is binarised to 0 or 255, and the binarisation error is distributed to its unvisited neighbours. The stencil weights used for distribution sum up to 1, thereby preserving the average grey value. This allows us to determine how many new mask pixels are added by rescaling the average grey value of the error map accordingly (see Belhachmi et al. [3]).

In order to preserve existing mask points, we always binarise them to 0. The accumulated error at these pixels is then transferred over to the next pixel in the scan order. Overall, this corresponds to skipping known pixels for the purpose of dithering. An additional modification is the use of serpentine scan order i.e., left to right for odd numbered rows, right to left for even numbered rows and top to bottom, since this improves the results [23]. See Algorithm 1 for details.

Densification by Dithering. Since error maps depend on the mask, adding new mask pixels will change the error map. Therefore, it has to be recomputed, which requires an inpainting. As exemplar-based inpaintings are computationally expensive, we insert multiple points at the same time and recompute the error map afterwards. In practice, we partition the desired number of mask pixels into n equally sized batches. This leads to the following iterative *densification by dithering (DbD)* approach (Algorithm 2). In each iteration, DbD performs three straightforward steps:

1. Inpaint with the current inpainting mask.
2. Compute the error map.
3. Apply modified Floyd-Steinberg dithering and update the mask.

We repeat the process until we reach the required density d. Note that in the first iteration, the inpainting mask is still empty and therefore, no error map can be computed. Instead of relying on error map dithering, we choose the initial set of known pixels randomly.

Nonlocal Pixel Exchange. Since DbD is a greedy algorithm that may get trapped in a bad local optimum, we can expect to improve the reconstruction quality by postprocessing the mask with a nonlocal pixel exchange (NLPE). For a random mask pixel, we consider a set of 10 randomly chosen non-mask pixels as candidates. As an acceleration strategy, we use the local error to identify the most promising candidate. Following Adam et al. [1] we then compute the global MSE that would result from a swap of the mask pixel with this non-mask pixel. To speed up the process further, we compute eight NLPE steps in parallel on the different threads of a multicore CPU and take the one with the smallest error.

Algorithm 1: Modified Floyd-Steinberg Dithering of the Error Map

---

Input: rescaled error map $e : \Omega \to [0, 255]$,
              binary inpainting mask $c : \quad \Omega \to \{0, 1\}$
Output: modified inpainting mask c

1 **for** *all pixels (i,j) scanned in serpentine order* **do**

2 | {*Binarise and set new mask pixel*}
3 | $old \leftarrow e_{i,j}$
4 | **if** $c_{i,j} = 0$ *and* $e_{i,j} > 127.5$ **then**
5 | | $e_{i,j} \leftarrow 255.0$
6 | | $c_{i,j} \leftarrow 1$
7 | **end**
8 | **else**
9 | | $e_{i,j} \leftarrow 0$
10 | **end**

11 | {*Compute binarisation error*}
12 | $\epsilon \leftarrow old - e_{i,j}$

13 | {*Diffuse binarisation error to unvisited neighbours*}
14 | **if** *scanning from left to right* **then**
15 | | $e_{i+1,j} \leftarrow e_{i+1,j} + \frac{7}{16}\epsilon$
16 | | $e_{i-1,j+1} \leftarrow e_{i-1,j+1} + \frac{3}{16}\epsilon$
17 | | $e_{i,j+1} \leftarrow e_{i,j+1} + \frac{5}{16}\epsilon$
18 | | $e_{i+1,j+1} \leftarrow e_{i+1,j+1} + \frac{1}{16}\epsilon$
19 | **end**
20 | **else** *scanning from right to left*
21 | | $e_{i-1,j} \leftarrow e_{i-1,j} + \frac{7}{16}\epsilon$
22 | | $e_{i+1,j+1} \leftarrow e_{i+1,j+1} + \frac{3}{16}\epsilon$
23 | | $e_{i,j+1} \leftarrow e_{i,j+1} + \frac{5}{16}\epsilon$
24 | | $e_{i-1,j+1} \leftarrow e_{i-1,j+1} + \frac{1}{16}\epsilon$
25 | **end**

26 **end**

This pixel exchange attempt is accepted if it leads to an MSE improvement. It appears natural to measure the amount of exchange attempts in terms of *cycles*: One cycle denotes as many exchange attempts as mask pixels $|\boldsymbol{K}|$. This ensures that on average each mask pixel is given a chance to be exchanged. As we consider eight parallel NLPE steps, this counts as eight exchange attempts.

5 Experiments

In this section, we evaluate the performance of DbD for the four test images shown in Fig. 2: *Synthetic* and *Barbara* with a mask pixel density of 5%, and *Trui* and *Baboon* with 10% density. For each of these test settings, we perform exemplar-based inpainting with random masks, DbD-optimised ones, and DbD

Algorithm 2: Densification by Dithering

Input: original image $f : \Omega \rightarrow [0, 255]$,
                desired pixel density d,
                number of iterations n
Output: final inpainting mask $c : \Omega \rightarrow \{0, 1\}$

1 Choose a random inpainting mask c with pixel density $\frac{d}{n}$.
2 **for** $i = 1$ *to* $n - 1$ **do**
3 $\quad$ Reconstruct with exemplar-based inpainting: $u := inpaint(c, f)$.
4 $\quad$ Compute the error map $e = (e_{i,j})$ via $e_{i,j} \leftarrow |f_{i,j} - u_{i,j}|^2$.
5 $\quad$ Rescale the error map to $\frac{255 \cdot d}{mean(e) \cdot n} \cdot e$.
6 $\quad$ Apply modified Floyd-Steinberg dithering on the error map e.
7 $\quad$ Update the current inpainting mask c by adding the dithered mask.
8 **end**

---

masks with NLPE. For DbD, we present results for $n = 10$ iterations, which was found empirically to give the best quality for modified Floyd-Steinberg dithering. NLPE is run for four cycles. The largest improvements by NLPE are achieved in the first cycle, while allowing for more cycles reduces the error only incrementally.

Fig. 2. Original images. *From left to right: Synthetic* (256×256), *Trui* (256×256), *Barbara* (512×512), and *Baboon* (256×256).

Figure 3 displays the reconstructed images and their MSE. It shows that both visually and quantitatively, DbD leads to substantially better inpainting results than random masks, and NLPE gives significant further quality improvements. The corresponding optimised masks are depicted in Fig. 4. For the *Synthetic* image, we observe that our algorithm correctly identifies that the pixels near texture boundaries are important to the exemplar-based inpainting operator. Thus, in contrast to random masks, many incorrect inpaintings at texture boundaries are avoided. For the *Trui* and the *Barbara* images, the recovery of highly textured regions is more faithful, since our data optimisation leads to higher pixel densities in these areas. Small-scale details such as the semantically important eye regions are not oversmoothed. The *Baboon* image requires a high, almost

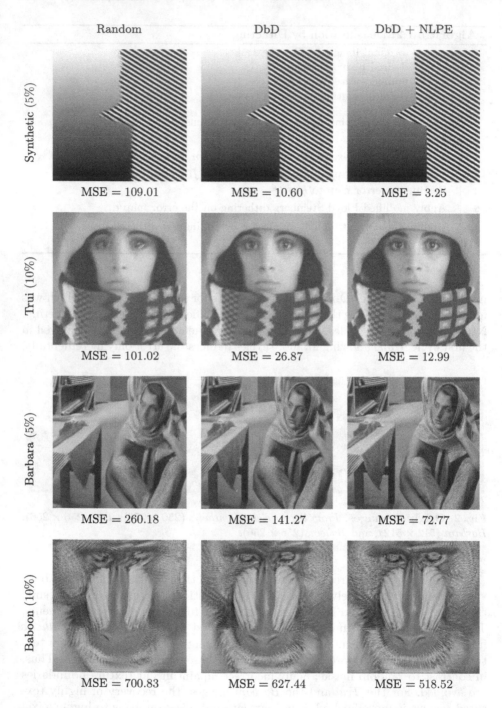

Fig. 3. Comparison of different data selection approaches. The densification by dithering (DbD) algorithm is run for $n = 10$ iterations, and we perform 4 cycles of nonlocal pixel exchange (NLPE).

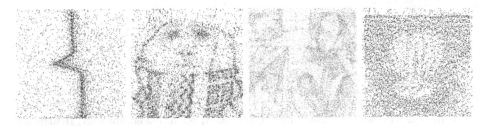

Fig. 4. Optimised inpainting masks. *From left to right:* Masks after DbD + NLPE optimisation for *Synthetic, Trui, Barbara,* and *Baboon.* The black pixels represent the known mask points.

equal mask density everywhere, such that the MSE is fairly high at 10% density, and improvements by DbD and NLPE are more moderate.

6 Conclusions and Outlook

We have presented the first paper that optimises data for exemplar-based inpainting. We achieve this with a novel approach called densification by dithering. It relies on iterative dithering of error maps for selection of important pixels. Further improvements are obtained with nonlocal pixel exchange. Our method shows that even for images which are heavily dominated by textures, a smart selection of the inpainting mask improves the quality substantially.

In our ongoing work, we explore the concept of error maps further and study more sophisticated dithering methods. Since we did not analyse the entropy of the resulting sparse data, the potential for compression applications is still unknown. Therefore, in the future we will also address this aspect and embed our method into a full image compression framework.

Acknowledgement. Part of our research has been funded by the ERC Advanced Grant INCOVID. This is gratefully acknowledged.

References

1. Adam, R.D., Peter, P., Weickert, J.: Denoising by inpainting. In: Lauze, F., Dong, Y., Dahl, A.B. (eds.) SSVM 2017. LNCS, vol. 10302, pp. 121–132. Springer, Cham (2017). https://doi.org/10.1007/978-3-319-58771-4_10
2. Arias, P., Facciolo, G., Caselles, V., Sapiro, G.: A variational framework for exemplar-based image inpainting. Int. J. Comput. Vis. **93**(3), 319–347 (2011)
3. Belhachmi, Z., Bucur, D., Burgeth, B., Weickert, J.: How to choose interpolation data in images. SIAM J. Appl. Math. **70**(1), 333–352 (2009)
4. Bertalmio, M., Sapiro, G., Caselles, V., Ballester, C.: Image inpainting. In: Proceedings of SIGGRAPH 2000, pp. 417–424. ACM Press, New Orleans, July 2000
5. Chen, Y., Ranftl, R., Pock, T.: A bi-level view of inpainting-based image compression. In: Kúkelová, Z., Heller, J. (eds.) Proceedings of Computer Vision Winter Workshop, Křtiny, Czech Republic, February 2014

6. Criminisi, A., Pérez, P., Toyama, K.: Region filling and object removal by exemplar-based image inpainting. IEEE Trans. Image Process. **13**(9), 1200–1212 (2004)
7. Efros, A.A., Leung, T.K.: Texture synthesis by non-parametric sampling. In: Proceedings of IEEE International Conference on Computer Vision, vol. 2, pp. 1033–1038. IEEE, Kerkyra, September 1999
8. Facciolo, G., Arias, P., Caselles, V., Sapiro, G.: Exemplar-based interpolation of sparsely sampled images. In: Cremers, D., Boykov, Y., Blake, A., Schmidt, F.R. (eds.) EMMCVPR 2009. LNCS, vol. 5681, pp. 331–344. Springer, Heidelberg (2009). https://doi.org/10.1007/978-3-642-03641-5_25
9. Floyd, R.W., Steinberg, L.: An adaptive algorithm for spatial greyscale. Soc. Inf. Disp. **17**(2), 75–77 (1976)
10. Galić, I., Weickert, J., Welk, M., Bruhn, A., Belyaev, A., Seidel, H.P.: Image compression with anisotropic diffusion. J. Math. Imaging Vis. **31**(2–3), 255–269 (2008)
11. Guillemot, C., Meur, O.L.: Image inpainting: overview and recent advances. IEEE Signal Process. Mag. **31**(1), 127–144 (2014)
12. Heeger, D.J., Bergen, J.R.: Pyramid-based texture analysis/synthesis. In: Proceedings of SIGGRAPH 1995, pp. 229–238. ACM Press, Los Angeles, July 1995
13. Hoeltgen, L., Setzer, S., Weickert, J.: An optimal control approach to find sparse data for laplace interpolation. In: Heyden, A., Kahl, F., Olsson, C., Oskarsson, M., Tai, X.-C. (eds.) EMMCVPR 2013. LNCS, vol. 8081, pp. 151–164. Springer, Heidelberg (2013). https://doi.org/10.1007/978-3-642-40395-8_12
14. Hoffmann, S., Mainberger, M., Weickert, J., Puhl, M.: Compression of depth maps with segment-based homogeneous diffusion. In: Kuijper, A., Bredies, K., Pock, T., Bischof, H. (eds.) SSVM 2013. LNCS, vol. 7893, pp. 319–330. Springer, Heidelberg (2013). https://doi.org/10.1007/978-3-642-38267-3_27
15. Liang, L., Liu, C., Xu, Y.Q., Guo, B., Shum, H.Y.: Real-time texture synthesis by patch-based sampling. ACM Trans. Graph. **20**(3), 127–150 (2001)
16. Mainberger, M., et al.: Optimising spatial and tonal data for homogeneous diffusion inpainting. In: Bruckstein, A.M., ter Haar Romeny, B.M., Bronstein, A.M., Bronstein, M.M. (eds.) SSVM 2011. LNCS, vol. 6667, pp. 26–37. Springer, Heidelberg (2012). https://doi.org/10.1007/978-3-642-24785-9_3
17. Masnou, S., Morel, J.M.: Level lines based disocclusion. In: Proceedings of 1998 IEEE International Conference on Image Processing, Chicago, IL, vol. 3, pp. 259–263, October 1998
18. Peter, P., Kaufhold, L., Weickert, J.: Turning diffusion-based image colorization into efficient color compression. IEEE Trans. Image Process. **26**(2), 860–869 (2017)
19. Peter, P., Weickert, J.: Compressing images with diffusion- and exemplar-based inpainting. In: Aujol, J.-F., Nikolova, M., Papadakis, N. (eds.) SSVM 2015. LNCS, vol. 9087, pp. 154–165. Springer, Cham (2015). https://doi.org/10.1007/978-3-319-18461-6_13
20. Schmaltz, C., Gwosdek, P., Bruhn, A., Weickert, J.: Electrostatic halftoning. Comput. Graph. Forum **29**(8), 2313–2327 (2010)
21. Schmaltz, C., Peter, P., Mainberger, M., Ebel, F., Weickert, J., Bruhn, A.: Understanding, optimising and extending data compression with anisotropic diffusion. Int. J. Comput. Vis. **108**(3), 222–240 (2014)
22. Schönlieb, C.B.: Partial Differential Equation Methods for Image Inpainting. Cambridge University Press, Cambridge (2015)
23. Ulichney, R.: Digital Halftoning. The MIT Press, London (1987)

Fast Light Field Inpainting Propagation Using Angular Warping and Color-Guided Disparity Interpolation

Pierre Allain[1(✉)], Laurent Guillo[2], and Christine Guillemot[1]

[1] Inria Rennes Bretagne-Atlantique,
263 Avenue du Général Leclerc, 35042 Rennes, France
pierre.allain@inria.fr
[2] University of Rennes 1, CNRS, Irisa, 35000 Rennes, France

Abstract. This paper describes a method for fast and efficient inpainting of light fields. We first revisit disparity estimation based on smoothed structure tensors and analyze typical artefacts with their impact for the inpainting problem. We then propose an approach which is computationally fast while giving more coherent disparity in the masked region. This disparity is then used for propagating, by angular warping, the inpainted texture of one view to the entire light field. Performed experiments show the ability of our approach to yield appealing results while running considerably faster.

1 Introduction

As the capture of 4D light fields from real scenes is gaining in popularity, the need for efficient editing tools is expected to rise as well. However, the very large volume of data which they represent, as well as the need to maintain texture and structure consistency across views, raise challenging complexity issues in these processing tasks.

In this paper, we focus on light field inpainting for applications such as object removal. Although the problem of image inpainting has already been extensively studied, only a few methods in the literature address the specific case of 4D light fields. One can find some work on the related topic of multiview image inpainting, however the methods as in [1,2] are designed for captures with large baselines and do not generalize well for dense light fields.

It is only recently that methods have been proposed for dense light fields inpainting. The authors in [3] extend 2D patch-based methods to 4D patches and ensure consistency by minimizing a 4D patch bi-directional similarity measure. This method progresses patch per patch in a greedy fashion and suffers from a high computational complexity. In addition, the global consistency of the entire light field is not guaranteed. In [4], the central view is first edited using a 2D

C. Guillemot—This work was supported by the EU H2020 Research and Innovation Programme under grant agreement No. 694122 (ERC advanced grant CLIM)

© Springer Nature Switzerland AG 2018
J. Blanc-Talon et al. (Eds.): ACIVS 2018, LNCS 11182, pp. 559–570, 2018.
https://doi.org/10.1007/978-3-030-01449-0_47

patch-based method, and the offsets between the filled patches and their best match in the known region are propagated to the other views. But the above methods suffer from high complexity and/or from angular view coherency.

Inpainting first one view (in general the center one) and then coherently propagating the inpainted texture to the other views is an interesting approach as it allows using any 2D inpainting method on the center view to have a good initialization. This is related to the general problem of edit propagation [5]. In this category of methods, the approach in [6] inpaints the center view using a 2D patch-based method and the other views are inpainted by searching for the best matching patches in the inpainted center view. However, because of the greedy nature of the algorithm and the fact that the views are processed separately, the inpainting results may contain angular inconsistencies. The authors in [7] exploit smoothness priors in the epipolar plane images (EPI) to coherently propagate information across views using PDE-based diffusion, while a method based on matrix completion is proposed in [8], assuming a low rank prior for the light field data.

One critical issue for propagating texture to all views is the estimation of disparity in the masked region. In [9], the disparity map is inpainted using a linear interpolation followed by variational regularization. However, an interpolation per EPI may lead to spatial inconsistencies in the disparity maps. In [7], dominant orientations of known pixels are computed as weighted averages of structure tensors across all the views of a given slice. Dominant orientations of unknown pixels are found using a super-pixel guided interpolation.

In this paper, we focus on the fast propagation of the inpainted texture across the entire light field. For sake of computational efficiency, we consider a simple warping strategy which requires knowledge of scene geometry in the inpainted region, that is to say the disparity at this location, as shown in Fig. 1. We revisit the problem of disparity estimation using smoothed structure tensors. We show that, while smoothing generally increases spatio-temporal coherence of estimated disparity fields, it may also lead to misalignment between color and disparity, and in turn to incoherent disparity in the masked region. A direct structure tensor computation is used instead and it is shown that, even if the disparity field is much noisier, this noise does not affect the inpainted disparity values as it turns out to be automatically removed via the color-guided interpolation. This simple approach reduces computational complexity while improving disparity coherency in the masked region. The inpainted disparity allows us to perform a simple angular warping to propagate inpainted texture across the whole light field.

In summary, the contributions of the paper are as follows: (1) A fast disparity estimation of a masked area guided by an inpainted view. (2) The ability of applying inpainting on any view of the light field with corresponding disparity at angular position, since disparity is computed exactly on the view (no averaging). (3) A direct propagation of inpainted area over the whole light field using a 2D forward warping method. (4) An overall speed increase as compared to other methods.

The paper is organized as follows: Sects. 3 and 4 describe the selected disparity computation method and the method for interpolating the disparity inside the inpainted area. Section 5 presents the method used for propagating a view across the light field, and in Sect. 6 we discuss some experimental results.

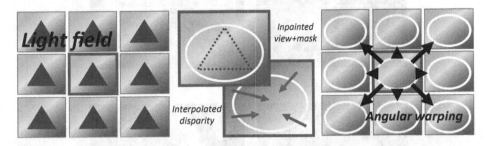

Fig. 1. Overview of the method. The red triangle represents inpainted object in the scene. Corresponding mask standing for inpainted area is delimited by the white line. Scene geometry (depth) behind the mask is unknown. Disparity is therefore interpolated according to subapertures color features. (Color figure online)

2   Disparity Estimation

Let $L(x, y, u, v)$ denote the 4D representation of a light field, describing the radiance of a light ray parameterized by its intersection with two parallel planes [10], and where (u, v) denote the angular (view) coordinates and (x, y) the spatial (pixel) coordinates. While disparity estimation methods based on structure tensors [11] usually operate on EPI smoothed with a 2D Gaussian kernel, we consider instead a direct computation of the structure tensors on the 4D light field L as

$$\mathbf{T}_{xu} = \nabla_{xu} L \otimes \nabla_{xu} L^\mathsf{T} * G_\sigma(x, y, u, v), \tag{1}$$

where $\nabla_{xu} L = \begin{pmatrix} \partial_x L \\ \partial_u L \end{pmatrix}$, denotes the (x, u) gradient of L, and where $G_\sigma(x, y, u, v)$ is a Gaussian kernel smoothing along the four light field dimensions.

The orthogonal eigenvectors θ_+ and θ_- with respective eigenvalues λ_+ and λ_- (where $\lambda_+ > \lambda_-$) of $\mathbf{T}_{xu}$ give a robust computation of the dominant gradient orientations. We are interested in the eigenvector θ_- with the smallest eigenvalue which corresponds to the isophote lines whose slopes give the amount of disparity between the views. Same applies for computing the vertical disparity with the tensor $\mathbf{T}_{yv}$.

While the use of a smoothing kernel yields disparities with enhanced spatial coherence, this also leads to mis-alignment (some cross overs) between the disparity estimates and objects contours (Fig. 2). We will see in the next section that such cross overs can produce bad estimates of disparity inside the inpainted area.

(a)   (b)

(c)   (d)

Fig. 2. Disparity estimate of one light field view (a), with black lines accounting for original image's segmentation of dominant shapes. (b) Using Sobel gradient and smoothing tensor with a Gaussian kernel of size 7. (c) Without gradient or tensor smoothing. (d) Without smoothing + TVL1. Despite being nosier, estimate without smoothing better fits object contours.

The structure tensor in the approach is therefore computed with no smoothing, and in order to remove most of outliers estimates, i.e. reducing the noise, we apply a Total Variation (TV-L1) denoising [12] on the disparity estimate:

$$D = \frac{1}{2}\left(\frac{\theta_{xu-}\overrightarrow{x}}{\theta_{xu-}\overrightarrow{u}} + \frac{\theta_{yv-}\overrightarrow{y}}{\theta_{yv-}\overrightarrow{v}}\right), \qquad (2)$$

searching to optimize:

$$\tilde{D} = arg\,min_{\tilde{D}}||\nabla\tilde{D}||_1 + \lambda||\tilde{D} - D||_2, \qquad (3)$$

where λ is the total variation regularization parameter and was set to 0.5 in our experiments. This value yields a good trade-off regarding smoothness of the regularization.

We finally obtain a disparity estimate specific to inpainted view, being spatially accurate regarding scene contours, and in a very fast way. In the next section, we will address the problem of both filling inpainted mask and dealing with remaining noise.

3   Superpixel Guided Disparity Interpolation

In order to propagate the inpainted area across all views, we need a disparity information inside the mask that does not exist in the light field scene, since inpainting a view can be seen as modifying depth geometry in the inpainted area. As proposed in [7], we use a superpixel segmentation of the inpainted view to guide disparity inpainting, making the assumption that disparity (i.e. depth) is homogeneous in local regions having the same color. The computed superpixels $S = \{S_s\}$, with $\{s \in \mathbb{N} | s \leq Q\}$, are first merged such as any of them lying even partly inside the masked area has at least K pixels outside the mask, i.e. known data. The number of superpixels is defined such as $Q = \frac{1}{\gamma}P$, P being the number of pixels in each subaperture. In our experiments γ was set to 1000 pixels per superpixel, which allows to capture main structures of the image, and K to 20, so that a certain area of each superpixel exists outside the mask.

The disparity $\hat{D}$ for any pixel at position x inside the mask and belonging to a superpixel S_s is estimated by interpolating known disparity values $\tilde{D}$ of $W \leq K$ closest pixels outside the mask but from the same superpixel as:

$$\hat{D}(x) = \frac{\Sigma_i^W w_{s,i}(x, y_i)\tilde{D}(y_i)}{\Sigma_i^W w_{s,i}(x, y_i)}, \tag{4}$$

where $y_i \in S_s$ is a pixel position outside the mask. We also set W to 20 in our experiments. The weights are defined as:

$$w_o(x, y_i) = exp\left(-\frac{(d_s(x, y_i) - \mu_s(x))^2}{\sigma_s(x)^2}\right), \tag{5}$$

$$\text{with:}\quad d_s(x, y_i) = \|x - y_i\|^2 + (I(x) - I(y_i))^2 f_s \tag{6}$$

accounting for the spatial distance between the two pixels (inside and outside the mask) and of the color proximity of the two pixels in the inpainted view.

Knowing that the spatial distance $d_s(x, y_i)$ can greatly vary for the different pixels x within a given superpixel, we apply in Eq. (5) a centering of weights around the minimum distance as:

$$\begin{cases} \mu_s(x) = \min(\{d_{s,i}\}), \\ \sigma_s(x)^2 = \beta \text{Var}(\{d_{s,i}\}). \end{cases} \tag{7}$$

Hence, pixels with known disparity and closest to the border of the mask have a higher weight. The scaling between spatial and color distance in Eq. (6) is defined by a function specific to superpixel geometry as $f_s = \alpha \frac{\text{Var}(\{x\})}{\text{Var}(\{I(x)\})}$, $x \in S_s$ to compensate for higher spatial and color variance inside some superpixels. In our experiments, we set $\alpha = 10^{-5}$ and $\beta = 1$ which yielded a good trade-off for interpolation smoothness and coherence with outer mask disparity.

Superpixel guided interpolation has proved to be effective for disparity reconstruction [7]. It is nonetheless very sensitive to disparity cross overs. In Fig. 3a, we can see that the disparity interpolation uses in some cases irrelevant information because of mask and segmentation configuration. In Fig. 3b, this reconstruction error is strongly alleviated because of the disparity cross over avoidance introduced in Sect. 2.

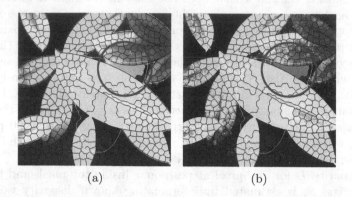

(a)   (b)

Fig. 3. Disparity estimate of inpainted view with focus on cross over issue, white line accounting for inpainting mask boundaries. (a) As in [7] for a vertical EPI., (b) Our method. Highlighted superpixel mostly use information close to the leaf due to mask configuration. Disparity retrieval is therefore extremely sensitive to estimate around object edges.

As for the rest of the inpainted area, interpolation weights guided by superpixels prove to be enough for obtaining an homogeneous disparity reconstruction despite noisy inputs, in the same fashion as in [13]. Finally, we obtain a consistent disparity estimate inside the mask which is derived from the inpainted view only and not by using information from the other views, with a few computations and approximations.

4 Inpainting Propagation

The computed disparity in the masked area is used to propagate the inpainted area of one view to all the other views of the light field using an angular forward warping. Instead of propagating information on one angular dimension in an EPI framework, we chose to warp directly in directions (u, v). Two main advantages of such approach are as follows. First, we do not need to repeat the whole pipeline of disparity computation and inpaint propagation for each EPI stripe $N+1$ times (for a light field of size $N \times N$ views). And second, direct warping is very fast as compared to diffusion.

However, warping has well known drawbacks. Backward warping implies to know disparity in each destination view, i.e. computing $N^2 - 1$ disparity estimates, which is time consuming. Forward warping on the other hand only needs disparity at the origin of propagation, but comes with ghosting and cracking effects. We propose a solution to tackle the latter issues in the scope of fast light field inpainting.

Handling Ghosting Artefacts: Ghosting effects, which are classical in depth image-based rendering, arise when multiple pixels with different depth and color converge to a same one in the warped image, resulting in a mean value. Ghosting artifacts are often avoided by detecting depth discontinuities, in order to separate the boundary layer (containing pixels near the contours of an object) from the main layer (containing pixels far from the contours) [14]. The main layer is first projected then the boundary layer is added everywhere it is visible (i.e. where its depth value is smaller than the main layers one). Ghosting artifacts can also be avoided by estimating background and foreground contributions in the rendered view with the help of advanced matting techniques [15], or using a confidence measure for each projected pixel as in [16].

For sake of reduced complexity, we consider instead a simple disparity mean filter to classify the projected pixels within the mask area as locally belonging to the foreground or the background. This simple mean depth classification approach gives satisfactory results in the case where the number of depth layers does not exceed two. However, in the case of more complex scenes with a higher number of layers, approaches such as z-buffering would be needed to avoid ghosting artefacts. Nonetheless, one can note that for dense light fields with small disparities, as those considered here, such case remains marginal.

Let us suppose that two pixels coming from two different objects with different depths are projected on the same pixel, when doing the texture propagation. Instead of handling a z-buffer to order the projected pixels from foreground to background, thus saving computationally expensive sorting operations, we compute the mean disparity of pixels projected on the same point. We then classify pixels having a disparity higher than the mean as foreground occluding pixels, and those having a disparity smaller than the mean as occluded pixels. In order to obtain such mean disparity, we simply warp the disparity by itself, see Fig. 4. The color is then warped with pixels classified as occluding pixels overlaying those classified as occluded pixels. This simple method also prevents aliasing of warping by still averaging occluding pixels that converge to the same point. This method is directly applied in the two angular dimensions (u, v) of light field, which also provides better continuity of propagation as compared to a sequential one dimension u, v projection.

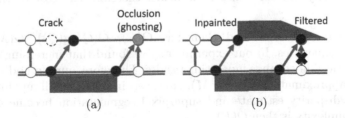

Fig. 4. Forward warping in 1D. Circles stand for image pixels. Disparity is represented in red. (a) Base algorithm, (b) Modified algorithm with warping of disparity by itself. (Color figure online)

Handling Cracks: Cracking effects, oppositely, arises where a destination pixel does not receive any information due to a local divergence of disparity. This issue is dealt with the help of unsupervised inpainting as in [17] and yields satisfying results for tested light fields.

Eventually, the computational cost of inpainting propagation for one view accounts for two forward warpings and an inpainting of cracks. The later is negligible provided cracks are small, which was the case in our tests.

5   Computational Complexity Analysis

Algorithmic complexity is presented for the differents stages of the method. First, disparity estimate obtained by Eqs. (1 and 2) is computed all over the view at which the inpainting is assigned. The estimate is regularized using Eq. (3). Both algorithms have $O(P)$ complexity.

Disparity Interpolation Guided by Inpainted View: SLIC algorithm [18] used for superpixel segmentation has $O(P)$ complexity. Fast implementations of the algorithm exist, and make in practice superpixel segmentation computational cost to be low. Merging superpixels with respect to the mask has an approximate complexity $O(R\log(R))$, with R being the number of superpixels not containing at least K pixels outside the mask. This process therefore strongly depends on image segmentation and mask characteristics, as well as on value K. Assuming R is mostly dependant on the number of superpixels Q, we obtain $O(Q\log(Q))$ complexity for the merging step. In our experiments, for a total number of superpixels around 400 (depending on the dataset), computation took in most complex cases only a few seconds. Disparity interpolation inside the mask defined at Eq. (4) has complexity $O(MW)$, with M being the number of pixels in the mask. It is, in most cases, more costly than the last two processes.

Inpainting Propagation: As presented in Sect. 4, two forward warpings of inpainted view are performed in our method. Each of them as $O(N^2M)$ complexity. The second one also involve unsupervised inpainting of cracks. This last step has $O(V)$ complexity, whith V being the number of *cracked* pixels. Their number is usally low for dense light fields, therefore the cost of this step is neglectible.

Overall: The global method then has $O(P)$, $O(Q\log(Q))$, $O(MW)$ and $O(N^2M)$ complexities. In our experiments, we found that increasing Q and W does not noticeably change inpainting results. Therefore, our method's complexity can be approximated by $O(N^2M)$. However, in case a small inpainting mask is applied, disparity estimate and superpixel segmentation become dominant, and the complexity is then $O(P)$.

6   Experimental Results

Figure 5 shows an inpainting result in comparison with method [6]. The experiments were performed on real light fields captured by camera Lytro 1.0. To alleviate color fluctuations between views, we first perform a histogram matching between the central and remaining views. The algorithm takes as input the mask of the area to be inpainted in the whole light field and an inpainted view (usually the central one), which here has been inpainted using the patch-based method of [19]. We compare our method to [7], for which we perform 100 iterations of diffusion in the reference algorithm. Computations for the later and ours were performed using a C++ implementation on a i7-6600U without any parallelization and the execution times are presented in Table 1.

Table 1. Execution times. Our new method performs with an average of ~ 21 times faster than proposed in [7], and ~ 2.2 times than [8]. Regarding the later, our method performs faster for larger inpainting masks. As compared to [6], our experiments on Lytro Illum light fields ran around ~ 1.3 times faster, without GPU implementation.

Dataset	N	Resolution	Mask	[7]	[8]	Our method
Bee1	9	625*434	4.4%	1'29"	-	4"6
Bee2	11	625*434	1.6%	1'17"	4"2	5"3
Figurines	11	625*434	17.4%	3'39"	19"3	7"6
Fruits	9	625*434	4.8%	1"38	-	4"6
Totoro Park	7	379*379	17.2%	50"	9"2	2"5
Totoro Waterfall	7	379*379	24%	1'2"	9"1	3"
Tape	7	379*379	17.1%	1"12	-	2"6
Still life	9	768*768	4.6%	4'2"	13.2"	11"9
Butterfly	9	768*768	21.6%	6'17"	45"	19"3

Our method performs the fastest inpainting for almost every dataset tested. Because the mask applied on light field *Bee2* has small size, method of [8] is faster. Indeed, the singular value decomposition used has a $O(N^4M)$ complexity linearly dependent on mask's size, whereas our method has $O(P)$ complexity in this case.

Let us note that from a memory allocation perspective, our method consumes only a few calculation images. For comparison [8] requires approximately $8M(N^2 + 400)$ floating point numbers to store in memory.

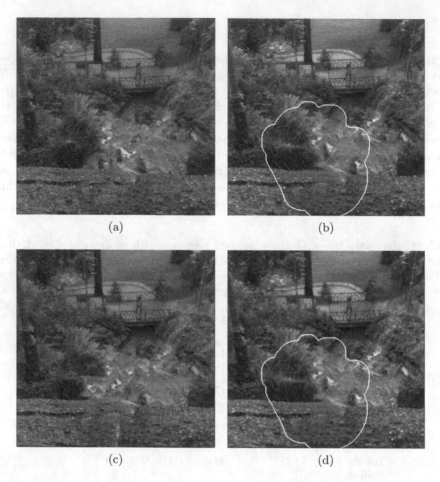

(a)   (b)

(c)   (d)

Fig. 5. Inpainting of *Totoro waterfall* light field. (a, b) Top left view. (c, d) Bottom right view. (a, c) With method proposed by [6]. See the incoherence of the texture in the mask area. (b–d) Using our method. Views are more coherent along angular coordinates inside the inpainting area with the proposed algorithm.

More results with videos showing the inpainting consistency across views are provided at https://www.irisa.fr/temics/demos/lightField/InpaintFast/index. html, and Fig. 6 shows additional results comparison to state of the art methods.

Experiments show very similar results regarding refocusing on focus plane of removed object and its corresponding background, while running considerably faster than other methods.

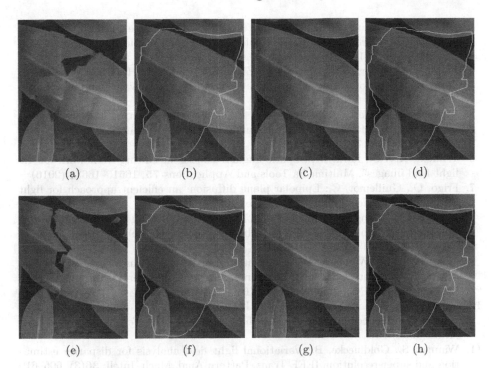

Fig. 6. Inpainting of *Butterfly* light field. (a–d) Top left view. (e–h) Bottom right view. (a, e) With method proposed by [6]. See the incoherence of the texture in the mask area. (b, f) With method proposed by [7]. See misalignment of the edges of the central leave. (c, g) With method proposed by [8] as reference result. (d, h) Using our method which result matches reference.

7   Conclusion

In this paper we have presented a new fast approach for inpainting light fields. Our method is based on a minimum of relevant computation steps. By first obtaining a coarse disparity estimate at inpainted view coordinates in angular space, we are able to interpolate the disparity at inpainted object's location. This allows a fast propagation of inpainted view using specifically adapted warping techniques. Results quality match state of the art, while running most of the time significantly faster. As a future work, our algorithm would highly benefit of parallel programming like GPU implementation.

References

1. Baek, S.-H., Choi, I., Kim, M. H.: Multiview image completion with space structure propagation. In: IEEE Conference on Computer Vision and Pattern Recognition (CVPR), June 2016
2. Thonat, T., Shechtman, E., Paris, S., Drettakis, G.: Multi-view inpainting for image-based scene editing and rendering. In: Fourth International Conference on 3D Vision (3DV), October, pp. 351–359 (2016)

3. Chen, K.-W., Chang, M.-H., Chuang, Y.-Y.: Light field image editing by 4D patch synthesis. In: IEEE International Conference on Multimedia and Expo, ICME, June, pp. 1–6 (2015)

4. Zhang, F.L., Wang, J., Shechtman, E., Zhou, Z.Y., Shi, J.X., Hu, S.M.: Plenopatch: patch-based plenoptic image manipulation. IEEE Trans. Vis. Comput. Graph. **23**(5), 1561–1573 (2016)

5. Jarabo, A., Masia, B., Gutierrez, D.: Efficient propagation of light field edits. In: Proceedings of the V Ibero-American Symposium in Computer Graphics, ser. SIACG 2011, pp. 75–80 (2011)

6. Williem, K.W.S., Park, I.K.: Spatio-angular consistent editing framework for 4D light field images". Multimedia Tools and Applications **75**, 16615–16631 (2016)

7. Frigo, O., Guillemot, C.: Epipolar plane diffusion: an efficient approach for light field editing. France, London, September 2017

8. Pendu, M.L., Jiang, X., Guillemot, C.: Light field inpainting propagation via low rank matrix completion. IEEE Trans. Image Process. **PP**(99), 1 (2018). https://doi.org/10.1109/TIP.2018.2791864

9. Goldluecke, B., Wanner, S.: The variational structure of disparity and regularization of 4D light fields. In: IEEE Conference on Computer Vision and Pattern Recognition (CVPR), June, pp. 1003–1010 (2013)

10. Gortler, S., Grzeszczuk, R., Szeliski, R., Cohen, M.: The lumigraph. In: 23rd Annual Conference on Computer Graphics and Interactive Techniques, pp. 43–54. ACM (1996)

11. Wanner, S., Goldluecke, B.: Variational light field analysis for disparity estimation and super-resolution. IEEE Trans. Pattern Anal. Mach. Intell. **36**(3), 606–619 (2013)

12. Rudin, L.I., Osher, S., Fatemi, E.: Nonlinear total variation based noise removal algorithms. Phys. D: Nonlinear Phenom. **60**(1), 259–268 (1992)

13. Min, D., Choi, S., Lu, J., Ham, B., Sohn, K., Do, M.N.: Fast global image smoothing based on weighted least squares. IEEE Trans. Image Process. **23**(12), 5638–5653 (2014). https://doi.org/10.1109/TIP.2014.2366600

14. Zitnick, C., Kang, S., Uyttendaele, M., Winder, S., Szeliski, R.: High-quality video view interpolation using a layered representation. ACM Trans. Graph. **23**, 600–608 (2004)

15. Hasinoff, S., Kang, S., Szeliski, R.: Boundary matting for view synthesis. Comput. Vis. Image Underst. **103**, 2232 (2006)

16. Jantet, V., Guillemot, C., Morin, L.: Joint Projection Filling method for occlusion handling in Depth-Image-Based Rendering. 3D Research (2011). http://vincent.jantet.free.fr/publication/jantet-11-3DResearch.pdf, https://hal.archives-ouvertes.fr/hal-00628019

17. Telea, A.: An image inpainting technique based on the fast marching method. J. Graph. Tools **9**(1), 23–34 (2004). https://doi.org/10.1080/10867651.2004.10487596

18. Achanta, R., Shaji, A., Smith, K., Lucchi, A., Fua, P., Susstrunk, S.: Slic superpixels compared to state-of-the-art superpixel methods. IEEE Trans. Pattern Anal. Mach. Intell. **34**(11), 2274–2282 (2012)

19. Daisy, M., Tschumperlé, D., Lézoray, O.: A fast spatial patch blending algorithm for artefact reduction in pattern-based image inpainting. In: SIGGRAPH Asia 2013 Technical Briefs, ser. SA 2013, pp. 8:1–8:4. ACM, New York (2013). https://doi.org/10.1145/2542355.2542365

Fusing Omnidirectional Visual Data for Probability Matching Prediction

David Valiente[1](✉), Luis Payá[1], Luis M. Jiménez[1], Jose M. Sebastián[2], and Oscar Reinoso[1]

[1] Systems Engineering and Automation Department, Miguel Hernández University, 03202 Elche, Alicante, Spain
{dvaliente,lpaya,luis.jimenez,o.reinoso}@umh.es
[2] Centre for Automation and Robotics (CAR), UPM-CSIC, Technical University of Madrid, C/ José Gutiérrez Abascal, 2, 28006 Madrid, Spain
jsebas@etsii.upm.es

Abstract. This work presents an approach to visual data fusion with omnidirectional imaging in the field of mobile robotics. An inference framework is established through Gaussian processes (GPs) and Information gain metrics, in order to fuse visual data between poses of the robot. Such framework permits producing a probability distribution of feature matching existence in the 3D global reference system. Designed together with a filter-based prediction scheme, this strategy allows us to propose an improved probability-oriented feature matching, since the probability distribution is projected onto the image in order to predict relevant areas where matches are more likely to appear. This approach reveals to improve standard matching techniques, since it confers adaptability to the changing visual conditions by means of the Information gain and probability encodings. Consequently, the output data can feed a reliable visual localization application. Real experiments have been produced with a publicly-available dataset in order to confirm the validity and robustness of the contributions. Moreover, comparisons with a standard matching technique are also presented.

Keywords: Omnidirectional image · Visual data fusion
Feature matching · Mobile robotics

1 Introduction

Visual localization represents an essential aspect in mobile robotics. Typically, widely acknowledged sensory data such as laser and sonar, have been extensively used [10,14], being normally applied to mapping models [5,6]. Nonetheless, in the last decade, the use of visual sensors have experienced a significant growth due to their numerous advantages. Amongst others, they provide ability to encode large amount of information in an unique snapshot, lightness, power economy, low price, and good ratios in the computation-accuracy balance. All in all, they have

© Springer Nature Switzerland AG 2018
J. Blanc-Talon et al. (Eds.): ACIVS 2018, LNCS 11182, pp. 571–583, 2018.
https://doi.org/10.1007/978-3-030-01449-0_48

emerged as a promising alternative to former sensors, in order to complement current localization data approaches [8,13].

General classifications for localization approaches respond to the processing method and the treatment of the input data. Specifically, visual approaches have extensively relied on the explicit correspondence to the physical environment, commonly known as feature matching [17]. In this work we rely on an omnidirectional camera, since it confers a wide field of view that is able to acquire large scenes in a single image.

The main issue related to these methods has to do with the correct data association [15]. Finding reliable visual matches [4,17] implies a non-trivial procedure which has been widely studied [1,22]. Unfortunately, the output localization estimations tend to suffer from the effect of false positives, commonly induced by noise sources. Despite the fact that different contributions have been proposed in this sense [7,11], this work proposes a self-consistent approach. Focusing on a former development [23], we intended to improve the design in order to avoid computation overload, which typically entails the traditional methods.

To that purpose, we exploit the omnidirectional visual data to be fused at each estimation step without compromising the implementation with additional nor external outlier rejection modules. We rely on Gaussian processes (GP) [19], as the Bayesian technique to infer a distribution for the probability of feature matching existence, in the global reference system. Next, taking the most of a filter-based prediction stage, we are able to predict relevant areas on images at $t+1$, from where visual matches are more likely to appear, in terms of probability. Furthermore, an Information gain metric, Kullback-Leibler (KL) [12], is also configured to account for the visual changes on the scene, and hence the probability distribution is adaptively modulated, according to the current uncertainty in the visual scene.

The structure of this paper is devised as follows: Sect. 2 describes the main characteristics of the omnidirectional vision system and the localization method. Section 3 concentrates on the main contributions regarding the design and implementation of the visual data fusion to obtain the probability distribution for feature matching prediction. Section 4 presents real data results to evaluate the suitability of the approach. Section 5 extracts the principal conclusions derived from this research work.

2 Vision System

The real equipment is presented in Fig. 1. In particular, Fig. 1(a) represents the vision system, which is constituted by a hyperbolic mirror assembled with a Charge-Coupled Devide (CCD) camera. Figure 1(b) depicts the projection model for the omnidirectional image, where the hyperbolic focal, F, coincides with the camera center. A 3D point, $Q(x, y, z)$, generates a mirror projection P, which is normalized as S on a central unitary sphere, and finally projected onto the image as $p(u, v)$. According to [9], we can estimate a Taylor expansion through a calibration toolbox [20] to project Q onto p as follows:

$$\lambda p = \lambda \left[\frac{p}{a_0 + a_2||p||^2 + \ldots + a_n||p||^n} \right] = MQ \tag{1}$$

being $M = [R|T] \in \mathbb{R}^{3\times4}$ the projection matrix, R a rotation matrix $\in \mathbb{R}^{3\times3}$ and T a translation $\in \mathbb{R}^3$, between the camera and the global reference system.

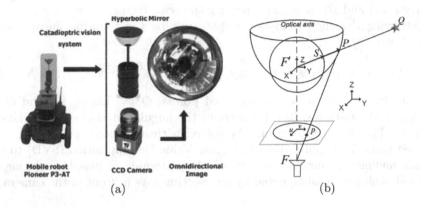

Fig. 1. (a) P3-AT mounted with the CCD DMK21BF04 camera and the Eizho Wide 70 hyperbolic mirror; (b) Omnidirectional camera projection model.

2.1 Omnidirectional Visual Localization

The localization model is conceived by means of the epipolar constraint [9]. In previous works [23, 24] we implemented the adaption of this constraint to the omnidirectional geometry of our vision system. Thus the motion transformation between two poses of the robot, from where associated images are captured, can be univocally determined by this constraint:

$$Q'^T E Q = 0 \tag{2}$$

being $E_{3\times3}$ [16] the essential matrix, and a feature matched pair, expressed in the first image, Q, and in the second image, Q', after normalization. E can be decomposed into a rotation $R_{3\times3}$ and translation $T = [t_x, t_y, t_z]$, through the skew symmetric $[T]_x$ [9]. Therefore a 2D motion transformation $\in XY$, between two poses, can be recovered as an angular movement (β, ϕ), up to a scale factor:

$$
E = [T]_x R = \begin{bmatrix} 0 & 0 & \sin(\phi) \\ 0 & 0 & -\cos(\phi) \\ -\sin(\phi) & \cos(\phi) & 0 \end{bmatrix} \begin{bmatrix} \cos(\beta) & -\sin(\beta) & 0 \\ \sin(\beta) & \cos(\beta) & 0 \\ 0 & 0 & 1 \end{bmatrix}
$$
$$
= \begin{bmatrix} 0 & 0 & \sin(\phi) \\ 0 & 0 & -\cos(\phi) \\ \sin(\beta - \phi) & \cos(\beta - \phi) & 0 \end{bmatrix} = \begin{bmatrix} 0 & 0 & e_1 \\ 0 & 0 & e_2 \\ e_3 & e_4 & 0 \end{bmatrix} \tag{3}
$$

being $\bar{e}_i = [e_1, e_2, e_3, e_4]$ the elements of E. Figure 2 depicts an example of this angular motion transformation (β, ϕ). The equivalence between the robot reference system, in Fig. 2(a), and the image reference system, in Fig. 2(b), can be noted. A 3D point $Q(x_Q, y_Q, z_Q)$, is represented in both, the 3D robot reference system and the 2D image reference system, while the robot moves between poses $A(x_A, y_A, z_A)$ and $B(x_B, y_B, z_B)$, being ρ the scale factor.

Retrieving the angular motion implies solving the resulting linear system from (2) and (3):

$$D_{mx4} \cdot \bar{e}_i = \begin{bmatrix} x_0 z_1 & y_0 z_1 & z_0 x_1 & z_0 y_1 \end{bmatrix} \begin{bmatrix} e_1 & e_2 & e_3 & e_4 \end{bmatrix}^T = \bar{0} \ \forall i \in [1, \dots, N] \quad (4)$$

Accordingly, for each m-pair of matched points, $Q^T = (x_0, y_0, z_0)$ and $Q' = (x_1, y_1, z_1)$, the m-th row in (4) constrains the angular motion through the elements $\bar{e}_i$. This linear system may be solved with a reduced set of $m_{min} = 4$ matched pairs. The application of Singular Value Decomposition (SVD) to (4), permits finding the angular pair (β, ϕ). A quaternion of possible solutions is obtained, which is disambiguated by intersecting rays in front of the camera.

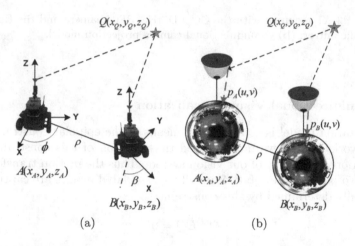

(a)   (b)

Fig. 2. Omnidirectional visual localization between poses A and B. (**a**) robot reference system; (**b**) camera reference system. Projections, $p_A(u, v)$ and $p_B(u, v)$, are indicated.

3   Visual Data Fusion

This section introduces the implementation of the principal contributions conducted in this work. The main objective is to obtain a model which encodes visual changes in the environment in terms of probability of feature matching occurrence in the 3D global reference system. In contrast to our previous works [23, 24], we rely on a global distribution which fuses visual data to infer the probability of feature matching existence and thus predicting its appearance, rather than computing multi-scale distributions for each possible matching candidate. Figure 3 illustrates this idea, along the navigation of the robot between poses, $\bar{x}_t = (\mathrm{x}_t, \mathrm{y}_t, \theta_t)^T$, from where omnidirectional images were captured, $\bar{x}_n = (\mathrm{x}_n, \mathrm{y}_n, \theta_n)^T$. The 3D global space is encoded in terms of probability of feature matching existence, represented by color spheres. This design is extended in order to predict areas on the images, where visual data is more likely to appear, in a more reliable and less computationally demanding scheme. These areas can be projected onto the image captured by the robot, in a single step-process.

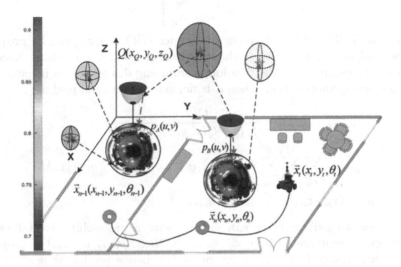

Fig. 3. Robot navigation example in an office-like scenario along three poses: x_{n-1}, x_n and $\bar{x}_t$. The 3D probability distribution of feature point's existence permits associating visual feature points with specific probability, indicated with spheres.

3.1 3D Probability Inference and Motion Prediction

A Bayesian regression technique such as GP [19] permits inferring probabilities for a sensory data distribution. Its notation responds to:

$$f(x) \sim \mathcal{GP}[m(x), k(x, x')] \tag{5}$$

where the GP function is expressed as $f(x) \equiv f[Q(x, y, z)]$, to represent the probability of feature matching existence over a 3D point $Q(x, y, z)$, in the global reference system. Thus $f(\cdot) \in [0,1]$, whereas $m(x)$ and $k(x, x')$ are the mean and covariance, respectively. It is worth noticing that the GP infers a fused probability distribution, taking the matching between images at each time step. Notice that the data have to be expressed in the 3D global reference system, X_o, rather than in the current robot reference system, X_t, through a rotation R, translation T, and scale factor ρ, as described in Sect. 2.1:

$$X_o = \rho T + R X_t \tag{6}$$

Finally, a motion prediction is applied to project the probabilities of feature matching occurrence onto the next image in $t+1$, before producing any matching. In this sense, a filter-based prediction stage, such as the Extended Kalman Filter (EKF), allows us to define a predicted motion, from the current pose, $\bar{x}_t$, to the next, $\hat{\bar{x}}_{t+1}$, by means of a predicted rotation and translation, $\hat{R} \equiv N(\hat{\beta}, \sigma_\beta)$ and $\hat{T} \equiv N(\hat{\phi}, \sigma_\phi)$. Notice that the EKF's innovation, S_t, permits establishing σ_β and σ_ϕ:

$$S_t = \begin{bmatrix} \sigma_\phi^2 & \sigma_{\phi\beta} \\ \sigma_{\beta\phi} & \sigma_\beta^2 \end{bmatrix} \tag{7}$$

Now the probability of feature point's existence, $f[Q(x, y, z)]$, can be projected onto the next image, according to the described prediction procedure. Nonetheless, before projecting such distribution, a sampling discretization is required in order to ease its computation. Then it is normalized and sampled as:

$$f_{norm}[Q(x, y, z)] = \frac{f[Q(x, y, z)]}{\iiint_V f[Q(x, y, z)]\, dx\, dy\, dz} \tag{8}$$

$$p_{norm} = \sum_{x_m}^{M} \sum_{y_m}^{M} \sum_{z_m}^{M} f_{norm}(x_m, y_m, z_m) = 1 \qquad m \in [1, M] \tag{9}$$

$$p(x_m, y_m, z_m) \equiv f_{norm}(x_m, y_m, z_m) \tag{10}$$

Thus the resulting distribution corresponds with a probability-normalized 3D sampled space, represented by $p(x_m, y_m, z_m)$. Next, $p(x_m, y_m, z_m)$ is projected onto the next image in $t + 1$, by means of (1), hence producing $p(u_m, v_m)$ in the image reference system. A real example is depicted in Fig. 4. Figure 4(a) presents the values of the normalized sampled distribution, once projected onto the image frame, $p(u_m, v_m)$ in 2D. Figure 4(b) presents $p(u_m, v_m)$ in 3D, being the Z-axis the values for the probability. Specific probability ranges may be chosen as $p \in [p_{max} - p_{min}]$ for narrower bounds.

3.2   Probability Matching

Considering that we are able to obtain the probability of feature point's existence onto the image frame, $p(u_m, v_m)$, and to predict feature matching, the last stage is designed to obtain the final probability-oriented feature matching. To that aim, we evaluate the visual descriptors associated to the set of feature points in $t+1$, $q(u, v)$, which are spatially arranged in concordance with the location of the probability distribution, $p(u_m, v_m)$, on the image. That constraint is established by the pixel location of $p(u_m, v_m)$, namely, (u_m, v_m). A reliable solution to assess the spatial arrangement is to use the Mahalanobis metric [18], with a well-accepted confidence metric. A set of matching candidates $\in q(u, v)$, can be more precisely obtained by narrowing the probability range $p(u_m, v_m) \in [p_{min} - p_{max}]$, to meet the $\chi(dof)$ distribution, being dof the degrees of freedom, stated by the dimensionality of the problem, that is $dof = dim(u, v) = 2$.

$$||(u_m, v_m) - q(u, v)|| \leq \chi[dim(u, v)] \tag{11}$$

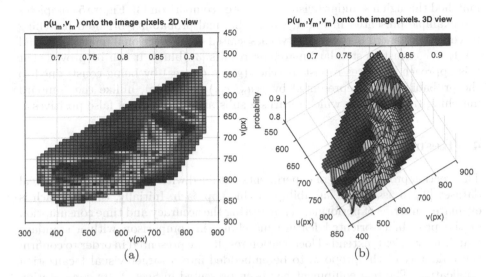

Fig. 4. Projection of $p(x_m, y_m, z_m)$ onto the image pixel axes as $p(u_m, v_m)$. (**a**) 2D view; (**b**) 3D view. (Color figure online)

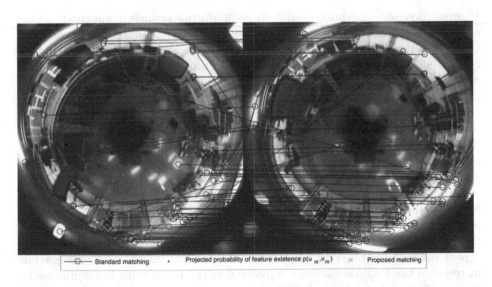

Fig. 5. Probability-oriented matching approach (green) is compared with standard matching (blue) and the projected probability of feature existence $p(u_m, v_m)$ (red). (Color figure online)

In this manner, a set of matching candidates is obtained, which are finally matched through a standard visual descriptor comparison [3]. Figure 5 completes the example presented in Fig. 4, where the final probability-oriented matching (green) is presented between the two associated omnidirectional images, in t and $t + 1$, in contrast to standard matching results [3] (blue). It can be proved that this approach produces robust matches (green crosses) by being constrained to the probability area represented by $p(u_m, v_m)$ (red dots), unlike the standard matching (blue circles), which returns a substantial amount of false positives.

4   Results

This section provides a set of experiments obtained with a publicly available real dataset [2], to assess the suitability of this approach. Initially, the approach is evaluated in terms of performance, by testing the accuracy and time consumption of the probability-oriented feature matching, in comparison with a standard matching [3]. Next, extended localization results are presented in order to confirm the reliability of the proposal to be embedded into a target visual localization application. The real equipment has been presented in Sect. 2. Its computation characteristics are 2×1.7 Ghz and 2 Gb RAM. The real dataset consists of indoor office and laboratory-like spaces with 1450 images, over a 174 m path.

4.1   Matching Results

Figure 6 presents matching results to compare the proposed approach and a standard matching technique. The number of features (left-side Y-axis) and the size of the total probability distribution, $p(x_m, y_m, z_m)$ (right-side Y-axis), is plotted versus the minimum range of the probability of feature existence, p_{min} (X-axis), that is, $p(u_m, v_m) \in [p_{min} - 1]$. Two sample distances between images, $d_1 = 0.25$ m and $d_2 = 1$ m, are respectively presented in Fig. 6(a) and (b). After inspecting the figures, an evident outcome it is observed: higher distances between images make both approaches return a lower number of matching points. Also, higher values of p_{min} restrict the number of matches. Nonetheless, our approach provides a more stable drop of matches with p_{min}. Moreover, even when the size of $p(x_m, y_m, z_m)$ is low, the proposal maintains a well ratio between matching candidates and final matches. However, these results are not complete for extracting further insights into the benefits of the approach. In this sense, Fig. 7 deals with the accuracy of this method. More specifically, Fig. 7(a) shows the percentage of false positive matches obtained with the standard matching (grey) and with the proposal (dark green). Figure 7(b) shows similar results, but in terms of the localization error, expressed as the mean error in angular localization of the robot (β, ϕ), described in Sect. 2.1. It may be noted that both error metrics decrease up to an intermediate value of p_{min} and then rise slightly. This is due to the fact that high values of p_{min} may restrict the probability of feature existence to a very narrow area on the image, and thus this is not desirable under certain circumstances in which the robot would avoid discovering new visual spaces over the image. Consequently, it is worth tuning $p \in [0.65 - 0.75]$, and then modulating its limits within that range, according to an adaptive informative metric, expressed as the KL divergence [12], together with the standard entropy [21]. Such scheme accounts for relevant changes in the visual appearance of the scene, and so adapts the current probability range in order to widen the visual areas from where matches are more likely to appear.

Lastly, Fig. 8 presents the computation efforts required by the standard matching and the proposed approach. The computation time has been divided into several contributions for obtaining, respectively: (i) feature matching; (ii) matching candidates; (iii) final localization estimation; Fig. 8(a) and (b) compare respectively, results obtained with a standard matching (grey), and the proposed approach (green), for distances d_1 and d_2, versus p_{min}. Again, it is immediately confirmed that both methods require less computation efforts when the number of matches is lower. Besides this, our approach reveals more efficient results than the standard matching.

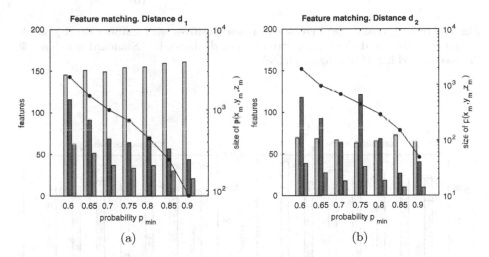

Fig. 6. Left axes: number of matches vs. p_{min}. Right axes: size of $p(x_m, y_m, z_m)$ (log) vs. p_{min}. ●— size $[p(x_m, y_m, z_m)]$. Distance between images (**a**) d_1; (**b**) d_2. Legend: ▪ Standard matching; ■ Proposed matching candidates; ▪ Proposed final matching. (Color figure online)

4.2 Localization Results

A final target application is also evaluated under the framework of visual localization. The same real dataset has been used in order to produce the results shown in Fig. 9. Figure 9(a) depicts the trajectory of the robot, estimated with the proposed probability-oriented matching (green), embedded into a visual localization system. The same estimation is obtained with a standard matching technique

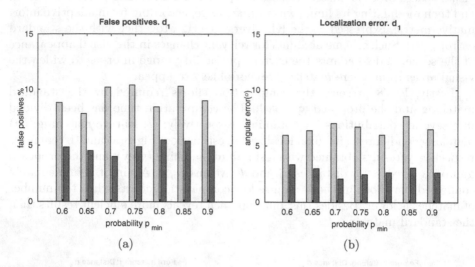

Fig. 7. Accuracy results. (**a**) Percentage of false positives with distance d_1; (**b**) Localization error (in β and ϕ) vs. p_{min} with distance d_1. Legend: ▨ Standard matching; ■ Proposed matching. (Color figure online)

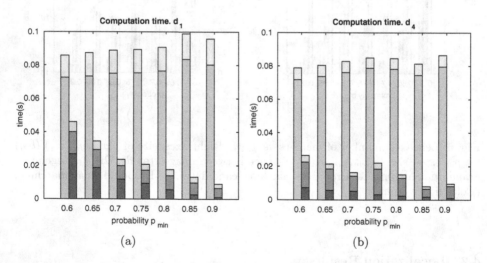

Fig. 8. Computation time vs. p_{min}. (**a**) distance d_1; (**b**) distance d_4. Legend: ■ Standard matching: matching computation; ▨ Standard matching: localization computation; ■ Proposed matching: candidates' computation; ▨ Proposed matching: matching computation; ▨ Proposed matching: localization computation. (Color figure online)

(grey). The ground truth is also indicated in dark. Figure 9(b) presents the evolution of the Root Mean Square Error (RMSE) along the navigation, for both approaches. It can be easily confirmed that our approach returns more reliable and robust results in terms of accuracy on the estimated localization.

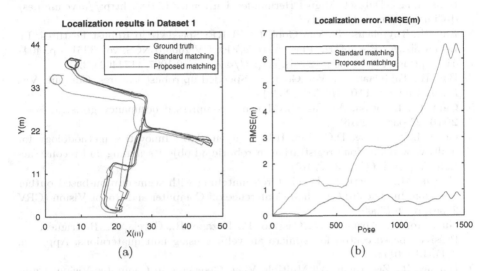

(a)   (b)

Fig. 9. Localization results. (a) Localization estimation obtained with: ground truth (black), standard matching (grey) and the proposed matching (green); (b) RMSE(m) in the localization estimation with: standard matching (grey) and the proposed matching (blue). (Color figure online)

5   Conclusions

This work proposes a reliable probability-oriented feature matching prediction, with visual data fusion achieved with omnidirectional imaging, in the field of mobile robotics. Such contribution permits obtaining predicted areas on the next image in $t + 1$, from where matches are more likely to appear. Furthermore, the approach accounts for the changing visual conditions of the scene, thus providing an adaptive and dynamic solution. Its design and implementation have been conducted according to a Bayesian inference framework, established by GP, together with Information gain metrics which adaptively modulate the ranges of the probability distribution that eventually map relevant areas on the image for the feature matching occurrence. Real results obtained with a publicly-available dataset have confirmed the suitability of the approach and compared it with results produced by a standard matching technique. These results have been extended to a visual localization approach. Again, the robustness and the improved performance of the approach have been confirmed.

References

1. Abduljabbar, Z.A.: SEPIM: secure and efficient private image matching. App. Sci. **6**(8), 213 (2016)
2. ARVC: Automation, Robotics and Computer Vision Research Group: Omnidirectional Image Dataset. Miguel Hernandez University (2018). http://arvc.umh.es/db/images/innova_trajectory/
3. Bay, H., Tuytelaars, T., Van Gool, L.: SURF: Speeded Up Robust Features. In: Leonardis, A., Bischof, H., Pinz, A. (eds.) ECCV 2006. LNCS, vol. 3951, pp. 404–417. Springer, Heidelberg (2006). https://doi.org/10.1007/11744023_32
4. Bay, H., Tuytelaars, T., Van Gool, L.: Speeded up robust features. Comput. Vis. Image Underst. **110**, 346–359 (2008)
5. Cain, C., Leonessa, A.: FastSLAM using compressed occupancy grids. J. Sens. **2016**, 23 pages (2016)
6. Chen, L.C., Hoang, D.C., Lin, H.I., Nguyen, T.H.: Innovative methodology for multi-view point cloud registration in robotic 3d object scanning and reconstruction. App. Sci. **6**(5), 132 (2016)
7. Gerrits, M., Bekaert, P.: Local stereo matching with segmentation-based outlier rejection. In: The 3rd Canadian Conference on Computer and Robot Vision (CRV 2006), p. 66 (2006)
8. Guerrero, M., Abaunza, H., Castillo, P., Lozano, R., Garcia, C., Rodriguez, A.: Passivity-based control for a micro air vehicle using unit quaternions. App. Sci. **7**(1), 13 (2017)
9. Hartley, R., Zisserman, A.: Multiple View Geometry in Computer Vision. Cambridge University Press, New York (2004)
10. Kim, Y., Lee, H., Park, C., Choi, S.: A study for optimum survey method of underwater structure using the dual sonar sensor. J. Sens. **2018**, 10 pages (2018)
11. Kitt, B., Geiger, A., Lategahn, H.: Visual odometry based on stereo image sequences with RANSAC-based outlier rejection scheme. In: 2010 IEEE Intelligent Vehicles Symposium, pp. 486–492, June 2010
12. Kulback, S., Leiber, R.A.: On information and sufficiency. Ann. Math. Stat. **22**, 79–86 (1951)
13. Lai, Y.C., Ting, W.O.: Design and implementation of an optimal energy control system for fixed-wing unmanned aerial vehicles. App. Sci. **6**(11), 369 (2016)
14. Li, J., Zhong, R., Hu, Q., Ai, M.: Feature-based laser scan matching and its application for indoor mapping. Sensors **16**, 1265 (2016)
15. Li, Y., Li, S., Song, Q., Liu, H., Meng, M.H.: Fast and robust data association using posterior based approximate joint compatibility test. IEEE Trans. Ind. Inform. **10**(1), 331–339 (2014)
16. Longuet-Higgins, H.C.: A computer algorithm for reconstructing a scene from two projections. Nature **293**(5828), 133–135 (1985)
17. Lowe, D.: Distinctive image features from scale-invariant keypoints. Int. J. Comput. Vis. **60**(2), 91–110 (2004)
18. McLachlan, G.: Discriminant Analysis and Statistical Pattern Recognition. Wiley Series in Probability an Statistics. Wiley, Hoboken (2004)
19. Rasmussen, C.E., Williams, C.K.I.: Gaussian Processes for Machine Learning. Adaptive Computation and Machine Learning series, Massachusetts Institute of Technology (2006)
20. Scaramuzza, D., Martinelli, A., Siegwart, R.: A toolbox for easily calibrating omnidirectional cameras. In: IEEE IROS, China, pp. 5695–5701 (2006)

21. Shannon, C.E.: A mathematical theory of communication. SIGMOBILE Mob. Comput. Commun. Rev. **5**(1), 3–55 (2001)
22. Shuang, Y., et al.: Encoded light image active feature matching approach in binocular stereo vision. In: IFOST, pp. 406–409 (2016)
23. Valiente, D., Gil, A., Payá, L., Sebastián, J.M., Reinoso, O.: Robust visual localization with dynamic uncertainty management in omnidirectional SLAM. App. Sciences **7**(12), 1294 (2017)
24. Valiente, D., Reinoso, Ó., Gil, A., Payá, L., Ballesta, M.: Omnidirectional localization in vSLAM with uncertainty propagation and Bayesian regression. In: Blanc-Talon, J., Penne, R., Philips, W., Popescu, D., Scheunders, P. (eds.) ACIVS 2017. LNCS, vol. 10617, pp. 263–274. Springer, Cham (2017). https://doi.org/10.1007/978-3-319-70353-4_23

Derivative Half Gaussian Kernels
and Shock Filter

Baptiste Magnier$^{(\boxtimes)}$, Vincent Noblet, Adrien Voisin, and Dylan Legouestre

IMT Mines Alès, LGI2P, 6. avenue de Clavières, 30100 Alès, France
Baptiste.Magnier@mines-ales.fr,
{Vincent.Noblet,Adrien.Voisin,Dylan.Legouestre}@mines-ales.org

Abstract. Shock filter represents an important family in the field of nonlinear Partial Differential Equations (PDEs) models for image restoration and enhancement. Commonly, the smoothed second order derivative of the image assists this type of method in the deblurring mechanism. This paper presents the advantages to insert information issued of oriented half Gaussian kernels in a shock filter process. Edge directions assist to preserve contours whereas the gradient direction allow to enhance and deblur images. For this purpose, the two edge directions are extracted by the oriented half kernels, preserving and enhancing well corner points and object contours as well as small objects. The proposed approach is compared to 7 other PDE techniques, presenting its robustness and reliability, without creating a grainy effect around edges.

Keywords: Half Gaussian Kernels · Shock filter · PDE

1   From Heat Equation to Anisotropic Diffusion

Since 1960, digital images may simply be deblurred by combining the difference between an original image I_0 and ΔI: a blurred version of this same image. Usually, ΔI corresponds to a blur process equivalent to the heat equation or a convolution of I_0 with an isotropic Gaussian. This original theory proposed by Gabor is proportional to using the Laplacian operator [7]. Thus, a simplest manner to remove blur in an image remains the equation:

$$\frac{\partial I}{\partial t} = I_0 - \alpha \cdot \Delta I, \tag{1}$$

where t represents the time or the observation scale and $\alpha < 1$ is a little scalar to control the deblurring. This process is equivalent to the inverse heat equation. However, this technique is not stable because the procedure blows up after several iterations and generates an unusable image [7]. To improve Eq. 1, rather than applying a global operator on all the image, the main idea is to iterate local operator at level of each pixel. Nonlinear Partial Differential Equations (PDEs) may achieve this task [2,7], practicing anisotropic diffusions of pixel informations

© Springer Nature Switzerland AG 2018
J. Blanc-Talon et al. (Eds.): ACIVS 2018, LNCS 11182, pp. 584–597, 2018.
https://doi.org/10.1007/978-3-030-01449-0_49

in the image. Indeed, PDEs belong to one of the most important part of mathematical analysis and are closely related to the physical world. In this context, images are considered as evolving functions of time and a regularized image can be seen as a version of the original image at a special scale.

2   On Existing Shock Filters

The main PDEs for image deblurring are presented in this section. In order to regularize images by controlling the diffusion, the pioneer work of Perona and Malik on anisotropic diffusion has been one of the most influential paper in the area [17]. The proposed model is described by the following equation:

$$\frac{\partial I}{\partial t}(x,y,t) = \text{div}\left(g\left(|\nabla I|\right) \cdot |\nabla I|\right), \tag{2}$$

where $|\nabla I|$ represents the modulus of the gradient with mask of type $[-1\ 1]$ and g a decreasing function satisfying $g(0) = 1$ and $g(+\infty) = 0$; this function may be:

$$g\left(|\nabla I|\right) = e^{-\left(\frac{|\nabla I|}{K}\right)^2}, \tag{3}$$

with $K \in \mathbb{R}_*^+$ a constant that can be assimilated to a gradient threshold or a diffusion barrier, slowing down diffusion near edges, where $|\nabla I|$ is large. Moreover, by developing Eq. 2, it is well known that the diffusion moves backward when $|\nabla I| > K$, creating time-reverse equation (deblurring effect) called a shock filter.

Original Shock Filter: In the PDE framework, the seminal contribution [16] is equivalent to Eq. 1, the 2D formulation of the original shock filter is:

$$\frac{\partial I}{\partial t} = -\text{sign}(\Delta I) \cdot |\nabla I|. \tag{4}$$

This PDE produces a dilation/erosion for each pixel, creating a high sensitivity to noise pixels, so a number of improvements have been proposed.

Shock Filter Involving Gaussian: In order to be more robust to noise, the Gaussian function G_σ may be convolved with $I_{\eta\eta}$, the second directional derivative of the image in the gradient direction, where σ represents the standard deviation of the Gaussian. Coupling diffusion ($I_{\xi\xi}$ term [2,7,13,19]) and shock filter, Alvarez and Mazorra (AM) in [3] proposed the following equation:

$$\frac{\partial I}{\partial t} = C_\xi \cdot I_{\xi\xi} - \text{sign}(G_\sigma * I_{\eta\eta}) \cdot |\nabla I|, \tag{5}$$

where $|\nabla I|$ is the modulus of the gradient with a 3×3 mask (as Sobel masks), C_ξ denotes a control function of the diffusivity, as in Eq. 3, $I_{\xi\xi}$ denotes the second derivative in the orthogonal direction of η, i.e., the edge direction. It corresponds to a pure diffusion in the contour directions, called the curvature equation.

Some evolutions of this approach have been proposed, as in [8], where an isotropic diffusion (ΔI) is applied concerning small gradients. Fu et al. [5] proposed a PDE technique that weights the diffusion and shock terms in accordance with threshold values of the gradient magnitude:

$$\begin{cases} \frac{\partial I}{\partial t} = c_1 \cdot I_{\xi\xi} - \text{sign}(G_\sigma * I_{\eta\eta}) \cdot |\nabla I|, & \text{if } |\nabla I| > T_1 \\ \frac{\partial I}{\partial t} = c_1 \cdot I_{\xi\xi} - c_2 \cdot \text{sign}(G_\sigma * I_{\eta\eta}) \cdot |\nabla I|, & \text{if } T_1 > |\nabla I| > T_2 \\ \frac{\partial I}{\partial t} = \Delta I = I_{\xi\xi} + I_{\eta\eta} & \text{elsewhere,} \end{cases} \quad (6)$$

with $c_1 = \frac{1}{1+\zeta_1 \cdot I_{\xi\xi}^2}$, ($\zeta_1 \in \mathbb{R}^+$), and $c_2 = |\text{th}(\zeta_2 \cdot I_{\eta\eta})|$, ($\zeta_2 \in \mathbb{R}^+$). The authors divided the image into three-type regions by its smoothed gradient magnitude. For high gradients (generally boundaries), a shock-type backward diffusion is performed in the gradient direction for the deblurring. For medium gradients (such as textures and details), a soft shock-type backward diffusion is performed, as Eq. 5. As far as small gradients are concerned(such as inside different areas or flat regions), an isotropic diffusion is applied (heat equation).

On the other hand, Bettahar [4] proposed a combination of nonlinear reaction-curvature diffusion and shock filter. The gradient modulus of $f(s) = f\frac{1}{1+s^2/k}$ applied on the gradient magnitude $|\nabla(f(|\nabla I_\sigma|))|$ favors or inhibits the shock filter in accordance with the gradient magnitude.

The main problem of the based-Gaussian models is the creation of homogeneous blobs in flat noisy regions, as illustrated in Figs. 1(b), 5(b) and 6(b). Moreover, after a certain number of iterations a corner smoothing could be created, as show for the extremity of the star in Fig. 5(b).

Complex Shock Filter: A different solution proposed by Gilboa et al. [6] is to change the *sign* function (cf. Eq. 4) in order to take into account both the 2nd order direction of the second derivative and its magnitude. Denoting $Im(I)$, the imaginary part of I, by using $\frac{2}{\pi} \arctan(a \cdot Im(\frac{I}{\theta}))$, $a \in \mathbb{R}^+$ represents the parameter controlling the steepness of the slope of the 2nd order derivative near 0. When $\theta \in \mathbb{R}^+_*$ tend to 0, $Im(\frac{I}{\theta})$ may be considered as the smoothed 2nd order derivative of I. The complex shock filter is described by the following equation:

$$\frac{\partial I}{\partial t} = -\frac{2}{\pi} arctan(a \cdot Im(\frac{I}{\theta})) \cdot |\nabla I| + \Lambda \cdot I_{\eta\eta} + \tilde{\Lambda} \cdot I_{\xi\xi}, \quad (7)$$

where Λ is a complex diffusion term regularizing the noise and indicating inflection points. $\tilde{\Lambda}$ is a real scalar parameter which correspond to the amount of diffusion in level-set direction. Using Eq. 7, the regions close to contours where the 2nd order derivative has a higher magnitude, i.e., inflection points will not have equal weights. This translates into a higher deblurring speed near edges and contours than in the flat regions of the image.

The complex shock has been improved in [9] to correct its main default: a weak edges enhancement. Denoting $\frac{\partial I}{\partial t} = I_t$ and $Re(I)$ the real part of I, the mathematical expression of this PDE can be decomposed in two expressions:

$$Re(I_t) = \frac{2}{\pi}arctan(a.Im(\frac{I}{\theta})).f_1 \cdot |\nabla I| - sign(Re(I_{\eta\eta})).f_2 \cdot |\nabla I|$$

$$+f_1 \cdot (Re(\Lambda) \cdot Re(I_{\eta\eta}) - Im(\Lambda) \cdot Im(I_{\eta\eta}) + \tilde{\Lambda} \cdot Re(I_{\xi\xi})) \quad (8)$$

$$Im(I_t) = Im(\lambda) \cdot Re(I_{\eta\eta}) - Re(\lambda) \cdot Im(I_{\eta\eta}) + \tilde{\Lambda} \cdot Im(I_{\xi\xi}). \quad (9)$$

As this filter is an improvement of the Eq. 7, it possesses the same variables. However the 2 new functions f_1 and f_2 are the filter core, and control the transition between an entirely complex and an entirely real filter. For N iterations, they are define as: for $i = 0, 1, ..., N - 1$ and $T_{l_1}, T_{s_1}, T_{l_2}, T_{s_2} \in (0, N - 1)$,

$$f_1(T_{l_1}, T_{s_1}) = \begin{cases} 1, & i < T_{l_1} \\ 1 - \frac{i-T_{l_1}}{T_{s_1}-T_{l_1}}, & \text{elsewhere,} \\ 0, & i > T_{s_1} \end{cases} \quad f_2(T_{l_2}, T_{s_2}) = \begin{cases} 0, & i < T_{l_2} \\ \frac{i-T_{l_1}}{T_{s_2}-T_{l_2}}, & \text{elsewhere.} \\ 1, & i > T_{s_2} \end{cases}$$

When $f_1 = 1$ and $f_2 = 0$ this technique behaves exclusively as a complex shock filter. In the case where $f_1 = 0$ and $f_2 = 1$ it correspond to an edge enhancement filter by using the classic shock filter. Elsewhere, it corresponds to a mix of both.

Despite the originality of complex shock filter and its evolutions, the main drawback of this techniques concerns noisy images. Namely, in the presence of noise, the deblurring process creates images looking grainy a level of edges (cf. Sect. 4). In other words, edges are not straight, but rather serrated boundaries.

(a) Noisy image   (b) Fu filter [5]   (c) Weickert [22]   (d) Eq. 17

Fig. 1. Restoration of a real highly noisy image.

Structure Tensor Based Approaches Shock Filter. The coherence enhancing shock filter has been developed by Weickert [22] with the shock filter theory via structure tensor based approach. Thus, the corresponding PDE:

$$\frac{\partial I}{\partial t} = -sign((G_\sigma * I)_{\omega_+\omega_+}) \cdot |\nabla I| \quad (10)$$

allows shock filtering along the direction of the eigenvector ω_+ corresponding to the largest eigenvalue of the structure tensor: $J_\rho(\nabla I_\sigma) = G_\rho * (\nabla I_\sigma \cdot \nabla I_\sigma^T)$, where the parameters ρ and σ represent standards deviations of the Gaussian.

Tschumperlé and Deriche proposed a different form of the diffusion shock filter coupling especially for enhancement of color images [19]. This model can be generalized to gray level images using the structure tensor:

$$\frac{\partial I}{\partial t} = \alpha \cdot (I_0 - I) + c_{\omega_-} \cdot I_{\omega_- \omega_-} + c_{\omega_+} \cdot I_{\omega_+ \omega_+} - (1 - g(\mathcal{N})) \cdot sign((G_\sigma * I)_{\omega_+ \omega_+}) \cdot |\nabla I| \quad (11)$$

where $\alpha \in \mathbb{R}_*^+$, and $c_{\omega_{|/}}$ are two decreasing functions: $c_\omega(\mathcal{N}) = \frac{1}{\sqrt{1+\mathcal{N}^2}}$ and $c_{\omega_+}(\mathcal{N}) = \frac{1}{1+s^2}$, with $\mathcal{N} = \sqrt{\lambda_+ + \lambda_-}$. The shock is controlled by $g(\mathcal{N})$ (cf. Eq. 3), where λ_+ and λ_- denote the two eigenvalues tied to the eigenvectors ω_+ and ω_- respectively, related to the tensor J_ρ.

These tensorial techniques behave like a contrast enhancing shock filter, they enhance well strip structures as in Fig. 6(c), however they create artificial lines or may engender an undesirable grainy effect around edges when dealing with highly noisy images, see Fig. 1(c). Notwithstanding a long-term and large scientific effort for PDE-based methods, a satisfactory solution has not been found yet.

3   A New Half Kernel Based Shock Filter

As stated in the methods above, image derivatives $(I_{\xi\xi}, I_{\eta\eta})$, and the gradient are crucial information for the image enhancement via PDEs. The more these features are accurate and robust against noise, the more the diffusion process is efficient. This section presents an oriented half kernels-based technique which enables a gradient computation and an estimation of edge directions modulo 2π. Thereafter, these components are useful for the creation of image derivatives and represent fundamental information tuning the proposed PDE.

3.1   Derivative Half Gaussian Kernels

Gaussian kernels are used in a large part of Shock filter for their efficiency in edge detection. Nevertheless weaknesses can be noticed at level of corners and small objects present in the image (as stated above about shock filters involving Gaussians). Edge detection techniques using elongated kernels are efficient at detecting large linear structures correctly [18], but small structures are considered to be noise and their edges are not extracted. Consequently, the accuracy of the detected edge points decreases strongly at corner points and for non-straight object contour parts. To bypass this undesirable effect, an anisotropic edge detection method is developed in [14] and deeper evaluated in [10]. Indeed, the proposed technique is able to detect crossing edges and corners due to two elongated and oriented filters in two different directions. The main idea is to "cut" the anisotropic Gaussian kernel using a Heaviside function (see Fig. 2(c)). The half kernel (HK) can be implemented using Gaussian and its first derivative:

- a half smoothing part: $\mathcal{G}(s) = H(s) \cdot e^{\frac{s^2}{2 \cdot \mu^2}}$, with H the Heaviside function,
- a derivative part of the filter: $\mathcal{G}_1(s) = s \cdot e^{\frac{s^2}{2 \cdot \sigma^2}}$, with $\sigma, \mu \in \mathbb{R}_+^*$ and $s \in \mathbb{R}$.

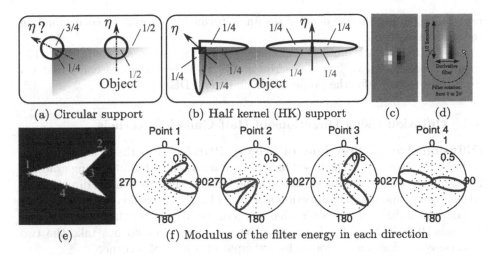

(a) Circular support   (b) Half kernel (HK) support   (c)   (d)

(e)   (f) Modulus of the filter energy in each direction

Fig. 2. Representation of filter supports concerning edges and corners. (c)Derivative Gaussian kernel. (d) 2D half kernel. (d) Selection of points before applying the HK_θ.

For signal and image processing, s belongs to integers. Figure 2(d) shows an example of a HK, built with the two function $\mathcal{G}_1$ at the horizontal and $\mathcal{G}$ at the vertical respectively. In order to create an elongated filter, the support of the half smoothing part must be higher than the derivative support i.e., $\mu > \sigma$. Then, the HK is rotated in several directions θ from 0 to 2π (bilinear rotation) to obtain a rotated version of the filter HK_θ. The convolution of the image I with HK_θ (i.e., $I * HK_\theta$) allows to compute a derivative information at each desired angle (illustrated in Fig. 2(e)–(f)). In order to better understand this technique to extract edge, the filter support of a HK on a straight contour is equivalent to 1/2 on both sides of the edge, illustrated in Fig. 2(b). On the contrary, for a corner point with a $\pi/2$ radian angle, the support of the half filter remains 1/2 on both sides of the edge, whereas it is around 1/4 and 3/4 concerning other filter supports [10] (cf. Fig. 2(a)). Eventually, HK corresponds to an oriented filter derivative, so its responses are either positive, or negative. For each pixel of coordinates (x, y), the gradient $|\nabla I|$ corresponds to the maximum value minus the minimum value of $I * HK_\theta$ among all the θ directions:

$$
\begin{cases}
|\nabla I|(x,y) = \max_{\theta \in [0,2\pi[} I * HK_\theta(x,y) - \min_{\theta \in [0,2\pi[} I * HK_\theta(x,y) \\
\theta_1(x,y) = \mathrm{argmax}_{\theta \in [0,2\pi[}(I * HK_\theta(x,y)) \\
\theta_2(x,y) = \mathrm{argmin}_{\theta \in [0,2\pi[}(I * HK_\theta(x,y)) \\
\eta(x,y) = \frac{\theta_1 + \theta_2}{2} \\
\beta(x,y) = \begin{cases} |\theta_1(x,y) - \theta_2(x,y)|, & \text{if } |\theta_1(x,y) - \theta_2(x,y)| \leqslant \pi \\ 2\pi - |\theta_1(x,y) - \theta_2(x,y)| & \text{elsewhere.} \end{cases}
\end{cases} \tag{12}
$$

Note that the matlab code of the HK is available on mathworks website [11], involving different HK filters. Once $|\nabla I|$, θ_1 and θ_2 have been obtained, η corresponds to the gradient orientation i.e., to the bisector between these two directions, thus the oriented second derivative $I_{\eta\eta}$ may be created (Fig. 3(d)). Also, as represented in Fig. 4(c), β denotes the angle formed by θ_1 and θ_2, corresponding to an angular sector.

These directions are useful and efficient for image restoration via PDE [13], corner detection [1] or image descriptor [20]. All these entities computed via Eq. 12 are required for the proposed shock filter PDE.

3.2 Diffusion Using Directions of Half Gaussian Kernels

Diffusion Along Directions of Edges: PDE-based methods are a mature technique. Therefore, PDE-based image regularization techniques using gradient intensities or tensorial diffusion smooth the image either both in the directions of edges ξ or ω_- and gradient direction η or ω_+. This process is doable by combining the current diffused image with oriented second derivative of the image ($I_{\xi\xi}$ and $I_{\eta\eta}$, see PDE in Sect. 2). However, all these approaches do not take the two directions of edges into account, for example at a level of a corner.

Similar to the curvature equation $I_{\xi\xi}$, the original idea developed in [12] is to smooth anisotropically at level of edges in two diffusion directions θ_1 and θ_2:

$$\frac{\partial I}{\partial t} = \frac{\partial^2 I}{\partial \theta_1 \partial \theta_2} = I_{\theta_1 \theta_2}. \qquad (13)$$

In [12], the two directions necessary for the diffusion are not computed by derivative half Gaussian kernels, only the diffusion process imports here. Indeed, contrary to previous methods, $I_{\theta_1 \theta_2}$ is computed by interpolations in the θ_1 and θ_2 directions, from 0 to 2π, with one and only one direction for each angle. As illustrated in Fig. 4(e), for each pixel, the smoothing is doable in 3×3 mask with the two pixels interpolated by the unweighted average in the desired directions. Yet, applying this PDE in an image using θ_1 and θ_2 issued from Eq. 12 is not stable and $I_{\theta_1 \theta_2}$ creates an effect of dilation/erosion close to edge points, because θ_1 and θ_2 are not well directed. Figure 3(b) illustrates this undesirable effect.

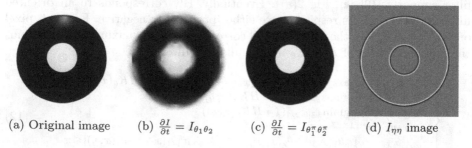

(a) Original image   (b) $\frac{\partial I}{\partial t} = I_{\theta_1 \theta_2}$ (c) $\frac{\partial I}{\partial t} = I_{\theta_1^\pi \theta_2^\pi}$ (d) $I_{\eta\eta}$ image

Fig. 3. A diffusion along θ_1 and θ_2 directions creates a blur effect around edges.

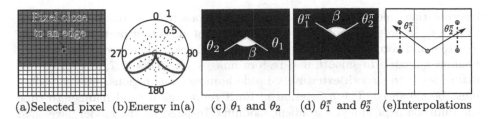

(a)Selected pixel (b)Energy in(a) (c) θ_1 and θ_2 (d) θ_1^π and θ_2^π (e)Interpolations

Fig. 4. Mirror rotation of θ_1 and θ_2 directions to avoid a blurring effect in the diffusion process. (a): Pixel close to an edge. (b-c): Diffusion directions for the selected point.

A Mirror Rotation in Order to Enhance the Diffusion: The PDE presented in Eq. 13 may be controlled by the gradient magnitude, for example, but an undetermined quantity of blur will be created. Consequently, instead of diffusing alongside the directions θ_1 and θ_2, it was proposed in [15] to diffuse in the opposite directions, i.e., the directions $\theta_1^\pi = \theta_1 - \pi$ and $\theta_2^\pi = \theta_2 - \pi$. Figure 4 illustrates these modifications where β keeps the same value. The new diffusion PDE becomes:

$$\frac{\partial I}{\partial t} = \frac{\partial^2 I}{\partial \theta_1^\pi \partial \theta_2^\pi} = I_{\theta_1^\pi \theta_2^\pi}. \tag{14}$$

Simply by a mirror effect, this diffusion offers a better preservation than the original as seen in Fig. 3(c), after many iterations without any control (!). Using the original diffusion the white pixels near the border are attracted by the color of the border which causes the appearance of grey blotches. By inverting the direction of the diffusion, those pixels are not attracted anymore and the disk is preserved, while few pixel of noise have disappeared in the middle of the image. The purpose of getting a stable PDE is to control the diffusion of Eq. 14 before developing a new combination with a shock filter scheme.

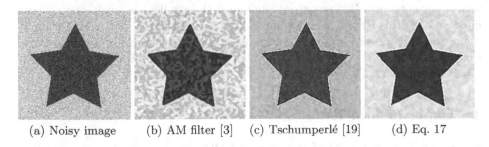

(a) Noisy image   (b) AM filter [3]   (c) Tschumperlé [19]   (d) Eq. 17

Fig. 5. Image restoration. For (b), $\sigma = 1$, (c) $\sigma = 1$, $\rho = 3$ and (d) $\sigma = 1$, $\mu = 3$.

A Diffusion Tuned by the Gradient and the Directions of Edges: The presented schemes in Eqs. 13 and 14 do not control the diffusion, they correspond to linear models. In other words, the pixel intensities are propagated involving interpolations along (θ_1, θ_2) or $(\theta_1^\pi, \theta_2^\pi)$ directions with a constant speed (cf. Fig. 4(e)). In order to avoid an over-smoothing of important parts of the image, as edges and corners, the diffusion must be nonlinear. There are many existing nonlinear models to smooth and restore images via PDEs [2,7,13,17,19]. The main idea is to smooth extensively inside homogeneous regions, whereas edges must be enhanced. The g function in Eq. 3 represents a good manner to tune the diffusion depending on gradient modulus. Inspired by the model developed in [13], a control function is proposed here to decrease the diffusion both at edge and corner points (when the angle between θ_1^π and θ_2^π is not open: $\beta < \pi$):

$$C_{(k,h)}(|\nabla I|, \beta) = \frac{1}{2} \cdot e^{-\left(\frac{|\nabla I|}{k}\right)^2} + \frac{1}{2} \cdot e^{-\left(\frac{(\pi - \beta)}{(\pi \cdot h)}\right)^2} \quad \text{with} \quad (k, h) \in]0; 1]^2 \quad (15)$$

Thus, Eq. 14 becomes:

$$\frac{\partial I}{\partial t} = C_{(k,h)} \cdot I_{\theta_1^\pi \theta_2^\pi}. \quad (16)$$

The smoothing process is driven by the gradient magnitude and β. The $C_{(k,h)}$ function ensures the diffusion preserving edges and corners and enables an extensive smoothing inside homogeneous regions. In case of a small gradient and a β angle close to π, the considered pixel will be largely diffused. If the gradient is important and the β angle is small, smoothing is weak and preserves boundaries.

A New Shock Filter Scheme: As stated in Sect. 2, corner rounding is the main weakness of the shock filter. To correct this problem a control function is inserted for the shock filter term, inversely to the control of $I_{\theta_1^\pi \theta_2^\pi}$. Thus, Eq. 16 evolves into:

$$\frac{\partial I}{\partial t} = C_{(k,h)} \cdot I_{\theta_1^\pi \theta_2^\pi} - (1 - C_{(k,h)}) \cdot sign(I_{\eta\eta}). \quad (17)$$

Consequently, the smoothing process and the shock are both tuned by $C_{(k,h)}$. When $|\nabla I| \approx 0$, it corresponds to homogeneous regions, $C_{(k,h)} \to 1$, and shock term does not work. If β is a close angle and the gradient is weak, the process must smooth this region; it is doable when $k > h$ in Eq. 16. When $|\nabla I| \approx 1$, it refers to pixels of (or close to) contours, $C_{(k,h)} \to 0$, thus edges are enhanced. Figure 5(d) illustrates the enhancement of an image containing corners (with $k = 0.1$, $h = 0.05$).

4   Experimental Results

Several application results of the proposed PDE are presented in this section. These results are compared to the different approaches described above. The parameters used for all the methods are also detailed. As far as the evaluation process is concerned, the Structural Similarity Index Measure (SSIM) [21] is reported and plotted as a function of the number of iterations for each method.

The parameters for the half kernels are $\sigma = 1$ for the derivative and $\mu = 3$ for the smoothing during all the experiments. Such a set of parameters enables

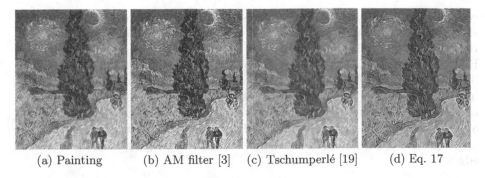

(a) Painting   (b) AM filter [3]   (c) Tschumperlé [19]   (d) Eq. 17

Fig. 6. Comparition of several PDE schemes on a picture of Vincent van Gogh paint. For (b), $\sigma = 1$, (c) $\sigma = 1$, $\rho = 3$ and (d) $\sigma = 1$, $\mu = 3$.

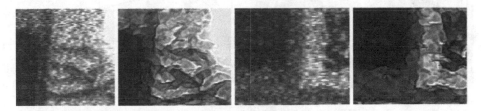

Fig. 7. Enlargement of parts extracted from images (a) -left- and (d) -right- in Fig. 1.

an enhancement of small objects and thin structures. With the enlargement of these two parameters, the scale effect will lead only to the enhancement of the large structures or dominant edges. The first result presented in Fig. 5 shows the behavior of the proposed PDE method with an image containing pointed corners. Even though the original image does not contain blur, edges remain sharp after 50 iterations whereas the noise disappears completely. For the second original image in Fig. 6, stripes are created by a brush on a painting. This result is somewhat reminiscent of tensorial approaches [19,22]. After 50 iterations, results depict the coherence of the proposed PDE scheme because the lineaments, trees and characters are preserved and enhanced. With the same number of iterations, boundaries in the image presented in Fig. 1 are well enhanced while corners are highly improved, as detailed in Figs. 7(b) and (d).

To shed light on the effectiveness of the proposed shock filter, two tested images are noised by adding random Gaussian noise and blurred using a convolution with a Gaussian of standard deviation of $\sigma = 1$, cf. Figs. 8(a)-(b) and 9(a)-(b). Then they are independently treated by Perona-Malik (PM), Alvarez-Mazorra (AM), Gilboa, Luduzan, Fu, Tschumperlé and Bettahar methods. The parameters for methods involving Gaussians is $\sigma = 1$ and $\rho = 3$ for tensorial PDEs. For complex shock filters, $\theta = 0.001$, $a = 2$ and $\hat{\lambda} = \lambda = 0.1$. In order to compare objective performance of these approaches, the SSIM is calculated for each iteration; presented images are tied to the best score of the measure. As shown in Figs. 8(c) and 9(c), the SSIM curve of the proposed scheme is above the others (called Half-filter), translating a better image enhancement than other PDE methods. The degraded images contain a strong noise and small objects have to be restored. As an example, the gray structure in the middle of the tripod in Fig. 8(a) is well enhanced by the new filter. Indeed, this object is recognizable and realistic, contours are straight. Details are also enhanced like the camera and the cameraman's lock of hair. Finally, by using the new PDE, the corners and edges of Fig. 9(b) are well deblurred, without creating a grainy effect around edges.

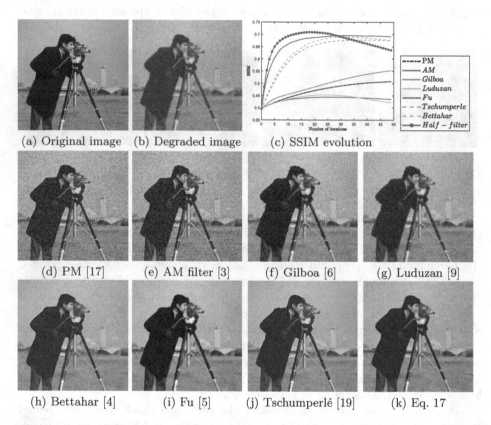

(a) Original image   (b) Degraded image   (c) SSIM evolution

(d) PM [17]   (e) AM filter [3]   (f) Gilboa [6]   (g) Luduzan [9]

(h) Bettahar [4]   (i) Fu [5]   (j) Tschumperlé [19]   (k) Eq. 17

Fig. 8. Restoration a blurred ($\sigma = 1$) and noisy image, 272×272, in (b), SNR = 13.1 dB.

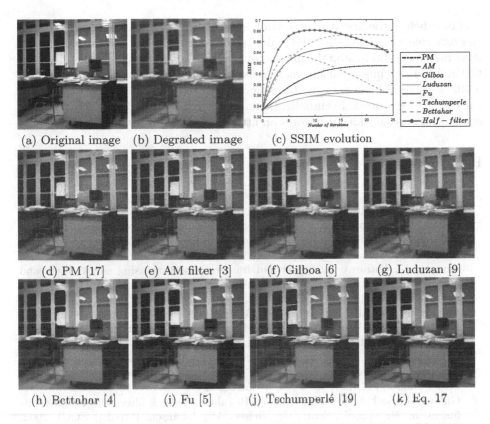

(a) Original image   (b) Degraded image   (c) SSIM evolution

(d) PM [17]   (e) AM filter [3]   (f) Gilboa [6]   (g) Luduzan [9]

(h) Bettahar [4]   (i) Fu [5]   (j) Tschumperlé [19]   (k) Eq. 17

Fig. 9. Restoration a blurred ($\sigma = 1$) and noisy image, 256×256, in (b), SNR $=$ 13.53 dB.

5 Conclusion

A new shock-filter for digital image enhancement has been presented in this paper. The principal characteristic of this filter is the use of half Gaussian kernels to detect a gradient modulus. This kind of kernels allows the filter to detect corners and their associated directions with accuracy. Furthermore, the shock-diffusion strategy depends on the value of the gradient modulus and the two directions of the estimated edges. By applying a mirror rotation of these directions, a new diffusion process via PDE is proposed and improves the shock-filter efficiency. Experiments on blurred and noisy images show the efficiency in removing blur and noise while enhancing important features like edges and also corners or small objects. The presented experimental results and images resulting of this PDE tied to the best score of the SSIM illustrates the relevance of the proposed image enhancement method. Future works will include automatic diffusion that would lead to an unsupervised restoration. In other words, it constrains to stop the PDE after a number of iterations allowing the enhancement of an image. This is doable by learning or statistic approaches estimating the level of noise and blur before the first iteration. On the other hand, an unsupervised measure (for example a method estimating if flat regions are really flat) on the restoration level may be useful to stop the diffusion process. The adaptation of the proposed method to color image enhancement remains a challenge without creating color artifacts, several manners must be explored for the diffusion process and overall the deblurring process. Eventually, this paper is not a breakthrough, but it adds some contribution, since it devotes an attention to the diffusion directions.

References

1. Abdulrahman, H., Magnier, B., Montesinos, P.: Oriented asymmetric kernels for corner detection. In: IEEE EUSIPCO, pp. 778–782 (2017)
2. Aubert, G., Kornprobst, P.: Mathematical Problems in Image Processing: Partial Differential Equations and the Calculus of Variations, vol. 147. Springer, New York (2006).https://doi.org/10.1007/978-0-387-44588-5
3. Alvarez, L., Mazorra, L.: Signal and image restoration using shock filters and anisotropic diffusion. SIAM J. Numer. Anal. **31**(2), 590–605 (1994)
4. Bettahar, S., Lambert, P., Stambouli, A.B.: Anisotropic color image denoising and sharpening. In: IEEE ICIP, pp. 2669–2673 (2014)
5. Fu, S., Ruan, Q., Wang, W., Chen, J.: Region-based shock-diffusion equation for adaptive image enhancement. In: Huang, D.S., Li, K., Irwin, G.W. (eds.) Advances in Machine Vision, Image Processing, and Pattern Analysis. Lecture Notes in Control and Information Sciences, vol. 345, pp. 387–395. Springer, Heidelberg (2006). https://doi.org/10.1007/978-3-540-37258-5_133
6. Gilboa, G., Sochen, N.A., Zeevi, Y.Y.: Regularized shock filters and complex diffusion. In: Heyden, A., Sparr, G., Nielsen, M., Johansen, P. (eds.) ECCV 2002. LNCS, vol. 2350, pp. 399–413. Springer, Heidelberg (2002). https://doi.org/10.1007/3-540-47969-4_27
7. Guichard, F., Moisan, L., Morel, J.-M.: A review of PDE models in image processing and image analysis. J. Phys. IV **12**(1), 137–154 (2002)
8. Kornprobst, P., Deriche, R., Aubert, G.: Image coupling, restoration and enhancement via PDE's. In: IEEE ICIP, pp. 458–46 (1997)
9. Ludusan, C., Lavialle, O., Terebes, R., Borda, M.: Morphological sharpening and denoising using a novel shock filter model. In: Elmoataz, A., Lezoray, O., Nouboud, F., Mammass, D., Meunier, J. (eds.) ICISP 2010. LNCS, vol. 6134, pp. 19–27. Springer, Heidelberg (2010). https://doi.org/10.1007/978-3-642-13681-8_3
10. Magnier, B.: An objective evaluation of edge detection methods based on oriented half kernels. In: Mansouri, A., El Moataz, A., Nouboud, F., Mammass, D. (eds.) ICISP 2018. LNCS, vol. 10884, pp. 80–89. Springer, Cham (2018). https://doi.org/10.1007/978-3-319-94211-7_10
11. Magnier, B.: Matlab code of "Edge detection methods based on oriented half kernels". https://fr.mathworks.com/matlabcentral/fileexchange/66853-edge-detection-methods-based-on-oriented-half-kernels?s_tid=srchtitle
12. Magnier, B., Montesinos, P., Diep, D.: Texture removal by pixel classification using a rotating filter. In: IEEE ICASSP, pp. 1097–1100 (2011)
13. Magnier, B., Montesinos, P.: Evolution of image regularization with PDEs toward a new anisotropic smoothing based on half kernels. In: IS&T/SPIE Electrical Imaging, International Society for Optics and Photonics, p. 86550M (2013)

14. Montesinos, P., Magnier, B.: A new perceptual edge detector in color images. In: Blanc-Talon, J., Bone, D., Philips, W., Popescu, D., Scheunders, P. (eds.) ACIVS 2010. LNCS, vol. 6474, pp. 209–220. Springer, Heidelberg (2010). https://doi.org/10.1007/978-3-642-17688-3_21

15. Montesinos, P., Magnier, B.: Des filtres anisotropes causaux pour une diffusion non contrôlées. In: GRETSI (2017)

16. Osher, S., Rudin, L.I.: Feature-oriented image enhancement using shock filters. SIAM J. Numer. Anal. **27**(4), 919–940 (1990). ISSN 0036-1429

17. Perona, P., Malik, J.: Scale-space and edge detection using anisotropic diffusion. IEEE TPAMI **12**, 629–639 (1990)

18. Püspöki, Z., Martin, S., Sage, D., Unser, M.: Transforms and operators for directional bioimage analysis: a survey. In: De Vos, W., Munck, S., Timmermans, J.P. (eds.) Focus on Bio-Image Informatics, vol. 219, pp. 69–93. Springer, Cham (2016). https://doi.org/10.1007/978-3-319-28549-8_3

19. Tschumperlé, D., Deriche, R.: Diffusion PDE's on vector-valued images: local approach and geometric viewpoint. IEEE Signal Process. Mag. **19**(5), 16–25 (2002)

20. Venkatrayappa, D., Montesinos, P., Diep, D., Magnier, B.: A novel image descriptor based on anisotropic filtering. In: Azzopardi, G., Petkov, N. (eds.) CAIP 2015. LNCS, vol. 9256, pp. 161–173. Springer, Cham (2015). https://doi.org/10.1007/978-3-319-23192-1_14

21. Wang, Z., Bovik, A., Sheikh, H., Simoncelli, E.: Image quality assessment: from error visibility to structural similarity. IEEE TIP **13**(4), 600–612 (2004)

22. Weickert, J.: Coherence-enhancing shock filters. In: Michaelis, B., Krell, G. (eds.) DAGM 2003. LNCS, vol. 2781, pp. 1–8. Springer, Heidelberg (2003). https://doi.org/10.1007/978-3-540-45243-0_1

Scanner Model Identification of Official Documents Using Noise Parameters Estimation in the Wavelet Domain

Chaima Ben Rabah[1,2]($\boxtimes$), Gouenou Coatrieux[2], and Riadh Abdelfattah[1,2]

[1] COSIM Lab, University of Carthage, Higher School of Communications of Tunis, El Ghazala city Ariana 2083, Ariana, Tunisia
chaima.benrabah@supcom.tn
[2] LaTIM Inserm UMR1101, IMT Atlantique, Technopôle Brest-Iroise, CS 83818, 29238 Brest Cedex 3, France

Abstract. In this article, we propose a novel approach for discerning which scanner has been used to scan a particular document. Its originality relates to a signature extracted in the wavelet domain of the digitized documents where the acquisition noise specific to a scanner is located in the first subbands of details. This signature is an estimate of the statistical noise model which is modeled by a General Gaussian distribution (GGD) and whose parameters are estimated in the HH subband by maximizing the likelihood function. These parameters constitute a unique identifier for a scanner. For a given image, we propose to identify its origin by minimizing the Kullback-Leibler divergence between its signature and those of known scanners. Experiments conducted on a real scanned-image database, developed for the validation of the work presented in this paper, show that the proposed approach achieves high detection performance. Total of 1000 images were used in experiments.

Keywords: Digitized documents · Scanner identification
Image forensics · Authenticity · Wavelet transform

1 Introduction

The use of scanners as an interface between the real world and the digital world makes it easier and faster to store and share documents. It reduces costs and will not disappear any time during the foreseeable future [1]. Companies, banks and many other institutions rely on digitization to save hundreds of thousand dollars annually in printing, shipping and labor costs. However, if access to information is facilitated, security issues are meanwhile increased. Communication networks such as Internet allow to send a digital document more easily than its paper version. Therefore, questions can be raised about the reliability of a scanned document especially when it is considered as proof (e.g. contract) or as a certificate (e.g. medical certificate).

© Springer Nature Switzerland AG 2018
J. Blanc-Talon et al. (Eds.): ACIVS 2018, LNCS 11182, pp. 598–608, 2018.
https://doi.org/10.1007/978-3-030-01449-0_50

Various tools have been proposed to provide proof of the origin of a document like digital signatures [2], Watermarking [3] and digital content forensics [4].

The first ones allow a person who scanned a document to 'sign' it. The signature protects the document in terms of integrity (any change will result in a different signature) and also non-repudiation. However, once the signature is deleted, it is difficult to ensure the integrity of the data and its origin. Watermarking is an alternative. It consists in hiding a message in the scanned document to prove its integrity and authenticity, by modifying it imperceptibly. The message is hidden in the pixels of the image. One of the drawbacks of this solution is that the watermark should be embedded in the document during its acquisition.

Digital content forensics [4] demonstrates an increasing interest in this field. It seeks to detect if an image has been modified or to prove its origin (i.e. Identification of the system that acquired it) without prior information of the image. In other words, no signature or other ancillary data is shared with the image contrary to the other techniques.

Digital forensics techniques rely on the analysis of traces intrinsically tied to the digital image by the acquisition device to form a fingerprint that can be used either to identify its origin or to detect changes.

Several methods are available for source scanner identification of a particular test image. These methods can be divided into five classes. The main idea of the first class aim at extracting significant features from scanned images and use correlation [5] or machine learning techniques [6–8] for classification. However, the difference between methods lies in the feature selection process. In fact, the acquisition process of scanner leaves some specific noises due to imperfect manufacture of the various components that constitute it. These noises have unique statistical characteristics for a scanner and introduce the same noisy pattern into the images [9]. However, this type of approaches suffers from the fact that such noise is almost absent in the saturated regions of an image. So, it may not work properly for text documents. This is why Khanna et al. [10] developed an approach that uses texture analysis to identify the scanner used to scan a text document. The proposed features used are Gray-Level Co-Occurrence Matrix (GLCM) features from groups of 'e' s and GLCM features and isotropic Gray-Level Difference Histogram (GLDH) features from blocks of the image. This method may fail to work for different font shapes and sizes. Therefore, Joshi et al. [11] present methods for authenticating scanned text documents using Local tetra patterns based features which succeeded to overcome this weakness. The third class includes methods that exploit as a single signature of a scanner the positions of dust and scratches or other debris that leave traces on the scanner stage and are measurable from the acquired image [12,13]. This type of footprint fails to evolve over time since traces may disappear or appear. A recent method [14] for scanner model identification which we will assign to the fourth category is based on the characterization of differences between holograms. First, it identifies capturing conditions by measuring correlations of spatial distribution of brightness. Then, the classification of scanners is made with correlations of color distributions.

The last type of approaches was introduced by Choi *et al.* [15]. This method introduces a new type of noise related to scanners called spectral noise which can be seen only in the frequency domain. They made a line reference pattern for each scanner and compared it with reference patterns of test images using the Euclidean distance. To enhance characteristics of this special noise, a refining process is applied to scanned images. This method uses the Euclidean distance to compare each reference pattern with the spectral noise from test images.

Our approach belongs to the last category. The solution that we propose differs from the method of Choi *et al.* [15] in two essential points.

First, we work in the wavelet domain and take advantage of the fact that the subbands of details of the first level of decomposition of an image comprise the noise related to scanning. In each subband, the wavelet coefficients follow a generalized Gaussian distribution whose parameters (position, scale factor, etc.) can constitute a signature. Another point of differentiation is how we identify a scanner based on the Kullback-Leibler divergence, an information-based measure of disparity among probability distributions which is suitable in our case.

The rest of the paper is organized as follows. Section 2 presents the scanner architecture, how it works as well as the noise that can result from sensor defects. In Sect. 3, we describe the method that we proposed. Experimental results are shown in Sec. 4. Section 5 draws the concluding remarks and the next challenges.

2   Scanner and Sensor Noise

2.1   Architecture and Operation

The basic principle of a scanner is to convert into a digital format a document which can be a color/grayscale image, a black and white text document or a composition of them.

The architecture of such a system varies a little from one device to another. Figure 1 illustrates the basic components constituting a scanner. The document to be scanned is placed over the glass plate and then isolated from the outside light by a cover (except in case of bulky object - e.g. book). Under the glass, mirrors, lenses and the charge-coupled device (CCD) image sensors are attached to a stabilizer bar which, driven by a motor, will scan the document at a steady pace. When scanning, the lamp illuminates the document and the light is reflected by a set of inclined mirrors. Each mirror is slightly curved so as to focus the image it reflects on a smaller surface. The image reflected by the last mirror is concentrated by a lens in a filter before meeting the CCD sensors. The latter converts the light into electrical signals according to their intensity in order to be able to digitize the document using an analog-to-digital converter (ADC). The digital document is then transmitted to a computer and may receive some post-processing operations.

2.2   Sensor Noise

Scanners can be identified using an inherent mark of scanning: the noise effect which is due to its imperfect manufacturing process. Two types of noise are distinguished in scanned documents. The first one called the chrominance noise results from the defects of the CCD sensors. It includes point defects, pixels or groups of gray-scale pixels saturated (black or white), as well as quantization, shot and thermal noise [16]. The second type of noise corresponds to the noise of luminance, it is also called 'Fixed Pattern Noise' (FPN), and refers to spatial patterns that do not change significantly from one image to another.

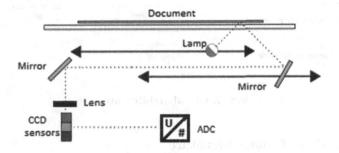

Fig. 1. Scanner components

This noise consists of two sub-components: the DSNU (for 'Dark Signal Non-Uniformity') and the PRNU (for 'Photo Response Non-Uniformity'). DSNU is defined as nonuniformity in the dark image. It can be measured by subtracting the dark image of the scanner, i.e. the image obtained without activating the scanner's light source [17]. The PRNU models the light responses of pixels.

3   Proposed Method

The main goal of the proposed solution is to authenticate digital documents by associating the document with its source device. In scanner identification techniques, the input is the questioned document and the output is a decision indicating the scanner model that acquired it.

In this paper, we propose a method for estimating the noise of the sensor as a whole in the wavelet domain in order to identify the scanner at the origin of a scanned document. Since, in practice, most documents scanned are official documents which contain mostly text with different font sizes and colors and may also contain pictures like logo and forms, we focus on these documents and show that our method is independent of the content of the digitized document. We will summarize below the process of extracting the signature of a document, a signature that will be used later in a scanner identification scheme as illustrated in Fig. 2.

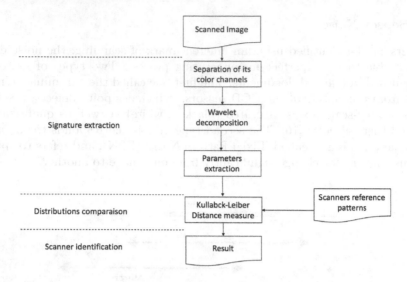

Fig. 2. Global architecture

3.1   Extraction of Image Signature

This step consists of extracting a signature from a given document, a characteristic signature of the scanner that captures it as summarized in Fig. 3.

Initially, the first decomposition level of the wavelet transform of each color channel (Red R, Green G, Blue B) of a given image I is calculated to get 4 subbands: Low-Low (LL), Low-High (LH), High-Low (HL), and High-High (HH).

The proposed method is applied to the RGB colormap since the scanner uses individual R, G and B sensors to scan the document. In this work, we choose to use the Symlet wavelet which showed good performance in features extraction compared to the other wavelet families particularly for printed documents.

The features of this wavelet make it a good candidate as a denoising tool [18]. In order to constitute our signature, we rely on the fact that the distribution of wavelet coefficients in a subband of details is close to a generalized Gaussian [19]:

$$f(x) = \frac{\beta}{2\alpha\Gamma(\frac{1}{\beta})} \exp\left[-(\frac{|x-\mu|}{\alpha})^{\beta}\right] \tag{1}$$

where α, β, μ and $\Gamma(.)$ represent respectively the parameters of scale, shape, location and the known Gamma function.

We propose to estimate these parameters for the HH subband since the wavelet coefficients in this subband are dominated by noise which is none other than sensor noise while the other subbands contains more image details and less noise. These parameters will constitute our signature.

Motivated by our observations, and to decrease the number of parameters to be estimated, we simply approximate the value of μ with zero. As a result, the α and β parameters suffice to characterize the distribution of noise in a subband.

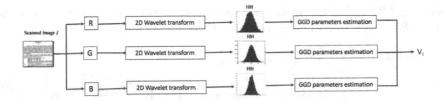

Fig. 3. Signature construction for a query document

We propose to calculate these parameters using the maximum likelihood estimation (MLE) method with the Newton-Raphson's iterative algorithm [20].

On this basis, we obtain a pair of estimated parameters per color channel, for the HH wavelet subband.

The characteristic vector of noise V_I of a document have this form:

$$V_I = \begin{bmatrix} V_I^R \\ V_I^G \\ V_I^B \end{bmatrix} = \begin{bmatrix} \left(\hat{\alpha}_R^{HH}, \hat{\beta}_R^{HH}\right) \\ \left(\hat{\alpha}_G^{HH}, \hat{\beta}_G^{HH}\right) \\ \left(\hat{\alpha}_B^{HH}, \hat{\beta}_B^{HH}\right) \end{bmatrix} \tag{2}$$

In our approach, the scanner signature is the median of the signatures of several images scanned with it. The identifier of a scanner S from the signatures of N images is then such that:

$$\begin{aligned} V_S^R &= median(V_{S,i}^R, i = 1..N) \\ V_S^G &= median(V_{S,i}^G, i = 1..N) \\ V_S^B &= median(V_{S,i}^B, i - 1..N) \end{aligned} \tag{3}$$

where $V_{S,i}^R$, $V_{S,i}^G$ and $V_{S,i}^B$ are the features vectors of the HH subband of the R, G and B color channels of the ith image scanned with the scanner S, respectively.

3.2 Scanner Model/Brand Identification

Identifying the origin of a digitized document is based on the comparison between the signature of several scanners and the signature extracted from a questionned document.

Here we made the choice to use the Kullback-Leibler divergence (KLD) [21].

It is an appropriate measure for comparing statistical distribution, with parameters which are images signatures in our case.

The 'source' scanner of a document will be the one that minimizes this measure. Since the coefficients of the detail subbands of the first level of the wavelet decomposition of an image follows generalized Gaussian distribution, the KLD of the color channel j is given by

$$KLD(\alpha_Q^j, \alpha_S^j, \beta_Q^j, \beta_S^j) = \log\left(\frac{\beta_S^j \alpha_Q^j \Gamma\left(\frac{1}{\beta_Q^j}\right)}{\beta_Q^j \alpha_S^j \Gamma\left(\frac{1}{\beta_S^j}\right)}\right) + \left(\frac{\alpha_S^j}{\alpha_Q^j}\right)^{\beta_Q^j} \frac{\Gamma\left(\frac{\beta_Q^j+1}{\beta_S^j}\right)}{\Gamma\left(\frac{1}{\beta_S^j}\right)} - \frac{1}{\beta_S^j} \tag{4}$$

where $j \in \{R, G, B\}$. The global distance between the query image Q and the scanner S is given by summing along the three color channels

$$
\begin{aligned}
\text{KLD}_{Q,S} &= \sum_{j=1}^{3} KLD(\alpha_Q^j, \alpha_S^j, \beta_Q^j, \beta_S^j) \\
&= \sum_{j=1}^{3} \left[\log\left(\frac{\beta_S^j \alpha_Q^j \Gamma\left(\frac{1}{\beta_Q^j}\right)}{\beta_Q^j \alpha_S^j \Gamma\left(\frac{1}{\beta_S^j}\right)} \right) + \left(\frac{\alpha_S^j}{\alpha_Q^j}\right)^{\beta_Q^j} \frac{\Gamma\left(\frac{\beta_Q^j+1}{\beta_S^j}\right)}{\Gamma\left(\frac{1}{\beta_S^j}\right)} - \frac{1}{\beta_S^j} \right]
\end{aligned}
\tag{5}
$$

4 Experimental Evaluation

In order to empirically evaluate the effectiveness of our proposed method, we tested it on five different scanners of various native resolutions as shown in Table 1.

Table 1. Scanners used in experiments

Id	Model \ Brand	Resolution
S1	Canon Lide 220	4800×4800
S2	Epson Perfection V39	4800×4800
S3	Epson Perfection V370 Photo	4800×9600
S4	Epson Perfection V550 Photo	6400×9600
S5	HP Scanjet Pro 2500 F1	1200×1200

First, we created a test database of 100 documents of varied content scanned with each scanner. These documents contains black and white text or colored text. Some of them includes figures and tables. The digital images obtained were saved in TIFF format, that is to say in a format of raw data without loss of information, with a resolution of 300 dpi in order to constitute a database of 500 images. This resolution is the commonly used in practice [22]. 50 out of 100 images were chosen randomly for generating each scanner fingerprint, the remaining 50 images were used to evaluate the detection performance.

The training and testing phases are repeated 5 times to obtain the final performance measures.

We give in Fig. 4, the distributions of the couples $(\widehat{\alpha}_R^{HH}, \widehat{\beta}_R^{HH})$ of the HH subband of the red color channel for all test images.

It can be seen that these parameters discriminate well the images according to the scanner that acquired it which proves that the noise is different from one scanner to another.

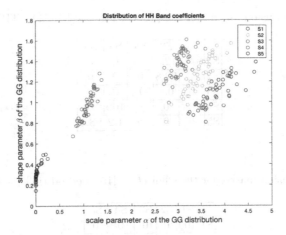

Fig. 4. Distribution of couples $(\widehat{\alpha}_R^{HH}, \widehat{\beta}_R^{HH})$ of the HH subband of the red color channel for different scanners (Color online figure)

We study the performance in terms of classification accuracy by measuring the percentage of images from scanner model correctly identified. We obtain a matrix as shown in Table 2 and a total accuracy 95.6%. Notice also that these detection results remain the same regardless of the content of the scanned document.

Moreover, the proposed method shows greater accuracy in discriminating different scanners comparing to the last and the most powerful existing method [15] which uses spectral noise in the frequency domain.

As it can be seen in Table 3, Choi *et al.* [15] method is not able to correctly discriminate scanners of the same brand.

In the following, we investigate the effect of lossy compression on the reliability of our scanner source identification technique as it is likely that in practice scanned documents will be stored in the JPEG format. Another dataset consisting of single JPEG compressed images is created by compressing the uncompressed previously generated dataset with a quality factor QF equal to 75. Table 4 shows average accuracy of the dedicated classifier for scanned documents saved in JPEG format.

The proposed method maintains an average classification accuracy of 67.6% despite the decline in performance. This demonstrates that our identification method is robust to JPEG compression.

The relative decline of accuracy in this experiments is due to the introduction of the JPEG quantization noise in the HH subband which deviates significantly the sensor noise characteristics from the original one. The introduced noise can be suppressed by tresholding the wavelet coefficients. We consider this issue as a future work.

Table 2. Confusion matrix for the proposed method (in%) - TIFF images

	S1	S2	S3	S4	S5
S1	87.6	12.4	0	0	0
S2	7.2	92.8	0	0	0
S3	0	0.8	99.2	0	0
S4	0	0	0	100	0
S5	0	0	0	1.2	98.8

Table 3. Confusion matrix for the Choi *et al.* [15] method (in%) - TIFF images

	S1	S2	S3	S4	S5
S1	100	0	0	0	0
S2	0	1.2	0	98.8	0
S3	0	0	100	0	0
S4	0	0.8	0	99.2	0
S5	0	0	0	0	100

Table 4. Confusion matrix for the proposed method (in%) - JPEG (QF = 75) images

	S1	S2	S3	S4	S5
S1	44.4	31.6	8	17.2	0
S2	3.2	71.6	25.2	0	0
S3	2.4	20	75.6	6	0
S4	18.8	0	0.8	80.4	0
S5	0	0	0	13.6	86.4

5 Conclusion

In this article, we presented a new approach to identify the scanner that was used to scan a document. We rely on a signature extracted from the wavelet decomposition of the image which highlights the statistical properties of the scanning noise of a scanner. As shown, the statistical properties of the scanning noise in the wavelet domain is specific for each scanner. The experimental results show a good identification accuracy independently of the content of the scanned documents. As future work, we will investigate the performance of using our proposed method against several post processes such as, contrast stretching and sharpening. We will also focus on discriminating higher number of scanners and particularly scanners of the same make and model.

Acknowledgements. This work was financially supported by the "PHC Utique" program of the French Ministry of Foreign Affairs and Ministry of higher education and research and the Tunisian Ministry of higher education and scientific research in the CMCU project number 17G1405.

References

1. de Laubier, C.: La difficile quête du zéro papier. http://www.lemonde.fr/economie/article/2017/09/03/la-difficile-quete-du-zero-papier_5180409_3234.html
2. Ferguson, N., Schneier, B., Kohno, T.: Cryptography Engineering: Design Principles and Practical Applications. Wiley, Hoboken (2011)
3. Qadir, M.A., Ahmad, I.: Digital text watermarking: secure content delivery and data hiding in digital documents. IEEE Aerosp. Electron. Syst. Mag. **21**(11) (2006)
4. Swaminathan, A., Min, W., Ray Liu, K.J.: Digital image forensics via intrinsic fingerprints. IEEE Trans. Inf. Forensics Secur. **3**(1), 101–117 (2008)
5. Gloe, T., Franz, E., Winkler, A.: Forensics for flatbed scanners, in security, steganography, and watermarking of multimedia contents IX. Int. Soc. Opt. Photonics **6505**, 65051I (2007)
6. Gou, H., Swaminathan, A., Min, W.: Robust scanner identification based on noise features. In: Security, Steganography, and Watermarking of Multimedia Contents IX. International Society for Optics and Photonics, vol. 6505, p. 65050S (2007)
7. Gou, H., Swaminathan, A., Min, W.: Intrinsic sensor noise features for forensic analysis on scanners and scanned images. IEEE Trans. Inf. Forensics Secur. **4**(3), 476–491 (2009)
8. Khanna, N., Mikkilineni, A.K., Delp, E.J.: Scanner identification using feature-based processing and analysis. IEEE Trans. Inf. Forensics Secur. **4**(1), 123–139 (2009)
9. Khanna, N., Mikkilineni, A.K., Chiu, G.T.C., Allebach, J.P., Delp, E.J.: Scanner identification using sensor pattern noise. In: Security, Steganography, and Watermarking of Multimedia Contents IX. International Society for Optics and Photonics, vol. 6505, p. 65051K (2007)
10. Khanna, N., Delp, E.J.: Source scanner identification for scanned documents. In: First IEEE International Workshop on Information Forensics and Security, WIFS 2009, pp. 166–170. IEEE (2009)
11. Joshi, S., Gupta, G., Khanna, N.: Source classification using document images from smartphones and flatbed scanners. In: Rameshan, R., Arora, C., Dutta Roy, S. (eds.) NCVPRIPG 2017. CCIS, vol. 841, pp. 281–292. Springer, Singapore (2018). https://doi.org/10.1007/978-981-13-0020-2_25
12. Dirik, A.E., Sencar, H.T., Memon, N.: Flatbed scanner identification based on dust and scratches over scanner platen. In: IEEE International Conference on Acoustics, Speech and Signal Processing, ICASSP 2009, pp. 1385–1388. IEEE (2009)
13. Elsharkawy, Z.F., Abdelwahab, S.A., Dessouky, M.I., Elaraby, S.M., Abd El-Samie, F.E.: Identifying unique flatbed scanner characteristics for matching a scanned image to its source. Digit. Image Process. **5**(9), 397–403 (2013)
14. Sugawara, S.: Identification of scanner models by comparison of scanned hologram images. Forensic Sci. Int. **241**, 69–83 (2014)
15. Choi, C.-H., Lee, M.-J., Lee, H.-K.: Scanner identification using spectral noise in the frequency domain. In: 2010 17th IEEE International Conference on Image Processing (ICIP), pp. 2121–2124. IEEE (2010)

16. Findlater, K.M., et al.: A CMOS image sensor with a double-junction active pixel. IEEE Trans. Electron Devices **50**(1), 32–42 (2003)
17. Khanna, N., et al.: A survey of forensic characterization methods for physical devices. Digit. Investig. **3**, 17–28 (2006)
18. Daubechies, I.: Ten Lectures on Wavelets, vol. 61. SIAM (1992)
19. Chang, S.G., Yu, B., Vetterli, M.: Adaptive wavelet thresholding for image denoising and compression. IEEE Trans. Image Process. **9**(9), 1532–1546 (2000)
20. Verbeke, J., Cools, R.: The newton-raphson method. Int. J. Math. Educ. Sci. Technol. **26**(2), 177–193 (1995)
21. Kullback, S., Leibler, R.A.: On information and sufficiency. Ann. Math. Stat. **22**(1), 79–86 (1951)
22. Cochran, M.: A proposed standard procedure to define minimum scanning attribute levels for hard copy documents. In: 2014 47th Hawaii International Conference on System Sciences (HICSS), pp. 2036–2043. IEEE (2014)

Relocated Colour Contrast Occurrence Matrix and Adapted Similarity Measure for Colour Texture Retrieval

Hela Jebali[1]([✉]), Noel Richard[2]([✉]), Hermine Chatoux[2], and Mohamed Naouai[1]

[1] Faculty of Science of Tunis, University campus el Manar, Tunis, Tunisia
jebalii.hela@gmail.com
[2] University of Poitiers, XLIM UMR CNRS 6172, Poitiers, France
noel.richard@univ-poitiers.fr

Abstract. For metrological purposes, distance between texture images is crucial. This work study as a pair the couple texture feature/similarity measure. Starting from the Colour Contrast Occurrence Matrix (C_2O) definition, we propose an adapted similarity measure improving texture retrieval. In a second step, we propose a modified version of the C_2O definition including the texture's colour average inside a modified similarity measure. Performance in texture retrieval is assessed for four challenging datasets: Vistex, Stex, Outex-TC13 and KTH-TIPS2b databases facing to the recent results from the state-of-the-art. Results show the high efficiency of the proposed approach based on a simple pair feature/similarity measure facing to more complex approaches including Convolutional Neural Networks.

1 Introduction

Texture provides a semantic description to characterize the similarity between images. Although, different efficient texture descriptors have been developed in the literature. *Cooccurrence* matrix and *Run-Length Matrix* (RLM) [28] are good descriptors of texture content, but are not suitable for direct use with a distance or similarity measure. The *Local Binary Pattern* [21] attribute retains only a very reduced part of texture information about spatial-frequency composition. Yet, it is suitable for a distribution operation with matching metrics of similarities. Nevertheless the pair attribute and similarity measure is very efficient for texture classification or discrimination.

The transition to colour images is not straightforward with the vector aspect of the pair descriptor/measure of similarity. Marginal and/or crossed marginal (channel by channel) approaches can produce simplest descriptors but sliced the difficulty of the vector data management to the similarity measure function. It exists few vector expressions of texture features, among them the *Colour Contrast Occurrence* (C_2O) matrix [22] proposes a vector descriptor producing a distribution of colour differences present in an image or region of interest. In previous work, we shown the metrological potential of the C_2O [22], but in the

© Springer Nature Switzerland AG 2018
J. Blanc-Talon et al. (Eds.): ACIVS 2018, LNCS 11182, pp. 609–619, 2018.
https://doi.org/10.1007/978-3-030-01449-0_51

same time the performances in texture classification were not as expected. In this work, we study the impact of the similarity metric on classification performance, and we propose an improved version of the C_2O definition more adapt to texture image retrieval in databases.

In the following, we first describe the Colour Contrast Occurrence (C_2O) construction (Sect. 2). Then, we develop the question of similarity measurement between trichromatic distributions (Sect. 3). Considering the invariance of the C_2O to the average lightning change, we propose the Relocated Colour Contrast Occurrence matrix (RC_2O) and an adapted expression of the similarity (Sect. 3.4). In a first experiment, similarity measure performances are studied from some images of classic image databases (KTH, Stex). Then we compare performances of the RC_2O using the proposed similarity metrics with a recent of state of the art approaches for Vistex, Stex, Outex-TC13 and KTH-TIPS2b databases (Sect. 4.2).

2   Color Contrast Occurrence Matrix

The C_2O texture feature is inspired from the first Julesz conjecture[1] [12]. The initial idea was to define a compact probability density function adapted to direct distance/similarity measurements between features.

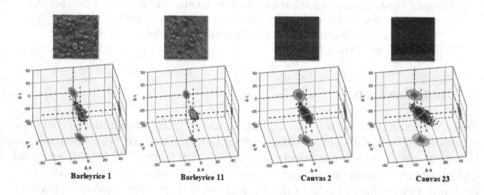

Fig. 1. Some color textures from Outex with their C_2O distributions ($d = 1, \theta = 0$) (Color figure online)

The colour contrast occurrence matrix proposed in [17] expresses the probability $\Lambda(\overrightarrow{c_i c_j})$ to find a specific local colour difference $\overrightarrow{c_i c_j} = \overrightarrow{\chi}$. This probability is assessed between two pixels separated by a vector defined by a spatial distance d and an orientation θ. The colour difference is defined in spherical coordinates

---

[1] The first Julesz conjecture states that preattentive discrimination of textures is possible only for textures that differ on second-order correlation statistics.

by a perceptual color difference measurement $\Delta E_{\overrightarrow{\chi}}$ established in CIELAB by two angles (α, β).

$$\overrightarrow{\Lambda(c_i, c_j)} : prob(\overrightarrow{\Lambda(c_i, c_j)} = \overrightarrow{\Lambda_\chi})$$
$$\text{With } \| \overrightarrow{\Lambda(c_i, c_j)} \| = \Delta E_\chi \tag{1}$$
$$\text{and } \llcorner(\overrightarrow{Oa}, \overrightarrow{c_i c_j}) = (\alpha, \beta)$$

Concretely, the C_2O matrix is a probability density function expressed in a $(\Delta_L, \Delta_a, \Delta_b)$ space. When texture is stationary, the C_2O matrix is centred at the origin in a dense distribution (examples in Fig. 1).

3 Similarity Measure Between C_2O matrix

3.1 Indirect Measurement

The main objective of the C_2O construction was to obtain a compact 3-dimensional probability distribution functions associated to an adapted distance or similarity measure. The initial construction in [17] was associated to a quantification using a spiral path in order to obtain a texture signature $Sig_{C_2O}(I)$ (Eq. 2 and Fig. 2). Then the distance between two textures is obtained using a L_2 norm between texture signatures.

$$Sig_{C_2O}(I) = \Delta_i \alpha_j \beta_k$$
$$= prob(\Delta_i \leq \| \overrightarrow{\Lambda(c_i, c_j)} \| < \Delta_j + \Delta E_{step})$$
$$\text{With} \quad \frac{\pi}{2n_\alpha}(j) \leq \alpha < \frac{\pi}{2n_\alpha}(j) \tag{2}$$
$$\text{and} \quad 0 \leq \beta < \frac{\pi}{n_\beta}(k+1)$$

where ΔE_{step}, n_α and n_β represent respectively the contrast norm, chromatic and lightning steps.

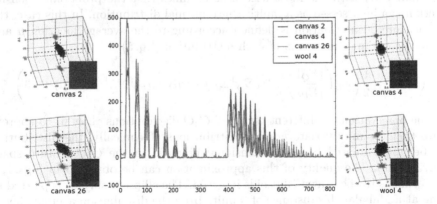

Fig. 2. C_2O signature proposed in [17] for some Outex textures. The signature is not compact and present periodic patterns due to the distribution's quantification using a spiral path from the distribution center (Eq. 2).

3.2   Measurement Using Kullback Leibler Divergence

Considering C_2O matrix as a probability density function, *f-divergence* are well adapted to assess the similarity between them. Inside this group, the Kullback-Leibler divergence is the most used. In [8], the authors analysed and compared several similarity functions between colour distributions. In this work, they shown the Kullback-Leibler divergence takes into account all colour changes between the two considered distributions, whatever their magnitude. This divergence is also used for texture discrimination as in [18] using features based on energy from multi-scale Gabor-filters.

The Kullback-Leibler measure of information $KL(P/Q)$ assess the quantity of information lost when Q is used to estimate P (Eq. 3) [14].

$$KL(P/Q) = \int_{-\infty}^{\infty} p(x) \ln(\frac{p(x)}{q(x)}) dx \qquad (3)$$

where p and q are the respective densities of P and Q. In order to define a similarity measure, the Kullback-Leibler divergence is defined as the sum of the measure of information of P relative to Q and the measure of information of Q relative to P.

$$div_{KL}(P, Q) = KL(P/Q) + KL(Q/P) \qquad (4)$$

The direct application of the Kullback-Leibler divergence is costly. In order to solve this problem, Wang and Goldberger proposed methods accelerating this process but loosing in precision [10, 27].

3.3   Estimation of the KL-Divergence

Due to the difficulty and cost of a direct processing, adaptations of the Kullback-Leibler measure of information were proposed for different types of distributions. For example, Mathiasan [18] proposed a dedicated estimation for Gamma distribution. Obviously, the most used case assumes that the probability density function can be assessed as a multivariate normal distribution. In this case, the KL measure of information is defined according to the average location μ and variance-covariance matrix Σ of each distribution (Eq. 5).

$$KL(P/Q) = \frac{1}{2}\left(\log\left(\frac{|\Sigma_Q|}{|\Sigma_P|}\right) + tr(\Sigma_Q^{-1}\Sigma_P) + (\mu_Q - \mu_P)^t \Sigma_Q^{-1}(\mu_Q - \mu_P) - 3 \right) \quad (5)$$

The observations of different textures' C_2O distributions allow us to approximate them as multivariate normal distributions. Technically, the distribution can be approximated by a 3-dimensional ellipsoid in the C_2O feature space $(\Delta L, \Delta a, \Delta b)$. The quality of this approximation can be observed in Figs. 1, 4 and 5 thanks to the drawn ellipsoids in green. The ellipsoid surface is plotted as being at a Mahalanobis distance of 3 units from the distribution average, hence 99.7% of the samples must be considered as being inside this ellipsoid.

3.4   Relocated Colour Contrast Occurrence

The C_2O matrix describes the internal colour variations of texture. When the texture is stationary, the C_2O is a distribution centred at the origin of the representation space. For application in Quality Control, all textures belonging to the same category have a similar colour average. But in order to assess the performance of the pair feature/similarity measure, we search to adapt the purpose to texture classification from image database. In this case, the colour average of each image is different. As shown in a lot of studies on texture discrimination, the colour average can be important for the texture retrieval or classification [25].

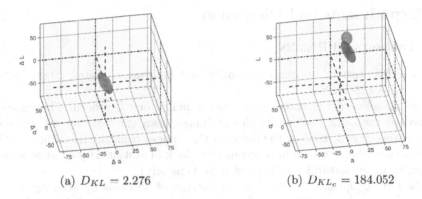

(a) $D_{KL} = 2.276$ (b) $D_{KL_c} = 184.052$

Fig. 3. Ellipsoid modelling of C_2O and the proposed RC_2O distributions $(d - 1, 0 - 0)$ from two Vistex textures (*Fabric* and *Flowers* presented in Fig. 4). As observed the RC_2O distribution will induces a bigger difference between the two textures.

We define the *Relocated Colour Contrast Occurrence Matrix* (RC_2O) as being the distribution of the C_2O relocated at the image colour disribution's average in the *CIELAB* colour space. Then the similarity measurement between two textures is assessed with the similarity between the two C_2O distributions taking into account their colour average difference.

As the colour average and the C_2O distribution are not calculated in parallel, we propose an easy way to process the similarity between the RC_2O of two textures (Eq. 6). The new measure of information, for the RC_2O, $KLc(S_1, S_2)$ is estimated from the estimation of the KL measure of information (Eq. 5).

$$KLc(S_1, S_2) = \tfrac{1}{2}\left(\log\left(\frac{|\Sigma_2|}{|\Sigma_1|}\right) + tr(\Sigma_2^{-1}\Sigma_1) \right.$$
$$\left. + (\mu_{S_1} + \mu_1 - \mu_{S_2} - \mu_2)^t \Sigma_2^{-1}(\mu_{S_1} + \mu_1 - \mu_{S_2} - \mu_2) - 3 \right) \tag{6}$$

where μ_i and Σ_i represent respectively the mean and the variance-covariance matrix of the C_2O descriptor of the image i, and μ_{S_i} represents the average coordinate of the image i in the *CIELAB* space. The similarity measure between two image thanks to the Relocated Colour Contrast Occurrence Matrix is defined by the sum of the two KL measures of informations:

$$d_{KLc} = KLc(S_1, S_2) + KLc(S_2, S_1) \tag{7}$$

Figure 3 presents the similarity assessed by the initial writing form using the C_2O distribution (Figs. 3(a) and 4(a)) or using the Relocated Colour Contrast Occurrence Matrix (Figs. 3(b) and 4(b)). As observed, the second solution induces the biggest measure of difference than the first one, only based on the differences of the local colour variations.

4   Experiments and Discussion

4.1   Dataset and Results

For experimentation, we use four challenging databases: Vistex, Stex, Outex-TC13 and KTH-TIPS2b.

The classification is based on a nearest neighbour approach. The image is affected to the same class as its closest image from the learning set. So, each image category is defined by the union of the images from the learning set. This construction allows to take into account the lack of stationarity of some images coming from the natural world (vegetation typically).

The Outex [2] database is a large collection of 68 classes. Following Arvis's process [5] for the classification, each colour image was divided into 20 sub-images of 128×128 pixels so the total of 1360 images. We take 50% of them for the training and the remaining 50% for validation.

The Fig. 1 shows some Outex textures with their C_2O distribution. Visually, we can notice the great similarity between the two classes *Barleyrice 1* and *Barleyrice 11*. This similarity is validated firstly by their C_2O distribution and then by the estimated distance $D_{KL} = 2.052$ (see Table 1). On the other hand the *Barleyrice 11* classe is far from the *Canvas 2* texture in the feature space, that is expressed by the difference value $D_{KL} = 90.435$.

Table 1. Inter and intra-class D_{KL_c} distance for images of Outex database.

Outex	Barleyrice 1	Barleyrice 11	Canvas 2	Canvas 23
Barleyrice 1	0	2.052	50.634	39.782
Barleyrice 11	2.052	0	90.435	72.330
Canvas 2	50.634	90.435	0	17.233
Canvas 23	39.782	72.330	17.233	0

Vistex test [3] is built of 54 colour images, whose initial size are 512×512 pixels, these images are acquired using uncontrolled conditions. As the Outex case, each colour image was divided into 16 sub-images of 128×128 pixels, resulting 432 images. The Fig. 4 presents some classes from Vistex with their C_2O distributions while Table 2 shows the inter and intra-class D_{KL_c} distance between them. The inter class distance is still zero while the intra-class distance varies according to the similarity between classes. The smallest distance is between the *Fabric* and the *Food* (5.025) which shows their big resemblance.

Table 2. Inter and intra-class D_{KL_c} distance for images of Vistex database.

Vistex	Fabric	Flowers	Food	Leaves
Fabric	**0**	184.052	5.025	81.305
Flowers	184.052	**0**	99.175	24.311
Food	5.025	99.175	**0**	34.327
Leaves	81.305	24.311	34.327	**0**

The Stex database (Salzburg texture images) [1] is composed by 476 color texture images, whose acquisition conditions are not defined. Each colour image is split up into 16 non-overlapping 128×128 sub-images.

The KTH-TIPS2b [7] database includes 11 different texture classes with four samples of each. This database contains in total number of 4752 imges, which captured at nine different scales, four illumination conditions and three poses. The Fig. 5 presents a Lettuce-leaf texture from the KTH-TIPS2b with different scale variations (scale 1, scale 6 and scale 9). We can show the highest similarity measurement between all C_2O distributions whose is almost stable for the two scales 6 and 9. The inter-class estimated distances are presented in Table 3, all values are very weak which confirms the similarity between these textures. Therefore we can validate the C_2O robustness to scale variation.

Table 3. Inter-class D_{KL_c} distance for Lettuce-leaf texture from KTH-TIPS2b database with different scale variation.

Lettuce-leaf	Scale 1	Scale 6	Scale 9
Scale 1	**0**	5.201	4.323
Scale 6	5.201	**0**	0.696
Scale 9	4.323	0.696	**0**

4.2   Performance in Classification

In this section, we present the obtained classification rates for our proposition based on the *Relocated Colour Contrast Occurrence* matrix RC_2O_{DKL} in comparison to results recently published (Table 4).

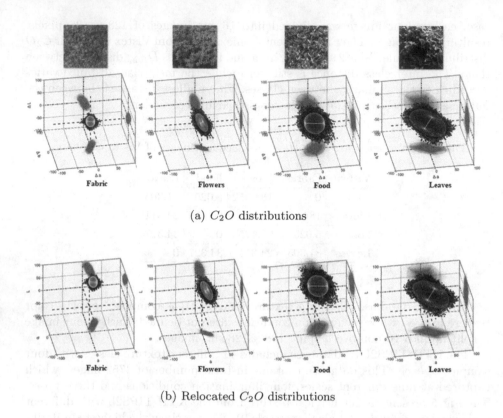

(a) C_2O distributions

(b) Relocated C_2O distributions

Fig. 4. Some color textures from Vistex with their: (a) C_2O distributions ($d = 1, \theta = 0$) and (b) Relocated C_2O distributions. (Color figure online)

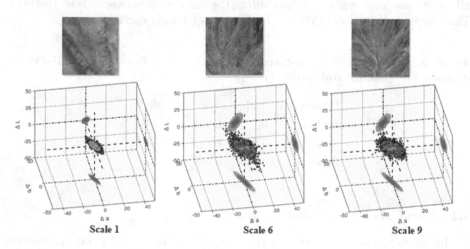

Fig. 5. Lettuce-leaf texture from KTH-TIPS2b database with their C_2O distributions ($d = 1, \theta = 0$) with different scale variation (Color figure online)

Table 4. Comparison of the classification rates (%) achieved on Vistex, Stex and Outex-TC13 and multi-scale KTH-TIPS2b databases. The proposed approach RC_2O_{KL} obtains good performances facing the initial C_2O and more complex approaches using Convolutional Neural Networks.

Methods	Vistex	Stex	Outex TC13	KTH TIPS2b
LED+ED [20]	94.70	80.08	76.67	-
CLP [23]	97.70	83.90	82.10	-
LSTM network [6]	99.09	-	94.7	-
3D-ASDH [24]	-	-	**95.8**	91.3
SMGD [16]	97.5	77.6	89.7	-
GLACI [15]	88.42	41.91	87.67	-
CCMA [11]	91.89	57.58	83.67	-
BF+CLBP+S/M+ScatNet [19]	-	-	-	78.09
FC-CNN+FV-CNN [26]	-	-	-	83.3
STD [4]	98.89	-	90.3	-
Compact DITC [13]	-	-	-	69
3D connectivity [9]	-	-	84.26	83.54
C_2O [17]	99.3	76.65	82.64	-
RC_2O_{DKL}	**100**	**87.24**	92.4	**96.21**

Firstly, it can be observed that for three of the four datasets, the RC_2O_{DKL} have the better classification rate: Vistex (100%), Stex (87.24%) and KTH-TIPS2b (96.21%). For the Outex database, the greatest classification rate is obtained by the *3D-ASDH* method (95.8%) [24] and the third highest classification rate of (92.4%) is achieved by our proposed approach after the LSTM network [6].

Secondly, we compare the gain of the proposed contribution to the initial C_2O construction. For the Outex database, the content is essentially manufactured with reduced local colour intra-variations, but important colour variations between the textures. In this case, the good-classification score gain is around 12%, in accordance with the information captured by the proposed RC_2O_{DKL}. The other image databases have more images from the natural world, local variations (so the textured aspect) are more important than in the Outex case therefore the gain is reduced but always positive.

It is interesting to note that when the image database content becomes more complex with a bigger number of images from the natural world, the different approaches present smallest good-classification scores (case of Stex image database). In this case, an important part of images is not stationary.

To finish result analysis, we also note that the good performances of the pair RC_2O feature/similarity measure are obtained compared to some approaches based on Deep-Learning solutions. These results show the interest of an accurate texture feature associated to an efficient similarity measurement function that obtain a good classification score on a basic classification scheme (1-NN).

5 Conclusion

In this paper, we proposed a new texture feature associated with an adapted similarity measure: the Relocated Colour Contrast Occurrence matrix associated to a Kullback-Leibler divergence. This texture feature characterizes the local colour difference processed in a perceptual colour space and take into account the colour average of the texture. The proposed construction is fully in accordance with the first Julesz conjecture. In order to consider the non-stationarity of some categories, each texture is considered as a collection of Relocated Colour Contrast Occurrence matrix.

The presented results show that the proposed texture feature is robust to different image changes and very efficient for texture image classification. The Relocated Colour Contrast Occurrence matrix presents the best classification score in three cases out of four, and is the third for the last one. These results will be extended to more image databases in order improve the ranking of this pair texture feature/similarity measure.

References

1. Salzburg texture image database stex, Department of Computer Sciences. http://www.wavelab.at/sources/STex
2. University of Oulu, Outex texture database. http://www.outex.oulu.fi
3. VisTex Vision Texture Database, Vision and Modeling Group, MIT Media Laboratory (1995). http://vismod.media.mit.edu/vismod/imagery/VisionTexture
4. Alvarez, S., Vanrell, M.: Texton theory revisited: a bag-of-words approach to combine textons. Pattern Recogn. **45**(12), 4312–4325 (2012)
5. Arvis, V., Debain, C., Berducat, M., Benassi, A.: Generalization of the co-occurrence matrix for colour images: application to colour texture segmentation. Image Anal. Estereol **23**, 63–72 (2004)
6. Byeon, W., Liwicki, M., Breuel, T.: Texture classification using 2D LSTM networks. In: IEEE 22nd International Conference on Pattern Recognition (ICPR) (2014)
7. Caputo, B., Hayman, E., Fritz, M., Eklundh, J.: Classifying materials in the real world. Image Vis. Comput. **28**, 150–163 (2010)
8. Chatoux, H., Richard, N., Lecellier, F., Fernandez-Maloigne, C.: Différence entre distributions couleur. ORASIS: 16ème journées francophones des jeunes chercheurs en vision par ordinateur, June 2017
9. Florindo, J.B., Landini, G., Bruno, O.M.: Three-dimensional connectivity index for texture recognition. Pattern Recogn. Lett. **84**, 239–244 (2016)
10. Goldberger, J., Gordon, S., Greenspan, H., et al.: An efficient image similarity measure based on approximations of kl-divergence between two gaussian mixtures. ICCV **3**, 487–493 (2003)

11. Hauta-Kasari, M., Parkkinen, J., Jaaskelainen, T., Lenz, R.: Genaralized coocurrence matrix for multispectral texture analisis. In: 13th International Conference on Pattern Recognition I, August 1996
12. Julesz, B.: Texture and visual perception. Sci. Am. **212**, 38–48 (1965)
13. Khan, F.S., Anwer, R.M., van de Weijer, J., Felsberg, M., Laaksonen, J.: Compact color-texture description for texture classification. Pattern Recogn. Lett. **51**, 16–22 (2015)
14. Kullback, S., Leibler, R.: On information and sufficiency. Ann. Math. Statist. **22**(1), 79–86 (1951)
15. Mäenpää, T., Pietikäinen, M.: Classification with color and texture: jointly or separately? Pattern Recogn. **37**, 1629–1640 (2004)
16. Maliani, A.D.E., Hassouni, M.E., Berthoumieu, Y., Aboutajdine, D.: Color texture classification method based on a statistical multi-model and geodesic distance. J. Vis. Commun. Image Represent. (2014)
17. Martnez, R., Richard, N., Fernandez, C.: Alternative to colour feature classification using colour contrast ocurrence matrix. In: Proceedings SPIE 9534, Twelfth International Conference on Quality Control by Artificial Vision, 30 April 2015
18. Mathiassen, J.R., Skavhaug, A., Bø, K.: Texture similarity measure using kullback-leibler divergence between gamma distributions. In: Heyden, A., Sparr, G., Nielsen, M., Johansen, P. (eds.) ECCV 2002. LNCS, vol. 2352, pp. 133–147. Springer, Heidelberg (2002). https://doi.org/10.1007/3-540-47977-5_9
19. Nguyen, V.L., Vu, N.S., Phan, H.H., Gosselin, P.H.: An integrated descriptor for texture classification. In: 23rd IEEE International Conference on Pattern Recognition (ICPR) (2016)
20. Pham, M.T., Mercier, G., Bombrun, L.: Color texture image retrieval based on local extrema features and riemannian distance. J. Imaging **3**(4) (2017)
21. Porebski, A., Vandenbroucke, N., Hamad, D.: LBP histogram selection for supervised color texture classification. In: ICIP, pp. 3239–3243 (2013)
22. Richard, N., Ivanovici, M., Bony, A.: Toward a metrology for non-uniform surface using the complexity notion. In: 4th CIE Expert Symposium on Colour and Visual Appearance, Czech Republic, Prague, pp. 40–50 September 2016
23. Richard, N., Martnez, R., Fernandez, C.: Colour local pattern: a texture feature for colour images. J. Int. Colour Assoc. **16**, 56–68 (2016)
24. Sandid, F., Douik, A.: Robust color texture descriptor for material recognition. Pattern Recogn. Lett. **80**, 15–23 (2016)
25. Mangijao, S., Hemachandran, K.: Content-based image retrieval using color moment and gabor texture feature. IJCSI Int. J. Comput. Sci. **9**, 299–309 (2012)
26. Song, Y., Li, Q., Feng, D., Zou, J.J., Ca, W.: Texture image classification with discriminative neural networks. Ann. Math. Statist. **2**(4), 367–377 (2016)
27. Wang, Q., Kulkarni, S., Verdú, S.: Divergence estimation for multidimensional densities via k-nearest-neighbor distances. IEEE Trans. Inf. Theory (55) (2009)
28. Xiaoyan, S., Shao-Hui, C., Jiang, L., Frederic, M.: Automatic diagnosis for prostate cancer using run-length matrix method. In: Medical Imaging. Procceding of SPIE 7260 (2009)

I-HAZE: A Dehazing Benchmark with Real Hazy and Haze-Free Indoor Images

Cosmin Ancuti[1]([⊠]), Codruta O. Ancuti[1], Radu Timofte[3,4],
and Christophe De Vleeschouwer[2]

[1] MEO, Universitatea Politehnica Timisoara, Timişoara, Romania
cosmin.ancuti@upt.ro
[2] ICTEAM, Universite Catholique de Louvain, Ottignies-Louvain-la-Neuve, Belgium
[3] ETH Zurich, Zürich, Switzerland
[4] Merantix GmbH, Berlin, Germany

Abstract. Image dehazing has become an important computational imaging topic in the recent years. However, due to the lack of ground truth images, the comparison of dehazing methods is not straightforward, nor objective. To overcome this issue we introduce I-HAZE, a new dataset that contains 35 image pairs of hazy and corresponding haze-free (ground-truth) indoor images. Different from most of the existing dehazing databases, hazy images have been generated using real haze produced by a professional haze machine. To ease color calibration and improve the assessment of dehazing algorithms, each scene includes a MacBeth color checker. Moreover, since the images are captured in a controlled environment, both haze-free and hazy images are captured under the same illumination conditions. This represents an important advantage of the I-HAZE dataset that allows us to objectively compare the existing image dehazing techniques using traditional image quality metrics such as PSNR and SSIM.

1 Introduction

Limited visibility and reduced contrast due to haze or fog conditions is a major issue that hinders the success of many outdoor computer vision and image processing algorithms. Consequently, automatic dehazing methods have been largely investigated. Oldest approaches rely on atmospheric cues [11,26], multiple images captured with polarization filters [27,30], or known depth [20,34]. Single image dehazing, meaning dehazing without side information related to the scene geometry or to the atmospheric conditions, is a complex mathematically ill-posed problem. This is because the degradation caused by haze is different for every pixel and depends on the distance between the scene point and the camera. This dependency is expressed in the transmission coefficients, that control the scene attenuation and amount of haze in every pixel. Due to lack of space, we refer the reader to previous papers for a formal description of a simplified but realistic light propagation model, combining transmission and airlight to describe how haze impacts the observed image.

© Springer Nature Switzerland AG 2018
J. Blanc-Talon et al. (Eds.): ACIVS 2018, LNCS 11182, pp. 620–631, 2018.
https://doi.org/10.1007/978-3-030-01449-0_52

Hazy images

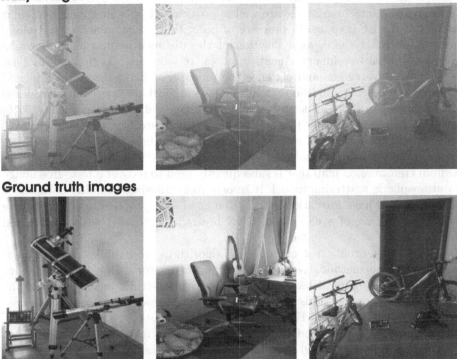

Ground truth images

Fig. 1. I-HAZE dataset provides 35 set of hazy indoor images and the corresponding ground truth (haze-free) images.

Single image dehazing directly builds on this simplified model. It has been addressed recently [3,5,12–14,16,22,32–34], by considering different kinds of priors to estimate the transmission. The method of Fattal [13] adopts a refined image formation model that accounts for surface shading in addition to the transmission function. This allows to regularize the transmission and haze color estimation problems by searching for a solution in which the resulting shading and transmission functions are locally statistically uncorrelated. Tan's approach [32] assumes the atmospheric airlight to be the brightest pixel in the scene, and estimate the transmission by maximizing the contrast. The maximal contrast assumption has also been exploited in [34], with a linear complexity in the number of image pixels. The dark channel priors, introduced in [16], has been proven to be a really effective solution to estimate the transmission, and has motivated many recent approaches. Meng et al. [24] extends the work on dark channel, by regularizing the transmission around boundaries to mitigate its lack of resolution. Zhu et al. [39] extend [24] by considering a color attenuation prior, assuming that the depth can be estimated from pixel saturation and intensity. The color-lines, introduced in [14], also exploit the impact of haze over the color channels distribution. Berman et al. [6] adopt a similar path. They observe that the colors of a haze-free image are well approximated by a limited set of tight clusters in the RGB space. In presence of haze, those clusters are spread along lines in the RGB space, as a function of the distance map, which allows to recover the haze free image. Ancuti et al. [3] rely on the hue channel analysis to identify hazy regions and enhance the image. For night-time dehazing and non-uniform lighting conditions, spatially varying airlight has been considered and estimated in [1,23]. Fusion-based single image dehazing approaches [5,10] have also achieved visually pleasant results without explicit transmission estimation and, more recently, several machine learning based methods have been introduced [7,28,33]. DehazeNet [7] takes a hazy image as input, and outputs its medium transmission map that is subsequently used to recover a haze-free image via atmospheric scattering model. It resorts to synthesized training data based on the physical haze formation model. Ren et al. [28] proposed a coarse-to-fine network consisting of a cascade of CNN layers, also trained with synthesized hazy images.

Despite this prolific set of dehazing algorithms, the validation and comparison of those methods remains largely unsatisfactory. Due to the absence of corresponding pairs of hazy and haze-free ground-truth image, most of the existing evaluation methods are based on non-reference image quality assessment (NR-IQA) strategies. For example, in [15], the assessment simply relies on the gradient of the visible edges. Other non-reference image quality assessment (NR-IQA) strategies [25,29] have been used for dehazing assessment. A more general framework has been introduced in [9], using subjective assessment of enhanced and original images captured in bad visibility conditions. Besides, the Fog Aware Density Evaluator (FADE) introduced in [10] predicts the visibility of a hazy/foggy scene from a single image without corresponding ground-truth. Unfortunately, due to the absence of the references (haze-free), none of these quality assessment approaches has been commonly accepted by the dehazing community.

Due to the practical issues associated to the recording of reference and hazy images under identical illumination condition, all existing data-sets have been built on synthesized hazy images based on the optical model and known depth. The work [35] presents the FRIDA dataset designed for Advanced Driver Assistance Systems (ADAS) that is a synthetic image database (computer graphics generated scenes) with 66 roads synthesized scenes. In [2], a dataset of 1400+ images of real complex scenes has been derived from the *Middlebury*[1] and the *NYU-Depth V2*[2] datasets. It contains high quality real scenes, and the depth map associated to each image has been used to yield synthesized hazy images based on Koschmieder's light propagation model [21]. Recently in [4] it has been introduced O-HAZE a similar dataset with I-HAZE that considers outdoor scenes.

---

[1] http://vision.middlebury.edu/stereo/data/scenes2014/.

[2] http://cs.nyu.edu/~silberman/datasets/nyu_depth_v2.html.

Complementary to the existing dehazing datasets, in this paper we contribute I-HAZE[3], a new dataset containing pairs of real hazy and corresponding haze-free images for 35 various indoor scenes. Haze has been generated with a professional haze machine that imitates with high fidelity real hazy conditions. A similar dataset (CHIC) that contains only two quite similar scenes was introduce in [19]. However, I-HAZE dataset provides significantly more scenes with a larger variety of object structures and colors. Another contribution of this paper is a comprehensive evaluation of several state-of-the-art single image dehazing methods. Interestingly, our work reveals that many of the existing dehazing techniques are not able to accurately reconstruct the original image from its hazy version. This observation is founded on SSIM [36] and CIEDE2000 [31] image quality metrics, computed using the known reference and the dehazed results produced by different dehazing techniques. This observation, combined with the release of our dataset, certainly paves the way for improved dehazing methods.

2   I-HAZE Dataset

This section describes how the I-HAZE dataset has been produced. All the 35 scenes presented in the I-HAZE correspond to indoor domestic environments, with objects with different colors and specularities. Besides the domestic objects, all the scenes contain a color checker chart (Macbeth color checker). We use a classical Macbeth color checker with the size 11 by 8.25 in. with 24 squares of painted samples (4×6 grid).

After carefully setting each scene, we first record the ground truth (haze-free image) and then we immediately start the process of introducing haze in the scene. Therefore, we use two professional fog/haze machines (LSM1500 PRO 1500 W) that generate a dense vapor. The fog generators use cast or platen type aluminum heat exchangers, which causes evaporation of the water-based fog liquid. The generated particles (since are water droplets) have approximately the same diameter size of $1-10\,\mu$ as the atmospheric haze. Before shooting the hazy scene, we use a fan that helps to obtain in a relatively short period of time a homogenous haze distribution in the entire room (obviously that is kept isolated as much as possible by closing all the doors and windows). The entire process to generate haze took approximately 1 min. Waiting approximately another 5–10 min, we obtain a homogenous distribution of the haze. The distances between the camera and the target objects range form 3 to 10 m. The recordings were performed during the daytime in relatively short intervals (20–30 min per scene recording) with natural lightning and when the light remains relatively constant (either smooth cloudy days or when the sun beams did not hit directly the room windows).

To capture haze-free and hazy images, we used a setup that includes a tripod and a Sony A5000 camera that was remotely controlled (Sony RM-VPR1). We acquired JPG and ARW (RAW) 5456×3632 images, with 24 bit depth. The cameras were set on manual mode and we kept the camera still (on a tripod)

---

[3] http://www.vision.ee.ethz.ch/ntire18/i-haze/.

over the entire shooting session of the scene. We calibrate the camera in haze-free scene, and then we kept the same parameters for the hazy scene. For each scene, the camera settings have been manually calibrated by adjusting manually the aperture (F-stop), shutter-speed (exposure-time), ISO speed and the white-balance. Setting the three parameters aperture-exposure-ISO has been realized using both the built-in light-meter of the camera and an external exponometer Sekonic. For the white-balance we used the gray-card, targeting a middle gray (18% gray). The calibration process is straight-forward, since it just requires to set the white-balance in manual mode and to place the gray-card in front of the subject. In practice, we placed the gray-card in the center of the scene, two meters away from the camera. In addition, since each scene contains the color checker, the white-balance can be manually adjusted (a posteriori) using specialized software such as Adobe Photoshop Lightroom.

3   Quantitative Evaluation

In this section, before presenting the dehazing techniques used in our evaluation we briefly discuss the optical model assumed in most of the existing dehazing techniques.

Mathematically the dehazing optical model is expressed by the image formation model of Koschmieder [21]. Based on this model, due to the atmospheric particles that absorb and scatter light, only a certain percentage of the reflected light reaches the observer. The light intensity $\mathcal{I}$ of each pixel coordinate x, that passes a hazy medium is expressed as:

$$\mathcal{I}(x) = \mathcal{J}(x)\ T(x) + A_\infty\ [1 - T(x)] \tag{1}$$

where the haze-free image is denoted by $\mathcal{J}$, T is the *transmission* (depth map) and A_∞ is the atmospheric light.

The atmospheric light A_∞ is a color constant and represents the principal source of the additive color shifting that distorts the hazy images. Assuming an homogeneous medium [21] the transmission map T can be expressed as:

$$T(x) = e^{[-\beta\ d(x)]} \tag{2}$$

where β is a medium coefficient due to the light scattering and d is the distance between the sensor and the target scene.

He et al. [16,17] introduced the popular dark-channel prior, an extension of the dark object assumption [8]. Their proposed dehazing algorithm exploits the observation that the majority of local regions (except the sky, or hazy regions) include some pixels that are characterized by a very low value in at least one color channel. This helps in roughly estimating the transmission map of the hazy images. Further refinement of the transmission map can be obtained based on an alpha matting strategy [16], or by using guided filters [17]. In our assessment, we employ the dark channel prior refined based on the guided filter [18].

Meng et al. [24] extend the dark channel prior [16]. It estimates the transmission map by formulating an optimization problem that embeds the constraints imposed by the scene radiance and a weighted L1-norm based contextual regularization, to avoid halo artifacts around sharp edges.

Fattal [14] introduced a method that exploits the observation that pixels of small image patches typically exhibit a one-dimensional distribution in RGB color space, named color lines. Since the haze tends to move color lines away from the RGB origin, an initial estimation of the transmission map is obtained by computing the lines offset from the origin. The final transmission map is refined by applying a Markov random field, which filters the noise and other artifacts resulting from the scattering.

Cai et al. [7] introduced an end-to-end system built on CNN that learns the mapping relations between hazy and haze free corresponding patches. To train the network they syntheses hazy images based on Middlebury stereo dataset. The method employs a non linear activation function that uses a bilateral restraint to improve convergence.

Ancuti et al. [1] introduced a local airlight estimation applied in a multi-scale fusion strategy for single-image dehazing. The method has been designed to solve the problems associated to the scattering effect, which is especially significant in the nigh-time hazy scenes. The method, however, generalizes to day-time dehazing.

Berman et al. [6] introduced an algorithm that also uses local airlight estimation, and extends the color consistency observation of Fattal [14]. The algorithm builds on the observation that the colors of a haze-free image can be approximated by a few distinct tight color clusters. Since the pixels of a cluster are spread on the whole image, they are affected differently by haze, to be spread along an elongated line, named haze-line. Those lines convey information about the transmission in different regions of the image, and are thus used for transmission estimation.

Ren et al. [28] adopts a multi-scale convolutional neural network to learn the mapping between hazy images and their corresponding transmission maps. The network is trained based on synthetic hazy images, generated by applying a simplified light propagation model to haze-free images for which corresponding depth maps are known.

4   Results and Discussion

In Fig. 2 are shown seven hazy images (first column), the corresponding ground truth (last column) and the dehazing results yielded by the specialised techniques of He et al. [16], Meng et al. [24], Fattal [14], Cai et al. [7], Ancuti et al. [1], Berman et al. [6] and Ren et al. [28].

Qualitatively, the well-known method of He et al. [16] yields results that visually seems to recover the structure, but suffers from color shifting in the hazy

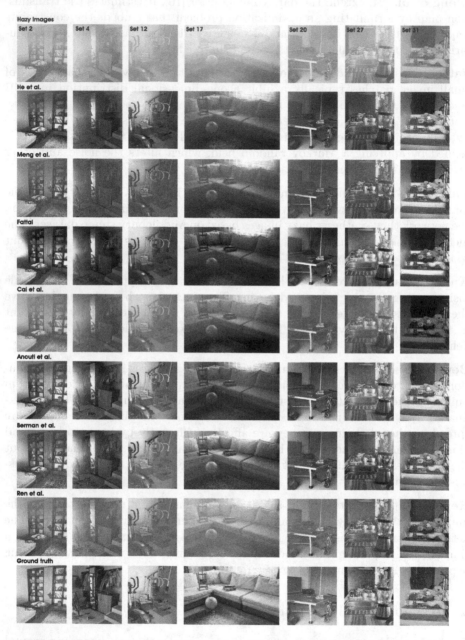

Fig. 2. Comparative results for seven I-HAZE sets of images. The first row shows the hazy images and the last row shows the ground truth. The other rows from top to bottom show the results yielded by the methods of He et al. [16], Meng et al. [24], Fattal [14], Cai et al. [7], Ancuti et al. [1], Berman et al. [6] and Ren et al. [28].

regions due to the poor airlight estimation. This happens when the scenes contains lighter color patches in the close-up regions or small reflections. The Meng et al. [24] approach, built also on dark channel prior, as expected, generates similar results as He et al. [16]. It presents an improved alternative filtering of the transmission for artifacts reduction and a more precise airlight estimation. The results of color-lines method of [14] suffers from unpleasing color shifting while the results of Ancuti et al. [1] are more accurate due to the local airlight estimation strategy. This approach also is less prone to introduce structural artifacts due to the multi-scale fusion strategy. The results of Berman et al. [6] are yielding visually compelling results and although it is built on a haze-line strategy due to its locally estimation of the airlight, shown to introduce less color shifting than [14]. In addition to the algorithms built on priors, deep learning techniques produce less visual artifacts. The results of Ren et al. [28] generates visually more compelling results in comparison with the ones generated by DehazeNet of Cai et al. [7].

Table 1. Quantitative evaluation. We randomly selected seven set of images from our I-HAZE dataset and we compute the CIEDE2000 and SSIM indexes between the ground truth images and the enhanced results of the evaluated techniques. The hazy images, ground truth and the results are shown in Fig. 2.

		Set 2	Set 4	Set 12	Set 17	Set 20	Set 27	Set 31
He et al. [16]	CIEDE2000	15.66	20.77	15.79	20.28	24.84	18.03	10.45
	SSIM	0.78	0.63	0.74	0.58	0.62	0.78	0.82
Meng et al. [24]	CIEDE2000	13.58	16.58	16.39	20.01	19.57	17.25	16.12
	SSIM	0.82	0.70	0.78	0.66	0.79	0.82	0.79
Fattal [14]	CIEDE2000	19.85	20.92	18.97	26.77	23.43	15.88	16.24
	SSIM	0.61	0.57	0.70	0.44	0.54	0.82	0.66
Cai et al. [7]	CIEDE2000	9.81	19.97	17.32	18.91	24.04	20.86	24.66
	SSIM	0.85	0.59	0.76	0.56	0.61	0.63	0.41
Ancuti et al. [1]	CIEDE2000	14.07	14.47	13.23	21.90	15.76	10.52	7.95
	SSIM	0.82	0.74	0.84	0.62	0.76	0.88	0.88
Berman et al. [6]	CIEDE2000	14.13	17.60	9.57	17.51	16.01	16.87	11.24
	SSIM	0.80	0.69	0.83	0.71	0.81	0.79	0.81
Ren et al. [28]	CIEDE2000	9.91	18.06	11.81	12.15	10.74	13.78	11.12
	SSIM	0.85	0.64	0.83	0.67	0.90	0.87	0.87

To quantitatively evaluate the dehazing methods described in the previous section, we compare directly their outcome with the ground-truth (haze free) images. Besides the well known PSNR, we compute also the structure similarity index SSIM [37] that compares local patterns of pixel intensities that have been normalized for luminance and contrast. The structure similarity index yields values in the range $[-1,1]$ with maximum value 1 for two identical images. Additionally, to evaluate also the color restoration we employ the CIEDE2000 [31,38]. Different than the earlier measures (e.g. CIE76 and CIE94) that shown shown important limitations to resolve the perceptual uniformity issue, CIEDE2000 defines a more complex, yet most accurate color difference algorithm. CIEDE2000 yields values in the range $[0,100]$ with smaller values indicating better color preservation.

Table 2. Quantitative evaluation of all the 35 set of images. In this table are shown the average values of the SSIM, PSNR and CIEDE2000 indexes over the entire dataset using the small size version of the images.

	SSIM	PSNR	CIEDE2000
He et al. [16]	0.711	15.285	17.171
Meng et al. [24]	0.750	14.574	16.834
Fattal [14]	0.574	12.421	21.385
Cai et al. [7]	0.771	16.983	12.991
Ancuti et al. [1]	0.770	16.632	14.428
Berman et al. [6]	0.767	15.942	14.629
Ren et al. [28]	0.791	17.280	12.736

Table 1 presents a detailed validation based on SSIM and CIEDE2000 for seven randomly selected images of the I-HAZE dataset that are shown in the Fig. 2. In Table 2 are presented the average values over the entire dataset of the SSIM, PSNR and CIEDE indexes. To avoid the parameter tweaking of some of the tested methods, the quantitative results have been generated using the small size of the images (we kept the original image proportion and the images have been resized to a maximum size of 800 pixels).

From these tables, we can conclude that the methods of Berman et al. [6], Ancuti et al. [1] and Ren et al. [28] performs the best in average when considering the SSIM, PSNR and CIEDE indexes. A second group of methods including Meng et al. [24] and He et al. [16], perform relatively well both in terms of structure and color restoration.

In general, all the tested methods introduce structural distortions such as halo artifacts close to the edges, that are amplified in the faraway regions. Moreover, due to the poor estimation of the airlight and transmission map from the hazy image, some color distortions may create some unnatural appearance of the restored images.

5   Conclusions

In this paper we introduced a novel dehazing dataset named I-HAZY. It consists of 35 image pairs of hazy and corresponding haze-free (ground-truth) indoor images. I-HAZE has been generated using real haze produced by a professional haze machine with both haze-free and hazy images captured under the same illumination conditions. Compared with previous dehazing datasets I-HAZE has an important advantage since it allows to objectively compare the existing image dehazing techniques using traditional image quality metrics. Additionally, we perform a comprehensive evaluation of several state-of-the-art single image dehazing methods. In summary, there is not a single technique that performs the best for all images. The relatively low values of SSIM, PSNR and CIEDE2000 measures prove once again the difficulty of single image dehazing task and the fact there is still much room for improvement.

References

1. Ancuti, C., Ancuti, C.O., Bovik, A., Vleeschouwer, C.D.: Night time dehazing by fusion. In: IEEE ICIP (2016)
2. Ancuti, C., Ancuti, C.O., Vleeschouwer, C.D.: D-HAZY: A dataset to evaluate quantitatively dehazing algorithms. In: IEEE ICIP (2016)
3. Ancuti, C.O., Ancuti, C., Hermans, C., Bekaert, P.: A fast semi-inverse approach to detect and remove the haze from a single image. In: Kimmel, R., Klette, R., Sugimoto, A. (eds.) ACCV 2010. LNCS, vol. 6493, pp. 501–514. Springer, Heidelberg (2011). https://doi.org/10.1007/978-3-042-19309-5_39
4. Ancuti, C.O., Ancuti, C., Timofte, R., Vleeschouwer, C.D.: O-HAZE: a dehazing benchmark with real hazy and haze-free outdoor images. In: IEEE CVPR, NTIRE Workshop (2018)
5. Ancuti, C., Ancuti, C.: Single image dehazing by multi-scale fusion. IEEE Trans. Image Process. **22**(8), 3271–3282 (2013)
6. Berman, D., Treibitz, T., Avidan, S.: Non-local image dehazing. In: IEEE International Conference on Computer Vision and Pattern Recognition (2016)
7. Cai, B., Xu, X., Jia, K., Qing, C., Tao, D.: Dehazenet: an end-to-end system for single image haze removal. IEEE Trans. Image Process. **25**, 5187–5198 (2016)
8. Chavez, P.: An improved dark-object subtraction technique for atmospheric scattering correction of multispectral data. Remote Sens. Environ. **24**, 459–479 (1988)
9. Chen, Z., Jiang, T., Tian, Y.: Quality assessment for comparing image enhancement algorithms. In: IEEE Conference on Computer Vision and Pattern Recognition (2014)
10. Choi, L.K., You, J., Bovik, A.C.: Referenceless prediction of perceptual fog density and perceptual image defogging. IEEE Trans. Image Process. **24**, 3888–3901 (2015)
11. Cozman, F., Krotkov, E.: Depth from scattering. In: IEEE Conference on Computer Vision and Pattern Recognition (1997)
12. Emberton, S., Chittka, L., Cavallaro, A.: Hierarchical rank-based veiling light estimation for underwater dehazing. In: Proceedings of British Machine Vision Conference (BMVC) (2015)
13. Fattal, R.: Single image dehazing. SIGGRAPH **27**, 72 (2008)
14. Fattal, R.: Dehazing using color-lines. ACM Trans. Graph. **34**, 13 (2014)

15. Hautiere, N., Tarel, J.P., Aubert, D., Dumont, E.: Blind contrast enhancement assessment by gradient ratioing at visible edges. J. Image Anal. Stereol. **27**, 87–95 (2008)
16. He, K., Sun, J., Tang, X.: Single image haze removal using dark channel prior. In: IEEE CVPR (2009)
17. He, K., Sun, J., Tang, X.: Single image haze removal using dark channel prior. IEEE Trans. Pattern Anal. Mach. Intell. **33**, 2341–2353 (2011)
18. He, K., Sun, J., Tang, X.: Guided image filtering. IEEE Trans. Pattern Anal. Mach. Intell. (TPAMI) **6**, 1397–1409 (2013)
19. El Khoury, J., Thomas, J.-B., Mansouri, A.: A color image database for haze model and dehazing methods evaluation. In: Mansouri, A., Nouboud, F., Chalifour, A., Mammass, D., Meunier, J., ElMoataz, A. (eds.) ICISP 2016. LNCS, vol. 9680, pp. 109–117. Springer, Cham (2016). https://doi.org/10.1007/978-3-319-33618-3_12
20. Kopf, J., et al.: Deep photo: model-based photograph enhancement and viewing. Siggraph ASIA ACM Trans. Graph. (2008)
21. Koschmieder, H.: Theorie der horizontalen sichtweite. In: Beitrage zur Physik der freien Atmosphare (1924)
22. Kratz, L., Nishino, K.: Factorizing scene albedo and depth from a single foggy image. In: ICCV (2009)
23. Li, Y., Tan, R.T., Brown, M.S.: Nighttime haze removal with glow and multiple light colors. In: IEEE International Conference on Computer Vision (2015)
24. Meng, G., Wang, Y., Duan, J., Xiang, S., Pan, C.: Efficient image dehazing with boundary constraint and contextual regularization. In: IEEE International Conference on Computer Vision (2013)
25. Mittal, A., Moorthy, A.K., Bovik, A.C.: No-reference image quality assessment in the spatial domain. IEEE Trans. Image Process. **21**, 4695–4708 (2012)
26. Narasimhan, S., Nayar, S.: Vision and the atmosphere. Int. J. Comput. Vis. **48**, 233–254 (2002)
27. Narasimhan, S., Nayar, S.: Contrast restoration of weather degraded images. IEEE Trans. Pattern Anal. Mach. Intell. (2003)
28. Ren, W., Liu, S., Zhang, H., Pan, J., Cao, X., Yang, M.-H.: Single image dehazing via multi-scale convolutional neural networks. In: Leibe, B., Matas, J., Sebe, N., Welling, M. (eds.) ECCV 2016. LNCS, vol. 9906, pp. 154–169. Springer, Cham (2016). https://doi.org/10.1007/978-3-319-46475-6_10
29. Saad, M.A., Bovik, A.C., Charrier, C.: Blind image quality assessment: a natural scene statistics approach in the DCT domain. IEEE Trans. Image Process. **21**, 3339–3352 (2012)
30. Schechner, Y.Y., Narasimhan, S.G., Nayar, S.K.: Polarization-based vision through haze. Appl. Opt. **42**, 511–525 (2003)
31. Sharma, G., Wu, W., Dalal, E.: The CIEDE2000 color-difference formula: Implementation notes, supplementary test data, and mathematical observations. Color Res. Appl. **30**, 21–30 (2005)
32. Tan, R.T.: Visibility in bad weather from a single image. In IEEE Conference on Computer Vision and Pattern Recognition (2008)
33. Tang, K., Yang, J., Wang, J.: Investigating haze-relevant features in a learning framework for image dehazing. In IEEE Conference on Computer Vision and Pattern Recognition (2014)
34. Tarel, J.P., Hautiere, N.: Fast visibility restoration from a single color or gray level image. In: IEEE ICCV (2009)

35. Tarel, J.P., Hautire, N., Caraffa, L., Cord, A., Halmaoui, H., Gruyer, D.: Vision enhancement in homogeneous and heterogeneous fog. IEEE Intell. Transp. Syst. Mag. **4**, 6–20 (2012)
36. Wang, Z., Bovik, A.C.: Modern Image Quality Assessment. Morgan and Claypool Publishers, New York (2006)
37. Wang, Z., Bovik, A.C., Sheikh, H.R., Simoncelli, E.P.: Image quality assessment: from error visibility to structural similarity. IEEE Trans. Image Process. **13**, 600–612 (2004)
38. Westland, S., Ripamonti, C., Cheung, V.: Computational Colour Science Using MATLAB, 2nd edn. Wiley, New York (2005)
39. Zhu, Q., Mai, J., Shao, L.: A fast single image haze removal algorithm using color attenuation prior. IEEE Trans. Image Process. **24**, 3522–3533 (2015)

Author Index